QUASARS

INTERNATIONAL ASTRONOMICAL UNION
UNION ASTRONOMIQUE INTERNATIONALE

QUASARS

PROCEEDINGS OF THE 119TH SYMPOSIUM OF THE
INTERNATIONAL ASTRONOMICAL UNION,
HELD IN BANGALORE, INDIA,
DECEMBER 2–6, 1985

EDITED BY

G. SWARUP

and

V. K. KAPAHI

Tata Institute of Fundamental Research, Bangalore, India

D. REIDEL PUBLISHING COMPANY

A MEMBER OF THE KLUWER ACADEMIC PUBLISHERS GROUP

DORDRECHT / BOSTON / LANCASTER / TOKYO

Library of Congress Cataloging in Publication Data

International Astronomical Union. Symposium (119th : 1985 : Bangalore, India)
    Quasars : proceedings of the 119th Symposium of the International Astronomical
Union, held in Bangalore, India, December 2–6, 1985.

    At head of title: International Astronomical Union.
    Includes indexes.
    1. Quasars—Congresses. I. Swarup, G. (Govind), 1929–   . II. Kapahi,
V. K. III. International Astronomical Union. IV. Title.
QB860.I58   1985   523   86–13759

ISBN-13: 978-90-277-2298-0   e-ISBN-13: 978-94-009-4716-0
DOI: 10.1007/978-94-009-4716-0

Published on behalf of
the International Astronomical Union
by
D. Reidel Publishing Company, P. O. Box 17, 3300 AA Dordrecht, Holland

Softcover reprint of the hardcover 1st edition 1986

Sold and distributed in the U.S.A. and Canada
by Kluwer Academic Publishers,
101 Philip Drive, Assinippi Park, Norwell, MA 02061, U.S.A.

In all other countries, sold and distributed
by Kluwer Academic Publishers Group,
P. O. Box 322, 3300 AH Dordrecht, Holland

TABLE OF CONTENTS

(Titles of invited papers are given in capital letters)

II. CONTINUUM EMISSION

III. SPECTRAL LINE STUDIES

V. COSMOLOGICAL STUDIES, CLUSTERING, ISOTROPY ETC

PREFACE

Quasars were discovered by Maarten Schmidt in 1963 when he interpreted the optical spectrum of the stellar object identified with the radio source 3C 273 to imply an unusually high redshift (by standards of those days) of about 0.16. The discovery posed a severe theoretical problem in understanding the tremendous amount of energy released in a rather small volume. Over the last two decades the quasar phenomenon has come to be recognized as an extreme manifestation of a range of violent activity occurring in active galactic nuclei. It is generally believed that the source of the energy is the gravitational field associated with massive objects, perhaps black holes of $10^8 - 10^9$ solar masses. Although many spectacular observational and theoretical results have been obtained and about 3000 quasars have been catalogued through radio, optical and X-ray means, yet the quasar phenomenon is still not well understood. Symposium 119 was the first IAU symposium, devoted exclusively to quasars.

Held at Hotel Ashok in Bangalore, India from December 2 to 6, 1985 soon after the XIX General Assembly of IAU in New Delhi, the symposium was attended by 200 scientists from 21 countries. The wide range of topics concerning quasars covered during the symposium included continuum properties, structure and morphology (in the radio, IR, optical, UV, and X-ray region of the spectrum), emission and absorption line studies, nature and models of the prime movers, cosmological evolution and implications, studies of clustering and pairing, as well as the use of quasars as probes of the intervening medium.

The scientific program consisted of 23 invited reviews, 24 contributed papers and 87 poster presentations. In view of the large number of contributions arrangements were made to keep all the posters on display throughout the symposium. Authors of posters dealing with topics covered on the first day were asked to make a very brief oral summary of their work. For the remaining days the posters were divided into 12 groups according to the similarity of topics covered by them. Posters in each group were briefly summarized in the relevant scientific session by one person (generally one of the authors or Chairman of the session), followed by a discussion between the authors and participants. This arrangement was by and large quite successful and provided an opportunity for every poster to be discussed.

With only one or two exceptions, authors of all the papers presented at the symposium have been quite prompt in sending us their camera-ready manuscripts which are reproduced in these proceedings. Some of the manuscripts had however to be retyped in order to ensure better reproduction. Thanks to the cooperation of many participants and speakers, we have also succeeded in reproducing most of the discussion that took

place during the symposium from written versions of the questions that
were submitted to the speakers for filling up the answers.

In keeping with the tradition of IAU symposia, some cultural and
social events were also organised during the symposium, giving an opportunity to the participants from abroad to experience the many facets
of India. One important event was the special 3-day tour both before and
after the symposium, that was taken by many of the participants. The
tour took them to the ancient stone-carved temple at Somnathpur, the old
princely city of Mysore, the wild life sanctuary at Mudumalai and to the
Ooty Radio Telescope located in the picturesque Nilgiri Hills of South
India. The trip into the wild life preserve became particularly interesting when some daring astronomers got out of the bus to photograph a
herd of wild elephants. They were charged at by the elephants but were
fortunately able to rush back to safety just in time !

Apart from the IAU, the symposium was sponsored by the Tata
Institute of Fundamental Research, the Departments of Science and
Technology and of Atomic Energy of the Government of India and by The
Raman Research Institute. We thank all of them for their generous
support.

We are grateful to a large number of colleagues in the Tata Institute
of Fundamental Research and the Raman Research Institute who contributed
to the successful organization of the Symposium. Special thanks are due
to Mr. T.M.Sahadevan and Ms Moxa Halesh of the TIFR Centre, Bangalore
for their tremendous help in the organisation work. We also thank
Ms. N.N.Shantha for cheerfully doing most of the typing and
Mr. C.Ramachandra Rao for taking all the pictures of the participants
that appear in the proceedings.

                                        G. Swarup
                                        V.K. Kapahi

SCIENTIFIC ORGANIZING COMMITTEE

Biermann, P.
Jauncey, D.L.
Kellermann, K.I.
Khachikian, E.
Menon, T.K.
Rees, M.J.
Schmidt, M.
Setti, G.
Swarup, G. (Chairman)

LOCAL ORGANIZING COMMITTEE

Gopal-Krishna
Kapahi, V.K. (Chairman)
Kembhavi, A.K.
Kulkarni, V.K.
Marar, T.M.K.
Prabhu, T.P.
Srinivasan, G.

LIST OF PARTICIPANTS

ABRAMOWICZ,M.A., International Centre for Advanced Studies,Trieste,Italy
AGRAWAL,P.C., Tata Institute of Fundamental Research,Bombay 400005,India
ALIGHIERI,S.di S., ST-ECF, ESO, Garching bei Munchen, West Germany
ALLADIN,S.M., CASA, Osmania University, Hyderabad 500007, India
ALLOIN,D., Observatoire de Meudon, Meudon Principal Cedex, France
ANANTHAKRISHNAN,S., Radio Astronomy Centre(TIFR),Ootacamund 643001,India
ANANTHARAMIAH,K., Raman Research Institute, Bangalore 560080, India
ANDERSON,M., Institute of Astronomy,Madingley Road, Cambridge CB3 OHE,UK
BAILEY,J., Anglo Australian Observatory, Epping, NSW 2121, Australia
BALDWIN,J.E., Cavendish Laboratory,Madingley Road,Cambridge CB3 OHE, UK
BARR,P., European Southern Observatory, 6100 Darmstadt, West Germany
BARTEL,N., Centre for Astrophysics, 60 Garden Street, Cambridge, Ma,USA
BARTHEL,P.D., Owens Valley Radio Obs., Caltech, Pasadena, Ca 91125, USA
BARVAINIS,R., NRAO, Edgemont Road, Charlottesville, Va 22903, USA
BERGERON,J.A., Institute d'Astrophysique, 98 Bis.bd. Arago, Paris
BHATTACHARYA,D., Raman Research Institute, Bangalore 560080, India
BHATTACHARYYA,J.C., Indian Inst. of Astrophysics,Bangalore 560034,India
BHAVSAR,S., Raman Research Institute, Bangalore 560080, India
BIRKINSHAW,M., Dept. Astronomy, Harvard University, Cambridge, Ma, USA
BLADES,J.C., Space Telescope Science Institute, Baltimore, Md 21218,USA
BLANDFORD,R.D., Theoretical Astrophysics, Caltech,Pasadena,Ca 91125,USA
BOKSENBERG,A., Royal Greenwich Obs., Hailsham, E.Sussex BN27 1RP, UK
BRAMWELL,D., National Inst. for Telecomm. Research,Johannesburg,S.Africa
BREGMAN,J.N., NRAO, Edgemont Road, Charlottesville, Va 22903, USA
BRODIE,J.P., University of California, Sp.Sc.Lab, Berkeley,Ca 94720,USA
BURBIDGE,G.R., Univ. of California, San Diego, La Jolla, Ca 92093, USA
BURKE,B.F., Dept. of Physics, M.I.T., Cambridge, Ma 02139, USA
CAMPUSANO,L.E., University of Chile, Casilla 36-D, Santiago, Chile
CANIZARES,C.R., Mass. Inst. of Technology, Cambridge, Ma 02139, USA
CANNON,R., Royal Observatory, Blackford Hill, Edinburgh EH9 3HJ, UK
CAVALIERE,A.G., University of Roma, 00173 Roma, Italy
CHATTERJEE,T.K., CASA, Osmania University, Hyderabad 500007, India
CHITRE,S.M., Tata Institute of Fundamental Research,Bombay 400005,India
CHU,Y., Observatorio Astrofisico, 36012 Asiago (Vicenza), Italy
CLARKE,J.T., NASA, Marshall Sp. Flt. Centre, Alabama 35812, USA
CLAVEL,J., IUE Observatory, European Space Agency, 28080 Madrid, Spain
COHEN,M.H., Astronomy Dept., Caltech, Pasadena, Ca 91125, USA
COLEMAN,C., University of Oxford,Dept. of Astrophys., Oxford OX1 3RQ,UK
COLLIN-SOUFFRIN,S., DAF, Observatoire De Meudon, Meudon Cedex, France
COTTON,W.D., NRAO, Edgemont Road, Charlottesville, Va 22901, USA
COWSIK,R., Tata Institute of Fundamental Research,Bombay 400005, India
CRISTIANI,S., European Southern Observatory,Casilla 567,La Serena,Chile
D'ODORICO,S., ESO, D-8046 Garching bei Munchen, West Germany
DADHICH,N., Dept. of Mathematics, Univ. of Poona, Pune 411007, India
DAS,P.K., Indian Institute of Astrophysics, Bangalore 560034, India
DAVIS,R.J., NRAL, Jodrell Bank, Maccelesfield, Cheshire SK11 9DL, UK
DE BRUYN,G., Radiosterrenwacht, Post Box 2, Dwingeloo, The Netherlands
DE RUITER,H.R., Instituto di Radio Astronomia, 40126 Bologna, Italy
DRINKWATER,M., Inst. of Astronomy,Madingley Road,Cambridge, CB3 OHE, UK

DUFFETT-SMITH,P., Cavendish Lab., Madingley Road, Cambridge CB3 OHE, UK
DULTZIN-HACYAN,D., Observat.Astron.Nacional, 04510 Mexico, D.F., Mexico
ELVIS,M.S., Center for Astrophysics, 60 Garden Street, Cambridge,Ma,USA
FABBIANO,G., Centre for Astrophys., 60 Garden Street,Cambridge, Ma,USA
FASANO,G., Observatorio Astron. dell'Universita di Padova, Padova,Italy
FERRARI,A., Instituto di Fisica, Generale Dell. University,Torino,Italy
FILIPPENKO,A.V., Dept. of Astron., Univ. of California, Berkeley,Ca,USA
FOLEY,A.R., Radiosterrenwacht, Post Bus 2, Dwingeloo, The Netherlands
FRANCESCHINI,A., Inst. di Astron. dell Universita di Padova,Padova,Italy
GARILLI,B., Instituto di Fisica Cosmica-CNR, 20133 Milano, Italy
GHIGO,F.D., Dept. of Astronomy, Univ. of Minnesota, Minneapolis, Mn,USA
GHOSH,P., Tata Institute of Fundamental Research, Bombay 400005, India
GHOSH,T., TIFR Centre, Post Box 1234, Bangalore 560012, India
GIOIA,I.M., Centre for Astrophysics, 60 Garden Street,Cambridge, Ma,USA
GONDHALEKAR,P.M., Rutherford Appleton Lab, Chilton, Oxfordshire, UK
GOPAL-KRISHNA, TIFR Centre, Post Box 1234, Bangalore 560012, India
GOSSET,E., Inst. d'Astrophysique, Univ. de Liege, Cointe-Ougree,Belgium
GOWER,A.C., Dept. of Physics, Univ. of Victoria, Victoria, BC, Canada
GREEN,R.F., Kitt Peak National Observatory, Box 2673, Tucson, Az, USA
HARRIS,D.E., Centre for Astrophys., 60 Garden Street, Cambridge, Ma,USA
HE, XIANG-TAO, Beijing Normal University, Beijing,Peoples Rep. of China
HOAG,A.A., Lowell Observatory, Box 1269, Flagstaff, Az 86002, USA
HUNSTEAD, R.W., School of Physics, Sydney Univ., Sydney, Australia
HUNT,G., VLA, NRAO, P.O.Box 0, Socorro, New Mexico 87801, USA
HUTCHINGS,J.B., Dominion Astrophys. Observatory, Victoria, B.C., Canada
ILLARIONOV,A.F., Acad. of Sciences,Profsoyuznaia 88, Moscow 117810,USSR
IMPEY,C.D., Div. of Astron. & Phys., Caltech, Pasadena, Ca 91125, USA
INDULEKHA,K., Indian Inst. of Astrophysics, Bangalore 560034, India
IYER,B., Raman Research Institute, Bangalore, 560080, India
JOHRI,V.B., Dept. of Mathematics, Indian Inst. of Tech., Madras, India
JOLY,M., DAF., Observatoire de Meudon, 92195 Meudon Cedex, France
JOSHI,M.N., Radio Astronomy Centre (TIFR), Ootacamund 643001, India
KAPAHI,V.K., TIFR Centre, Post Box No. 1234, Bangalore 560012, India
KAPOOR,R.C., Indian Institute of Astrophysics, Bangalore 560034, India
KASTURIRANGAN,K., ISRO Satellite Centre, Bangalore 560017, India
KELLERMANN,K.I., NRAO, Edgemont Road, Charlottesville, Va 22903, USA
KEMBHAVI,A.K.,Tata Institute of Fundamental Research,Bombay 400005,India
KHACHIKIAN,E.Y., Byurakan Astrophys. Observatory, 378433 Armenia, USSR
KOCHHAR,R.K., Indian Institute of Astrophysics, Bangalore 560034, India
KRISHAN,V., Indian Institute of Astrophysics, Bangalore 560034, India
KRISHNAMOHAN,S., TIFR Centre, Post Box 1234, Bangalore 560012, India
KÜHR,H., Max-Planck Inst. fur Astronomie,D-6900 Heidelberg,West Germany
KULKARNI,V.K., Radio Astronomy Centre (TIFR), Ootacamund 643001, India
KUNDT,W., Institut fur Astrophysik, D-5300 Bonn, West Germany
KUNDU,M.R., Astronomy Program, Univ. of Maryland, College Park, Md, USA
KUNTH,D., Institute d'Astrophysique,98 Bis bd Arago, 75014 Paris,France
LAKSHMI,S., JAP., TIFR Centre, Post Box 1234, Bangalore 560012, India
LAWRENCE,C.R., OVRO, Caltech, Pasadena, Ca 91125, USA
LEAHY,J.E., NRAL, Jodrell Bank, Macclesfield, Cheshire, SK11 9DL, UK
LUMINET,J.P., Observatoire de Paris,92195 Meudon Principal Cedex,France
MACCACARO,T., Center for Astrophys., 60 Garden Street,Cambridge, Ma,USA

MACCAGNI,D., Instituto di Fisica Cosmica-CNR, 20133 Milano, Italy
MACHALSKI,J., Ul Mazowiecka 36/33, 30-019 Krakow, Poland
MALKAN,M., Univ. of California, Dept. of Astronomy, Los Angeles, Ca,USA
MALLIK,D.C.V., Indian Inst. of Astrophysics, Bangalore 560034, India
MANTOVANI,F., Instituto di Radioastronomia, I-40126 Bologna, Italy
MARAR,T.M.K., ISRO Satellite Centre, Bangalore 560017, India
MARGON,B.H., Astronomy Dept., University of Washington, Seattle, Wa,USA
MATVEYENKO,L.I., Sp.Res.Inst.,USSR Acad. of Sciences,117810 Moscow,USSR
McADAM,W.B., School of Physics, Sydney Univ., Sydney NSW 2006,Australia
MENON,T.K., University of British Columbia,Vancouver,B.C. V6T 1W5,Canada
MURDOCH,H.S., Astrophys. Dept., Sydney Univ., Sydney NSW 2006,Australia
MUTEL,R.L., Dept. of Phys. & Astron., Univ. of Iowa, Iowa City, Ia, USA
MORRIS,D., IRAM, BP391, Grenoble, Cedex 38017, France
NARAYAN,R., Steward Obs., Univ. of Arizona, Tucson, AZ 85721, USA
NARLIKAR,J.V.,Tata Institute of Fundamental Research,Bombay 400005,India
NITYANANDA,R., Raman Research Institute, Bangalore 560080, India
NOVIKOV,I.D., Space Research Inst., USSR Acad. of Sciences, Moscow,USSR
OWEN,F.N., NRAO, Edgemont Road, Charlottesville, Va 22901, USA
O'DEA,C.P., NRAO, Edgemont Road, Charlottesville, Va 22901, USA
PADRIELLI,L., Laboratoria di Radioastronomia, I-40126 Bologna, Italy
PARIJSKIJ,Y.N., Special Astrophysical Observatory, Kraj 357140, USSR
PEACOCK,J.A., Royal Observatory, Blackford HIll, Edinburgh EH9 3HJ, UK
PEARSON,T.J., Owens Valley Radio Observatory, Caltech, Pasadena, Ca,USA
PEREZ-FOURNON,I., Inst. de Astrofisica de Canarias, Univ. De La Laguna
    La Laguna (Tenerife), Spain
PERRY,J.J., Institute of Astronomy,Madingley Road,Cambridge CB3 OHE,UK
PETERSON,B.A., Australian National Univ., Woden ACT 2606, Australia
PETROSIAN,V., Inst. of Sp. Sc. & Astroph., Stanford Univ., Stanford,USA
PORCAS,R.W., Max-Planck Inst.fur Radioastron.,D-5300 Bonn 1,West Germany
PRABHU,T.P., Indian Institute of Astrophysics, Bangalore 560034, India
PUETTER,R.C., University of California, San Diego, La Jolla, Ca,USA
RADHAKRISHNAN,V., Raman Research Institute, Bangalore 560080, India
RAO,A.P., Radio Astronomy Centre (TIFR),P.B.No.8,Ootacamund 643001,India
RAO,U.R., Dept. of Space,Govt. of India,K.G.Road,Bangalore 560009,India
RAY,A., Tata Institute of Fundamental Research, Bombay 400005, India
REBOUL,H., Laboratoire d'Astronomie, U.S.T.L., Montpellier Cedex,France
REES,M.J., Institute of Astronomy, Madingley Road, Cambridge CB3 OHE,UK
ROBERTS,D.H., Physics Department, Brandeis University, Waltham, Ma, USA
ROBERTS,J.A., Radiophysics, CSIRO, P.O.Box 76, Epping NSW, Australia
SAIKIA,D.J., NRAL, Jodrell Bank, Macclesfield, Cheshire SK11 9DL, UK
SALTER,C., IRAM, 7 Nucleo Central, 18012 Granada, Spain
SALVATI,M., Instituto di Astrofisica Spaziale, I-00044 Frascati, Italy
SAPRE,A.K., Physics Dept., Ravishankar University, Raipur 492002, India
SARMA,N.V.G., Raman Research Institute, Bangalore 560080, India
SASLAW,W.C., Astronomy Dept., Univ. Virginia, Charlottesville, Va, USA
SASTRI,Ch.V., Indian Institute of Astrophysics, Bangalore 560034, India
SCHILD,R.E., Center for Astrophys., 60 Garden Street, Cambridge, Ma,USA
SCHILIZZI,R.T., Radiosterrenwacht, Postbus 2, Dwingeloo,The Netherlands
SCHMIDT,M., H.M.Robinson Lab. of Astrophys., Caltech, Pasadena, Ca, USA
SEGAL,I.E., Mass. Inst. of Tech., Dept.of Mathematics,Cambridge,Ma,USA
SEIELSTAD,G.A., NRAO, P.O.Box 2, Green Bank, West Virginia 24944, USA

SETTI,G., ESO, D-8046 Garching bei Munchen, West Germany
SHANBHAG,S., Tata Institute of Fundamental Research, Bombay 400005,India
SHANKS,T., Univ. of Durham, Department of Physics, Durham DH1 3LE, UK
SHASTRI,P., TIFR Centre, Post Box No.1234, Bangalore 560012, India
SHAVER,P.A., ESO, D-8046 Garching bei Munchen, West Germany
SHUKRE,C.S. Raman Research Institute, Bangalore 560080, India
SINGAL,A.K., Radio Astronomy Centre (TIFR), Ootacamund 643001, India
SINHA,R.P., SASC Technologies, 4400 Forbes Blvd., Lanham, Md 20706, USA
SIVARAM,C., Indian Institute of Astrophysics, Bangalore 560034, India
SMITH,M.G., UK Infrared Telescope,900 Leilani St., Hilo,Hawaii 56720,USA
SMITH,H.J., Astronomy Dept., University of Texas, Austin, Tx 78712, USA
SRINIVASAN,G., Raman Research Institute, Bangalore 560080, India
STABELL,R., Inst. of Theoret. Astrophys., P.B. 1029, Blindem,Oslo,Norway
STURROCK,P., Inst. of Sp.Sc. & Astrophys., Stanford Univ., Stanford, USA
SUBRAHMANYA,C.R., School of Phys.,Sydney Univ.,Sydney NSW 2006,Australia
SUBRAHMANYAN,R., Radio Astronomy Centre (TIFR), Ootacamund 643001,India
SUBRAMANIAN,K., Astronomy Centre, Univ. of Sussex, Brighton BNI 9GH, UK
SUKUMAR,S., Radio Astronomy Centre (TIFR), Ootacamund 643001, India
SVENSON,R., Nordita, Bledgamsvej 17,DK-2100 Kopenhagen, Denmark
SWARUP,G., Radio Astronomy Centre (TIFR), Ootacamund 643001, India
SWINGS,J.P., Inst. d'Astrophysique, Univ.de Liege,Cointe-Ougree,Belgium
THAKUR,R.K., Phys. Department,Ravishankar University,Raipur-452002,India
TRIMBLE,V., Astronomy Program, Univ. of Maryland, College Park, Md, USA
TRINCHIERI,G., Osservatorio Astrofisico di Arcetri, 50125 Firenze,Italy
TURNSHEK,D.A., Space Telescope Sc.Institute, Baltimore, Md 21218, USA
TYSON,J.A., Bell Labs, 600 Mountain Avenue, Murray Hill, New Jeresy,USA
UNWIN,S.C., Owens Valley Radio Observatory, Caltech, Pasadena, Ca, USA
VAGNETII,F., Instituto Astronomico, Via Lancisi 29, I-00161 Roma, Italy
VALTAOJA,E., Univ.of Turku,Dept.of Physical Sciences, Turku 50, Finland
VAN HEERDE,G.M., Sterrewacht Leiden, 2300 RA Leiden, The Netherlands
VATS,H.O., Physical Research Laboratory, Ahemedabad 380009, India
VEERARAGHAVAN,S., Astronomy Dept., Univ. of California,Berkeley, Ca,USA
VERON,M.P., Obs. de Haute-Provence, Saint-Michel-L'Observatoire, France
WAGH,S.M., Dept. of Mathematics, Univ. of Poona, Pune 411007, India
WALL,J.V., Royal Greenwich Observatory,Hailsham,East Sussex BN27 1RP,UK
WAMPLER,E.J., ESO, D-8046 Garching bei Munchen, West Germany
WANDEL,A., Astronomy Program, Univ. of Maryland, College Park, Md, USA
WHITE,G.L., Radio Physics CSIRO, P.O.Box 76, Epping NSW 2121, Australia
WILKES,B.J., Smithsonian Astrophys. Obs., Cambridge, Ma 02138, USA
WILKINSON,P.N., NRAL, Jodrell Bank, Macclesfield, Cheshire SK11 9DL, UK
WILLS,B.J., Dept. of Astronomy, Univ. of Texas, Austin, Tx 78712, USA
WILLS,D., Dept. of Astronomy, Univ. of Texas, Austin, Tx 78712, USA
WLERICK,G., Observatoire de Paris, 75014 Paris, France
VAN WOERDEN,H., Kapteyn Laboratory, P.B.800, Groningen, The Netherlands
WOLTJER,L., ESO, D-8046 Garching bei Munchen, West Germany
WORRALL,D., Centre for Astrophys., 60 Garden Street, Cambridge, Ma, USA
YEE,H.K.C., Univ. of Montreal, Dept. of Phys., Montreal, H3C 3J7,Canada
ZAMORANI,G., Instituto Di Radio Astronomia, I-40126 Bologna, Italy
ZENSUS,A., OVRO, Caltech 105-24, Pasadena, Ca 91125, USA
ZHOU YOU-YUAN, Centre of Theoretical Astrophysics, University of Science
    & Technology, Hefei, Peoples Rep. of China

"It might be logical to start off with definitions and taxonomy, but I doubt that this would really help. Let us just define a quasar as what Dr.Swarup and his co-organisers deem to be one for the purposes of this conference, but recall S.Toulmin's dictum that 'definitions are like belts: the shorter they are the more elastic they must be' ".

— Martin Rees (p.1)

Martin Rees, Govind Swarup and Smita Shanbhag

INTRODUCTORY LECTURE

Martin J. Rees
Institute of Astronomy
Madingley Road
Cambridge CB3 OHA
ENGLAND

It is now 22 years since quasars were discovered. When the ageing
veterans of those pioneering investigations think back over two decades
of boisterous debate, their reactions are probably rather mixed.
Wonderment at the range and variety of novel phenomena revealed in all
wavebands must be tinged with disappointment that we seem so slow in
grasping what is really going on. Our understanding has advanced
slowly, through many small steps — forward steps preponderating
(fortunately) over backward ones.

Only gradually did quasars become generally recognised as just
extreme instances of the bewildering variety of phenomena associated
with active galactic nuclei. This zoo of objects spawned a confusing
nomenclature: quasars, BL Lacs, blazars, optically violent variables
(OVVs), Seyferts 1 - 2, starbursts, liners, and warmers. It might be
logical to start off with definitions and taxonomy, but I doubt that
this would really help. Let us just define a quasar as what Dr Swarup
and his co- organisers deem to be one for the purposes of this
conference, but recall S. Toulmin's dictum that "definitions are like
belts: the shorter they are the more elastic they must be".

It is still premature to propose a multiparameter classification
scheme with a real physical basis. Meanwhile, we may as well stick
with the established nomenclature, "historical baggage" though it may
be. All that matters is that we should agree on what the words mean,
and be mindful that many characteristics (for instance, whether or not
images appear "stellar") depend on the sensitivity of our instruments,
being neither intrinsic nor fundamental.

Figure 1 depicts the main phenomena manifested by quasars, with
rough estimates (which in some respects are uncertain and model-
dependent) of the associated length scales. I went through the
provisional programme of this meeting trying to classify the papers in
terms of the relevant lengthscale, assigning this according to the
theoretical prejudices embodied in Figure 1. There are 23 papers
dealing with scales $\sim 10^{15}$ cms ("prime movers", X-ray and UV continuum), 48 papers with scales $\sim 10^{19}$ cms ((broad emission lines and
VLBI radio), 15 papers with $\sim 10^{23}$ cms (fuzz and extended radio

1

G. Swarup and V. K. Kapahi (eds.), Quasars, 1–13.
© *1986 by the IAU.*

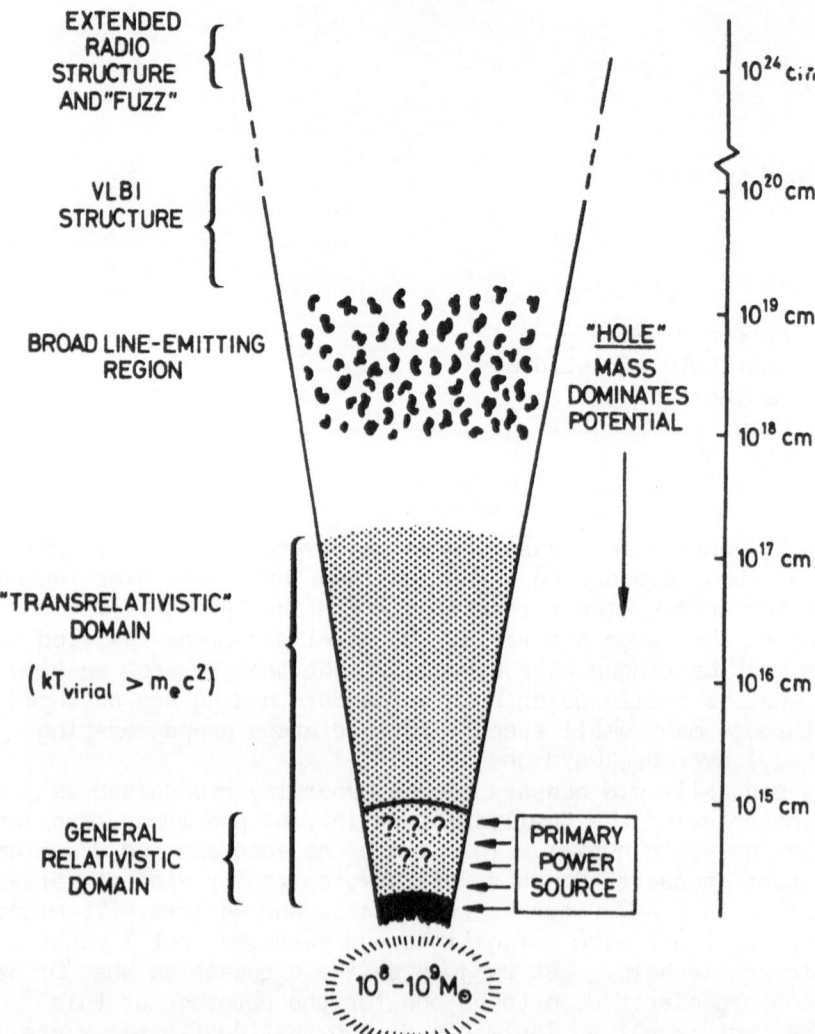

Figure 1. A schematic "slice" through a quasar, illustrating that observed phenomena span a range of almost $10^{10} - 1$ in lengthscale. If the primary power source involves a single mass of $10^8 - 10^9 M_\odot$, this object dominates the gravitational field (yielding a potential well of "$1/r$" form) out to a few parsecs. On larger scales, interaction of radiation and outflowing matter with the environment of the host galaxy is the determinant factor. General relativistic effects would be important within a few times r_s (where $r_s = 3 \times 10^{13}(M/10^8 M_\odot)$ cm); the primary power supply may be concentrated within this domain. Within a radius $(m_p/m_e) r_s \cong 10^3 r_s$, the virial temperature would exceed the rest mass energy of an electron.

structure); finally 30 papers address cosmological effects —
clustering, absorption lines, lensing, and cosmological evolution (i.e.
scales $\sim 10^{27}$ cm). Let us hope that this indeed proves to be the most
useless statistical data presented at this conference! It is my
prerogative as opening speaker to raise questions rather than
necessarily answer them. I shall do this by highlighting, in a
qualitative and subjective way, some issues which I hope we shall focus
on during the week. Let me apologise in advance for omitting references or attribution for the various ideas mentioned here. Useful compilations, and benchmarks against which we can assess current
progress, are the proceedings of five recent conferences concerned with
quasars: those held at Liège (Swings 1983), Munich (Brinkman and
Trumper 1984), Manchester (Dyson 1985), Santa Cruz (Miller 1985) and
Trieste (Giuricin et al. 1986).

THE EMISSION LINES ETC.

It is unsurprising to find so many papers dealing with the broad
emission lines - undoubtedly the most intensively studied aspect of
quasars. There is a substantial body of data with high signal-to-
noise, and modelling rests on the firm basis of conventional and
well-understood physics. Among the still-unsettled questions are the
following.

(a) Is photoionization by a UV continuum the only heat input?

(b) How well-defined is the radius of the broad line region and
the cloud density within it? The standard density normally quoted is
of order 10^{10} or 10^{11} cm^{-3}. But how much gas could be present with
densities below 10^9 cm^{-3} or above 10^{11} cm^{-3}? Maybe those theorists
who aim to show why clouds only condense at one characteristic radius
are trying harder than they need.

(c) What is the spatial configuration? Is the gas in an
approximately spherical distribution; or is it, on the other hand,
in some kind of thick disc, or in a jet-like outflow?

(d) What are the kinematics? Is the gas flowing out in a wind,
and if so what pushes it? Are clouds on gravitationally bound orbits?
Or are they falling inward, perhaps as part of some accretion flow
pattern?

Further issues, involving somewhat less direct inferences from
observations, concern the formation of the inferred clouds and filaments, and the relation of the broadline region to the VLBI radio
structure, which has comparable size.

Before "homing in" on the central engine ($\lesssim 10^{15}$ cm) note that many
complexities may occur on intermediate scales $10^{15} - 10^{18}$ cm. A massive
central object would dominate the gravitational field throughout this
volume. The spatial distribution of the material would be as intricate
as on any larger scale (though of course we have little hope of
resolving it directly). One would expect, moreover, a multiphase
structure: optically thick clouds at $10^4 - 10^5$ °K, embedded in plasma
where all the electrons may be marginally relativistic, and the ion
kinetic temperature $\gtrsim 1$ Mev. The density of > 1 Mev photons may be

high enough for $\gamma + \gamma \to e^+ + e^-$ collisions to maintain a pair density
at least comparable with the original electron density. Processes on
these length-scales may be crucial in determining the emergent continuum spectrum (including a UV bump?) and in establishing the collimation of jets. The continuum emission mechanism may be quite
different in a typical radio quiet quasar from in an OVV - only in the
latter objects do we really know (from polarization, variability etc)
that synchrotron-type radiation dominates.

The degree of <u>directionality</u> of the power output is another key
question. There is good evidence for relativistic beaming on scales 1
- 10 parsecs from VLBI measurements of superluminal sources. More
controversial, however, is the relevance of relativistic beaming to the
demarcation between radio loud and radio quiet quasars, and the issue
of whether BL Lacs and OVVs are beamed towards us. Quantitative
details of the beaming would be sensitive to the flow pattern, and to
shocks, etc., in the outflow. Beaming is directly observed on scales
of parsecs and larger; it is still unclear whether the collimation is
established only on that scale, or on the dimensions of the central
source (see Figure 1).

Some directionality could arise merely from non-spherical
geometry (even if there are no relativistic motions or large doppler
effects). For instance, optically thick discs would appear brighter
viewed face-on, and edge-on objects may appear faint because of
absorption in a larger-scale disc. These issues could be clarified by
seeking   correlations between emission line widths and other
properties.

THE CENTRAL POWER SUPPLY

At the first Texas conference, held in 1963, the implications of
quasars for theorists were already recognised. As T. Gold said in his
banquet speech on that occasion "Relativists and their sophisticated
work are not only magnificent cultural ornaments, but might actually be
useful to science!" (see Robinson <u>et al</u>. 1964). And one of the few
opinions about quasars where the consensus has been steady since 1963
is that the energy is gravitational in origin. (An un- weighted
majority vote is in itself, however, worth little in this subject!)
Conditions in the central continuum-emitting region are more extreme,
in terms of energy densities, etc., than in the emission line clouds
(though the physics is by no means as exotic as on the surfaces of
neutron stars, for instance).

The prime reason for invoking collapsed objects is that any
gravitationally-powered source which releases more than 1% of its rest
mass energy contracts unstoppably; and a collapsed object, once formed,
offers a more powerful and efficient power source than any precursor
system. So if the power is gravitational in origin, it is inconsistent <u>not</u> to envisage that collaped objects power the most powerful
active nuclei — quasars and radio galaxies in particular.

Irrespective of the details of the central mechanism, there are
simple thermodynamic constraints on the form of the primary emergent

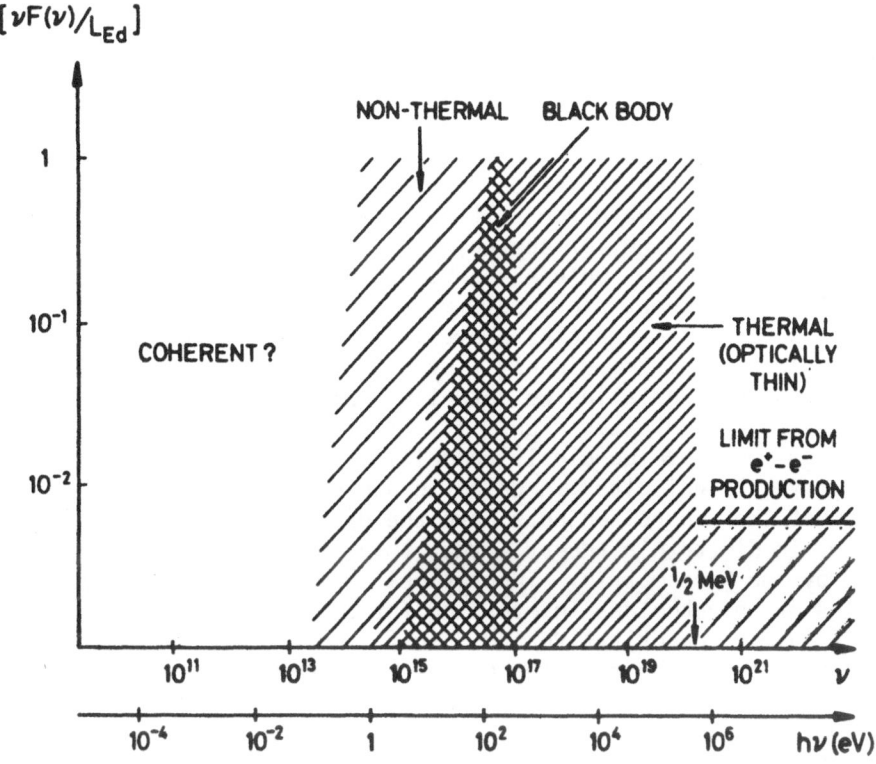

Figure 2. This diagram illustrates schematically the thermodynamic constraints on the radiation spectrum that could emerge directly from a source of luminosity $L_{Ed} \cong 10^{46}$erg s^{-1} and dimensions $\sim 10^{14}$cm. Black body radiation with $T_{bb} \cong 3 \times 10^{5}$ K (heaviest shading) would peak in the UV. Optically thin thermal radiation from plasma with $T > T_{bb}$ could emerge in the X-ray band, but absorption would still prevent a high fraction of the radiation from emerging in the optical band even though the Rayleigh-Jeans Law would be higher than for the black body by a factor T/T_{bb} (intermediate shading). Synchrotron self-absorption yields a cut-off in the infra-red, so incoherent non-thermal radiation is restricted to the part of the diagram that is lightly shaded. (This is calculated assuming a magnetic field strength $\sim 10^{4}$ G, the equipartition value expected in radial accretion flow). Absorption of γ-rays via $\gamma + \gamma \to e^{+} + e^{-}$ limits the luminosity above 1 Mev to $\lesssim 0.01$ L_{Ed}. The main direct output from the central source would therefore be: (i) Thermal radiation in the UV or X-ray bands, (ii) non- thermal radiation, anywhere from the near infra-red to hard X-rays (but few γ-rays) and (iii) a relativistic e^{+} - e^{-} pair outflow.

radiation. Figure 2 illustrates these for a typical quasar with M = 10^8 $M_\odot$, $L_{Ed} \cong 10^{46}$ erg s^{-1} and dimensions $\sim 10^{14}$cm (~ 3 r_s). If a luminosity L_{Ed} emerged with a black body spectrum, the temperature would be $\sim 3 \times 10^5$ K, the flux therefore peaking in the far ultra-violet; a luminosity $\sim L_{Ed}$ in the UV or X-ray band could be generated via an optically-thin thermal mechanism (e.g. Comptonised brems-sstrahlung); even if the radiation were non-thermal, involving rela-tivistic electrons, synchrotron self-absorption produces a cut-off in the infrared (implying, of course, that no radio emission can come from dimensions $\sim 10^{14}$cm unless a coherent mechanism is operative); the pair production constraint would prevent more than 0.01 L_{Ed} from emerging as γ-rays with energies above 1 Mev. These simple consid-erations would tell us immediately — even if we did not have other evidence — that the radiation we actually observe must be generated or reprocessed further out (see Figure 1).

Although there are genuinely compelling reasons for invoking generic "gravitational pits" in galactic nuclei, there is little evidence that the central objects have the distinctive properties of black holes in Einstein's theory — that they are, in other words, described by the Kerr metric. Theoretical models that postulate Kerr black holes are nonetheless interesting for two reasons. First, they may lead to a more quantitative and specific model for the form of the continuum radiation: the emergent energy could derive from accretion, or alternatively from electromagnetic effects that tap the holes' spin energy, converting it into non-thermal power and electromagnetic beams. Secondly, quasars may offer relativists a diagnostic that permits, for the first time, tests of general relativity beyond the weak-field approximation.   Such modelling also involves other poorly-understood complications — for instance, relativistic MHD and radiation processes in pair-dominated plasmas.

WHAT CAN WE REALISTICALLY EXPECT FROM THEORIES AND MODELS?

How will we ever really know whether there are black holes in AGNs? The evidence can never be more than circumstantial. But we should not be too downcast by that. After all, the evidence that the Sun is powered by nuclear fusion, a cherished dogma never seriously contested, is also really just circumstantial. However, the confrontation of theory with observations, indirect even for stars, is more ambiguous still for quasars: in stars, energy percolates to the observable surface in a relatively steady, symmetric, and well-understood fashion; but in galactic nuclei it is reprocessed into all parts of the electromagnetic spectrum on scales spanning many powers of 10, in a way that depends on poorly-known environmental and geometrical effects in the host galaxy.

The generic black hole model is not infinitely flexible, and not invulnerable. It could be refuted in at least three ways:

1. by finding very regular periodicities, particularly on timescales below 1 hour.

2. by showing that the central masses were $<<$ 10^6 $M_\odot$ in Seyferts or $\ll 10^8$ $M_\odot$ in radio galaxies;
 or 3. by developing a theory of gravity more convincing than general relativity which prohibits black holes.

But a clean-cut refutation, leading to abandonment of some theory, happens only rarely in astrophysics; many models have persisted unrefuted for a long time. A cynic might argue that they have survived only because they don't go beyond generalities, or else because their proponents have been adept, like the motor mechanics who must abound here in India, at replacing or modifying faulty parts to keep shaky old models "roadworthy". Such a cynical attitude is not necessarily justified, and to explain why I must digress briefly into methodology. The way we are told science is done is like this: the data suggest a model, which suggests further tests, whereby the original model is either refuted or refined. Such a simple scheme is realistic in, for instance, particle physics, where the fundamental entities may be exactly reducable to a few basic constants and equations. But other sciences deal with inherently complex phenomena and no theoretical scheme can be expected to account for every detail. In geophysics, for instance, the concepts of continental drift and plate tectonics have undoubtedly led to key advances; but they cannot be expected to explain the shape — the precise topography — of the continents. What we should aim to do, in our attempts to understand quasars, is focus on those features of the data which genuinely test crucial ideas, and not to be diverted into measuring or modelling something which is accidental or secondary.

The problem of interpretation can be illustrated by an (almost arbitrarily chosen) example: the data on continuum X-ray emission from quasars. As summarised in Figure 3, one can envisage various

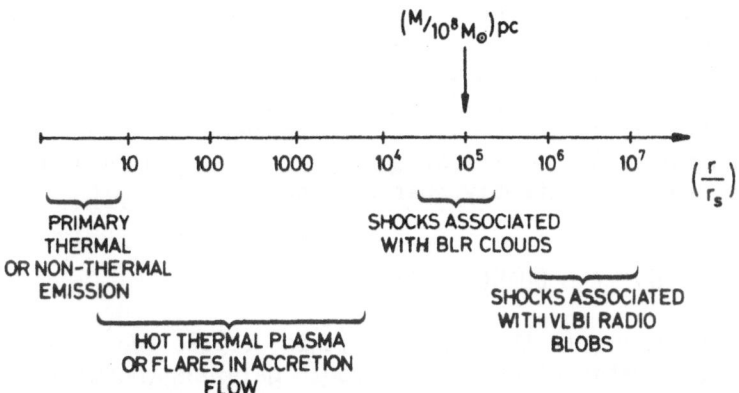

Figure 3. This diagram indicates that several different mechanisms, operating on quite different dimensions, could contribute to the observed X-ray continuum from quasars. The relative contribution from each mechanism could plausibly depend on radio properties, orientation, etc. (Absorption of soft X-rays by gas on kpc scales in the host galaxy could further complicate attempts to fit the data in terms of a few-parameter model.)

processes, on very different scales, any of which could contribute
significantly to the observed X-rays. This means that the <u>overall X-ray
spectrum</u> is unlikely to give a good fit to any simple model and is in
itself of limited interest. On the other hand, the detection of <u>large
amplitude rapid variability</u> in the X-rays would clearly pin down the
scale of the dominant emission, and thereby discriminate among the
candidate production mechanisms mentioned in Figure 3. Detection of
<u>line features</u> would also, obviously, be an important discriminant.

When the phenomena being studied are inherently complicated and
ambiguous, the confrontation between observation (or experiment) and
theory is indirect. It is mediated by what one might call a scenario",
like this:

In the case of quasars, elements of the scenario include the things
depicted in Figure 1 - clouds, inflow, winds, jets, radio blobs,
shocks, collapsed objects, and so forth. We need some such scenario in
our minds in order to interpret observations, and formulate ideas for
further observations. A quasar is too complicated, and involves far
too many parameters, to be quantitatively modelled all in one go; the
scenario nevertheless suggests what the essential ingredients of the
system really are, and which physical processes merit detailed
investigation by theorists. Among these would be photoionization
equilibrium, particle acceleration, electron-positron pairs,
relativistic MHD, and black holes.

The above diagram describes, I believe, how our subject actually
progresses. We start off with only a very vague picture of what might
be going on (or maybe several quite different alternatives); different
kinds of observations, and modelling of various individual ingredients,
play complementary roles in revealing inconsistencies in our original
viewpoint, and suggesting modifications, so that gradually a favoured
scenario comes into consistently sharper focus.

HOST GALAXIES, "FUZZ" AND QUASAR STATISTICS

The relation of quasars to their host galaxies raises two questions.
What is the relationship of quasars to radio galaxies (normally
ellipticals) and to Seyferts (normally spirals)? Also, does a close
companion or group membership enhance quasar activity? The latter has
often been claimed, but the mechanism whereby a companion can trigger
enhanced activity in the nucleus (rather than merely enhancing the star
formation in the body of the galaxy) is still quite unclear.

A more general issue deserving further attention is the
<u>environmental impact</u> of quasars. X-rays and UV from the quasar can
heat, and even expel, gas in a surrounding galactic disc. Shock waves

related to the extended radio structure could trigger star formation. For these reasons, and no doubt for others also, the observed fuzz is unlikely to resemble any normal undisturbed galaxies. Processes whereby the quasar reacts back on its surroundings may determine the life-cycle and the fuelling rate. Variability has of course been recorded over intervals from hours to years, but quasars may be equally variable on all longer timescales between 10 years and their total active lifetime. Moreover such variability could lead to transformations between different types of active galaxies (quasars $\rightarrow$ Seyferts, or quasars $\leftrightarrow$ BL Lacs).

Let us turn now from individual objects to the properties of the collectivity of quasars. The energy output from quasars is known to within a factor of order 2: it is $\sim 3000\ M_\odot c^2$ per Mpc^3, about half coming from quasars with apparent magnitudes in the range 19 - 21. Galaxies contribute $\sim 10^5$ in the same units, and the microwave background $\sim 7.5 \times 10^6$. So, even though quasars may influence their host galaxies, they are collectively rather modest contributors to the cosmic energy budget, because of their low space density. They may nevertheless have a crucial cumulative effect on the entire intergalactic medium, because their energy emerges largely in forms such as an ionizing continuum and high velocity jets and winds.

The prime era of quasar activity is at z = 2 or 3. It is from these redshifts that most of the quasar background light originates. The population thereafter decays on a timescale of order t_{Evo} = 2 x 10^9 years. This is, however, merely an upper limit to the lifetime t_Q of each object: many generations of individual quasars could be born, and could die, in the period over which the population declines. Several important numbers depend on what t_Q actually is: the mass and the number of quasar remnants, the ratio of their luminosity to the Eddington limit, and the issue of whether the broadline region can be gravitationally bound.

The following table brings out these points. It contrasts two hypotheses: (i) that there was, in effect, only one generation of quasars, which were long-lived and massive; versus (ii) that there were $\sim$ 50 generations of quasars, so that their individual masses (for a given efficiency) need not have built up to such high values, and quasar remnants would be more numerous.

(i)	(ii)
$t_Q \cong t_{Evo}$	$t_Q \cong 4 \times 10^7$ yrs $\cong 0.02\ t_{Evo}$
$M = 2.5 \times 10^9\ \varepsilon_{0.1}\ M_\odot$	$M = 5 \times 10^7\ \varepsilon_{0.1}\ M_\odot$
$L \ll L_{Ed}$	$L \simeq L_{Ed}\ \varepsilon_{0.1}$
Broad-line regions gravitationally bound	Broad-line region not gravitationally bound
Very massive remnants in $\sim$ 2% of galaxies	$\sim 10^8\ M_\odot$ remnants in most bright galaxies

In this rough table, $\varepsilon_{0.1}$ is the efficiency measured in units of 0.1.

Clearly the z-dependence of the quasar population is related to galactic evolution, though we cannot yet quantify how. What other handle do we have on what the universe was like at redshift z = 3, when it was perhaps only 1/8th its present age? Ordinary galaxies offer some clues. Cosmologists tell us that after the universe had expanded for about 10^6 years, the fireball cooled down below 3000°, and the primaeval black body radiation shifted into the infrared. The universe then experienced a "dark age", which persisted until the first bound systems formed, igniting localised nuclear or gravitational power sources. When this "first light" occurred depends on the specific cosmogonic scenario. It could be at redshifts much greater than 2, but need not be.

There are now, indeed, firmer reasons for believing that galaxy formation was still going on at z = 2 or 3. This reappraisal comes from evidence of dark halos around galaxies extending out to r $\cong$ 100 kpc ($r_{100} \cong 1$). If galaxies extended only out to 10 kpc, they could have formed on a timescale of 10^8 years, at z > 10. However, it is different if their massive halos extend out to $\gtrsim 100$ kpc. The discs of galaxies could have formed from tidally-torqued material falling in, conserving its angular momentum, from $r_{100} \cong 1$. The infall time is of order $10^9 r_{100}$ years, implying that infall from the halo, and disc formation, cannot be completed until the age of the universe is 2 x 10^9 years (i.e. z = 2 or 3) for $r_{100} \equiv 1$.

When the universe was $\lesssim 20\%$ its present age, the host galaxies, whose discs were still in the process of forming, would have been very different. We would expect more uncondensed or infalling gas to be present, implying that the fuzz should be more conspicuous, because it would more effectively scatter or reprocess the quasar light.

Is a high-z cutoff related to the formation of galaxies with well-defined nuclei? Or do higher z quasars exist, shrouded by gas or dust in the host galaxy and therefore unrecognised?

QUASARS AS PROBES

An aspect of quasar studies that has burgeoned over the last few years, partly because it is in no way stymied by our confusion about their intrinsic properties, is their use as probes of the intervening medium: absorption by gas along the line of sight, and gravitational lensing by compact or massive bodies. Quasars allow us to probe 80 or 90% of cosmic history — a period spanning the later stages of galaxy formation, and the contraction of clusters.

Absorption lines: intervening gas clouds

The statistics of the inferred absorbing clouds yield information on the z-dependence of the co-moving density of the clouds, the distribution of column densities, and velocity correlations. On the basis of such evidence one can begin to address the physics of individual clouds. In particular, how are they confined? Is it by the pressure of an external hot diffuse medium, or are they gravitationally bound?

What are the sizes and shapes of the clouds? How do they relate to galaxies or protogalaxies?

Gravitational lensing by galaxies and "dark matter"

Lensing by the overall mass concentrations in galaxies, halos, and clusters yields multiple images on scales $\sim$1 arcsec. Several such cases have been found over the last few years. Additionally, mini-lensing by individual compact objects of mass M_ℓ (stars, "Jupiters", or black holes in halos) yield images separated by scales $\sim 10^{-6} (M_\ell/M_\odot)^{\frac{1}{2}}$ arcsec. Even if these images cannot be optically resolved, they can contribute to variability and yield magnification.

Although lensing offers an important probe of the intervening medium, it is merely a confusing complication if we are trying to infer the intrinsic properties of quasars, such as the luminosity function. It is as though we view the remote universe through frosted glass. If the intrinsic luminosity function were very steep, magnified objects could preponderate in magnitude-limited surveys of high z quasars.

FINAL COMMENTS

These sketchy and impressionistic comments — less a smorgasbord, I fear, than a dog's breakfast — paid no heed to the historical development of quasar studies. But the way things actually happened was really far from optimal — we were given many clues, but in a confusing order. It's ungracious to the discoverers of quasars to lament, with hindsight, the insight and resourcefulness they displayed in 1963. But if quasars had only been discovered in (say) 1973 — when we already knew more about lower-level activity in galactic nuclei (Seyferts, etc), and when pulsars and X-ray binaries had convinced us that gravitational energy could be channelleld efficiently into radiation — quasars would never have seemed 'sui generis', and theoretical ideas would have evolved less waywardly to their present state.

We shall be reminded later this week that there is not yet complete unanimity on even the most basic aspects of quasars. Though most astrophysicists are impressed that the growing body of data, when "conventionally" interpreted, meshes consistently into an overall cosmological and cosmogonic scheme, a few are convinced that some apparent anomalies demand a new (non-cosmological) mechanism for redshifts. The "dissidents", professing surprise that their claims meet persistent scepticism, sometimes attribute this to a conservative intellectual bias in the astronomical community — a blinkered reluc-tance to entertain fundamentally new ideas. But we sceptics — some of us, anyway — have, if anything, the _opposite_ bias. The prospect that astronomers might discover some fundamentally new physics is a seductive one, and nothing could be more gratifying — and do more to render this a memorable conference — than the presentation of some novel and really clinching evidence for anomalies. Most of us might lay high odds against this happening; but let's hope nevertheless that the overall progress to be reported this week, along a broad front, will oblige Dr Woltjer drastically to update the draft of his concluding remarks which he doubtless already has on file.

REFERENCES

Brinkman, W. and Trumper, J. (eds) 1984, 'X-ray and UV emission from
     Quasars and Active Galactic Nuclei' (Publ. Max Planck Institut,
     München).
Dyson, J.E. (ed) 1985, 'Active Galactic Nuclei' (Manchester
     University Press).
Giuricin, G., Mardirossian, F. and Mezzetti, M. (eds) 1986, 'Structure
     and Evolution of Active Galactic Nuclei' (Reidel, Dordrecht).
Miller, J. (ed) 1985. 'Astrophysics of Active Galaxies and Quasi-
     Stellar Objects' (University Science Books, California).
Robinson, I., Schild, A. and Schucking, E.L. (eds) 1964, 'Quasi-
     Stellar Sources and Gravitational Collapse' (University of Chicago
     Press).
Swings, J.P. (ed) 1983, 'Quasars and Gravitational Lenses': Proc. 24th
     Liege Internat. Colloquium (Universite de Liège, Belgium).

DISCUSSION

Cowsik : Could you comment on Quasars as probes of dark matter ?

Rees : If the dark matter consists of individual compact objects - low mass faint stars or black holes - then "minilensing" is expected, as we shall I'm sure, learn from later speakers. The multiple images of QSOs which are interpreted as lensing by galactic-mass objects are of course a probe for massive halos in the lensing galaxies. Moreover, it seems that the M/L of some lensing objects is exceptionally high, and this raises the interesting question of whether there may be a population of "failed galaxies" - massive halos without luminous cores. Such objects are expected in some models of galaxy formation, and gravitational lensing would be an important probe.

Kembhavi : Couldn't a large number of quasars at high redshift shrouded by young galaxies produce bumps in the background radiation at various wavelengths ?

Rees : Yes. In particular, hard X-rays would not be absorbed by gas in a young galaxy. Even if we cannot directly detect objects of the highest z, background limits can certainly offer constraints on their collective properties.

Kundt : Could you comment on possible connections between QSO activity and the activity of young stellar objects (both classes of sources appear to produce supersonic jets.) ?

Rees : The velocities are of course much lower in the protostellar objects, may be because the gravitational potential wells are shallow. Your question also relates to the question of the collimation mechanism, which indeed may be similar in these two contexts. It's important to remember that we have no direct evidence that collimation in AGNs occurs on scales below about one parsec.

Burbidge : You have set the stage by talking about the "host galaxy". I would like to point out that there is no unambiguous evidence for galaxies around QSOs, and if they were there, the effect of the QSO might well make them abnormal. Thus in principle it may be difficult to prove that QSOs are embedded in host galaxies.

Rees : I agree.

Thakur : Validity of GTR has been tested and verified only in the weak field approximation but not for the strong fields. How can the models of the prime mover based on black holes be relied upon, especially when the singularity invovled does not appear to be physical and when the laws of physics do not hold good at the singularity ?

Rees : A motivation for developing detailed models is the hope that this may reveal a diagnostic of the precise metric, and thereby permit some test of GTR, which has indeed, as you imply, only so far been validated when fields are weak. (However, I don't think our ignorance about physics near the singularity is an impediment. This singularity lies within the event horizon, and is irrelevant to observational phenomena. Analogously, one has confidence in the theory of atomic structure, despite our ignorance of subnuclear structure within individual protons.)

Thakur : Accretion discs around black holes have been suggested as the sources of energy from quasars and AGNs. However, these models appear to give steady out-flow of energy whereas violent activities have been observed in the quasars and AGNs. How can those contrary facts be reconciled ?

Rees : Theorists have concentrated on stationary flow patterns in accretion flows. However, one would realistically expect the phenomena to be irregular on a whole range of timescales.

Segal : This was a very coherent and comprehensive account of quasar phenomena, but in part, and especially as regards evolution and differences between quasars at various redshifts, it appears as model-dependent. Is there any direct observational evidence for evolution, or is this simply a corollary to interpretation in a Friedman cosmology ?

Rees : It depends on assuming a Friedman-type model, as you rightly imply.

I

SURVEYS

"We must continue this vigorous search for quasars at all wavelengths if we are to gain an understanding of the overall population of quasars and an accurate appreciation of the full range of quasar activity."

- Malcolm Smith (p.28)

QUASAR SURVEYS

M. G. Smith
United Kingdom Infrared Telescope
Hawaii Headquarters
665 Komohana Street
Hilo, Hawaii, U.S.A.

ABSTRACT. A review is given of progress in surveys for quasars at frequencies from radio to x-ray. Radio results show evidence for a decline in the radio luminosity function for flat-spectrum radio sources at redshifts z > 2. The IRAS survey is uncovering hitherto unknown dusty Seyfert galaxies. Optical surveys, which yield the largest number of QSOs per square degree, may suffer from selection effects which depend on intrinsic luminosity, redshift, and spectral evolution - particularly above redshift 2. Below redshifts of about 2.3, the optical magnitude-redshift plane is being filled in to the point where the evolution of the luminosity function can be seen directly. The statistics of quasar pair separations provide the best evidence so far for quasar clustering.

The existence of many potentially significant selection effects means that a multi-frequency approach to quasar surveys is likely to prove essential to an understanding of the evolutionary behaviour of the quasar population as a whole.

1. INTRODUCTION

In his introductory remarks at this conference, Martin Rees outlined five main research areas being addressed with the products of current surveys for QSOs, viz: (a) the cosmological evolution of QSOs, (b) clustering of QSOs, (c) QSOs as background probes of intervening absorbing material, (d) gravitational lensing and (e) the intrinsic physical properties of QSOs. In this article, I will review progress in surveys for quasars, at frequencies from radio to x-rays, in the 2 years since my review at Liège (Smith 1983); that article (see also Smith 1978; 1981) gives a much more detailed account of the survey methods themselves. In the last two years, work has concentrated more on applying these existing basic survey techniques, rather than developing new ones. The other chapters in this book provide more detailed reviews of the physical properties of quasars.

17

G. Swarup and V. K. Kapahi (eds.), Quasars, 17–32.

2. RADIO SEARCHES

Following the discovery of PKS2000-330 at z=3.78 using radio search
techniques (Peterson et al. 1982), Dunlop et al. (1985) have found
that PKS 1351-018 has a redshift of 3.71. Their discovery is
important because PKS1351-018 is a member of a complete sample. The
lack of objects like it in an "intermediate" flux density sample (S >
0.1 Jy at 2.7GHz - Peacock and Gull (1981)) re-inforces a recent
suggestion by Peacock (1985) that the quasar radio luminosity function
for flat-spectrum sources declines for redshifts z > 2. Dunlop et al.
predict that there should be about 8 quasars in the main Parkes survey
with z > 4.

Peacock gives a more detailed account of his statistical studies
in a later chapter. Peacock has extended his free-form analysis that
he described at Liège (see also Peacock and Gull 1981) and concluded
that, for the class of compact flat-spectrum sources, there is a
redshift "cutoff". At redshift z=2 the radio luminosity function no
longer evolves, and at z=4 it is reduced by a factor > 3 of its peak
value. Peacock, Miller and Longair (1985) have concluded that the
apparent correlation between the radio and optical fluxes from quasars
is an artifact caused by the existence of two classes of quasar with
differing host galaxies; radio quiet QSOs exist in spiral galaxies,
while radio-loud quasars are found in elliptical galaxies. Miley and
de Grijp (1985) also note that "optically selected Seyferts are
associated predominately with spirals and radio selected ones occur
mainly in elliptical-type galaxies." The evolutionary properties of
optically selected QSOs may thus differ from those of radio quasars.
We will discuss the "cutoff" for optically selected QSOs in section 4.

3. INFRARED SEARCHES

Infrared observations have, for the first time, been used in a
successful search for high-redshift QSOs. Dunlop et al. (1985) used
measurements on the United Kingdom Infrared Telescope (UKIRT) to
establish the optical-infrared colour of the candidate identification
for the z=3.71 QSO PKS1351-018. They found that the "combined
properties of (i) compact, flat spectrum radio source, (ii) faint red
optical object, and (iii) blue optical-infrared colour ... made
1351-018 a prime candidate for a high redshift quasar and consequently
a high priority candidate for spectroscopy." Paul Hewett and I have
recently started to make measurements in the infrared of the "peculiar
red objects" he has discovered in his generalised pattern recognition
approach to IIIaF objective-prism spectra. We shall observe known
very high redshift quasars (z>3.5) that fall in his large samples to
see if their infrared properties can help distinguish between quasars
and other kinds of "peculiar red objects" (see discussion in section
4).

The IRAS point source catalogue was released in its final form at
about the time I got on the plane to come to this conference. This
IRAS "strong source" survey has a characteristic depth to redshifts of

about 0.03 and covers about 96% of the sky to flux levels about 0.5Jy at 12, 25 and 60 microns and to about 1.5 Jy at 100 microns for point sources; it has yielded detections of several hundred previously known active galactic nuclei (AGN). However, there are only about 20 matches with known objects having M(v) brighter than -23, and 9 with M(abs)<-25. None of the classical, luminous QSOs are infra-red selected - all are objects discovered earlier in other wavebands.

The infrared waveband offers an opportunity to combat some of the selection effects which particularly hamper optical searches. It is usually supposed that classical quasars are very luminous active galactic nuclei - AGN. ("Activity" here refers to sources of energy not traceable eventually to nuclear reactions in stars or to their remnants such as supernovae). Surveys identifying low-powered quasar-like activity in the nuclei of nearby galaxies have been highly successful in recent years. In his introductory remarks, Martin Rees compared estimates of the relative co-moving space densities of quasars and galaxies near redshifts of two; activity in the nuclei of a large percentage of nearby galaxies favours models in which frequent cycles of activity occur in any given quasar; luminosity evolution would be a gradual reduction in the mean intensity of the QSO cycles. If black holes are the eventual energy source for the quasar's energy output then, as Rees pointed out, less massive ones are needed to power a series of relatively short outbursts rather than a single gradual dimming maintained through much of the history of the universe. Surveys for low-luminosity quasar activity merit attention similar to that usually afforded the more luminous classical quasars.

Some sort of weak quasar-like activity probably exists in most bright spiral galaxies (see, e.g. Keel, 1983; 1985). At the recent Manchester conference on AGN, I presented a catalogue of composite nuclei, in which the degree of non-thermal activity becomes increasingly obvious (Smith, 1985). Penston and Perez (1984) have suggested that Seyfert 2s are merely "Seyfert 1s" whose central source is temporarily quiescent. Recent IRAS releases support the contention that 30-micron, flat-spectrum, thermal components (a million solar masses of warm dust at 80K from a region 10^2 parsecs across, with luminosities ~ 10^9 to 10^{11}Lo) are widespread, at least in Seyfert 2s. Miley and de Grijp (1985) speculate that these components may, in the spirit of the suggestion of Penston and Perez, be dominant in the "off" phase of Seyfert 1s.

The first active galaxy to be selected on the basis of its far-infrared emission, IRAS0421+040P06, has been discussed by Beichmann et al. (1985). They conclude that the detection of 25-micron radiation may prove to be a key method for selecting AGN from the IRAS sample. Miley and de Grijp (1985) elaborate on this theme and conclude from the IRAS strong source survey that the wide occurence of flat "warm" IR spectra can be used as an indicator of nuclear activity; hence "there should be about 1400 Seyferts present in the strong point-source survey and a factor ~ 4 more could be found if the complete survey data were co-added ... comparing this number of several thousand IR Seyferts in the IRAS survey data base with the 554 known active galaxies and the 2251 known QSOs tabulated by Veron-Cetty

and Veron (1984), one can safely predict that within the next few
years IRAS will have led to the discovery of a significant fraction of
all known AGNs." The IR Seyferts are systematically more distant than
the Markarian Seyferts. Some 5% of the IR Seyferts have z>0.1.

Narrow-emission-line galaxies with luminosities in excess of
10^{12}Lo have been discovered using IRAS. Such objects would be visible
as unresolved emission-line objects at redshifts comparable to those
of quasars. A comparison between these (red?) IRAS galaxies and the
(blue) highly luminous emission-line galaxies discovered by Terlevich
on United Kingdom Schmidt Telescope (UKST) objective-prism plates
should be made.

Possibly the closest quasar-like activity is invisible at optical
wavelengths. The centre of our galaxy suffers about 30 magnitudes of
visual extinction. Recent measurements by Geballe et al. (1984) on
UKIRT have revealed HI and He I lines with full widths at zero
intensity of 1500 km/s within a central region less than 0.1pc across.
Serabyn and Lacy (1985) have chosen to interpret these velocities,
along with their own measurements further from IRS16, in terms of
circular motions, and thus derive a central mass of several million
solar masses, all within this tiny volume; they take the measurements
of Geballe et al. as excellent evidence for a black hole at the centre
of our Galaxy.

Other interpretations of these high velocities are possible. An
infrared survey on UKIRT of the Galactic centre region by Gatley et
al. (1984) revealed evidence for a possible mass-loss bubble about 4pc
across, centred on IRS16. IRS16 was found to have colours distinct
from other infrared sources in the immediate vicinity, and a
luminosity of several million Lo. A ring of shock-excited molecular
hydrogen was found, again centred on IRS16. This suggestion of a
powerful outflow source led Geballe et al. to make the crucial
infrared line-profile measurements. Whatever IRS16 really is, its
concentrated high-luminosity source is unique in our Galaxy.

4. OPTICAL SEARCHES

4.1. Quasar-like activity in nearby galaxies

As mentioned earlier, statistics concerning the incidence of activity
in nearby galaxies provide indications of the properties and overall
evolution of quasars. Keel (1985) has defined a low-level active
nucleus as any nucleus with [NII]6563/Hα greater than 0.7. Elvis and
Lawrence have dubbed these objects "microquasars". Such objects may
occur in the nuclei of a large fraction of all spiral galaxies (Keel,
1983). Some of these objects have been detected in x-rays (e.g. Elvis
and van Speybroeck 1982); perhaps IRS16 is a very low-luminosity
object of this type. Moving up in luminosity to what Elvis and
Lawrence call "mini-quasars", we come to the Low Excitation Nuclear
Emission Line Regions, or LINERS - which are known to occur widely in
galaxies (Heckman 1980 a, b, c). Evidence has accumulated from
extensive survey work (see, e.g., the review by Keel, 1985) that "many

LINERS are powered by photoionisation from a flat-spectrum radiation
source similar to the very luminous ones in conventional active
nuclei." Several LINERS are known to be X-ray sources (Halpern and
Steiner, 1983). Many, though not all, LINER nuclei may be low-power
quasars.

Seyfert galaxies usually have [OIII]5007/Hβ greater than 3; the
ratio for LINERS is smaller (see e.g., Baldwin, Phillips and Terlevich
1981). From surveys of nearby galaxies (Stauffer 1982; Keel 1983;
Phillips, Charles and Baldwin 1983), 5% of luminous disk systems have
been found to contain a Seyfert nucleus, as diagnosed by highly
ionised gas (FeVII, NeV) consistent with an ionising continuum
extending to at least 100eV.

4.2. The optical luminosity function and the apparent lack of faint,
high-redshift quasars

Most of this conference will be dealing with more luminous objects
than those seen in the nuclear regions of nearby galaxies.
Considerable progress in optical selection of QSOs has been made
recently using multicolour techniques. Marshall's (1985) paper on
optically selected QSOs with z<2.2, and B<20 is a particularly
important characterisation of the quasars that have been discovered.
He concludes that the slope of the luminosity function for optically
selected quasars shows no significant relation to redshift (c.f. Veron
1983) and that "a power-law luminosity function that evolves
homologously in time describes the data well and suggests that the
same mechanism operates in all quasars to supply the gas and produce
the light."

The diagram shown by Koo (1983, Fig. 3) at Liège shows that
Marshall's power-law luminosity functions appear to flatten out for
fainter quasars. He concluded that quasars were brighter, but
possibly less numerous in the past, especially for redshifts greater
than 2.5. Koo (1985) has selected candidates from SA57 with B<22.6
using 4-metre photographic multicolour photometry, astrometry,
variability and CCD spectroscopy; candidates were chosen that had
star-like images and colours unlike normal stars. This combination of
techniques allowed Koo to filter out white dwarfs, hot subdwarfs and
narrow-emission-line galaxies. No quasars with z>2.54 were found
among the 8 confirmed quasars with B>21.5. Koo discusses the
possibility that higher-redshift quasars may be hiding in the region
of the two-colour plot occupied by most Galactic stars. From his
arguments alone, the numbers seem likely to be small. It is important
to find them - virtually nothing is known about the luminosity
function, spectroscopic properties, or redshift distribution of
quasars with z>3.4. Knowledge of the space density of such objects,
and subsequent detailed studies of the Lyman-alpha absorption systems
in the line of sight will provide strong constraints on the energy
input to, and the properties of the intergalactic medium at those
early epochs (see, e.g., Atwood, Baldwin and Carswell 1985).

Hewett has used plates from the United Kingdom Schmidt Telescope
(UKST) and the Automated Plate Measuring (APM) facility to obtain very

large numbers of low-resolution optical spectra. Hewett and Irwin
(1985) discuss their generalised pattern-recognition techniques for
classifying these spectra. An example of this work, as applied to
quasar research is given in Figure 1. The cluster of 43 large filled
circles at the left of the diagram corresponds to known quasars with
redshifts z<2.2. Five known quasars in this field with very high
redshifts, are also plotted as large filled circles. They are easily
distinguished from most of the known quasars, but are not so easy to
separate from the majority of objects found on the plate. Similar
stellar colours have been found by Dunlop et al. (1985) for
PKS1351-018. It is obvious from diagrams like this that simple colour
searches over a restricted wavelength range are very effective at
finding QSOs with z<2.2, but are not much good at z>2.9. I am
therefore still extremely cautious about claims of "cutoffs" in the
range 2.2<z<2.9! The next "cutoff" one hears about comes at around
3.3<z<3.5. IIIaJ, the emulsion which has yielded the most
high-redshift QSOs discovered so far, cuts off at 5350A, the
wavelength of Lyman-alpha at z=3.4. How did Kodak know?

     Schmidt, Schneider and Gunn (1986) report a striking paucity of
high-redshift, faint quasars (see also Schneider, Schmidt and Gunn
1984). They have surveyed a total of 0.91 sq. deg. in 113 fields at
galactic latitudes above 30 degrees, using a grism technique with the
PFUEI at the Palomar 5-metre telescope. They used well-defined
selection criteria to pick out a final list of 27 emission-line
objects from calibrated spectra for over 9,000 objects in the range
4500-7200A. 17 of these objects are emission-line galaxies with
0.04<z<0.31. The remaining 10 are quasars with 0.91<z<2.66. The
luminosity-function models of Schmidt and Green (1983) yielded good
agreement with the new survey data over the redshift range 0.7-2.7,
but predicted a yield of 40-120 quasars in the range 2.7<z<4.9. From
the absence of such detections in their survey, Schmidt, Schneider and
Gunn concluded that "quasars with an absolute magnitude of M(B) ~ -25
suffer a redshift cutoff near or below a redshift of 3."

     Published systematic searches for objects beyond redshift 3.4 -
the IIIaJ emulsion cutoff for Lyman-alpha - have thus proved largely
unsuccessful (see, e.g. Koo and Kron, 1980; Osmer 1982; Schneider,
Schmidt and Gunn 1984; Koo 1985). These surveys have examined small
areas of sky to faint limiting magnitudes, with a strong emphasis on
the discovery of objects with clear emission-line features. In
contrast, Hazard, McMahon and co-workers have discovered 5 quasars
with 3.4<z<3.7 on low-resolution objective-prism plates covering a
large area of sky to relatively bright magnitudes m(R)=18 (Hazard et
al. 1984; Hazard and McMahon 1985). Three of these objects lie within
a single UK Schmidt field, where a z=3.61 object had already been
discovered by Shanks et al. (1983). One of these objects does not
exhibit strong emission - it is a broad-absorption-line QSO. The
highest-redshift optically-selected QSO is 0055-2659, discovered by
Hazard and McMahon (1985) at z = 3.68. Dunlop et al. (1985) have
suggested, from the radio discoveries of objects like PKS2000-330 (z =
3.78) and PKS 1351-018 (z = 3.71) which have weak emission lines and
stellar colours, that it "therefore seems feasible that both

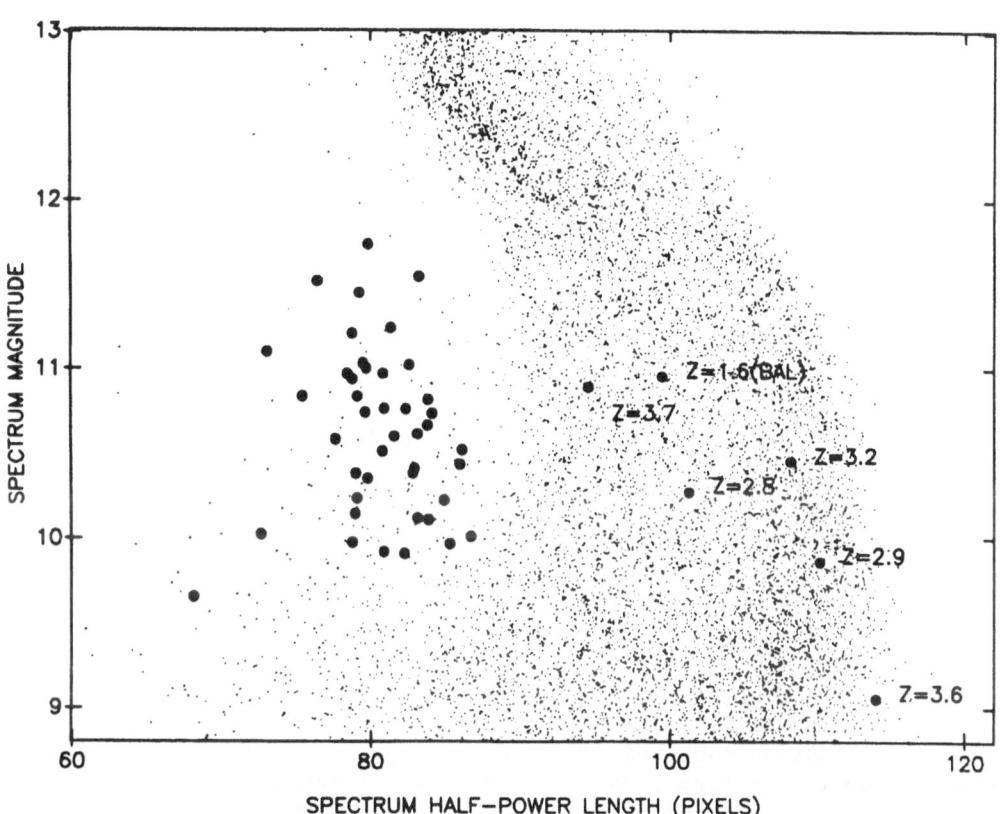

Figure 1. A colour-magnitude for ~10,000 star-like images from a
U.K. Schmidt field containing the South Galactic Pole. In this diagram, kindly supplied by Paul Hewett, we see that quasars with redshifts z<2.2 (the group of solid circles to the left of the diagram)
are bluer than the majority of objects in the field. Quasars of
higher redshifts (and at least some broad absorption-line QSOs) do
not stand out from galactic stars in diagrams of this type.

The units of magnitude are arbitrary, roughly linear, and cover
the range 16<m_B<20; brighter objects are nearer the top of the diagram. The abscissae measure colour; redder objects are nearer the
right of the diagram. The solid circles represent all the known,
confirmed, non-overlapped quasars in the central 22 sq. deg. of
this field-based on work by Clowes and Savage (1983); Shanks et al.
(1983); Boyle et al. (1985); Hazard and McMahon (1985).

prism-plate and colour selection of high-redshift quasar samples may
be biased towards bright, strong emission-line objects, in which case
it is not unexpected that both techniques should agree on the shape of
the quasar luminosity function at high redshift." Concerning the
widely-held view that quasars were fewer but brighter in the past
Dunlop et al. add that "This conclusion may indeed be correct but at
present does not seem fully justified."

4.3. Progress with the automatic measuring machines

Very rapid progress is now being made in the UK with the APM machine
in Cambridge and the COSMOS machine in Edinburgh. All the work to be
described in this section has appeared since 16 months ago when I was
able to state - at a conference on future applications of measuring
machines (Smith, 1984) - that "The challenge to the new high-speed
measuring machines is obvious. The only significant cosmological
results from the studies of optically selected quasars have so far
involved neither APM nor COSMOS. Most of the work on clustering of
optically selected quasars has been based on visually-selected samples
of poor quality. The majority of bright QSOs with z>2 selected for
absorption-line studies have been selected by eye. All the BAL QSOs
have been found by eye, not by machine. There is no reason for this
situation to continue. The hardware is ready for those with the
software."

Schmidt plates provide the most efficient means of amassing large
samples of quasars; large surface densities are particularly important
for studying the large-scale clustering of matter at early epochs.
Automated measuring machines do at last, in practice, provide the best
means for obtaining large, objective samples of these quasars (Clowes,
Cooke and Beard 1984; Hewett et al. 1984; Hewett 1984; Stobie 1984;
Boyle et al. 1985; Hawkins 1985). The highest redshift quasar
independently discovered using these machines is DHM 0054-284, found
at a redshift of z=3.61 by Shanks et al. (1983) using multicolour
searches with the COSMOS machine.

Hewett has used APM surveys of UK Schmidt objective-prism plates
to identify homogeneous samples of quasar candidates with a broad
range of spectral features. His search is not just limited to IIIaJ
plates, or to candidates showing strong emission. In one project - to
isolate very-high-redshift quasars - he has so far examined a total of
150,000 objects in 75 square degrees to m(R) = 19.25 and has analysed
his spectra in the following specific ways: (i) to identify objects
having properties similar to those of known high-redshift quasars,
(ii) to search for red spectra that can not be classified as normal
stars. This subset of "peculiar red objects" has been separated out
from the "main sequence" in Figure 1 by a generalised
cross-correlation method applied to mean galactic-star templates at
different magnitude levels; the subset does contain known
very-high-redshift quasars (z>3.4). Hewett's survey produces about 2
good candidates per square degree. As mentioned in section 3, he and
I hope to use infrared photometry to help separate out the very high
redshift quasars from this subset of peculiar red objects. Hewett is

naturally interested to check whether the four z>3.4 quasars in the one "South Galactic Pole" field are part of a statistical fluctuation and whether there is a population of high-redshift quasars with spectra that do not exhibit strong emission. Failure to detect such a population would help to extend Peacock's (1985) conclusions to a much broader class of quasar.

In a later chapter, Shanks et al. present their recent data on an important complete redshift survey of a sample of 200 ultraviolet-excess QSOs obtained from COSMOS measures of U and B plates taken with the UK Schmidt telescope, and present the 2-point correlation function for QSOs over a range of separations up to 500 h^{-1} Mpc. Redshifts for the entire sample were obtained using the AAT fibre-optics spectrograph.

The COSMOS machine at Edinburgh has also been used by Clowes (1985) to locate, automatically, a sample of 56 quasars and to obtain their redshifts. For a subset of objects selected to lie in the redshift range 1.8<z<2.2, apparently significant clustering was observed in two dimensions, which disappeared when the third dimension (redshift) was added. Clowes discusses two possible causes, viz: (i) obscuration, which hides quasars in the directions of dense foreground clusters of galaxies (Shanks et al. 1983) or (ii) a problem with spectral overlaps on objective-prism plates in regions where dense clusters of galaxies occur. Nevertheless, such studies from a single Schmidt field have low statistical significance. It is therefore necessary to use the machines to combine data from more than one field.

Hawkins (1985) has used the COSMOS machine to search two 18 sq. deg. fields, separated by 13 degrees, to look for large-scale fluctuations in QSO surface density. He used variability and ultraviolet excess as selection criteria. With careful calibration, and the use of COSMOS, "it now appears to be feasible to map quasar density" [fluctuations] "at a level of a few percent." He found no differences between the samples at a level of 6%. This is not small enough to be of interest in constraining the fluctuations of the microwave background, but is very much less than the large inhomogeneities being reported by some workers using subjective selection criteria. His conclusions do, however, depend on confirmation of his QSO candidates.

The best evidence so far for quasar clustering is that based on the statistics of quasar pairs, as presented in a later chapter by Shaver.

The APM laser-scanning machine at Cambridge has been used - primarily by Hewett, Hazard and collaborators - to produce very large lists containing hundreds of quasar candidates. Interesting spectra of objects from a number of these lists have been shown at various conferences by Hazard, in which he demonstrates, for example (Hazard 1985) the range of spectra one can encounter at z>2. He has not yet published his sample, nor his magnitude calibration, but reports detection of 36 confirmed QSOs with m(J)<19 in 4.3 sq. deg.; he concludes that "Either QSO densities are still seriously underestimated even for objects brighter than 19 mag., or there is

evidence here of significant clustering in scales of several degrees."
In my view, the most significant results of Hazard's visual searches
have been (i) the distinct possibility that the spectral properties of
QSOs evolve - see also Wampler and Ponz (1985) - and (ii) his
demonstration of the wide range of QSO spectra (see, e.g. Hazard 1984,
1985). As Hewett and Irwin (1986) point out, "It is now believed that
the fraction of quasars showing particular features may change
significantly with redshift. As a consequence, techniques that only
detect certain types of quasar may produce trends in the number
density as a function of redshift that are not representative of the
quasar population as a whole." In this paper, I am trying to add to
this point by stressing the value of multi-wavelength techniques for
exploring this diverse and possibly evolving range of QSO properties.

     In addition to the various cosmological studies, the high-speed
measuring machines should answer the top requirement for future work
on broad-absorption-line (BAL) QSOs suggested by Weymann, Turnshek and
Christiansen (1985; see also Smith 1984; Hazard et al. 1984). They
state "We need a much larger sample of Broad Absorption Line QSOs to
study.....We think that this will require a machine-selection
procedure, partly to ensure some objective and quantifiable set of
selection criteria, but also just because of the sheer amount of plate
material that must be examined. A study over a large area of sky,
yielding objects bright enough (e.g. 18-18.5) to be studied with good
signal-to-noise without huge investments of telescope time, is
strongly to be preferred over studies going much fainter over smaller
areas of sky." They, with Hewett, are currently undergoing a survey
using APM measures of Curtis Schmidt plates from Chile.

     Another application of the automatic machines to absorption-line
studies in QSOs is given in a poster paper by He et al. based on a
sample of objects selected by APM from a UK Schmidt field in Virgo
(see also He et al. 1984). Here the intention is to seek background
light sources to probe the halo regions of different kinds of galaxy
in Virgo. Virgo was chosen, as the whole field is covered by the
anticipated absorption cross-sections of Virgo galaxies. It is
important, however to cover a large area of sky - i.e. several Schmidt
fields - in order to be able to find quasars bright enough to allow
the absorption features to be studied at high spectral resolution with
the Hubble Space Telescope.

     A significant spin-off of Clowes' survey work with COSMOS has
been the discovery of 3 more close pairs of QSOs. These pairs are
being incorporated into the lists being used (see, e.g., Shaver and
Robertson 1984) to study absorptions in QSO halo regions.

     A large automated survey searching 150 square degrees of sky for
gravitational lenses has recently been completed by Webster, Hewett
and Irwin (1986); they used the APM facility to examine a sample of
3000 quasar candidates with 17<m(b)<20 for distorted image structures
indicative of gravitational lenses. They have explored the separation
range 1"<θ<1° in order to derive the probability of detecting a lens
of specified magnitude, component magnitude difference and component
separation. They find a remarkable lack of pairs with small

separations, contrary to the predictions of most existing lensing models (see also the paper by Shaver later in this book).

5.   SEARCHES AT X-RAY WAVELENGTHS

Elvis and Lawrence (1985) claim that "the intense output of x-rays is the only property shared by all 'active' objects." Many hundreds of quasars and active galactic nuclei (AGN) have now been discovered at x-ray wavelengths. Many of these discoveries have been made in the medium-sensitivity survey with the IPC on the Einstein (HEAO-2) satellite in the 0.1 - 4.5keV energy band. Later articles by, Gianni Zamorani and Tomasso Maccacaro provide more detailed results from the x-ray satellites, in particular from the medium sensitivity survey. I shall therefore not deal here with the soft x-ray surveys themselves.

For those quasars selected at harder x-ray wavelengths, and having 2-10 keV luminosities greater than 10^{44} erg/sec, the typical X-ray spectrum is a single power law $F(\nu) \propto \nu^{\alpha}$ of energy index $\alpha = -0.7$ over the range 0.1 - 120 keV, with very little low-energy absorption (e.g. Mushotsky et al. 1980; Mushotsky and Marshall 1980; Mushotsky 1982; Rothschild et al. 1983; Petre et al. 1984; Reichert et al. 1985). However, exceptions are known to exist. Furthermore, selection effects are almost always underestimated by workers in a new waveband. X-ray astronomers are not the first to believe that most quasars have spectra with the 'canonical' slope of -0.7!

Only fairly recently has stress been laid on the diversity in soft X-ray spectral slope - (Elvis and Lawrence 1985; Elvis, Wilkes and Tananbaum 1985). Indeed, they and Branduardi-Raymont et al. (1985) have raised the possibility of a 'break' in the soft X-ray spectrum of AGN below energies of about 0.5keV; the x-ray energy spectrum at higher energies has slope -0.7, whereas below 0.5 keV, the x-ray spectrum steepens to meet the short-UV upturn.

Although the tight relation found at energies above 0.5keV for x-ray selected quasars holds over 4 decades of luminosity, it does not apply so strikingly to quasars selected outside the x-ray window. Martin Elvis will show us a number of illustrations of the X-ray to IR continua of optically-selected quasars, which should help to illustrate this point; optically selected QSOs have a mean X-ray power-law index near 1.2. This revision of our ideas away from the idea of a "Universal" X-ray spectrum lead to consideration of different X-ray production mechanisms; indeed, it is quite possible that the x-ray production mechanism in most radio and x-ray selected quasars differs from that in most optically selected QSOs.

Piccinotti et al. (1982) showed that Seyfert galaxies contribute about 20% to the diffuse x-ray background. On the most commonly invoked cosmological models (e.g. Schmidt and Green 1983), quasars were more luminous and perhaps more numerous in the past, so their contribution to the x-ray background is likely to be substantial. Indeed, about the time of the launch of the "Einstein" satellite, many were expecting that the x-ray background would provide good, quantitative upper limits to the evolution of quasars (see, e.g.,

Setti and Woltjer 1979). Unfortunately, it proved hard to detect the
very numerous high-redshift optically selected QSOs with the Einstein
satellite (see e.g. Khembhavi and Fabian 1982; Margon, Chanan and
Downes 1982). Furthermore, since then, the exact relation between the
optical and x-ray properties of quasars has become less clear.

The parameter α(ox) generally used to predict X-ray luminosity
functions from optical ones (Tananbaum et al. 1979) depends on
measurement of f(v) at 2500A; this is in the region of the 'big bump'
discussed by Malkan (1983) and by Malkan and Sargent (1983), which
they attribute to black-body radiation from an accretion disk at a
range of temperatures. QSOs can dominate the x-ray background if
certain relations between <α(ox)> and optical luminosity and/or
redshift are found to hold. Kriss and Canizares (1985) find that the
dependence of <α(ox)> on redshift "is small and may be null"; they go
on to suggest that the CfA Medium Survey data need to be corrected for
the effects of low-luminosity, partially reddened and X-ray absorbed
objects. In his PhD thesis, Anderson (1985) concludes that "the
typical contributor to the X-ray background may well be a
moderate-redshift object, with moderate to low optical luminosity:
the high-redshift QSOs apparently have too low a surface density, and
because they are more optically luminous, have lower x-ray fluxes for
a given apparent magnitude as well. Nearby QSOs (z<0.5) are simply
not abundant enough to make a substantial contribution." In the
context of luminosity evolution, the X-ray luminosity evolves less
than the optical luminosity (see also Avni and Tananbaum 1982).

Immediately to either side of the x-ray band we have no survey
data. Although most of a quasar's luminosity could appear in the MeV
gamma-ray region (Bezler et al. 1984), the first trickle of data from
AGN in this waveband is not expected until after the launch of the
Gamma Ray Observatory in 1988. Few survey data for quasars are
expected from the X-ray Ultraviolet Explorer (XUVE) - also due for
launch in 1988 - as few quasars will be detectable through the local
interstellar medium in the 100A-1000A wavelength range of XUVE. IUE
has done magnificent work on studying individual AGN and bright
quasars, but has not resulted in the discovery of large samples of
quasars useable for statistical studies.

6.   CONCLUSIONS

We must continue this vigorous search for quasars at all wavelengths
if we are to gain an understanding of the overall population of
quasars and an accurate appreciation of the full range of quasar
activity. Ever since the announcement of a redshift "cutoff" at z>2.5
by Sandage (1972) and Schmidt (1972) and the subsequent discovery of
many objects at redshifts z>3 (e.g. Smith, 1975), there has been
considerable controversy as to the actual form of the evolution
functions at z>2. I have repeatedly stressed the sensitivity of our
conclusions to selection effects and the consequent importance of
understanding quantitatively the nature of any selection effects
involved. For example, there may be problems with optical surveys

which could hide a population of very high redshift quasars.
Evolution of the optical spectrum is possible, indeed likely. Optical
surveys for quasars use methods of selection which probably vary with
intrinsic luminosity and with redshift! Though often forgotten, it is
obviously necessary to deal with these effects quantitatively before
making any quantitative statements concerning the form of the
luminosity evolution, particularly at redshifts $z>2.5$. It also
appears likely that the radio and optical properties of quasars are
uncorrelated; a demonstration of a redshift cutoff for radio sources
is therefore not a sufficient measure of the switch-on time for all
quasars because the radio-loud quasars exist mainly in elliptical
galaxies whose evolutionary properties could quite easily differ
substantially from those of spirals.

     Much of the above assumes a Friedmann cosmology and that
redshifts are a reliable indicator of distance - assumptions which are
not proven, as Profs. Burbidge and Segal often reminded us during the
conference.

7. ACKNOWLEDGEMENTS

I wish to thank Ann Savage for her constructive comments on a draft
version of this manuscript.

REFERENCES

Anderson, S.F., 1985. Ph.D. Thesis, University of Washington: 'The
    X-ray Properties of High-Redshift Optically-Selected QSOs'.
Atwood, B., Baldwin, J.A. & Carswell, R.F., 1985. Ap.J., 292, 58.
Avni, Y. & Tananbaum, H., 1982. Ap.J. (Letters), 262, L17.
Baldwin, J.A., Phillips, M.M. & Terlevich, R., 1981. P.A.S.P., 93, 5.
Beichman, C., Wynn-Williams, C.G., Lonsdale, C.J., Persson, S.E.,
        Heasley, J.N., Miley, G.K., Soifer, B.T., Neugebauer, G.,
        Becklin, E.E. & Houck, J.R., 1985. Ap.J., 293, 148.
Bezler, M., Kendziorra, E., Staubert, R., Hasinger, G., Pietsch, W.,
        Reppin, C., Trümper, J., & Voges, W., 1984. Astron. and Ap.,
        136, 351.
Boyle, B.J., Fong, R., Shanks, T. & Clowes, R.G., 1985. M.N.R.A.S.,
        216, 623.
Branduardi-Raymont, G., Mason, K.O., Murdin, P.G. & Martin, C., 1985.
        M.N.R.A.S., 216, 1043.
Clowes, R.G., 1985. M.N.R.A.S., in press. 'Automated quasar
    detection in the SGP field: a clustering study'.
Clowes, R.G., Cooke, J.A. & Beard, S.M., 1984. M.N.R.A.S., 207, 99.
Clowes, R.G. & Savage, A., 1983. M.N.R.A.S., 204, 365.
Dunlop, J.S., Downes, A.J.B., Peacock, J.A., Savage, A., Lilly, S.J.,
        Watson, F.G. & Longair, M.S., 1985. Nature, in press.
        'Discovery of a Quasar with z = 3.71 and Limits on the Number of
        More Distant Objects.'
Elvis, M. & Lawrence, A., 1985. In 'Astrophysics of Quasars and
        Active Galactic Nuclei', Proc. 7th Santa Cruz Workshop on
        Astronomy and Astrophysics, (ed. J.S. Miller), p289.
Elvis, M. & van Speybroeck, L.P., 1982. Ap.J. (Letters), 257, L51.
Elvis, M., Wilkes, B.J. & Tananbaum, H., 1985. Ap.J., 292, 357.
Gatley, I., Jones, T.J., Hyland, A.R., Beattie, D.H. & Lee, T.J.,
        1984. M.N.R.A.S., 210, 565.
Geballe, T.R., Krisciunas, K., Lee, T.J., Gatley, I., Wade, R.,
        Duncan, W.D., Garden, R. & Becklin, E.E., 1984. Ap.J., 284, 118.
Grindlay, J.E., Steiner, J.E., Forman, W.R., Canizares, C.R. &
        McClintock, J.E., 1980. Ap.J., 239, L43.
Halpern, J.P. & Steiner, J.E. 1983. Ap.J. (Letters), 269, L37.
Hawkins, M.R.S., 1985. M.N.R.A.S., 216, 589.
Hazard, C., 1984. M.N.R.A.S., 211, 45P.
Hazard, C., 1985. In Proc. Manchester Conf. 'Active Galactic Nuclei',
        p1. (Ed. J.E. Dyson) University of Manchester press.
Hazard, C. & McMahon, R., 1985. Nature, 314, 238.
Hazard, C., Morton, D., Terlevich, R. & McMahon, R., 1984. Ap.J.,
        282, 33.
He, X.-T., Cannon, R.D., Peacock, J.A., Smith, M.G. & Oke, J.B., 1984.
        M.N.R.A.S., 211, 443.
Heckman, T.M., 1980 a. Astron. and Ap., 87, 142.
Heckman, T.M., 1980 b. Astron. and Ap., 87, 152.
Heckman, T.M., 1980 c. Astron. and Ap., 88, 365.

Hewett, P.C., 1984. In Proc. Astronomical Measuring Machines Workshop 'Astronomy from Measuring Machines', p5. (Royal Greenwich Observatory).

Hewett, P.C. & Irwin, M.J., 1986. Preprint, 'Finding high-redshift quasars using low-resolution spectra'.

Hewett, P.C., Irwin, M.J., Bunclark, P., Bridgeland, M.T., Kibblewhite, E.J., He, X.-T. & Smith, M.G., 1985. M.N.R.A.S., 213, 971.

Keel, W.C., 1983. Ap.J., Suppl., 52, 229.

Keel, W.C., 1983. Ap.J. 269, 466.

Keel, W.C., 1985. In 'Astrophysics of Quasars and Active Galactic Nuclei', Proc. 7th Santa Cruz Workshop on Astronomy and Astrophysics, (ed. J. S. Miller), p1.

Kembhavi, A.K. & Fabian, A.C., 1982. M.N.R.A.S., 198, 921.

Koo, D., 1983. In Proc. 24th Liège Astrophysical Symposium 'Quasars and Gravitational Lenses', p240. (Institut d'Astrophysique, Leige).

Koo, D., 1985. In Proc. Trieste Conf. 'The Structure and Evolution of AGNs', in press.

Koo, D. & Kron, R., 1980. P.A.S.P., 92, 537.

Kriss, G.A. & Canizares, C.R., 1985. Ap.J., 297, 177.

Lewis, D.W., MacAlpine, G.M., & Weedman, D.W., 1979. Ap.J., 233, 787.

Malkan, M.A., 1983. Ap.J., 268, 582.

Malkan, M.A. & Sargent, W.L.W., 1982. Ap.J., 254, 22.

Margon, B., Chanan, G.A. & Downes, R.A., 1982. Ap.J. (Letters), 253, L7.

Marshall, H., 1984. Ap.J., 283, 50.

Marshall, H., 1985. Ap.J., 299, 109.

Miley and de Grijp, R., 1985. In Proc. 1st IRAS Symposium, Noordwijk. 'IRAS Observations of Active Galaxies - a Review' (Space Telescope Science Institute Preprint #65).

Mushotsky, R.F., 1982. Ap.J., 256, 92.

Mushotsky, R.F. & Marshall, F.E., 1980. Ap.J., 239, L5.

Mushotsky, R.F., Marshall, F.E., Boldt, E.A., Holt, S.S., & Serlemitsos, P.J., 1980. Ap.J., 235, 377.

Osmer, P.S., 1982. Ap.J., 253, 28.

Peacock, J., 1985. Edinburgh Astronomy Preprint #12/85: 'The High-Redshift Evolution of Radio Galaxies and Quasars'.

Peacock, J.A. & Gull, S.F., 1981. M.N.R.A.S., 196, 611.

Peacock, J., Miller, L. & Longair, M.S., 1985. Edinburgh Astronomy Preprint #17/85. 'The Statistics of Radio Emission from Quasars'.

Penston, M.V. & Perez, E., 1984. M.N.R.A.S., 211, 33P.

Peterson, B.A., Savage, A., Jauncey, D.L., & Wright, A.E., 1982. Ap.J. (Letters), 260, L27.

Petre, R., Mushotsky, R.F., Krolik, J.H., & Holt, S.S., 1984. Ap.J., 280, 499.

Phillips, M.M., Charles, P.A. & Baldwin, J.A., 1983. Ap.J., 266, 485.

Piccinotti, G., Mushotsky, R.F., Boldt, E.A., Holt, S.S., Marshall, F.E., Serlemitsos, P.J., & Shafer, R.A., 1982. Ap.J., 253, 485.

Reichert, G.A., Mushotsky, R.F., Petre, R. & Holt, S.S., 1985.
     Ap.J., 296, 69.
Rothschild, R.E., Mushotsky, R.F., Baity, W.A., Gruber, D.E.,
     Matteson, J.L. & Peterson, L.E., 1983. Ap.J., 269, 423.
Schmidt, M. & Green, R.F., 1983. Ap.J., 269, 352.
Schmidt, M., Schneider, & Gunn, J.E., 1985. Preprint.
Schneider, D., Schmidt, M. & Gunn, J.E., 1984. B.A.A.S., 16, 488.
Setti, G. & Woltjer, L., 1979. Astron. and Ap., 76, L1.
Serabyn, E. & Lacy, J.H., 1985. Ap.J., 293, 445.
Shanks, T., Fong, R. & Boyle, B.J., 1983. Nature, 303, 156.
Shaver, P.A., & Robertson, J.G., 1983. Nature, 303, 155.
Smith, M.G., 1975. Ap.J., 202, 591.
Smith, M.G., 1978. Vistas in Astron., 22, 321.
Smith, M.G., 1983. In Proc. 24th Liège Astrophysical Symposium
     'Quasars and Gravitational Lenses', p4. (Institut
     d'Astrophysique, Leige).
Smith, M.G., 1984. In Proc. Astronomical Measuring Machines Workshop
     'Astronomy from Measuring Machines', p100. (Royal Greenwich
     Observatory).
Smith, M.G., 1985. Proc. Manchester Conf. 'Active Galactic Nuclei',
     p103. (Ed. J. E. Dyson - University of Manchester press.)
Stauffer, J.R., 1982. Ap.J., Suppl., 50, 517.
Stobie, R.S., 1984. In Proc. Astronomical Measuring Machines Workshop
     'Astronomy from Measuring Machines', p5. (Royal Greenwich
     Observatory.)
Tananbaum, H., Avni, Y., Branduardi, G., Elvis, M., Fabbiano, G.,
     Feigelson, E., Giacconi, R., Henry, J.P., Pye, J.P., Soltan, A. &
     Zamorani, G., 1979. Ap.J. (Letters), 234, L9.
Veron-Cetty, M.P. & Veron, P., 1984. 'A Catalogue of Quasars and
     Active Nuclei (2nd Edition)', Munich: ESO Sci. Rep. 4.
Wampler, E.J. & Ponz, D., 1985. Ap.J., 298, 448.
Webster, R.L., Hewett, P.C., & Irwin, M.J., 1986. Preprint. 'An
     Automated Survey for Gravitational Lenses'.
Weymann, R.J., Turnshek, D.A., & Christiansen, W.A., 1986. Ap.J., in
     press.

DISCUSSION

Burbidge : You suggested that perhaps the spectra of the QSOs changed systematically with redshift. I want to point out that the work on 3CR radio galaxies by Spinrad and others shows that for these objects the spectra systematically change as one goes to higher z.

Petrosian : It appears both from your review and Martin Ree's introductory lecture that the question of redshift cutoff will be discussed extensively in this Symposium. I would like to point out that in addition to the observational selection effects you mentioned and the environmental effects by Rees at high redshifts where the cutoff may or may not be present the cosmological model becomes important, even if one is limited to the Friedman-Lemaitre models of General Relativity.

COUNTS OF QUASARS AT FAINT MAGNITUDES: A NEW COMPLETE SAMPLE*

G. Zamorani[1], V. Zitelli[2] and B. Marano[2]

1. Istituto di Radioastronomia
    Via Irnerio 46, 40126 Bologna
    Italy

2. Dipartimento di Astronomia
    Via Zamboni 33, 40126 Bologna
    Italy

ABSTRACT. We present results on a new sample of optically selected
quasar candidates.   The "standard" multicolor technique for selecting
quasar candidates has been applied to all the objects brighter than J =
22.0 in a field of 0.69 square degrees. Additional candidates have been
selected from a search on grism plates obtained in the same area.
Spectroscopy for all the candidates brighter than J = 20.9 has provided
a sample of 22 confirmed quasars. The redshift distribution of these
objects is essentially flat from z = 0.6 up to z = 2.8. Three out of
the eight quasars with redshift larger than two were selected from the
grism plates and were missed by our color selection. This result,
although based on a small number of objects, suggests that the
luminosity functions computed in this redshift range from samples which
are only color selected might have to be increased by a factor up to
1.5.   On the other hand, these possible losses of the multicolor search
technique are negligible (of the order of 15% only) for the estimates of
the integral number counts at magnitudes of the order 20-21.

1. INTRODUCTION

The study of the basic evolutionary properties of quasars requires the
use of complete samples covering the largest possible regions of the
magnitude - redshift plane. At relatively bright magnitudes (J < 20.0)
various samples of UVX selected objects have recently become available
(Schmidt and Green 1983; Mitchell, Warnock and Usher 1984; Marshall et
al. 1984).   These samples, together with the time honored AB sample
(Braccesi, Formiggini and Gandolfi 1970), provide the basis for studies
of the quasar luminosity function and its evolution for redshifts less

---

*Based on observations collected at the European Southern Observatory,
La Silla, Chile.

G. Swarup and V. K. Kapahi (eds.), Quasars, 33–36.
© *1986 by the IAU.*

than 2.2 (see, for example, Marshall 1985). At fainter magnitudes and higher redshifts the available information from complete samples is still scanty (Koo, Kron and Cudworth 1985; see also Boyle et al. **1986**). For this reason, we started a program with the aim of selecting a sample of candidate quasars complete down to J = 22.0, for which we are now in the process of obtaining spectroscopic information.

2. THE SELECTION OF THE CANDIDATES

The observations and the data reduction are described in great detail in Marano, Zamorani and Zitelli **1985** and **1986** (Paper I and II). We have obtained various sets of deep exposures in the U(IIIaJ+UG1), J(IIIaJ+GG385) and F(IIIaF+GG495) bands at the prime focus of the ESO 3.6 m equipped with the triplet corrector. The whole useful field on six plates, two for each passband, was scanned with the ESO PDS microphotometer (50x50 micron aperture). The PDS scannings produced a working list of about 6000 objects complete to J = 22.0. After plates linearization and magnitude calibration, we have classified all the detected objects in three classes, "stellar", "intermediate" and "extended", on the basis of a shape parameter (see Paper I and II). For relatively bright magnitudes (J < 21.0) the "intermediate" objects would be commonly defined as "fuzzy". For each of the three classes we have then constructed the color – color diagrams (U–J vs. J–F), from which we have selected our quasar candidates following the "extended" color criterion suggested by Koo and Kron (1982).

In the same field we obtained also three grism plates at the ESO 3.6 m telescope. The wavelength range (3400 – 5300 A) is such that Lyα would not be detected at redshifts larger than about 3.3. The best two of these plates were visually inspected by us and additional quasar candidates from this inspection were added to the list of candidates selected by the color – color diagrams.

3. RESULTS OF SPECTROSCOPY

We have obtained slit spectroscopy for all the candidates brighter than J = 20.9. In the last spectroscopic run (November 1985) we have used the Eso Faint Object Spectrographic Camera which has become recently available at the ESO 3.6 m telescope (D'Odorico 1986). This camera allows to obtain in about 30 minutes good spectra (signal to noise of the order of 10) of objects in the magnitude range 20.5 – 21.0 (see Figure 1).

A summary of the results is as follows:
a) Down to J = 20.9 we have a sample of 22 confirmed quasars (broad emission lines objects). Ten of them are brighter than J = 20.0.
b) The redshift distribution is essentially flat in the redshift range 0.6 < z < 2.8, with eight objects with redshift larger than two.
c) Three of these 22 objects were selected only by the analysis of the grism plates. Each of them has a redshift larger than two. In order to select them also in the color – color diagram, it would have been

necessary to include a relatively large number of blue subdwarf stars in the sample of candidates. To check this result we are currently obtaining new PDS scannings of our plates with a smaller window (20x20 micron). This will allow us to improve our magnitude measurements and to test with more accuracy the position of these three objects in the color - color diagram.

d) One of the three objects selected only from the grism plates is the only Broad Absorption Line quasar in our sample (see Figure 1).

e) Six further objects, in the redshift range 0.1 - 0.5, showed only narrow emission lines . Five of them were classified as "intermediate" (see above) and were the most ultraviolet objects in this class.

4. CONCLUSION

Having complete spectroscopic information for all the quasar candidates brighter than J = 20.9, and taking into account the additional candidates with 20.9 < J < 21.0, we estimate a surface density of 43±8 quasars per square degree at J = 21.0. This number is in excellent

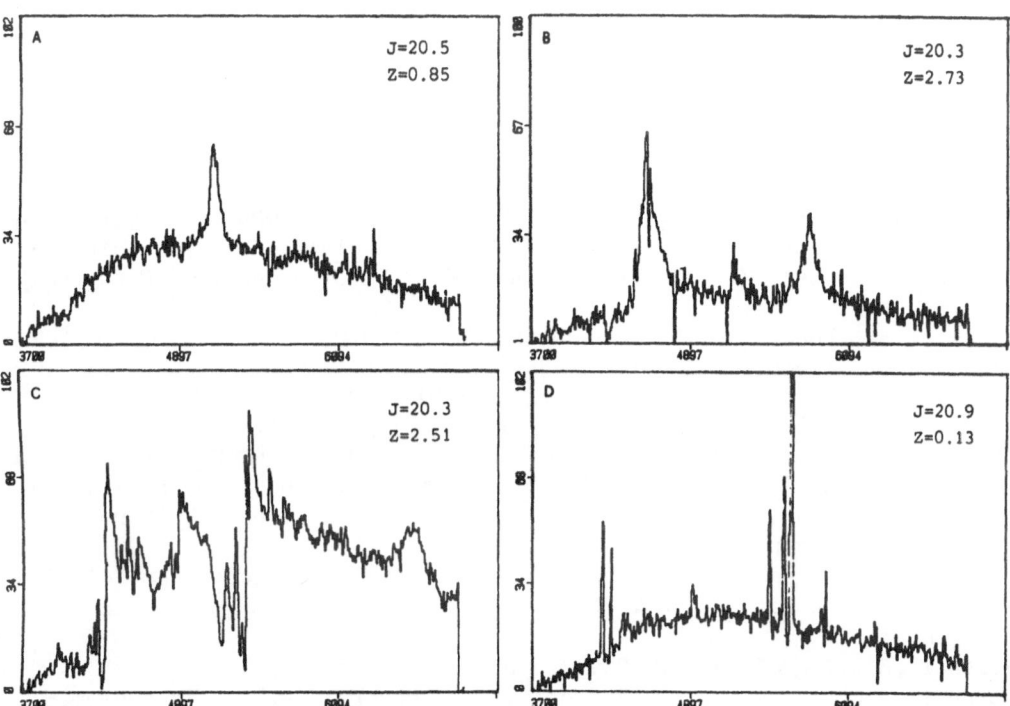

Figure 1. Spectra of four objects obtained with the EFOSC. The Y axis gives the sky subtracted intensity, not corrected for the instrumental response, in arbitrary units. Cosmic ray events have not been removed. Object c) is the only Broad Absorption Line quasar in our sample. Object d) is a low redshift Narrow Emission Line Object.

agreement with the results of similar surveys (Koo, Kron and Cudworth 1985; Shanks at al. 1986).

Our spectroscopic results on the grism selected candidates suggest, although on the basis of a few objects, that quasar searches based exclusively on color - color diagrams may be somewhat incomplete. While this incompleteness has no significant impact on the estimates of the integral number counts, it can be significant for redshifts larger than two.

In our sample, as in similar surveys, we do not find very high redshift quasars. The highest redshift among the 22 confirmed quasars is 2.73. This gives additional evidence against the existence of a large number of high redshift, faint quasars (see Osmer 1986 for a comprehensive review of the subject).

At the other extreme of the redshift range, we do not find any broad emission line object with redshift smaller than 0.6. On the other hand, at least six objects with prominent narrow emission lines (but this sample is probably incomplete) have been found in the same redshift range. This suggests that our selection, which includes also objects with "intermediate" shape in the list of candidates, is not strongly biased against low redshift objects. If this is the case, the absence of broad emission line objects at low redshift may be taken as an indication of a flattening of their luminosity function for absolute magnitudes fainter than about M_B = -23.0 (but see the discussion on completeness of the multicolor selection in Paper I).

REFERENCES

Boyle,B.J., Fong,R., Shanks,T., and Peterson,B.A. 1986, in Structure and Evolution of Active Galactic Nuclei, G. Giuricin, F. Mardirossian, M. Mezzetti, and M. Ramella eds., (Dordrecht: Reidel).

Braccesi,A., Formiggini,L., and Gandolfi,E. 1970, Astr.Ap., 5, 264.

D'Odorico,S. 1986, This Volume., p.57.

Koo,D.C., and Kron,R.G. 1982, Astr.Ap., 105, 107.

Koo,D.C., Kron,R.G., and Cudworth,K.M. 1985, preprint.

Marano,B., Zamorani,G., and Zitelli V. 1986, in Structure and Evolution of Active Galactic Nuclei, G. Giuricin, F. Mardirossian, M. Mezzetti, and M. Ramella eds., (Dordrecht: Reidel), in press.

Marano,B., Zamorani,G., and Zitelli V. 1985, Proceedings of the Convegno Nazionale ASTRONET 1984-1985, 1985, Mem. Soc. Astr. It., in press.

Marshall,H.L. 1985, Ap.J., 299, in press.

Marshall,H.L., Avni,Y., Braccesi,A., Huchra,J.P., Tananbaum,H., Zamorani,G., and Zitelli,V. 1984, Ap.J., 283, 50.

Mitchell,K.J., Warnock III,A., and Usher,P.D. 1984, Ap.J.(Letters), 287, L3.

Osmer,P.S. 1986, This Volume , p.447.

Schmidt,M., and Green,R.F. 1983, Ap.J., 269, 352.

Shanks,T., Boyle,B.J., Fong,R., and Peterson,B.A. 1986, This Volume ,p.37.

RESULTS FROM A FAINT QSO REDSHIFT SURVEY

T.Shanks, R.Fong and B.J.Boyle
Physics Department, University of Durham
Durham DH1 3LE, England

B.A.Peterson
Mount Stromlo Observatory, Woden, ACT 2606, Australia

ABSTRACT.

We have used the FOCAP fibre optic coupler at the Anglo-Australian Telescope (AAT) to measure redshifts for a complete sample of $\sim$ 170 B $\leqslant$ 21^m QSO's selected using the ultraviolet excess (UVX) criterion. We present preliminary estimates of the QSO luminosity function in discrete redshift ranges and show how these observations differentiate between models of QSO evolution. We have also investigated the clustering of QSOs in this complete sample by estimating the QSO 2-point correlation function and we use this to derive direct constraints on the homogeneity of the Universe at large scales.

1. BASIC DATA

The data on which this paper is based are COSMOS (Stobie et al. 1979) machine measurements of deep U and J U.K. Schmidt plates. We now have U and B photometry on 7 4°5 x 4°5 widely scattered high latitude fields. The B photometry is calibrated using a CCD to B $\sim$ 21^m. The U calibration goes less deep (B $\sim$ 18^{m}5) and here we have calibrated at the faint limit on the assumption that the U-B colours of halo stars do not change significantly between B = 18^{m}5 and B = 21^m. On the one field where we have U photometry to U = 20^m this assumption proved reasonable and so presumably it will hold good on our other high latitude fields. With the limits U-B $\leqslant$-0^{m}35 and B $\leqslant$ 21^m we find our samples produce approximately 100 UVX stars per square degree. As described later the spectroscopic surveys show that $\sim$ 40% of these stars are QSOs.

The initial use of these samples was to test the claim by Seldner and Peebles 1979 that high redshift QSO's in the Burbidge et al. 1977 catalogue are preferentially detected near low redshift Lick catalogue galaxies (Shanks et al. 1983). We have now checked the galaxy-QSO cross-correlation function on 5 fields with 'eyeballed' objective prism QSO surveys and on the above 7 COSMOS/UKST UVX catalogues (here restricted so that they are 70% QSO dominated). The result is that

37

G. Swarup and V. K. Kapahi (eds.), Quasars, 37–43.
© *1986 by the IAU.*

nowhere do we find a positive cross-correlation between galaxies and
QSO's and so we cannot confirm the Seldner and Peebles result. If
anything we find an anticorrelation, with the QSO density in both the
objective prism and the UVX samples being 30% lower in the vicinity of
average galaxy groups and clusters. This confirms our earlier result
from a smaller sample and the interpretation is that dust in low redshift galaxy clusters is obscuring line-of-sight QSO's at cosmological
distances. Because of the steepness of the QSO number counts at
magnitudes brighter than $B = 20^m$ the average amount of dust absorption
needed in each cluster is only $0^m.2$ in the B band.

2. THE SPECTROSCOPIC SURVEYS

We next proceeded to make spectroscopic surveys of subsets of the
above UVX samples. As shown by Véron (1983) the UVX technique selects
over 95% of QSO's in the redshift range $0 \leqslant Z \leqslant 2.2$. Because of this
completeness redshift surveys of UVX QSO's have great potential for
investigations of the QSO luminosity function and its evolution with
redshift.

We first made a $B \leqslant 19^m$ survey at the AAT using conventional
spectroscopic techniques and in 4 nights observing identified 23 QSO's
out of 60 UVX stars observed (see Boyle et al. 1985). However, our
observing efficiency improved dramatically with the advent of the
FOCAP fibre coupler (Gray 1983) at the AAT which enables 45 UVX stars
to be observed simultaneously. In 4 clear nights at the AAT we surveyed
~ 500 UVX $B \leqslant 21^m$ stars in 12 40 arcmin diameter fields, observing at
least 1 fibre field in 6 of our U.K. Schmidt fields. Of these 500
UVX stars, 170 proved to be broad line QSO's and we obtained unambiguous redshifts for 85% of these. Six QSO's showed broad absorption
lines. We also found 22 narrow emission line galaxies ($Z \leqslant 0.42$), 13
White Dwarfs with the rest of the sample being halo stars. The overall
QSO number redshift relation is reasonably flat between $0.4 \leqslant Z \leqslant 2.2$
with no peaks that are obviously due to redshift selection or any other
effect.

3. QSO COUNTS AND LUMINOSITY FUNCTION

Figure 1 shows the differential number counts of the spectroscopically confirmed QSO's ($Z \leqslant 2.2$) in our fibre survey, together with
the counts from several other authors. For $B \leqslant 19^m.5$ our n(m) relation
has a slope consistent with the steep $0^m.86$ slope fitted by Braccesi
et al. 1980 to other counts in this magnitude range. However, beyond
$B = 19^m.5$ our QSO counts turn over in reasonable agreement with the
fainter counts of Koo (1986). Cavaliere et al. (1983), Koo (1983) and
Marshall et al. (1983a) have suggested that such a feature may be inconsistent with a pure density evolution model for the QSO's and also
that the turnover in the counts may indicate a similar turnover in the
QSO luminosity function. As we shall see when we consider the QSO
luminosity function, we agree with this interpretation of the counts.
The integral sky densities of QSO's we find are 26 ± 3 deg^{-2} at

B = $20\overset{m}{.}5$ and 40 ± 4 deg^{-2} at B = $21\overset{m}{.}$ The field-to-field variation is
consistent with a Poisson error distribution, implying a fairly iso-
tropic distribution of QSO's on the sky.

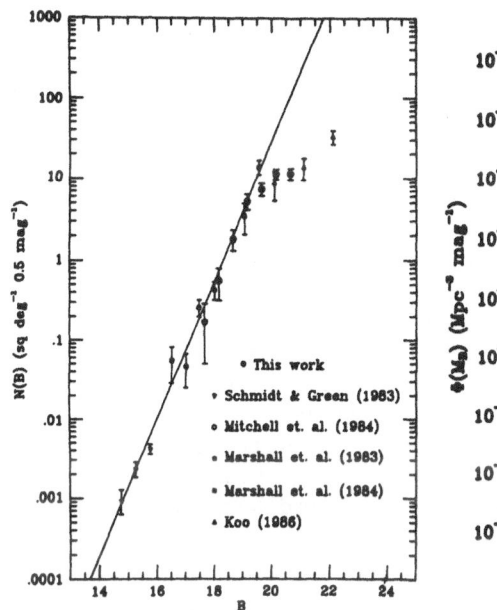

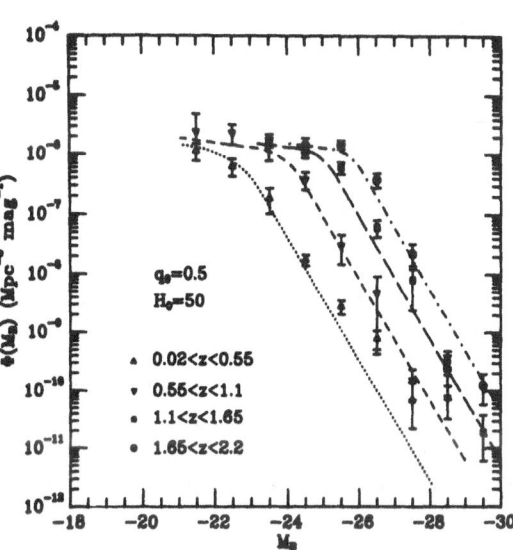

Fig. 1.   Differential number
count for spectroscopically
confirmed QSO's.

Fig. 2.   QSO luminosity functions
derived from our surveys combined
with those of Schmidt and Green
(1983), Marshall et al. (1983) and
Marshall et al. 1984.

Since we have a large number of QSO's we can directly estimate the
QSO luminosity function over a range of discrete redshift intervals.
Figure 2 shows the results for the luminosity function based on our own
and previous redshift surveys of UVX QSO's.   (The details of the
estimation procedure will be described elsewhere).   Bright surveys such
as that of Schmidt and Green (1983) dominate these estimates at high
luminosities whereas at fainter luminosities our own survey dominates.
We first note that at intermediate magnitudes there is good agreement
between our luminosity function results and the other authors'.
Secondly, in each of the 4 redshift bins in Figure 2, we see evidence
for a feature in the QSO luminosity function; in each bin the luminosity
function turns over to a flatter slope at faint absolute magnitudes.
Finally, the luminosity function is seen to move predominantly in the
luminosity direction with redshift, in the sense of increasing
luminosity towards higher redshift.

Before further discussing these results we note that it is an
intriguing coincidence in our data that the effect of evolution on the
QSO luminosity function in the range $0.4 \lesssim Z \lesssim 2.2$ is just enough to
cancel the effects of increasing distance.   It is this cancellation

which produces a turnover in the number counts as sharp as is observed, the feature in the luminosity function at each redshift always occurring in apparent magnitude at B $\sim 19^m.5$. This re-emphasises the already well known fact that the Hubble diagram of optically selected QSO's shows an extremely poor correlation between apparent magnitude and redshift.

The results in Figure 2 suggest that QSO evolution is well represented phenomenologically by pure luminosity rather than density evolution. We have also shown elsewhere (e.g. Boyle 1986 thesis) that our QSO luminosity function when extrapolated to z = 0 is not inconsistent with the luminosity function of nearby Seyfert galaxy nuclei as derived, for example, by Weedman (1986). The simplest model which is consistent with these observations is that all QSO's might have been created at a single epoch (z > 2) and that they dim in a manner independent of their luminosity to become the lower luminosity Seyfert galaxy nuclei which we see at the present day. However, this would imply that QSO's are long-lived and therefore have very large energy requirements. This could in the standard picture, result in some Seyfert nuclei containing very massive ($10^9 - 10^{10}$ M$_\odot$) black holes at the present time. Although this mass range is in good agreement with the inferred mass of the black hole in the galaxy NGC 4151 (Ulrich et al. 1984) it could still be that the simplest pure luminosity evolution model will turn out to be too naive an interpretation of our results. The future theoretical challenge is, therefore, to make luminosity function predictions for alternative physical models where the birth and death rates of short lived QSO's are significant to evaluate the real strength of the present evidence for a pure luminosity evolution model.

4.   THE SPATIAL CLUSTERING OF QSO's.

We finally used our redshift survey to estimate the spatial correlation function, $\xi_{qq}(r)$, for QSO's and this is shown in Figure 3. At small scales (< $10h^{-1}$ comoving Mpc) the statistics are poor but since we observe 11 QSO pairs in this range of separations whereas on a random hypothesis we expect only 4, there is tentative (3σ) evidence for QSO clustering. Even on a model where we assume QSO's cluster like galaxies and that the clustering pattern is stable at scales below $10h^{-1}$Mpc to z = 2.2 we expect to see only $\sim$ 5 QSO's and so there is also tentative evidence that QSO's may cluster more strongly than galaxies. It would be interesting if the QSO correlation function proved to be as high as the correlation function of rich galaxy clusters because this might indicate that QSO's are associated more with superclusters rather than clusters of galaxies. However, a larger QSO redshift survey will be required before any detailed conclusion can be drawn on the small scale clustering of QSO's.

At larger scales the QSO correlation function produces new limits on the homogeneity of the Universe to $1000h^{-1}$ Mpc. At present the results are consistent with a Poisson distribution for QSO's at all scales above $10h^{-1}$ Mpc. However, from other considerations the

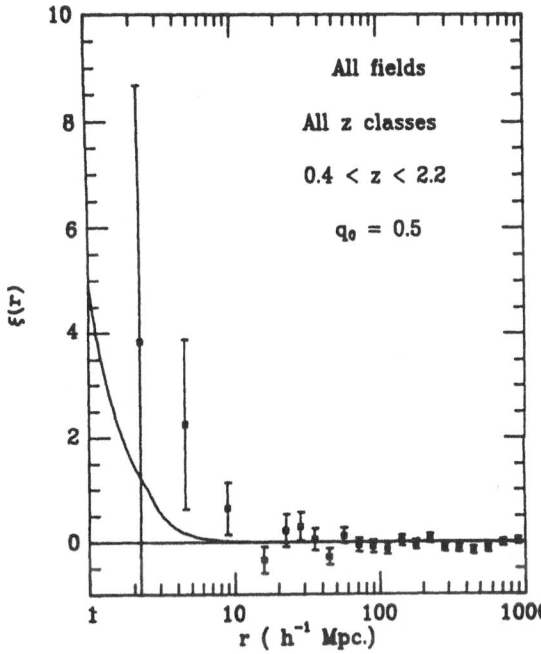

Fig. 3. QSO 2-point cor-
relation function, $\xi_{qq}(r)$. r
is a comoving coordinate computed assuming $q_0 = 0.5$. The
solid line is the expected
galaxy correlation function at
$z = 1.5$.

expected amplitude of any inhomogeneity in $\xi_{qq}(r)$ is still of the
order of the error bars in Figure 3. In a larger QSO survey it would
be interesting to see if there exist any weak features in ξ_{qq} which
are also detected in the galaxy correlation function at low redshift.
If so then it will not only be very important for theories of galaxy
formation but it could also afford a powerful test for q_0; at $z = 1.5$
any feature in ξ_{qq} moves by 50% in comoving separation between models
which assume $q_0 = 0.1$ and $q_0 = 0.5$ and so comparison between a feature's
location at high and low redshift produces a q_0 test. The determination
of the form of the QSO correlation function at both large and small
scales therefore forms a primary motivation for making further faint
QSO redshift surveys in the manner described here.

5. CONCLUSIONS

    Using our COSMOS/UKST UVX QSO surveys we have found evidence for
dust absorption of high redshift QSO's by low redshift galaxy groups
and clusters. Consideration of the number counts and luminosity
functions of our spectroscopially confirmed QSO's suggest a pure
luminosity evolution model which could imply that QSO's are long-lived.
Finally correlation analysis of the clustering of QSO's in our redshift
survey has shown tentative evidence for QSO clustering at scales up
to $10h^{-1}$ Mpc but a reasonably homogeneous QSO distribution at larger
scales.

REFERENCES

Boyle, B.J., Fong, R., Shanks, T., and Clowes, R.G., 1985, M.N.R.A.S.,
216, 623.

Braccesi, A., Zitelli, V., Bonoli, F., and Formiggini, L., 1980, Astr. Astrophys. <u>85</u>, 80.

Burbridge, G.R., Crowne, A.M., and Smith, H.E., 1977, Ap. J. Supp. <u>33</u>, 113.

Cavaliere, A., Giallongo, E., Messina, A., and Vagnetti, F., 1983, Ap. J. <u>269</u>, 57.

Gray, P.M., 1983, Proc. SPIE, <u>445</u>, 57.

Koo, D.C., 1983, in "QSO's and Gravitational Lenses", Univ. of Liege, p.240.

Koo, D.C., 1986, in "Structure and Evolution of Active Galactic Nuclei", Reidel: Dordrecht.

Marshall, H.L., Tananbaum, H., Zamorani, G., Huchra, J.P., Braccesi, A., and Zitelli, V., 1983, Ap. J. <u>269</u>, 42.

Marshall, H.L., Avni, Y., Braccesi, A., Huchra, J.P., Tananbaum, H., Zamorani, G., and Zitelli, V., 1983a, in "Quasars and Gravitational Lenses", Univ. of Liege, p.238.

Marshall. H.L., Avni, Y., Braccesi, A., Huchra, J.P., Tananbaum, H., Zamorani, G., and Zitelli, V., 1984, Ap. J. <u>283</u>, 50.

Mitchell, K.J., Warnock, A., Usher, P.D., 1984, Ap. J. Lett. <u>287</u>, L3.

Schmidt, M., and Green, R.F., 1983, Ap. J. <u>269</u>, 352.

Seldner, M., and Peebles, P.J.E., 1979, Ap. J. <u>227</u>, 30.

Shanks, T., Fong, R., Green, M.R., Clowes, R.G., and Savage, A., MNRAS 1983, <u>203</u>, 103.

Stobie, R.S., Smith, G.M., Lutz, R.K., and Martin, R., 1979, in "Image Processing in Astronomy", eds. G. Sedmak, M. Cappacioli, R.J. Allen, p.48.

Ulrich, M.H., Boksenberg, A., Bromage, G.E., Clavel, J., Elvius, A., Penston, M.V., Perola, G.C., Pettini, M., Snijders, M.A.J., Tanzi, E.G., and Tarenghi, M., 1984, MNRAS <u>209</u>, 479.

Veron, P., 1983, in "QSO's and Gravitational Lenses", Univ. of Liege, p.210.

Weedman, D.W., 1986, preprint.

DISCUSSION

Segal : The luminosity evolution estimated by Marshall et. al., which you cite, as well as that for the X-ray sample of AGNs of Maccacaro et. al., is closely similar to the simple difference between the non-evolutionary Friedman prediction and the Chronometric prediction, - though the latter has a smaller rms deviation in its magnitude-redshift residuals. Do you have any explanation for this ?

Shanks : I thought that Marshall found that the Chronometric model was inconsistent with V/V_{max} results obtained from the bright quasar samples.

Canizares : There appear to be statistically significant deviations from a powerlaw in the luminosity functions at the high luminosity end. Would you please comment on this ?

Shanks : In the magnitude range up to 3^m brighter than the knee in the luminosity function a single powerlaw provides an excellent fit at all redshifts. At higher luminosities the statistical errors are too large to allow any meaningful conclusions to be drawn.

Wampler : Did you correct your broad-band blue magnitudes for the emission lines. This could be done using your spectra.

Shanks : The magnitudes are corrected to the rest B band using a K-correction based only on a powerlaw continuum. In the redshift range of interest here I expect the contribution of the emission lines to the B magnitude to be small especially since our photometry is based on the very broad IIIaJ photographic blue band.

Rees : I would like to ask what inferences you think can be drawn from the clustering your data show. As you know, this is a potentially important discriminant among different cosmogonic schemes. In the "pancake" picture, quasars would be highly clustered at formation (though the clusters should become few and far between as one looks back to the highest z, when only the rare "high-σ" pancakes have collapsed); in a simple heirarchical clustering scheme, there would be less clustering, on a given comoving scale, at higher z; in schemes involving "biasing", the clustering may be enhanced and less sensitive to z.

Shanks : At small scales it is important to see whether the QSOs cluster as strongly as Abell galaxy clusters for this could imply that QSO's are associated with superclusters. At present the data are not inconsistent with this idea, although the errors are large and more data are needed. In bigger samples it will also be possible to divide the data by redshift to directly study the evolution of the QSO clustering which might also constrain Ω_o.

Vijay Kapahi and Geoff Burbidge

ULTRAVIOLET EXCESS QUASAR CANDIDATES IN LARGE FIELDS:
A STATUS REPORT CONCERNING THE NGC 450 AREA.

E. GOSSET, J. SURDEJ and J.P. SWINGS
Institut d'Astrophysique
Université de Liège
B-4200 COINTE-OUGREE
BELGIUM

ABSTRACT. New results are presented based on several statistical analyses of a 25 square degree field around NGC450. They include the detection of a clustering on a small scale ($\simeq 10$ arcminutes) and possible large scale inhomogeneities.

This paper is in fact a status report on a long term program initiated in 1981 with the aim of collecting new and extensive observing material on the background density, the distribution and the luminosity of quasars in fields near bright galaxies as well as far from them. This project is intended to give further evidence for or against the location of quasars in superclusters, in the vicinity of irregular galaxies, or nearby their companion(s) etc. Three large ($5^{o} \times 5^{o}$) fields have presently been surveyed: one near the irregular galaxy NGC450, another one near the amorphous galaxy NGC520 and the third one coinciding with the ESO field nb 300 at $\alpha = 3$ h., $\delta = -40^{o}$ which is essentially void of any particularly bright or active galaxy.

The surveys are carried out on U/B dual image ESO or Palomar Schmidt plates for which the exposure times were adjusted in order to obtain similar images in both the U and B filters for a U-B index of about -0.4 mag. Plates have been scanned at least four times by two different persons in an unbiased homogeneous and objective manner. Preliminary results have already been reported for the NGC 450 field (Swings et al.,1985, Rev. Mex. Astron. Astrof.,10, 91) and will not be repeated here.

One, two and three dimensional distributions of quasars and of quasar candidates are being investigated using several statistical methods. These are: Multiple Binning Analysis (MBA) with classical tests as well as with randomisation ones (Gosset and Louis, submitted to

G. Swarup and V. K. Kapahi (eds.), Quasars, 45–46.

Astrophysics and Space Science), Nearest Neighbours
Analysis (NNA), Correlation Function Analysis (CFA),
Power Spectrum Analysis (PSA) and the Extended Kolmogorov
Smirnov (EKS) test. We restrict ourselves here to outline
the main results relative to the two dimensional (α, δ)
distribution of objects in the field of NGC450. Three
samples have been considered: primary + secondary
candidates(140), primary candidates (94) and confirmed
quasars (60). All three samples deviate from uniformity
and randomness in the same manner, although not to the
same extent. MBA with the 4 within 16 randomisation test
as well as EKS, and possibly also the PSA, detect large
inhomogeneities with a scale of the order of the size of
the field. This cannot be attributed to a center-to-edge
effect. On the other hand, NNA reveals that the three
first nearest neighbours of an object are too close to
one another with respect to uniformity. This is confirmed
by the CFA which detects an overpopulation up to angular
distances of 10 arcminutes. PSA and MBA with the 4 within
16 randomisation test both detect a clustering with a
characteristic scale of about 10 arcminutes. The effect
is stronger for the confirmed quasars'sample than for the
two other ones. Thus, this 10 arcmin clustering can
provisionally be considered as being a characteristic of
the distribution of the quasars in the investigated
field. The significance level is of the order of .05. A
more detailed and complete analysis will be published
elsewhere.

It should be mentioned that low resolution
spectroscopy has already been performed for a fair number
of quasar candidates in the three fields mentioned above.
High resolution spectroscopy has been obtained for
several interesting targets: some results are presently
being reported concerning the doublet Q0107-025A, B
(Surdej et al, Astronomy and Astrophysics, in press) and
the triplet Q0118-031 A, B, C (Robertson et al, MNRAS, in
press). In addition, faint UBV photoelectric photometry
sequences up to B$\leqslant$19.5 mag. have now been obtained with
the ESO 1 m and Las Campanas 2.5 m telescopes in small
areas of the three different fields in order to establish
the values of the limiting magnitudes and of the U-B
thresholds.Grism plates of small areas are being studied
as well.

A more complete account of the surveys briefly
presented here is in preparation and will contain quasar
candidates, luminosities, U/B excesses, representative
spectra, line identifications, data on line strengths and
on redshifts. Furthermore, the results of the above-
mentioned statistical analyses of the distribution of the
different objects will be described and analyzed in
detail.

QSO SEARCH BY SLITLESS SPECTROSCOPY *

A. Hoag
Lowell Observatory
Mars Hill Road, 1400 West
Flagstaff, Arizona 86001
U.S.A.

Sandage and I have obtained "grism" plates that redundantly cover one square degree fields centered on ten Selected Areas. We used a grating-prism in the converging beam of the Mayall 4-meter telescope prime focus to photograph 1580 A·mm^{-1} slitless spectra to a limit of about B = 21.5 (cf. Hoag and Smith 1977).

Emission-line, Q, and ultraviolet excess, U, quasar candidates have been independently selected by two or more persons, and a final list for each plate has been prepared by coordinated review. Spectra of candidates were then measured microdensitometrically and analyzed by procedures described by Vaucher **et al.** (1982). Positions for identification have been determined by tying zero-order images into secondary standards set up by measures of images on plates taken with the A. Lawrence Lowell Astrograph.

The completeness of our lists as a function of magnitude depends not only on the quality of the plates but also on the intrinsic properties of the objects. To the limit we achieve in practice ($19.5 \leq B \leq 21.5$), we typically find the order of a dozen emission-line, and an equal number of ultraviolet excess quasar candidates per square degree. However, the range in the projected surface density of emission-line candidates is about a factor of three among the fields surveyed.

Results for a representative field are presented in Table 1. Lower-case letters following the serial number in the table refer to previous identifications as follows: a = U283, b = U310, and c = U314 from Usher (1981); d = 18, e = 22, and f = 5C7 from Kron and Chiu (1983). A colon following a Q class symbol indicates that only one emission line was detected, so that a redshift can be determined only by assuming a line identification. Redshifts are provisional and depend on the correctness of line identifications. Continuum magnitudes are estimated at $\lambda 1475$ A in the rest frame of the object or at $\lambda 4500$ A in the rest frame of the observer if the redshift is not determined.

---

* Discussion on p.61

G. Swarup and V. K. Kapahi (eds.), Quasars, 47–48.
© 1986 by the IAU.

Norman G. Thomas (Lowell) and Barbara G. Vaucher (Pennsylvania State University), Sister Mary Matthew (Mercyhurst College), Barry Feierman (Westtown School), and Gregory Berns (Princeton University) have shared in the work described.

Table 1. S.A. 57 Objects

Ser.	Class	R.A. (1950)	Dec.	Zp	m	Remarks
1	Q:	$13^h03^m48\overset{s}{.}0$	+29°46'04"	--	19.5	λ4740
2	Q	13 03 54.7	29 10 46	2.06	21:	λλ3700, 4800
3	Q	13 04 11.1	29 40 45	2.00	19.7	Lyα, CIV
4	Q:	13 04 20.0	29 34 20	--	21.0	λ3870
5	Q:	13 04 28.5	29 30 11	--	21.0	λ4800
6a	Q	13 04 32.3	29 20 34	2.11	20:	Lyα,CIV
7	U	13 04 41.5	29 17 32	--	--	Possible emission
8	Q:	13 04 56.1	29 25 42	--	20.0	λ4340
9	U	13 04 59.6	29 49 44	--	--	
10	Q:	13 05 27.1	29 24 31	--	20.0	λ4030
11	G	13 05 44.3	29 10 09	--	--	λλ4860, 6560
12b	U	13 05 45.8	29 50 48	--	--	
13d	Q	13 05 49.2	29 41 12	3.15	18.6	Lyα abs., Lyα
14c,e	U	13 05 53.3	29 35 17	--	--	
15	Q:	13 05 54.5	29 57 22	--	20.6	λ3600
16	Q	13 06 02.4	29 23 45	1.59	21.1	CIV, CIII]
17	Q:	13 06 02.5	29 07 53	--	20.7	λ3740
18	Q:	13 06 07.1	29 56 38	--	21:	λ3750
19f	Q:	13 06 26.9	29 28 43	--	20.5	λ3430
20	U	13 06 27.5	29 20 19	--	--	
21	U	13 06 42.1	29 33 28	--	--	
22	G	13 06 47.7	29 08 44	0.02	--	[OIII], Hα
23	U	13 07 17.6	29 29 33	--	--	
24	Q	13 07 50.3	29 50 41	2.22	19.6	Lyα, CIV
25	Q	13 07 54.4	29 50 52	1.75	19.0	Lyα, CIV
26	Q	13 07 58.4	29 42 55	2.21	18.9	Lyα, CIV
27	Q	13 08 07.8	29 37 21	1.36	20.1	CIV, MgII
28	Q	$13^h08^m41\overset{s}{.}8$	+29°42'08"	1.82	17.9	Lyα, CIV

References

Hoag, A. A., and Smith, M. G. (1977). Astrophys. J. **217**, 362.
Kron, R.G., and Chiu, L.-T. G. (1981). Publ. Astron. Soc. Pacific **93**, 397.
Usher, P. D. (1981). Astrophys. J. Suppl. **46**, 117.
Vaucher, B. G., Kreidl, T. J., Thomas, N. G., and Hoag, A. A. (1982). Astrophys. J. **261**, 18.

ABOUT THE QSOs CONTAINED IN THE CERRO EL ROBLE SURVEY AND THE DENSITY OF UVX-QSOs IN THE SGP FIELD

Luis E. Campusano[+]
Departamento de Astronomía, Universidad de Chile
Casilla 36-D, Santiago de Chile
+ Guest Investigator at Las Campanas Observatory

A PROMISING AREA: THE SOUTH GALACTIC POLE FIELD

A region containing the SGP, centered at α 00h 53m (1950) δ -28°03', is becoming a selected region for QSO research. Three lists of QSO candidates have been published for this field. One consists of candidates discovered visually on an objective prism plate, selected in a 25-deg^2 area and with B(lim) $\simeq$ 20 mag (Clowes and Savage, 1983; the CS sample). The other list came from visual inspection of U and B plates (UVX stars), covering a region of 44-deg^2 and with approximately the same limiting magnitude of the CS sample (Campusano and Torres, 1983; the CT sample). The third survey of QSO-candidates involved a machine selection of UVX stars (Shanks et al., 1983), whose published components correspond to two small areas of 1.6 and 8.2 deg^2 with B(lim) = 19 mag (Boyle et al., 1985).

In this contribution we briefly report some results from our spectroscopic survey of all the candidates in the central 25 deg^2 of the CT sample and with B $\simeq$ 19 mag, carried out with the 2.5 m telescope of Las Campanas Observatory. We have identified all the QSOs in this CT-

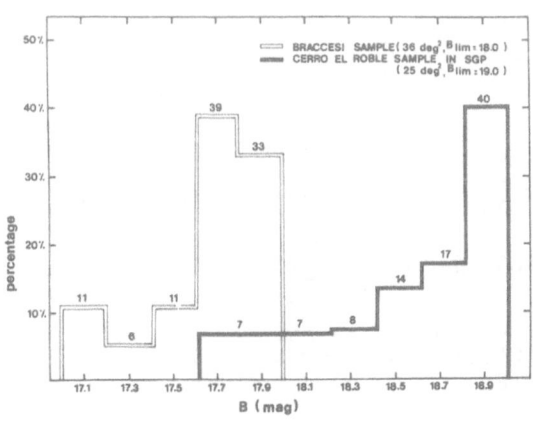

Figure 1. Magnitude distri - bution for two ultraviolet excess samples. Notice the differences in their shapes and the rather narrow magnitude range present in them (aprox. 1 mag). The Cerro El Roble sample has an estimated completeness of 50%.

G. Swarup and V. K. Kapahi (eds.), Quasars, 49–50.

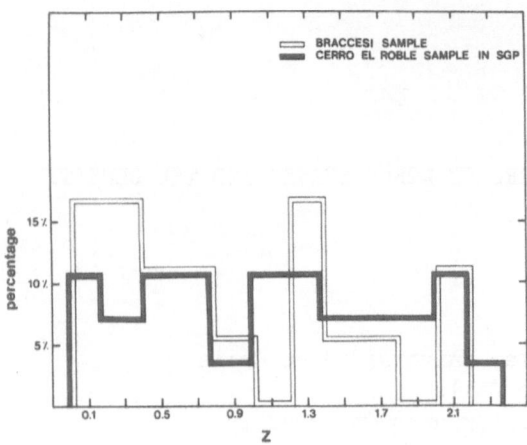

Figure 2. Redshift distri -
bution of all the optical QSOs
contained in Cerro El Roble and
Braccesi surveys defined in Fig.
1. The CT sample presents a
flatter distribution than the
one by Braccesi. The observed
upper redshift limit is compati
ble with the restrictions of
the selection criterion.

subset and determined their redshifts, and then compared the resulting
magnitude and redshift distributions with those corresponding to the
Braccesi UVX-QSO sample which has been adopted by Schmidt and Green (1983)
with B(lim) = 18 mag. The behaviour of these distributions can be seen
in Figs. 1 and 2.

PRESENTLY DERIVED QSO-DENSITY IN THE SGP (B $\lesssim$ 19 MAG)

In this derivation we consider only confirmed QSOs with slit spectrosco
py in the 25 deg^2 area. All candidates with B $\lesssim$ 19 mag in the CT sample
and most of the corresponding ones in the CS sample have been examined
spectroscopically. Then, from all known CT and CS QSOs, we get ρ (B $\lesssim$ 19
mag) = 1.8 deg^{-2}. With the six additional QSOs found by Boyle et al.,
the observed surface density increases to 2.0 deg^{-2}. Allowing for the
incompleteness of the CT sample and its restricted z range, we obtain
an estimate of 2.8 deg^{-2}. Boyle et al., from the identification of 4
UVX-QSOs in a 1.62 deg^2 area and the UVX-star density over the entire
plate, got a formal value of ρ (B $\lesssim$ 19 mag) = 3.3 deg^{-2}. Therefore, ap
proximately 40% of the UVX-QSOs probably remain to be identified in the
25-deg^2 SGP field. The consequences of the availability of such sample
for quasar research cannot be overemphasized.

REFERENCES

Boyle, B.J., Fong, R., Shanks, T. and Clowes, R.G.: 1985, Mon. Not. R.
    astr. Soc. 216, 623.
Campusano, L.E. and Torres, C.: 1983, Astron. J. 88, 1304.
Clowes, R.G. and Savage, A.: 1983, Mon. Not. R. astr. Soc. 204, 365.
Schmidt, M. and Green, R.G.: 1983, Astrophys. J. 269, 352.
Shanks, T., Fong, R., Green, M.R., Clowes, R.G. and Savage, A.: 1983,
    Mon. Not. R. astr. Soc. 203, 181.

A NEW CATALOGUE OF QUASI-STELLAR OBJECTS

A. Hewitt and G. Burbidge
Center for Astrophysics and Space Sciences
Mail Code: C-011
University of California, San Diego
La Jolla, California 92093

We have prepared a new catalogue of QSOs and BL Lac objects containing approximately 3400 entries. A complete update of the Hewitt-Burbidge (1980) catalogue has been made with approximately another 2000 objects with known redshifts added. The references to discovery, magnitudes, redshifts, color, spectra and polarimetry have been updated for the objects listed in 1980, and complete new references are included for the new objects. In addition to the basic optical information, the new catalogue also contains X-ray, radio and infrared information for all objects. Absorption redshifts are listed when they are available. A supplementary catalogue which is now in preparation will contain similar information for objects described variously as Seyfert galaxies, N systems and AGNs. In doubtful cases we have used the operational dividing line $z = 0.1$. All objects with $z < 0.1$ are put in the supplementary catalogue unless their discoverers have unambiguously defined them as QSOs. With approximately twice as many objects included it is interesting to note that:

 a) There are still very few genuine BL Lac objects, ~100.
 b) The largest number of additions has come from identifications using the objective prism-grism techniques.

In Figures 1 and 2 we show the redshift-apparent magnitude (Fig. 1) and the distribution of apparent magnitudes (Fig. 2). Comparison of Fig. 2 with the corresponding histogram in our 1980 paper shows that the studies have by now gone somewhat fainter.

In Figures 3, 4, and 5 we show the redshift distribution for all objects (Fig. 3), for the objects identified by objective prism techniques (Fig. 4), and for the objects identified by position and/or color alone (Fig. 5).

Figure 4 confirms completely that the objective prism methods discriminate in favor of QSOs with $z > 1.8$, though the peak close to $z = 0.3$ should be noted.

Figure 5 shows a fairly uniform distribution from $z = 0.1$ to $z = 2.0$ with peaks superimposed on it at $z = 0.3$, 0.6, 1.4, and 1.95. These peaks were first seen in small samples in 1968, and we believe that they are real and not an effect of spectroscopic selection.

The very steep fall off in the number of QSOs beyond about $z = 2.3$

51

G. Swarup and V. K. Kapahi (eds.), Quasars, 51–52.

is apparent in Figs. 3, 4, and 5 and we believe it to be real. Only
about 15% of the QSOs in the catalogue have z > 2.3.

We anticipate that this catalogue will be published in 1986. Tapes
and reprints of the catalogue will be available upon publication from
the authors. We are grateful for support from the National Science
Foundation through grant AST-84-17650.

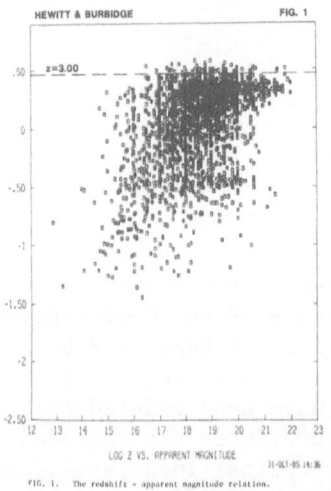

FIG. 1. The redshift - apparent magnitude relation.

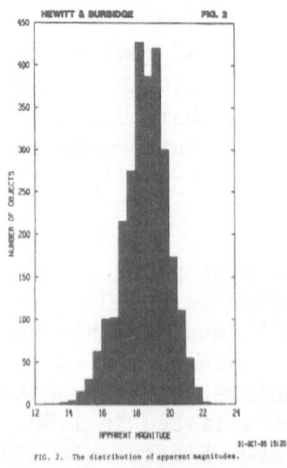

FIG. 2. The distribution of apparent magnitudes.

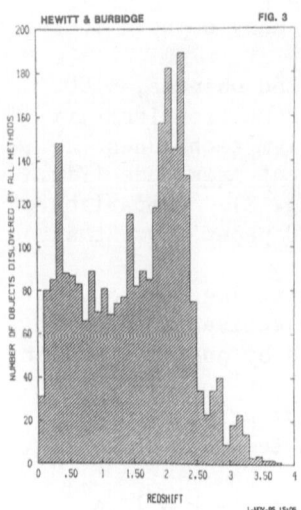

FIG. 3 Histogram of the redshift distribution for all
objects contained in the catalogue.

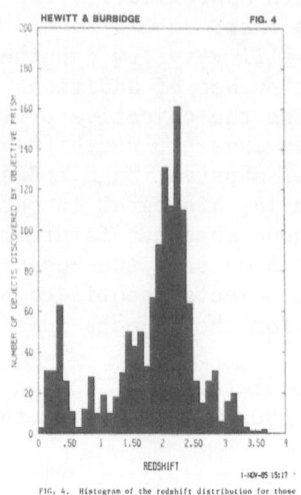

FIG. 4. Histogram of the redshift distribution for those
objects identified by objective prism, grism or
related techniques.

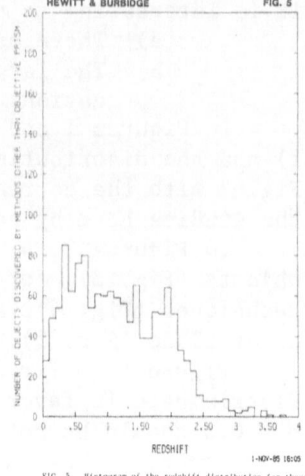

FIG. 5. Histogram of the redshift distribution for those
objects identified by position and/or color alone.

THE ASIAGO CATALOGUE OF QSOs EDITION 1985 *

C.Barbieri[1] , Y.Chu[2] , S.Cristiani[1] [3] , , A.Nota.[1]
1-Istituto di Astronomia,Università di Padova,Italia
2-Center for Astrophysics,University of Science &
  Technology of China
2-ESO, La Silla, Chile

A new Edition of the Asiago Catalogue of QSOs (Barbieri et
al. 1982) is available on magnetic tape. It contains essentially all the information of the previous editions updated
with the papers published before Dec.31, 1984. The catalog
should be essentially complete to that date in respect to
objects having slit spectra (2330 objects), while it is seriously incomplete regarding candidates discovered by objective prism or grism/grens surveys after 1982 (279 such objects included in the present version).
No attempt has been made to define precisely what a QSO is,
and actually several objects contained in the catalog have
been variously classified as Seyfert 1, or N galaxy. The collective term "Active Galactic Nuclei" is used more and more
by several authors.
Beyond updating, several additions and improvements have been
made to the catalog, which is now structured on fewer files
containing more information. One file contains the basic optical and X-ray data for 2610 objects, whilst radio data is
still entirely lacking (at the moment of this writing); a
second file contains the references ordered by numbers (a
FORTRAN 77 program is available to order them alphabetically);
a third file gives additional Notes for several objects.Other
files, such as the one giving synonyms or the one giving absorption lines, have been simply updated, but no change in
the file structure has been implemented. Among the improvements, the following are worth mentioning:

- two parameters have been added in the fundamental file for
each object. The first parameter confirms the availability
of slit spectra; the second parameter specifies in which way
the object was firstly identified. At the moment, three basic
identification modes have been parametrized, namely the opti-

---

* Discussion on p.61

G. Swarup and V. K. Kapahi (eds.), Quasars, 53–54.

cal identification of a Radio Source (1), the optical identification of an X-Ray source (3), and finally an entirely optical identification (2). This latter mode could be further
subdivided: identification from multicolour surveys (mostly
UV-excess, or UVX) from objective prism plates, from grism/
grens material, from optical variability. However in practice several candidates were identified from a combination say
of UVX and slitless characteristics, so that we decided not
to implement this finer subdivision (at any rate, provision
for it is maintained in the structure on the file).
The addition of a parameter giving the way the QSO was firstly discovered greatly helps the statistical examination of
the Catalogue. As examples of this improved analysis, we mention the distribution of magnitudes and redshifts separately
of optical, radio and X QSOs. For instance one sees that optically selected QSOs tend to be fainter than the radio ones,
which are abruptly cut around 19.5. Another interesting fact
is that radio QSOs do not show the peak in z around 2 which
is instead so prominent in optical candidates.
- a third parameter (characters BL) has been added to distinguish BL LAC objects

- the Reference numbering system has been expanded to include up to 9999 references (at the present moment some 1300
papers have been actually listed).

- the coding of the data adheres more closely to accepted
standards.

As part of the more general task to provide complete data
for each object, a particular effort has been made to cover
the optical variability. The result is a file that contains
for every object the reference numbers of papers giving data
about its optical variability (in general we selected the
papers that give original data).

Copy of the mag tape can be obtained from:
A.Nota, ST ScI, Homewood Campus, Baltimore, USA

Acknowledgement - This paper was written during a stay of
Y.C. at Asiago Observatory. The Section on QSO optical variability is being written by L.Mazzocco as part of his Thesis.

Reference

Barbieri, C., Capaccioli, M., Cristiani, S., Nardon, G.,
Omizzolo, A.1982, Mem.S.A.It.vol.53,nr.3.

THE AUTOMATIC DETECTION OF FAINT QUASARS FROM CFHT GRENS PLATES

C.J. Keable and R.G. Clowes
[1] Dept of Astronomy, University of Edinburgh
[2] Royal Observatory, Edinburgh
Blackford Hill, Edinburgh EH9 3HJ,
UNITED KINGDOM

ABSTRACT. The automatic quasar detection system (AQD, see Clowes, 1984) has been used to successfully select quasars from prism plates taken by the United Kingdom Schmidt Telescope. Surveys compiled using this technique have a magnitude limit of B-19.5. We present work here which extends the possible survey limit to B-21. This is done by putting plates taken with the Canada-France-Hawaii Telescope's grens (see Richardson, 1984, for a description) through the AQD measuring system. We will follow this work with a comparable photometric survey of the same field.

1. THE SELECTION OF CANDIDATES

This survey is of 0.7 square degrees centred on 10 40 0 +5 00 0 (1950). The plate material used for the grens part of the survey was a direct J plate from the UKST, paired with a CFHT grens plate. The multicolour survey uses direct UKST plates taken in the U,B,V,R and I bands. Thus the survey limit will be about B = 21, the plate limit of UKST direct plates. This limit is about a magnitude and a half fainter than that attainable from UKST objective prism plates.

It has been shown before (by Vaucher et al., 1982 and others) that it is possible to pick good quasar candidates from CFHT grens plates. However, such surveys suffer because of the eye's subjectivity in identifying features and inability to integrate across the spectrum to pick up real features on fainter objects.

An obvious route to take in the compilation of large, well defined, candidate lists is the use of a very fast microdensitometer. In this case, the COSMOS measuring machine at the Royal Observatory, Edinburgh (MacGillivray and Stobie, 1984, give an up to date description) was used to scan all plates used. Following our procedure developed for the analysis of UKST prism plates (see Clowes, Cooke and Beard, 1983 and Clowes, 1984) all spectra were then analysed to detect the presence of emission lines of equivalent width greater than a specified value.

G. Swarup and V. K. Kapahi (eds.), Quasars, 55–56.
© 1986 by the IAU.

The reason that grens plates have not been processed automatically before is that a grens, being a grating based instrument,
produces plates on which brighter objects have multiple images.   The
corrector system is a part of the instrument, and so no direct plates
can be taken with the same field effects.   As a result, we used direct
plates from the UKST to locate the emulsion cutoff position.   A
polynomial fitting process was developed to account for field effects
between the 2 plates.

2.   RESULTS

The survey has yielded about 100 objects down to a limit of B = 21, in
an area of 0.7 square degrees.   For the survey to be used to extend
the quasar luminosity function, for example, slit spectra of
candidates are required.   Very few candidates so far have slit
spectra, but Fig. 1 shows the slit spectrum of a faint quasar,
independently rediscovered by the survey.   There are 4 previously
known brighter (B < 19) quasars in the field;   1 was missed by the
survey, being overlapped by the image of a bright star.

3.   REFERENCES

Clowes, R.G. in 'Astronomy with Schmidt-Type Telescopes', 107, (1984).
Clowes, R.G., Cooke, J.A. and Beard, S.M., MNRAS 207, 99, (1983).
MacGillivray, H.T. and Stobie, R.S., Vistas in Astron. 27, 433,
        (1984).
Richardson, E.H., 'Can.J.Phys', 57, 1365, (1979).
Vaucher, B.G., Kreidl, T.J., Thomas, N.G. and Hoag, A.A., Ap.J. 261,
        18 (1982).

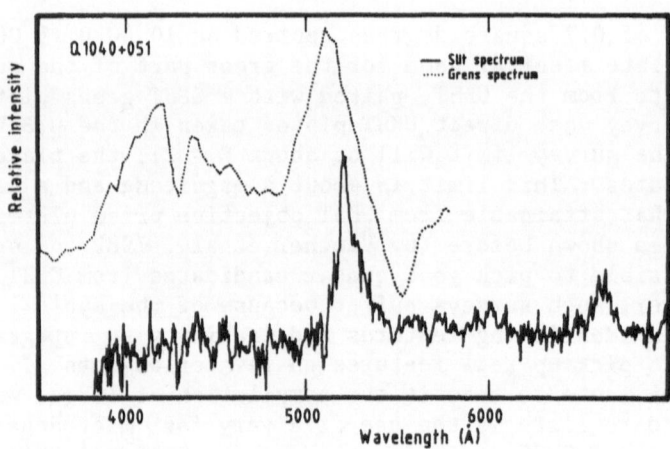

Fig. 1:   Spectra of a quasar discovered by S. D'Odorico (private
communication), of B ≅ 20.8, z ≅ 3.27.   This object was independently
selected by this survey because of the Ly α emission line.

OBSERVATIONS OF QSOs AND RELATED OBJECTS WITH EFOSC, THE ESO FAINT OBJECT SPECTROGRAPH AND CAMERA

S. D'Odorico[+], S. Cristiani[+], R.G. Clowes* and C.J. Keable*

[+] European Southern Observatory, Karl-Schwarzschild-Str. 2,
    D-8046 Garching bei München, Federal Republic of Germany
* Royal Observatory, Blackford Hill, Edinburgh EH9 3HJ, U.K.

EFOSC is a standard ESO instrument operating at the Cassegrain focus of the 3.6 m telescope since April 1st, 1985. A description of its optical design and operating modes is given in Enard and Delabre (1982) and Dekker and D'Odorico (1985). Briefly, it is a focal reducer with spectroscopic capability. The collimator produces a collimated beam with a diameter of 40 mm which passes through a filter and/or grism. The f/2.5 camera focusses the beam on the detector which is at present a thinned, back-illuminated RCA CCD with 320×512 pixels. The pixel size is 30 μm which corresponds to .675" on the sky. There are three remotely controlled wheels in the instrument: the aperture wheel in the focal plane of the telescope, with long slits of different widths, a filter and a grism wheel with 12 positions each. The instrument can be operated in four different modes: direct imaging, slit spectroscopy, grism or multiple object spectroscopy with specially made aperture plates.

In direct imaging, the field is $3\overset{.}{.}6 \times 5\overset{.}{.}7$. The global efficiency at 5500 Å (atmosphere + telescope + filter + instrument + detector) is 32%. A star of m_v = 25th is detected at a S/N ≈ 3 in a 15 m exposure with average seeing (FWHM = 1.5 arcsec).

In spectroscopy, observations of standard stars with a wide slit indicate a global efficiency of 30% at 5500 Å. 1 photon/Å/s is detected at this wavelength from an object of m_v = 18.

In its testing phase, EFOSC has been used for a number of faint QSO observations. In a grism CCD frame, a new QSO of m_v = 20.9 and z = 3.28 was discovered. Two QSO candidates from a CFHT grens plate searched with the AQD technique (Clowes et al., 1984) were confirmed by slit spectroscopy. These results will be reported in more detail elsewhere.

As an example of the performance of EFOSC, we show here a deep direct image (Fig. 1) and a spectrum of a faint galaxy (Fig. 2). These observations were obtained within a program to search for optical counterparts of the absorption systems seen in the high resolution spectra of the BL Lac object 0215+015.

REFERENCES

Clowes, R.G., Cooke, J.A., and Beard, S.M. 1984, M.N.R.A.S. 207, 99.
Dekker, H., and D'Odorico, S. 1985, ESO Operating Manual #4.
Enard, D., and Delabre, B. 1982, Proc. SPIE 445, 522.

G. Swarup and V. K. Kapahi (eds.), Quasars, 57–58.

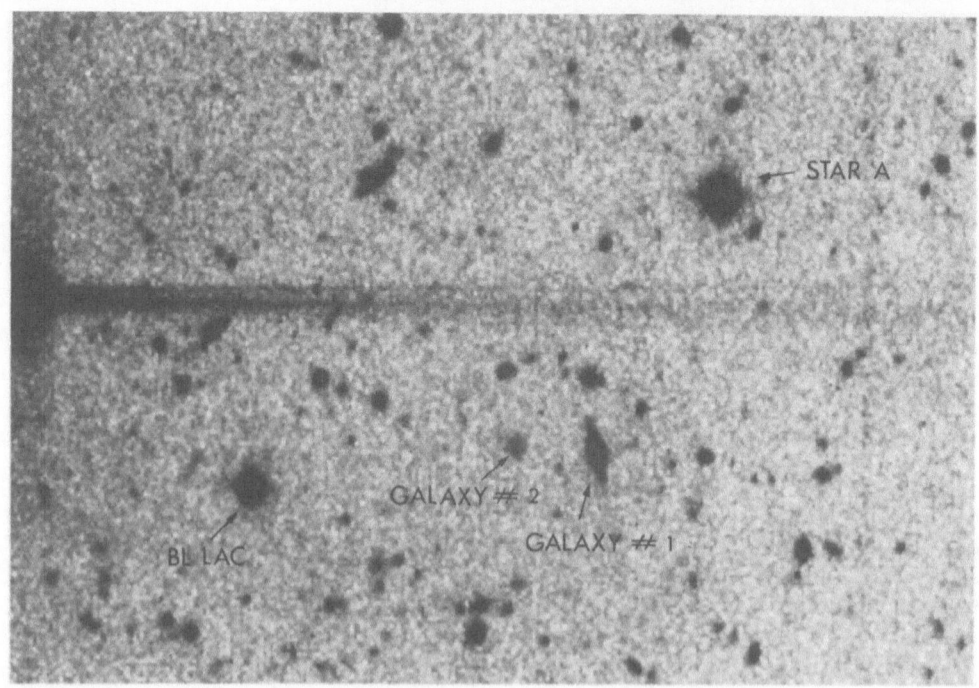

Figure 1. A 10 m exposure through a Gunn r filter of the field of
BL Lac object 0215+015, obtained on Sept. 23, 1985. The object
brightness has varied by more than two magnitudes in one year. From a V
exposure on the same night, the following magnitudes were derived:
BL Lac 17.3±.1, star A 16.25±.1, galaxy #1 20.3±.1, galaxy #2 22.4±.2.
North at the top, east on the side of the bright star.

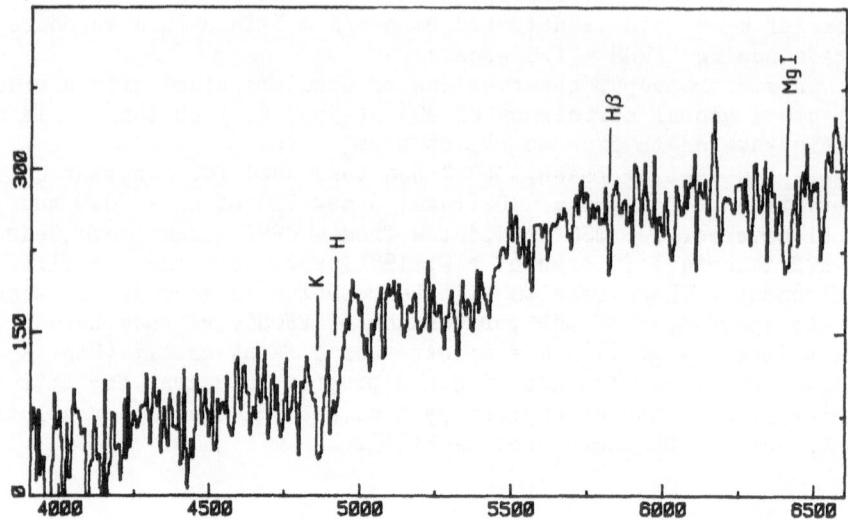

Figure 2. A sky subtracted spectrum of the galaxy #1 in relative
intensity units. The exposure time is 30 minutes and the resolution
15 Å. The absorption lines give a redshift z = 0.238±0.001.

PKS 2005-489: A VERY BRIGHT BL LAC OBJECT IN A NEARBY GALAXY[*]

J V Wall[1], I J Danziger[2], M Pettini[1], R S Warwick[3], W Wamsteker[4]

1 Royal Greenwich Observatory, Herstmonceux Castle, Hailsham, East Sussex BN27 1RP, U.K.
2 European Southern Observatory, Karl-Schwarzschild-Strasse 2, D-8046 Garching-bei-Munchen, Federal Republic of Germany
3 Department of Physics, University of Leicester, University Road, Leicester LE1 7RH, U.K.
4 IUE Observatory, European Space Agency, Villafranca, Madrid, Spain

The galaxy identified with the flat-spectrum radio source PKS 2005-489 has a bright stellar nucleus with V $\simeq$ 13 mag. Optical, UV and X-ray observations indicate variability and power-law continua in each of these wavebands, leading to the conclusion that PKS 2005-489 is one of the brightest BL Lac objects known.

The remarkable nature of the object is apparent in Fig 1. The image is clearly nebulous but diffraction spikes emerge from the centre of the image to indicate the stellar nature of the nucleus and its domination of the total light from the system. CCD photometry shows that the object varied by at least 0.5 mag over a year; in Aug 1982 V = 12.9 mag, putting PKS 2005-489 among the 3 or 4 brightest BL Lac objects known, brighter than others for which the nebulosity is clearly visible. Optical spectra obtained between 1980 and 1985 yield power-law continua with varying spectral indices ranging from α ($S_\nu \propto \nu^{-\alpha}$) = 0.3 to 1.6. None of these observations provides a convincing estimate of the redshift.

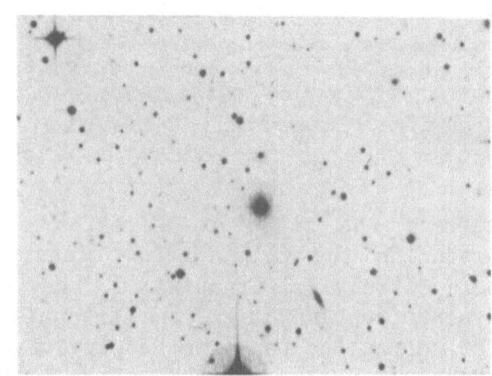

Fig 1. *The image of PKS 2005-489 as reproduced from a copy of the UKST survey. NE is up and left. The faint lenticular to the SW has z = 0.136; if the redshift of PKS 2005-489 is similar, the host galaxy would be exceptionally luminous at $M_B \sim -24$ mag ($H_o = 50$ km s^{-1} Mpc^{-1})*

Ultraviolet observations carried out with IUE (Fig 2) show a power-law spectrum with α = 1.4. All 3 interstellar lines marked in the figure have large equivalent widths ($W_\lambda \simeq$ 2 to 4 Å). The line-of-sight to PKS 2005-489, at its closest to the Galactic centre, passes 5.5 kpc below the plane. The strong absorption lines, similar to those

---

[*] Discussion on p.61

G. Swarup and V. K. Kapahi (eds.), Quasars, 59–60.
© *1986 by the IAU.*

found in halo stars with sight-lines passing <2 kpc from the centre,
indicate that the halo associated with the inner regions of the Galaxy
extends to several kpc from the plane.

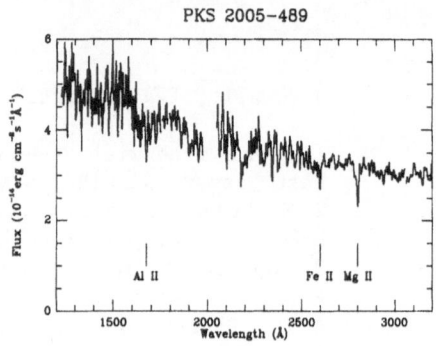

Fig 2. The UV spectrum of PKS 2005-489 from IUE observations in March 1984. There is no obvious 2200 Å feature, limiting E(B-V) to < 0.04 mag. The only definite features are absorption lines at low velocities, presumably formed in the ISM of our Galaxy.

X-ray observations from EXOSAT (September 1984) show an extremely
steep power-law spectrum ($\alpha \simeq 2.1$), even for a BL Lac object. One
month later the source was 2.5 times brighter and the spectrum had
hardened markedly. Long-term variability is indicated by the absence
of PKS 2005-489 from Uhuru, Ariel V, and HEAO-1 catalogues, its 1984
intensity being far above the limit of each of these. The X-ray
measurements are consistent with a column density for HI of $N_H \sim 6$ x
10^{20} cm^{-2}. Direct measurements of Galactic HI at 21 cm yield $N_{HI} = 4.7$
x 10^{20} cm^{-2}, and the column density intrinsic to the object must
therefore be small.

If the X-ray excess in the composite spectrum (Fig 3) is due to
the synchrotron self-Compton process, then the X-ray spectrum should

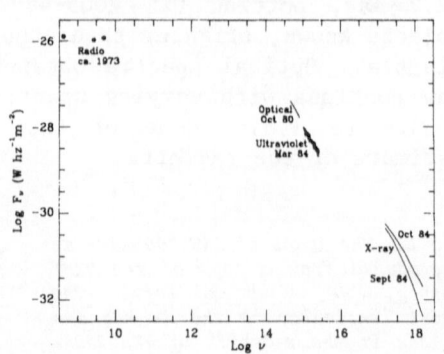

Fig 3. The composite (non-contemporaneous) spectrum of PKS 2005-489. There is a large apparent excess of X-ray emission over that predicted from extrapolation of the optical-UV spectrum, and this X-ray emission has a steeper spectrum than the optical-UV; in this respect the object differs from other BL Lacs with known X-ray excesses.

reflect the slope of a region of (presumably synchrotron) spectrum
producing the Comptonizing electrons. That it does not suggests a
more likely alternative: the entire optical to X-ray band is
synchrotron emission, the discontinuity between UV and X-ray regions
reflecting the non-simultaneity of observation. This interpretation is
supported by the variable slope in the optical region and the hardening
of the X-ray spectrum with increased X-ray output. It carries the
exciting implication that substantially greater UV (and perhaps
optical) output occurs than at the epochs observed. There are thus
substantial incentives for continued, additional and simultaneous
observations of PKS 2005-489, which may bear on the intrinsic nature of
emission, the nature of the surrounding nebulosity, and the nature of
the interstellar medium of our Galaxy.

DISCUSSION ON THE PAPER BY **HOAG** (p.47)

Hutchings : I would like to comment on the grens survey results from CFHT plates by D. Crampton. This technique is based on visual inspection but appears to be substantially more complete than other objective prism work in the z < 1.8 range. 80% of some 100 candidates have confirming slit spectra. The improvement in selection effects can be attributed to the dispersion and emulsion characteristics in the plates, and throw considerable doubt on the assumed number/redshift counts currently accepted.

DISCUSSION ON THE PAPER BY **BARBIERI** ET AL. (p.53)

Filippenko : A question to the Chairman, P.Veron : Do you plan to publish an update of your extensive catalogue, and will it be available on magnetic tape ?

Veron : Our quasar catalogue is continuously updated. The second edition has been published only about six months ago as ESO Scientific Report No. 3 and is available on magnetic tape. We will certainly publish a third edition, but certainly not before a year from now.

DISCUSSION ON THE PAPER BY **WALL** ET AL. (p.59)

Veron : Has this object been detected by IRAS ?

Wall : No. PKS 2005-489 is unlucky enough to lie in one of the small areas of low sensitivity in the IRAS survey; in any case, interpolation suggests that the IR flux density might be near the IRAS detection limit.

Burbidge : What do you know about radio variability in this object ?

Wall : Nothing; no radio monitoring has been carried out. It is certainly variable in the optical, UV, and X-ray bands.

Roberts : Is the stellar component in the centre of the fuzz ?

Wall : CCD frames and sky survey transparencies both indicate that it is. The unequal lengths of diffraction spikes for the image on the sky survey transparency suggest some non-symmetry in the underlying light distribution, but the effect is probably not significant.

G. Swarup and V. K. Kapahi (eds.), Quasars, 61.
© *1986 by the IAU.*

Janice and Bruce McAdam, Peter Shaver and
Dominique Radhakrishnan

II

CONTINUUM EMISSION

"But if quasars had only been discovered in (say) 1973-when we already knew more about lower-level activity in galactic nuclei (Seyferts etc), and when pulsars and X-ray binaries had convinced us that gravitational energy could be channelled efficiently into radiation - quasars would never have seemed 'sui generis', and theoretical ideas would have evolved less waywardly to their present state."

- Martin Rees (p.11)

"The success of this program is a result of the dedication by a large number of collaborators who operate nine different telescopes that obtain data from $10^8 - 10^{18}$ Hz (Table I)."

- Joel Bregman (p.65)

COMPACT CONTINUUM EMISSION FROM VIOLENTLY VARIABLE QUASARS AND BL LACERTAE OBJECTS

Joel N. Bregman
National Radio Astronomy Observatory
Edgemont Road
Charlottesville, VA   22903
USA

ABSTRACT.   A dozen BL Lacertae objects and violently
variable quasars were observed repeatedly during the last
six years by obtaining simultaneous measurements from the
radio through X-ray region. The relationship between the
radio-millimeter, infrared-ultraviolet, and X-ray regions
are determined from single epoch spectra and variability
measurements. The structure of the emitting regions and
the emission mechanisms are discussed. The results of this
program are summarized.

1.   INTRODUCTION

We report upon a program in which BL Lacertae objects and
violently variable quasars are observed simultaneously over
as much of the electromagnetic spectrum as possible. These
observations are designed to shed light on the connection
between various emitting regions, which are frequently
studied separately (e.g. radio, optical, X-ray regions),
and to determine the physical processes and conditions in
the emitting plasma. Of a dozen sources, most have been
observed at several epochs, providing important variability
information. The success of this program is a result of
the dedication by a large number of collaborators who
operate nine different telescopes that obtain data from 10^8
- 10^{18} Hz (Table I).

2.   SINGLE EPOCH SPECTRA

Multifrequency spectra have been obtained for the
violently variable quasars 1156+295 (Glassgold et al. 1983,
Wills et al. 1983), 3C 345 (Bregman et al. 1986), 3C 446
(Bregman et al. 1986) and the BL Lacertae objects IZw-187
(Bregman et al. 1982), 0735+178 (Bregman et al. 1984),
OJ 287 (Pollock et al. 1985), 1413+135 (Bregman et
al. 1981), OJ 049, 0215+015, OV-236, BL Lac, 1308+326,

65

G. Swarup and V. K. Kapahi (eds.), Quasars, 65–71.
© 1986 by the IAU.

TABLE I   MULTIFREQUENCY COLLABORATORS

Observers	Institution	Waveband
Dent, Balonek, Barvanis, O'Dea	Univ. Mass	1.5-90 GHz
Aller, Aller, Hodge	Univ. Mich.	4.8-24 GHz
Werner, Roellig	NASA/Ames	1mm - 1μm
Harvey	Univ. Texas	Far IR
Neugebauer, Soifer, Mathews, and Elias	Palomar Obs. and Caltech	100-0.55 μm
Rieke, Lebofsky, Rudy, and Wisniewski	Steward Obs.	10-0.36 μm
J.D. Bregman, Witteborn, Lester	NASA/Ames	14-8 μm
Impey, Williams, Brand	UKIRT, REO	2.2-1.2 μm
Hackwell	Univ. Wyoming	2.2-1.2 μm
A. Smith, Pollock, Webb, Pica	Univ. Florida	0.9-0.36 μm
Miller, Stephens	Lick Obs.	0.8-0.4 μm
Wills, Wills	Univ. Texas	0.8-0.35 μm
Bregman, Glassgold, Huggins, and Kinney	NYU and NRAO	0.12-0.32 μm
Ku	Columbia Univ.	0.2-4 keV
Elvis, Schwartz, Tannanbaum	CFA	0.2-4 keV
Pollock, Willmore	U. Birmingham	0.2-1 keV
McHardy	U. Leicester	0.2-1 keV
Maccagni	CNR, Milan	0.2-1 keV

and 1215+303 (in preparation). The radio spectra of these
sources, which are usually categorized as being "flat",
sometimes show curvature and structure. The flatness of
the spectrum appears to arise from several partially
emitting regions of differing opacities; this
interpretation is supported by VLBI observations (Cotton et
al. 1980) and flux variability measurements (Aller et
al. 1985). In nearly all of the sources we observed, the
flat radio spectrum steepens in the 30 - 1000 GHz region so
that it has a slope of approximately -1. We interpret this
turnover as the frequency at which all radio components
become transparent. The 100 - 2 μm region is usually well
described by a power law in the range -0.9 to -1.5. In the
optical and ultraviolet region, the spectrum sometimes
steepens, reaching slopes of -3 occasionally. Some
sources, such as IZw-187, have constant slopes from the
infrared through the ultraviolet (-0.9 in this case). The
other extreme is represented by sources in with the
spectrum steepens exponentially in less than a decade of
frequency space. This feature, first noticed in red
quasars (Rieke et al. 1977) and seen in sources such as
1413+135, occurs at higher frequencies as well (in the
ultraviolet in BL Lac, 3C 446), and is probably a common

phenomena. In addition, when blue bumps have been
detected, they occur only in violently variable quasars
such as 3C 345 (the blue bump that was discussed for
IZw-187 by Bregman et al. 1982 was not verified by more
accurate observations).

There is sometimes a smooth connection between the
IR-UV spectrum and the X-ray region, such as in IZw-187.
However, for the majority of cases, an extrapolation of the
IR-UV continuum to higher frequencies passes far below the
X-ray flux density. This implies that X-rays emission
arises either from a different physical process (thermal or
inverse Compton emission) or from a different region, or
both. The X-ray spectral slope of our sources, typically
-1, is generally steeper than the "universal" spectral
slope of -0.62 (e.g. Rothschild 1983), although no single
spectral slope characterizes quasars in the Palomar-
Green survey (Elvis et al., this volume).

The radiated power in our sample is dominated by the
contribution from the far infrared through optical
regions. This may be seen in sources such as 3C 345, where
the luminosity of the flat radio spectrum is comparable to
that from the X-ray region (if extrapolated to 165 keV with
a slope of -0.62) but only about a sixth of the luminosity
from the far infrared - ultraviolet region. The luminosity
in the blue bump is only about 0.65% of the total emitted
power.

When a comparison is made between BL Lacertae objects
and violently variable quasars, no obvious differences are
apparent in the spectral shape save for the blue bump and
the emission lines. One violently variable object, 3C 446,
had been suggested to be an object intermediate between BL
Lacertae objects and violently variable quasars. It indeed
has very small equivalent widths in the optical region, but
the spectrum is also quite steep and the equivalent widths
of the ultraviolet lines are more nearly like normal
quasars. We find that for the number of ionizing photons
emitted by 3C 446, its line strengths are perfectly normal
(e.g. its covering factor is like normal quasars). The
ratio of the line strengths to the number of ionizing
photons is also normal for the violently variable quasar
3C 345. Kinney et al. (1985) found that there was little
dispersion in this ratio in normal quasars and argued that
this implied that the ionizing radiation is not
relativistically beamed into a small opening angle. By
analogy, the ionizing radiation from the violently variable
quasars 3C 345 and 3C 446 is also not beamed.

3. VARIABILITY

Each of the sources were selected because they have
shown variability in the optical and radio wavebands. In

3C 345 and OJ 287, there appears to be a connection in the variation in these two wavebands. In 3C 345, infrared and optical monitoring data reveal five outbursts during a 16 year period. There are also five outbursts defined by the 15 GHz and 90 GHz data. The radio outbursts are either coincident with or follow the IR-optical outbursts by 1-2 years, and the 90 GHz outbursts precede the 15 GHz outbursts, typically by a year. The onset of outbursts as defined by the infrared or millimeter data coincide with the zero separation of individual superluminal components. One interpretation of these events is that the plasma responsible for optical outbursts moves outward, eventually becoming transparent at radio wavelengths and distinct as a VLBI component as it separates from the core.

The spectral variation in the IR-UV region during outbursts can differ from one object to another. During the dramatic outburst in 1156+295, the spectrum retained its shape as it dimmed (over a period of about a week). However, a month later when the source was fainter, the spectrum was steeper and daily flux measurements at that time also showed no change in the spectral shape. Flux variation in which the spectral shape is preserved is quite common and has been seen in 0735+178, OJ 287, and other objects. In 3C 345, spectral variation is dramatic and during a particular infrared outburst, the 10 μm flux changed less than at higher frequencies, so the spectrum became harder when the source brightened and softened when it faded. In the following year, a general dimming occurred that was barely perceptible at 100 μm (8% dimming) but became more dramatic at higher frequencies and decreased by a factor of 2.5 at 0.55 μm. This flux variation is most easily modeled by a hard and soft component, where the hard component appears and then fades.

The connection between X-ray variability and the emission at other wavelengths is especially complex. For IZw-187, the X-rays are probably just an extension of the IR-UV emission (the timescale and amplitude of variation is the same in all wavebands and the X-rays are smoothly connected to the IR-UV emission). However, in 0735+178, the optical flux changed by 350% while the X-ray emission was unchanged to within the uncertainties of the Einstein Observatory (30%; the X-rays may be consistent with the radio variation in which the fluxes changed by 5-40 %, depending upon the frequency). In 3C 345, the X-ray variation is not closely correlated with the IR-optical changes, but appears to be related to the radio flux variation. The sources 3C 446, and OJ 287 show optical and X-ray flux variations that seem to occur at nearly the same time and either with similar amplitudes or with greater

X-ray flux variation. However, 1156+295 showed an X-ray
outburst flanked by two optical outbursts, all of
comparable amplitude (it is impossible to tell whether the
X-ray outburst precedes or follows the optical outburst).
In the handful of sources studies, the assortment of X-ray
to optical behavior is astonishing. The results imply that
X-rays are produced in several regions of (and) by several
processes.

4. MODELING

The synchrotron-self-Compton model has frequently been
applied to data such as ours. This model can be applied to
our data when we assume (1) that the turnover in the
millimeter and submillimeter region is the frequency at
which the source becomes entirely transparent and (2) that
the X-rays are produced through the inverse Compton process
by the same electrons responsible for the submillimeter
flux. The model results indicate that the plasma causing
the submillimeter emission has a magnetic field of 0.1 -
100 G, a size of 10^{16} - 10^{17} cm, and a density of 10 -
10^4 cm^{-3} (assuming a minimum electron energy of 15 MeV).
Relativistic bulk motion is modest, with Lorentz factors of
1-3. These properties for the emitting plasma are
significantly different from those found in the GHz region,
where the sizes are greater by a hundredfold while the
densities and magnetic fields are 10^2-10^5 times smaller.
The variability data indicate that the optical,
ultraviolet, and (sometimes) the X-ray emitting regions may
be considerably smaller than the submillimeter region, with
more intense magnetic fields imbedded in a denser plasma.
Complex synchrotron-self-Compton models with special
geometries, variation of the plasma parameters with radius
and bulk acceleration of the plasma have been developed by
a number of workers and provide a more realistic approach
to interpreting the observations (Blandford and Konigl
1979; Marscher 1980, 1983; Konigl 1981, Reynolds 1982a,b).

5. REFERENCES

Blandford, R.D., and Konigl, A. 1979, Ap.J., 232, 34.
Bregman, J.N., Lebofsky, M.J., Aller, M.F., Rieke, G.H.,
  Aller, H.D., Hodge, P.E., Glassgold, A.E., and Huggins,
  P.J. 1981, Nature, 293, 714.
Bregman, J.N., Glassgold, A.E., Huggins, P.J., Pollock,
  J.T., Pica, A.J., Smith, A.G., Webb, J.R., Ku, W.H.-M.,
  Rudy, R.J., LeVan, P.D., Williams, P.M., Brand, P.W.J.L.,
  Neugebauer, G., Balonek, T.J., Dent, W.A., Aller, H.D.,
  Aller, M.F., Hodge, P.E. 1982, Ap.J., 253, 19.
Bregman, J.N., Glassgold, A.E., Huggins, P.J., Aller, H.D.,
  Aller, M.F., Hodge, P.E., Rieke, G.H., Lebofsky, M.J.,

Pollock, J.T., Pica, A.J., Leacock, R.J., Smith, A.G., Webb, J., Balonek, T.J., Dent, W.A., O'Dea, C.P., Ku, W.H.-M., Schwartz, D.A., Miller, J.S., Rudy, R.J., and LeVan, P.D. 1984, Ap.J., 276, 454.

Cotton, W.D. et al. 1980, Ap.J. (Letters), 238, L123.

Glassgold, A.E., Bregman, J.N., Huggins, P.J., Kinney, A.L., Pica, A.J., Pollock, J.T., Leacock, R.J., Smith, A.G., Webb, J.R., Wisniewski, W.Z., Jeske, N., Spinrad, H., Henry, R.B.C., Miller, J.S., Impey, C., Neugebauer, G., Aller, M.F., Aller, H.D., Hodge, P.E., Balonek, T.J., Dent, W.A., and O'Dea, C.P. 1983, Ap.J., 274, 101.

Kinney, A.L., Huggins, P.J., Bregman, J.N., and Glassgold, A.E. 1985, Ap.J., 291, 128.

Konigl, A. 1981, Ap.J., 243, 700.

Marscher, A.P. 1980, Ap.J., 235, 386.

Marscher, A.P., 1983, Ap.J., 264, 296.

Pollock, A.M.T., Brand, P.W.J.L., Bregman, J.N., and Robson, E.I. 1985, Apace Science Reviews, 40, 607.

Reynolds, S.P. 1982a, Ap.J., 256, 13.

Reynolds, S.P. 1982b, Ap.J., 256, 38.

Rieke, G.H., Lebofsky, M.J., Kemp, J.C., Coyne, G.V., and Tapia, S. 1977, Ap.J. (Letters), 218, L37.

Rothschild, R.E., Mushotzky, R.F., Baity, W.A., Gruber, D.E., Matteson, J.L., and Peterson, L.E. 1983, Ap.J., 269, 423.

Wills, B.J., Pollock, J.T., Aller, H.D., Aller, M.F., Balonek, T.J., Barvanis, R.E., Binzel, R.P., Chaffee, F.H., Dent, W.A., Douglas, J.N., Fanti, C., Garrett, D.B., Gregorini, L. Henry, R.B.C., Hill, R.E., Howard, R., Jeske, N., Kepler, S.O., Leacock, R.J., Mantovani, F., O'Dea, C.P., Padrielli, L., Perley, R., Pica, A.J., Puschell, J.J., Sanduleak, N., Shields, G.A., Smith, A.G., Thuan, T.X., Wade, C.M., Wasilewski, A.J., Webb, J.R., Wills, D., and Wisniewski, W.Z. 1983, Ap.J., 274, 62.

DISCUSSION

Abramowicz : What is the relative number of objects which show quasi periodic variations to those which do not ?

Bregman : To answer this question, one would need a complete sample of high polarization sources. We have not observed a complete sample, but have selected sources that are known to have outbursts and that are bright enough to be observed with the IUE.

Cowsik : What are the reasons to believe that the continuum radiation is generated through Synchrotron and Compton processes ?

Bregman : These sources are highly polarized in the IR-optical region and they vary rapidly, giving information on the size. As others have argued, these facts suggest synchrotron emission. There is little evidence that the X-ray emission in inverse Compton radiation and this must be regarded as an assumption.

McAdam : Could you clarify the correlation between optical-infrared bursts and radio variability - In your presentation you said 4 or 5 outburst ~ 1 year earlier. Yet your abstract states .. "usually followed by radio outbursts ... " Is the correlation significant ?

Bregman : The outbursts are easy to identify but there is sometimes a problem in determining the beginning of an outburst when the sampling rate is too low. I encourage you to look at the data, which will appear in the Ap. J., Feb. 15, 1986. The connection between IR-optical and radio outbursts seems quite clear and we plan to examine this more quantitatively with Fourier techniques.

Cohen : In Neugebauer's IR burst, a new hard component is needed because the radiative time scale is wrong for the existing soft component. In that case, how can the short time scale be explained for the hard component ?

Bregman : It is likely that the outburst occurs in a region much smaller than the size deduced from the multifrequency spectrum, which addresses the size of the emitting region where $\tau = 1$ (in the millimeter region in 3C345). The decay of the hard component may be due to adiabatic expansion as the plasma flows into a larger region.

X-RAY TO INFRARED CONTINUA OF OPTICALLY SELECTED QUASARS

Martin Elvis
Harvard-Smithsonian Center for Astrophysics
60 Garden Street
Cambridge, Massachusetts 02138 USA

ABSTRACT. X-ray to infrared continuum data for nine optically selected (PG) quasars show a form which can be simply described in terms of two components — a power law of slope ~1 joining smoothly the 1-10 μm infrared with the 0.1-10 keV x-ray points and superimposed on this, a 'big bump' of optical-uv emission. The 'big bump' can be interpreted as thermal emission from an accretion disk. In a unique case where this 'big bump' extends to soft x-rays the accretion disk parameters can be constrained interestingly.

1. INTRODUCTION

Wilkes and Elvis (this meeting) present the results of an survey of the soft x-ray spectra of quasars using the Einstein Observatory IPC. These results show a diversity of slopes in x-ray power-law index, contrasting with earlier work (Mushotzky 1984). To interpret these indices in physical terms it helps greatly to know how the x-ray continuum joins up with the ultraviolet to infrared (UVOIR) spectrum. We have therefore started a program to determine the UVOIR shape of all the quasars that have good IPC x-ray spectra. Here we show the results for the first nine quasars. They are all uv-excess selected quasars from the Bright Quasar Survey (Schmidt and Green 1983). A full report of these results will be presented in Elvis et al. (1986), Bechtold et al. (1986), and Czerny, B., et al. (1986, in preparation).

2. CONTINUUM DISTRIBUTION

The energy distribution (log νf_ν vs. log ν) for the best determined continuum distribution (PG1501+106, Mkn 841) is shown in Figure 1. The 'bow-tie' shape in the x-ray band shows the range of allowed slopes.

The luminosity distribution across the entire five decade range of frequency is remarkably uniform (a horizontal line in Figure 1 would have equal luminosities per decade). The energy distribution of Figure 1 is similar to that of all the other quasars in the distribution (except PG1211+143, see below). These continua could be interpreted in many ways, the simplest however is to divide the distribution into just 2 components: a power law of slope ~1.0 in the

G. Swarup and V. K. Kapahi (eds.), Quasars, 73–77.

Figure 1.

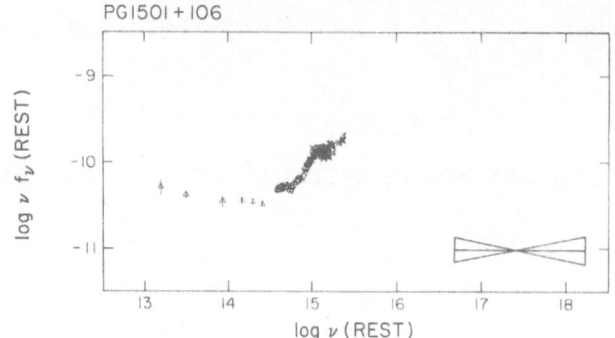

infrared extending without a break into the soft x-ray region; and a
'big bump' in the optical-ultraviolet superimposed on this power-law.
(The 'small bump' in Figure 1 has been shown convincingly to be due to
Balmer continuum and FeII emission, see Wills, Netzer, and Wills
1985.) This two component description is the same as that proposed
by Shields (1978) and Malkan and Sargent (1982) but extended now into
the 0.1-10 keV x-ray range.

One unique quasar is particularly important. PG1211+143 has a very
steep ($\alpha_E \sim 2.2$) soft x-ray spectrum and a flatter ($\alpha_E < 1.2$) hard x-ray
spectrum. No other quasar in the Wilkes and Elvis IPC survey has such
a steep soft spectrum. Its uniqueness may be related to its L_x/L_{opt}
ratio which is the largest in the PG sample (Tananbaum et al. 1986).
Figure 2 shows its x-ray to infrared energy distribution.

Figure 2.

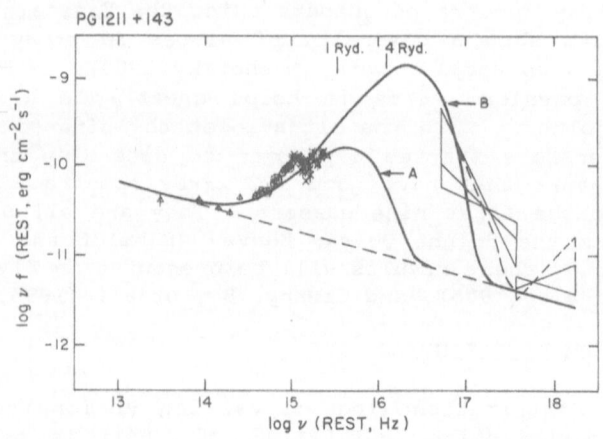

It is natural to apply the same 2-component decomposition as for
the other quasars by assuming the steep soft x-ray component to be the
high frequency extension of the 'big bump'.

3.   THE POWER-LAW

The data do not require that a single power-law extends from 10 μm to

10 keV. They are however consistent with this simple interpretation.
The coincidence of $\alpha_{IR} \approx \alpha_x \approx \alpha_{IR-x} \approx 1.0$ must certainly be explained
in some way.   In both the infrared and the x-ray bands there are
ambiguities which could easily remove this connection (e.g., IR dust
emission, two-component x-ray spectra).

If we adopt a single power-law as a working hypothesis there are
several mechanisms which could produce it: (1) Direct synchrotron
emission;   (2) 'Unsaturated Comptonization' of a arbitrary infrared
seed spectrum (e.g., Ipser and Price 1983);   (3) Inverse Compton
x-rays up-scattered from the 1-10 μm infrared;   (4) A pair-production
cascade spectrum (see Novikov, this meeting).   There is no immediate
way of deciding between these possibilities.   Infrared and x-ray
polarization and broad-band variability studies seem to offer the best
hope.

4. THE 'BIG BUMP'

Although the nature of the big bump is not decided the possibility
that it is due to the integrated thermal emission of an accretion disk
is an attractive one.   Since the steep soft x-ray spectrum of
PG1211+143 gives us a new constraint on the 'big bump' spectrum we
have taken the accretion disk idea as, again, a working hypothesis to
see what parameters it would imply for PG1211+143.

We used the simplest optically thick, geometrically thin disk
models at a constant cos i = 0.5 inclination.   This model turns out
not to be self-consistent but it does allow a first look at the
problem and possibilities.   More sophisticated models are now being
investigated by B. Czerny.

Figure 3.

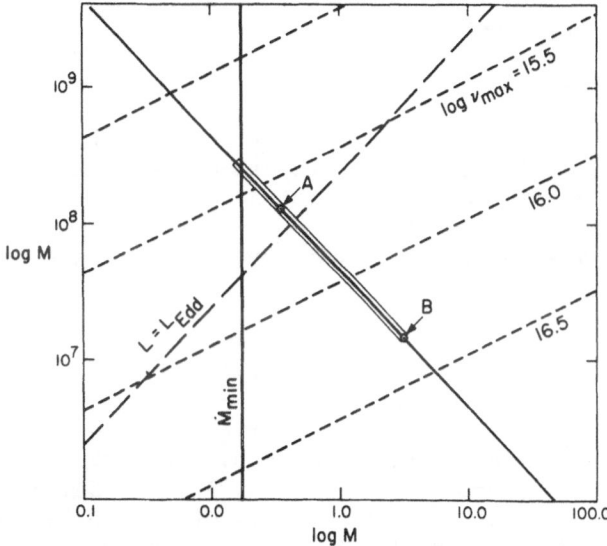

Figure 3 shows how the uv luminosity and the peak in νf_ν constrain
the central mass and accretion rate for PG1211+143.   The points
labelled A and B correspond to the models shown in Figure 2 and

illustrate the extreme range of allowed models. If the soft x-rays are really from the hot inner edge of a disk then model B is more appropriate. Model B however is ~20 times super-Eddington. Such a disk will be puffed-up by radiation pressure and so the thin disk approximation will not apply. However any model that is more realistic than a black body approximation will be less super-Eddington, probably by quite a large factor.

We can also use the optical-infrared break at ~1 μm to limit the outer radius of the disk to be at least ~.005 parsec. This may be interesting in limiting the value of the viscosity parameter α to prevent the break-up of the disk. It also makes clear that the 1-year timescale of ultraviolet variability which we see in the IUE data cannot be due to a global change in the accretion rate fed through the disk. This is simply because the viscous timescale at the outer radius we measure is already ~50,000 years. This imposes a low pass filter on any variation in accretion rate.

At the high energy end better x-ray spectra are sorely needed. Even the steep slope we measured is too flat to be explained with a simple disk model. We propose that the disk might be embedded in a hot (~80 keV) Compton scattering atmosphere. However the too flat slope might be an artifact of our low resolution ($\Delta E/E$ ~1) IPC spectrum.

4. CONCLUSIONS

Truly broad-band studies of quasars are clearly a valuable, even necessary, tool for understanding the quasar continua. We are now extending our study to include all quasars with good x-ray spectra. The new continua will include IRAS far-infrared data to look for the limits of the 'power-law' and accurate galactic N_H values to allow better x-ray slope determinations.

ACKNOWLEDGEMENTS

This work is part of a collaborative study involving many workers. I thank J. Bechtold, B. Czerny, R. Green, M. Schmidt, G. Neugebauer, B.T. Sorfer, K. Matthews, G. Fabbiano, and B. Wilkes for their great contributions to this study.

This work was supported in part by NASA Contract NAS8-30751.

REFERENCES

Bechtold, J., et al. 1986, Ap.J., submitted.
Czerny, B., et al. 1986, in preparation.
Elvis, M., et al. 1986, Ap.J., submitted.
Ipser, J.R. and Price, R.H. 1983, Ap.J., 267, 371.
Malkan, M.A. and Sargent, W.L.W. 1982, Ap.J., 254, 22.
Mushotzky, R.F. 1984, Advances in Space Research, 3, no. 10-13, 157.
Schmidt, M. and Green, R.F. 1983, Ap.J., 269, 352.
Shields, G.A. 1978, Nature, 272, 706.
Tananbaum, H., et al. 1986, Ap.J., submitted.

DISCUSSION

Bregman : Could the X-ray emission have varied and you have observed PG 1211+143 in a bright state ?

Elvis : Of course. We are monitoring PG 1211+143 with EXOSAT so we will soon know if it varies a lot. Whatever state it was in, it still had a very soft spectrum.

Burbidge : You have described ways in which you can explain the continuum observations by disk accretion models. Can you tell me which critical observations might rule out disk accretion models ?

Elvis : Strong variability on timescales much too short for dynamical timescales would be one way. Grey shift changes in the entire 'big bump' would, I think, be another. The constancy of the break point at 1 μm between IR power-law and the big bump may also be a problem since a disk should have very different relative amplitudes as a function of inclination angle.

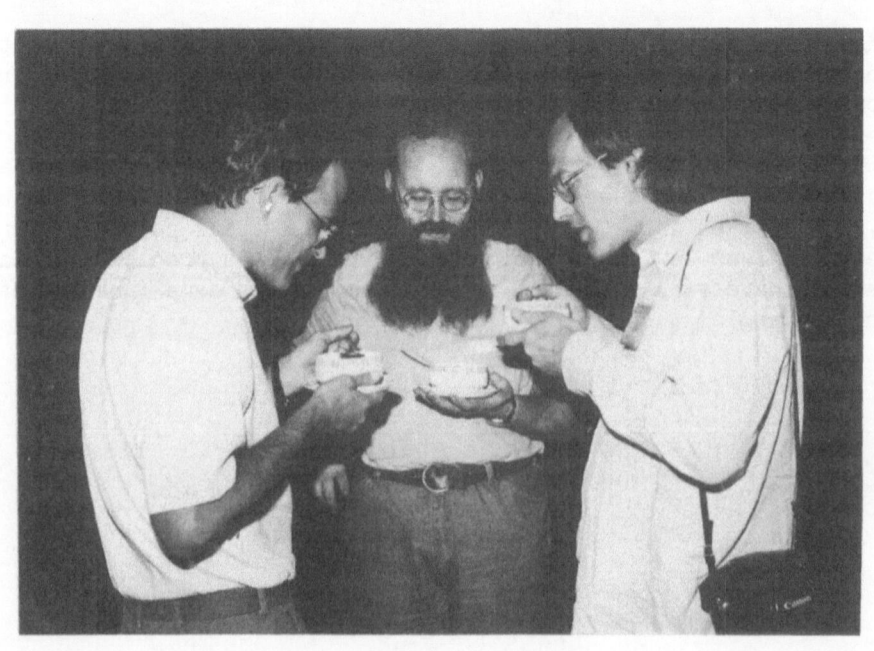

CONTINUUM MONITORING OF BRIGHT QUASARS

Esko Valtaoja
Turku University Observatory
SF-20500 Turku 50, Finland

ABSTRACT. Beginning in 1980, the fluxes of over 40 bright quasars, BL
Lacs and Seyferts have been monitored at radio and optical wavelengths
in a Finnish-Soviet collaboration. The chief aims of the program are to
provide information about the variability characteristics of active
nuclei over a wide range of timescales and to fill a gap in monitoring
programs at millimeter wavelengths. The results are also used as a basis
for various other investigations.

1. THE PROGRAM AND THE OBSERVATIONS

The monitoring program is a collaboration between Turku University Observatory (Finland), Helsinki University of Technology (Finland) and
Crimean Astrophysical Observatory (USSR). Additional measurements have
also been done at NRAO, West Virginia (USA) and at Sodankylä Geophysical
Observatory (Finland).

The principal radio telescopes used are the 13.7-m telescope in
Metsähovi, Finland (12, 22, 37, 77 and 90 GHz) and the 22-m telescope in
Simiz, Crimea (22 and 37 GHz). Observations with the 32-m telescope in
Sodankylä, Finland, have recently been initiated at 933 MHz. Polarimetry
is possible with the 933 MHz and 90 GHz receivers. The flexible sched-
uling of observing time has allowed us to investigate extensively also
shorter timescales (hours to weeks), which are not resolved in other
comparable monitoring programs, and for which rather little systematic
information exists, especially on frequencies above 15 GHz.

The radio sample consists of 42 bright (over 1 Jy) compact sources
known to be variable. Over 1/3 of the known blazars are included. About
25 sources have been monitored monthly or more often, mainly at 22 and
37 GHz with some additional measurements at 12 and 77 GHz. Monitoring
at 90 GHz has started recently with the new very sensitive receiver at
Metsähovi (Räisänen, 1983). The radio data up to June 1985, over 3000
measurements, has been reduced (Urpo et al., in preparation). Several
sources have also been monitored extensively with good time resolution
for rapid variability. Earlier results are described in Salonen et al.
(1983) and in Valtaoja et al. (1985).

G. Swarup and V. K. Kapahi (eds.), Quasars, 79–83.

The optical observations complement the radio monitoring. Results are described in Sillanpää et al. (1985, and in preparation).

2. DIFFERENT TYPES OF VARIABILITY

The observed variability behaviour at 22 and 37 GHz can be broadly divided into three distinct classes, which may have different underlying causes:

1) <u>Slow quasilinear trends</u> (timescale 10 years). This type is exemplified by NRAO 150, PKS 0735+17, 3C 273 and AO 1308+32, among others.

2) <u>Outbursts</u> (1 year). Prominent outbursts were seen in about 1/3 of the well-observed sources. Both the durations of the outbursts and their normalized lightcurves are very similar in the rest frames of the sources, independent of the source type; they seem to be well described by the canonical expanding source models with $T_b \sim 10^{12}$ K. There seems to be a tendency for the stronger outbursts to have more rapid rise times.

3) <u>Flickering</u> (0.1 year). Some sources show a rather chaotic behaviour on timescales of weeks or months, with apparently random flickering. This flickering may be superposed on other trends or outbursts. The amount of flickering may be quite different in sources of the same type. In BL Lac there is very little variation on monthly timescales (besides the outburst-related trends), while OJ 287 shows flickering down to at least daily timescales. This could indicate that the radiation from OJ 287 is beamed more directly towards us.

3. MAXIMUM AMPLITUDE OF VARIATIONS

Our coverage gives information about the variability characteristics of the sample over 4 orders of magnitude in time resolution (hours to 5 years). Of particular interest for theoretical models are the largest observed flux variations on different timescales. For each source in the sample, we have searched for the largest observed changes $\Delta S(max)$ within 0.1, 1, 10, 100 and 1800 days. (Note that this $\Delta S(max)$, which may represent a single event during five years of observing, will be much larger than the *average* variations on the timescale in question.) As Figure 1 shows, $\Delta S(max) \propto \Delta t^{0.3}$ both on the average for the sources in the sample as well as for the largest variations found in any QSO in the sample. A similar dependence for the rapidity of variations is also noticeable in the reported extreme flux changes (taken from Epstein et al. (1980) and Reich and Steffen (1982)). Since on shortest timescales the random statistical errors start to affect the data, possibly causing spurious large variations, the data for 3C 84 (several measurements) is also shown.

Since the brightness temperature $T_b \propto (\Delta S/\Delta t)^2$, our data indicates that $T_b \propto \Delta t^{-1.4}$, and the more rapid variations exceed the 10^{12} K limit in most of the sources.

We have made altogether about 130 flux measurements with $\Delta t < 6$ h, and over 400 measurements with $\Delta t < 30$ h to study short-timescale vari-

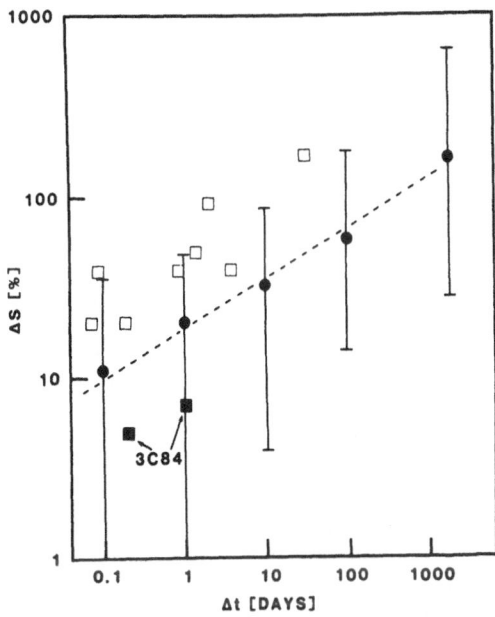

Figure 1. Largest observed variations on different timescales. In addition to the sample average, both the largest and the smallest variations found in any of the sample sources are shown. Open squares are extreme variations reported in the literature. The dashed line is the dependence $\Delta S(\max) \propto \Delta t^{0.3}$.

ations. To see whether the sample sources on the average are variable from day to day we have plotted all the observed flux changes $\Delta S(1,2)$ in units of $\sigma(1,2) = \{\sigma(1)^2+\sigma(2)^2\}^{1/2}$. The observed spread in ΔS is consistent with no average daily variability larger than about $\sigma(1,2)/2 \simeq 4\%$ (a conservative upper limit).

Even if the sources are generally not variable on daily timescales, some individual large changes may still occur. It may be significant that many of the most dramatic flux changes reported in the literature are sudden drops from a constant level with subsequent recovery; after careful consideration we have removed several such occurrences from our data as being possibly due to some undetected errors resulting in abnormally low fluxes. The largest remaining variations in our data are a 36 % (3.3σ) drop in 2 hours in OJ 287 and a 49 % (3.1σ) drop in one day

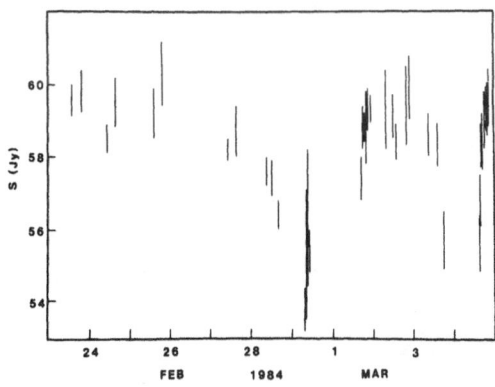

Figure 2. Rapid variability in 3C 84 at 12 GHz in Feb-Mar 1984.

in 4C 38.41. As Figure 1 shows, most of the sources have had at least
one episode of daily variability with $\Delta S > 10\ \%$, but many of these may
simply be the tail end of the gaussian distribution of errors. The
largest well-documented daily changes are comparable to those shown in
Figure 2 for 3C 84.

4. FUTURE WORK

The radio monitoring will be continued mainly at 37 and 90 GHz. Of special interest is the relation between mm- and optical wavelengths. Attempts to link cm- and optical data have remained rather inconclusive;
we hope to establish better correlations with simultaneous 90 GHz and
optical flux and polarization measurements.

Simultaneous UBVRI photopolarimetry (Korhonen et al., 1984) of
blazars has recently been added to the monitoring program to study rapid
variability and the little known frequency dependent polarization. The
most interesting observation so far concerns OJ 287, where our observations, combined with those of Sitko et al. (1985), show transient FDP
occurring in connection with a sudden drop in optical flux, preceded by
linear rotation of the position angle during the previous week and a
sudden change in P.A. simultaneously with the drop.

REFERENCES

Epstein, E.E., et al., *Astron. J.* 85, 1427 (1980)
Korhonen, T., et al., *ESO Messenger* 38, 20 (1984)
Reich, W., Steffen, P., *Astron. Astrophys.* 113, 348 (1982)
Räisänen, A., et al., *Electr. Lett.* 19, 892 (1983)
Salonen, E., et al., *Astron. Astrophys. Suppl.* 51, 47 (1983)
Sillanpää, A., et al., *Astron. Astrophys.* 147, 67 (1985)
Sillanpää, A., et al., in preparation
Sitko, M.L., et al., *Astrophys. J. Suppl.* (in press, 1985)
Urpo, S., et al., in preparation
Valtaoja, E., et al., *Nature* 314, 148 (1985)

DISCUSSION

Petrosian : Have you seen any periodic variability ?

Valtaoja : Our best candidate is OJ 287 with a possible 15.7 min period. We cannot do absolute flux measurements on so short timescales but must instead use Fourier analyses to search for periodicities in long continuous observing runs, and try to identify all the extrinsic factors that could cause apparent periodicities. After that it becomes a question of statistical significance of peaks in Fourier spectra. We estimate that the probability for the 15.7 min period being spurious is less than one percent.

"Most quasar optical light curves appear essentially the same whether displayed inverted or even with time-reversal."

- Harlan Smith and Ronald Angione (p.88)

0.3 TO 100 μm CONTINUA OF SEYFERT 1 GALAXIES[*]

G. Fabbiano, M. Elvis, N. Carleton, S.P. Willner
Harvard-Smithsonian Center for Astrophysics
60 Garden Street
Cambridge, MA 02138, USA

A. Lawrence   and   M.J. Ward
Queen Mary College   Institute of Astronomy
London, England   Cambridge, England

The questions we wish to address with this work are: What is the shape of the continuum spectrum of an active nucleus? Can we identify thermal and non-thermal components? Previous work on this subject studied different non-overlapping samples in narrow frequency bands. A significant step forward was taken by Malkan and Sargent (1982) who studied the IR (10μm) through UV continuum of a small sample of objects. We report here results for the first part of the study of the complete continuum for a hard X-ray selected sample of 37 Seyfert 1 galaxies. This selection criterion ensures a sample unbiased for optical-IR studies, as the hard X-rays are not affected by reddening (Lawrence and Elvis 1982). The data consist of our own 1-20 μm ground-based IR observations, 100 and 60 μm IRAS fluxes and optical UBVR photometry.

We found that the continuum spectra can be simply classified in three families by means of IR (60μm/12μm) and optical (1.2μm/0.3μm) colors (Figure 1). Class A galaxies, or "bare AGN" are non-reddened and do not present any significant amount of cool dust or galaxy IRAS component. These galaxies also tend to be brighter in X-rays, in agreement with the conclusions of Lawrence and Elvis (1982) and Mushotzky (1982) and have lower Balmer decrements than the other two classes. Class B galaxies are "reddened AGN". Class C galaxies are both reddened and present large "galaxy disk" cool dust components. Class B and C galaxies are low luminosity X-ray objects.

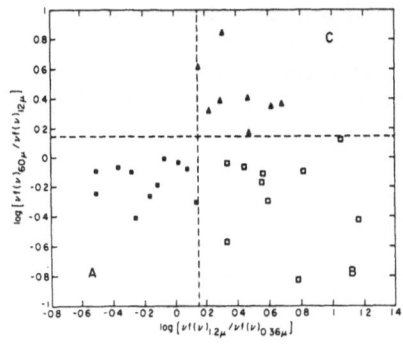

Figure 1.
The Classification.

[*] Discussion on p.93

G. Swarup and V. K. Kapahi (eds.), Quasars, 85–86.

To study the continuum, we excluded Class C galaxies and we restricted ourselves to the $100\mu m$ to $1\mu m$ region, to minimize the effects of reddening. We also made the working hypothesis that the spectra can be represented by an "underlying power law", plus IR excesses and we define this power law as the highest possible power law component allowed by the data. This hypothesis seems to have some meaning. We found that the power law fits for Class A galaxies (bare AGN) cluster tightly around a slope of 0 (in $\nu f \nu$) (or 1 in $f\nu$). The allowed power laws for Class B (reddened AGN) are instead more spread. Since a flat power law is suggested by the IR continuum, we define P_{below} as the power below a maximum power law $\alpha=0$ ($\nu f \nu$) and we explore the possibility that P_{below} could be a good estimate of the non-thermal continuum. Malkan (1984) found a tight correlation between the $3.5\mu m$ and the hard X-ray emission that led him to suggest that the $3.5\mu m$ emission is mainly non-thermal. If we use P_{below} instead of $L(3.5\mu m)$, we find an even tighter correlation (Figure 2).

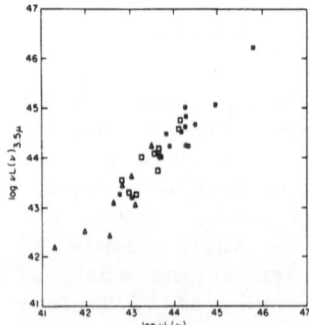 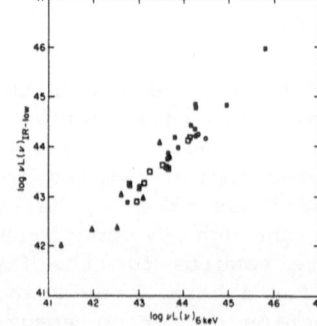

Figure 2.
Left: The Malkan
       correlation.
Right: Correlation
       of P_{below}
       with hard
       x-rays.

Therefore P_{below} is a __better__ estimator of the non-thermal continuum than the $3.5\mu m$ emission. Finally, where does the thermal component originate from? We addressed this question by studying correlations of broad emission lines (Hα) which originate within 0.1 pc of the nucleus and narrow emission lines ([OIII]), which originate from farther out (~100 pc) with $20\mu m$ and $3.5\mu m$ emission. We find that Hα is strongly correlated with the $3.5\mu m$ emission, while [OIII] is correlated instead with the $20\mu m$ emission: the hotter dust is nearer to the nucleus.

Concluding, using broad band photometry over a wide baseline covering 2-1/2 decades in frequency we can isolate "bare" nuclei, we can define thermal and non-thermal continuum components contributing to various degrees in different objects and we can have a first cut at the study of the dust distribution. We now have a new tool to explore further the temperature and spatial distribution of the IR emitting dust by studying the "thermal" component. This is work for the future, but we can already say that this distribution is not smooth and must have "temperature gaps" since we can define a meaningful non-thermal component. We plan to expand this work by including radio, UV, and X-ray data.

Lawrence, A. and Elvis, M, 1982, Ap.J., **253**, 410.
Malkan, M.A. 1984
Malkan, M.A. and Sargent, W.L.W. 1982, Ap.J., **254**, 22.
Mushotzky, R. 1982, Ap.J., **256**, 92.

PHYSICAL INTERPRETATIONS FROM QUASAR LIGHT CURVES -- 3C273

Harlan J. Smith   McDonald Observatory, U. of Texas
Ronald J. Angione   San Diego State Univ. California

Light curves of quasars give information on time scales (hence limits on size of principal luminous source), and on phase of variation (hence evidence bearing on the distance of surrounding regions or shells).

Insights derived from photometry as to the physical mechanisms involved have been slower to come. Some information lies in the amplitude of variation. The fractional amplitudes of most quasars are so big as to rule out the hypothesis that the luminosity comes from a very large number of separate accreting or exploding sources. There is evidence for inverse correlation of amplitude with redshift or with quasar luminosity (some of this can be accounted for by time dilation associated with large z). The correlation is consistent with models in which a modest number of variable sources contribute to the output of a quasar, but single-source models can also produce such data. Unambiguous detections of stable periodicities would be extremely important but have been elusive. Finally, at least in principle, some clues as to the character of the source may be present in the form of the variations.

3C273 has the longest relatively well-sampled run of historical photometric data. Our previously published light curves for this object included visual estimates from photographic plates, and comprised only mean points over lunations for 100-day intervals. Figure 1 presents for the first time the definitive light curve covering nearly a century and including objectively determined individual points. Prior to 1963 these come entirely from iris photometry of the Harvard historical plate collection; most of the remainder are photoelectric. Two cautions are in order: The four highest points are from plates taken with the smallest cameras having apertures of only several cm and very short focal lengths. Quasar images are very small and usually near the plate limit for these cameras, hence are unusually sensitive to occasional grain-clumpings or other accidents enlarging the image. These bright points could well be spurious, in which case there is no excess of positive over negative "flaring". Also, within a given year may be individual points having standard deviations as high as 0.2 mag; this exaggerates somewhat the appearance of sharp annual drops and rises.

G. Swarup and V. K. Kapahi (eds.), Quasars, 87–88.

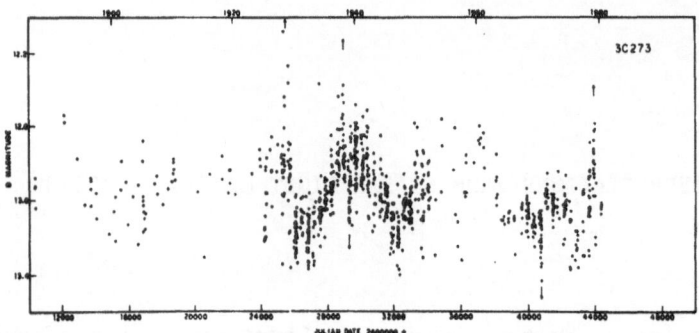

    Period analysis. Angione and Smith (AJ) have treated this light
curve using Deeming's method of power spectrum analysis applicable to
unequally spaced data. The Harvard data alone give only a single and
highly significant period of about 16 years, but this is heavily
weighted by the fact that 90% of the Harvard points lie in the interval 1928-1955 where the nearly sinusoidal variation is strongly marked.
If a true period, it should be present in the preceding and subsequent
data. However, a separate analysis excluding the 1928-1955 interval
shows only a weak 20-year period (largely due to alias) and a possibly
significant 6-year period. This means either that we are being fooled
by random variations or that real quasi-periodicity may be involved.
The latter interpretation is consistent with current models of an
accretion disk around a massive black hole, in which e.g. different
transient disk pulsation periods ranging from days to decades may be
present, and/or variable obscuring clouds may be orbiting at different
distances from a small bright nuclear region.

    Character of the Variations. Most quasar optical light curves
appear essentially the same whether displayed inverted or even with
time-reversal. Angione and Smith (1972) showed that, for 22 quasars
with relatively rich historical data, any selected rate of change of
brightness was effectively equally represented among increases and
decreases. If true flaring were an important phenomenon, we might
expect excess of rapid rises and of peaks standing out from the general
run of brightness. But even a casual inspection of the 3C273 light
curve shows one can make as good a case for anti-flares (or "dropouts").
This stupendously powerful central source of 3C273 must somehow be able
equally to appear to increase or decrease by up to nearly a factor of
two on times scales ranging from weeks to decades, yet normally remaining within rather tight limits. Irregular feeding of the monster may
well play a role. Here again, obscuration of the central source by
electron-scattering clouds orbiting in or above an accretion disk at
radii in the range of 10^{15} to 10^{17} cm offers a mechanism intrinsically
able to reproduce the character of the observed light curve, including
the wide range of time scales, the symmetrical rise and fall of brightness, and the presence of transient quasi-periodicities. On this picture, quasars observed relatively pole-on should in general suffer less
obscuration, hence have larger absolute magnitudes, be seen on the
average at greater red-shifts, and have smaller amplitudes of variation than would the majority of quasars seen relatively edge-on.

VARIABILITY IN THE OPTICAL CONTINUUM OF TYPE 1 SEYFERT NUCLEI[*]

H.R. de Ruiter[1] and J. Lub[2]

1 Istituto di Radioastronomia, Bologna, Italy
2 Sterrewacht Leiden, The Netherlands

1. INTRODUCTION

In 1979 we started a monitoring program of Seyfert galaxies, using various telescopes at ESO (La Silla, Chile). At present this program is still continuing, but here we discuss only the photometric data obtained in the period 1979-1982.

One of the aims of the program is to study the variability of the optical continuum radiation coming from active galactic nuclei and to identify the components that are responsible for the variability.

Although by no means statistically complete, our observing sample should be representative of Seyfert galaxies in general. We selected the Seyferts that were known in 1979 and satisfied the following criteria:
i) visual magnitude brighter than 15,
ii) declination below about 15 degrees,
iii) right ascension in the range 22 to 13 hours. Objects outside this range, but satisfying the other two conditions were observed in one period, August 1980.

Ideally an object is observed each year on 14 consecutive nights with the Walraven photometer attached to the 90 cm Dutch telescope at La Silla. We thus get an idea of the long term variations but also of the day-to-day variability.

The Walraven system has a number of advantages over the standard UBV photometry: it has five passbands (observed simultaneously), out of which three are in the range 3000 to 4000 Å. These passbands are relatively narrow (about 200 Å). Because of this the Walraven photometer is very suitable for studying the time-evolution of the so called 3000 Å bump in low redshift AGN's. More details on the Walraven photometry can be found in de Ruiter and Lub (1986).

2. VARIABILITY

We present the results concerning the period December 1979 till December 1982.

---

[*] Discussion on p.93

G. Swarup and V. K. Kapahi (eds.), Quasars, 89–90.

Out of eleven type 1 Seyferts that were observed regularly, ten showed significant variations on timescales of one year, from 0.1 to 0.6 mag. in the visual (the strongest variability was observed in Fairall 9 and 3C 120). The only object that remained constant, NGC 7603, is known to exhibit strong variations in the strength of the Balmer lines (Tohline and Osterbrock, 1976). The four type 2 Seyferts included in our program all remained constant, with an upper limit to variations of the order of 0.05 mag. This same upper limit applies to short term variations, i.e. in none of the Seyferts did we find significant changes on timescales of one day to one week. This tells us, incidentally, that such short term variability must be very rare.

3. COMPOSITION OF A SEYFERT TYPE 1 SPECTRUM

For each of the ten Seyferts that had varied we selected the phases when the object was brightest and faintest and compared the corresponding five-point spectra after correction for interstellar extinction and emission lines (whose strengths are known since we also have spectra obtained simultaneously with the photometry). Subtraction of the "faint" from the "bright" spectrum then gives the variable components. Assuming that the variability is caused by a Balmer continuum of constant temperature that changes its intensity and a power law component with variable spectral index, good fits to the difference spectra can be obtained. The procedure failed only for AKN 120: this object varied in all five passbands by the same factor.

The Seyfert spectra will also contain constant components; a normal galaxy spectrum will contribute a significant fraction of the total intensity in the V and B bands. Fitting the one constant and two variable components to the five-point spectra, in most objects a constant excess was left in the W band (at about 3200 Å). We identify this excess with the UV bump seen in AGN's below 3000 Å; it may be the black body radiation discussed by Malkan and Sargent (1982).

In summary, we suggest that the optical variability in the continuum of type 1 Seyfert spectra is caused by two components; the first is a Balmer continuum (with a temperature around 10000 K) that changes in intensity, the second is a power law with variable spectral index. In all cases the power law flattens when the object becomes brighter.

In order to reproduce a Seyfert spectrum we also need a normal galaxy component, which is important in the visual and to a lesser extent in the blue, and an ultraviolet component, only important in the W passband.

REFERENCES

de Ruiter, H.R., Lub, J.: 1986, Astron.Astrophys.Suppl., in press
Malkan, M.A., Sargent, W.L.W.: 1982, Astrophys.J. **254**, 22
Tohline, J.E., Osterbrock, D.E., 1976, Astrophys.J. **210**, L117

VLA OBSERVATIONS OF RAPID VARIABILITY IN OJ287 [*]

David H. Roberts,[1] John W. Dreher,[2] and Joseph Lehár[1,2]
[1]Department of Physics, Brandeis University
Waltham, MA 02254 USA
[2]Department of Physics, Massachusetts Institute of Technology
Cambridge, MA 02139 USA

ABSTRACT. Using the VLA at 6 cm, we have detected non-periodic variations in the powerful BL Lacertae object OJ287 on timescales between minutes and hours at levels of the order of one percent. Brightness temperatures inferred from causality arguments range from 10^{16} to 10^{20} K. No periodic component was found to a limit of 0.1 %.

We observed OJ287 with the Very Large Array at 5, 15, and 22 GHz several times in early 1983. Here we discuss only the 5 GHz data which were taken coincident with the 22 GHz observations of Valtaoja et al. (1985); a complete discussion is given in Dreher, Roberts, and Lehár (1986). The array was divided into two independent subarrays which observed OJ287 and the calibration source 0839+187 for 6 min 20 s each in two alternating sequences 180 degrees out of step. The visibility data for each subarray and circular polarization were separately corrected for the amplitude and phase gains of each antenna, using 3C286 to set the amplitude scale of the calibrator and phase calibrating each source on itself. Time series of the flux densities of OJ287 and 0839+187 were derived for each antenna from the instantaneous power gains. These were averaged over the antennas and polarizations, filtered of bad data points, and rescaled so that the average was 1000. Errors were assigned to each data point from the internal consistency of the averages. The results are given in Figure 1, and show OJ287 to decrease in flux by about 1.8 % in an irregular way over a 7 hour time span. This corresponds to a formal time scale $\tau = S/|dS/dt| < 16$ days, a short but not atypical value for active sources. The more rapid "dip" of ~ 0.5 % which occurs over ~ 15 min near $t = 23.89$ days has $\tau < 2$ days. These variations were seen in both polarizations of all antennas and appear highly significant in terms of the known errors. We feel, therefore, that they are real, but experience using the VLA to measure fluxes to an accuracy of < 1 % is as yet too limited to be absolutely certain.

Simple causality arguments which limit the size R of a source to $R < c\tau/(1+z)$ may be used to infer lower limits to the brightness

---

[*] Discussion on p.93

G. Swarup and V. K. Kapahi (eds.), Quasars, 91–92.

temperature of OJ287 of 1×10^{16} K for a 1.8% change in 7 hours and
6×10^{17} K for a 0.5 % change in 15 minutes. If the varying component
δS is physically distinct from the steady component S, its brightness
temperature is increased by a factor $(S/\delta S)$, leading to 6×10^{17} K and
1×10^{20} K. Such extraordinary inferred brightness temperatures would
seem to require coherent emission, unprecedented Doppler boosting, or
strong centimeter-wavelength scattering. In order to search for
periodic variations in OJ287, we formed a discrete Fourier transform
of the time series and removed the effects of the data window with a
one-dimensional complex CLEAN (Lehár 1985; Roberts, Lehár, and Dreher,
in preparation). The resulting distribution of harmonic amplitudes
is shown in Figure 2. There is no peak near the previously-reported
period of 943 s. Tests with artificial signals enable us to place a
conservative upper limit of 0.1 % on any harmonic variation.

We thank A. G. de Bruyn and K. Sowinski for helpful discussions.

Dreher, J.W., Roberts, D.H., and Lehár, J. 1986, to be published.
Lehar, J. 1985, Bachelor's Thesis, Physics Dept., Brandeis University.
Valtaoja, E. et al. 1985, Nature, 314, 148.

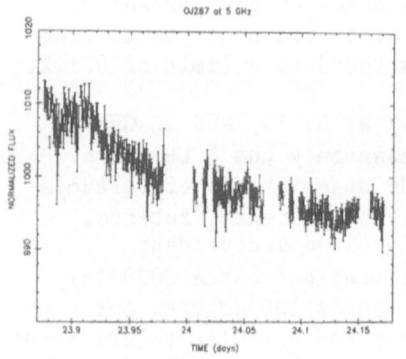 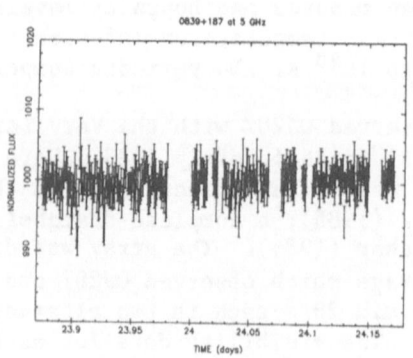

Figure 1. Flux histories of OJ287 and 0839+187 on 1983 May 23-24.

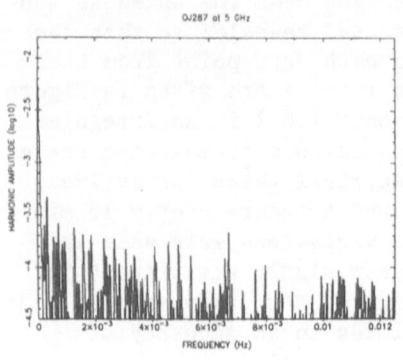 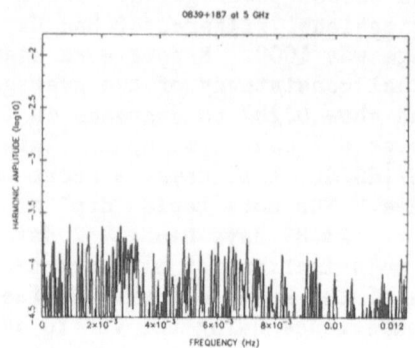

Figure 2. Harmonic amplitudes of the time series in Figure 1, on a
logarithmic scale. A conservative upper limit on harmonic variability
of OJ287 from these data is 0.1 % of the mean.

DISCUSSION ON THE PAPER BY **FABBIANO** ET AL. (p.85)

Filippenko : When comparing broadband colors of Seyfert 1 nuclei, how did you account for the very different effective apertures of the IRAS, near-IR, and optical data ?

Fabbiano : We subtracted the stellar contribution by using CCD images at R, and M31 bulge colors when possible, otherwise assuming a de Vaucoulurs profile for the bulge and using Young's (1976?) tables. We did not correct the IRAS data. The 60 and 20 μ IRAS and ground based points are typically within 20% of each other, and this would not affect our conclusions. The four IR points (100 μ , 50 μ .....) are definitely indicative of galaxy disk emission in low luminosity objects.

DISCUSSION ON THE PAPER BY **DE RUITER** & **LUB** (p.89)

Khachikian : Have you any data about Sy 2 Galaxies ?

De Ruiter : We have observed six type 2 Seyferts, of which four regularly in the period 1979-1982. We did not find any variations and stopped observing the type 2 Seyferts after 1982.

DISCUSSION ON THE PAPER BY **ROBERTS** ET AL. (p.91)

Dultzin-Hacyan : The variability can be strongly frequency dependent. We published optical data for periodicity in OJ 287, they appeared in Nature, simultaneously with the Finnish group results, and we are pretty sure of our results.

Roberts : The results I presented pertain to λλ6, 2, and 1.3 cm only. A coordinated world-wide program to observe OJ287 at all wavelengths from optical through decameter radio would help sort out the existence, variability, and frequency dependence of any periodic variations in OJ287.

Rees : The magnetic field within (say) 10^{14}cm of an accreting massive black hole could be 10^4-10^6 gauss. The variable radiation observed could then be at the Larmor frequency. We know that coherent emission at the cyclotron frequency is more readily produced than in the case of synchrotron radiation by ultra-relativistic particles. The inferred brightness temperatures could then arise if the source contained a large number of regions, each large enough to act coherently, and the overall properties of the entire system varied quasi-periodically. To explain the lack of high overall circular polarization, the field must be disordered,

93

G. Swarup and V. K. Kapahi (eds.), Quasars, 93–94.
© *1986 by the IAU.*

or the plasma a mixture of electrons and positrons. The main problem
with this interpretation is that the postulated coherent emission would
be vulnerable to incoherent synchrotron absorption and induced Compton
scattering by plasma further out.

RADIO EMISSION FROM OPTICALLY SELECTED QUASARS

K. Kellermann and R. Sramek
National Radio Astronomy Observatory

D. Shaffer, Interferometrics Inc.

R. Green, Kitt Peak National Observatory

M. Schmidt, California Institute of Technology

I. INTRODUCTION

Shortly after their discovery, it was realized that most quasars are not found in radio source catalogues, and a number of searches were initiated to try to detect weak radio emission from "optically selected quasars." See Table 1.

The earliest observations, all made using single-dish radio telescopes, were unsuccessful due to inadequate sensitivity. By the early 1970's much greater sensitivities were available with the use of interferometer systems and large single dishes operating at shorter wavelengths to reduce the effects of confusion. Flux densities of the order of 10 mJy could be reached. At this level, some 10% to 20% of optically selected quasars could be detected as radio sources. It was noticed by several investigators that about half of the quasars detected were already catalogued radio sources, and that the number of radio detections increased only slowly as the flux density limit was decreased. (Fanti et al. 1977, Sramek and Weedman 1980, Strittmatter et al. 1980, Shaffer et al. 1982. Several authors also commented that among quasars brighter than 17th magnitude, the fraction of radio detections was apparently higher than for the fainter quasars (e.g., Smith and Wright 1980, Condon, et al. 1980).

Since quasars are generally thought of as active objects, the apparent absence of radio emission from most quasars has been a subject of much discussion. The simplest and most trivial interpretation, of course, is that the radio luminosity function of quasars is broad, and because of their generally large distance, only the most luminous appear as radio sources. It is also possible that there may be two fundamentally different types of quasars; those which are radio loud and those which are radio quiet. Variability may also explain the lack of observed strong radio emission from most quasars. But known variable radio sources generally do not vary by more than a factor of two or three; so it seems unlikely that variability alone can account

95

G. Swarup and V. K. Kapahi (eds.), Quasars, 95–102.
© *1986 by the IAU.*

TABLE 1

λ	$B_{\ell im}$	S_{min}	n_{obs}	n_{det}	%	Reference
70	18	500	15	0	0	Bracessi et al. 1967, Nuovo Cimento 49B, 151
2,4	18	1000	2	0	0	Weinreb and Shapiro 1966, Ap. J. 143, 598
6	18	5	12	0	0	Pauliny-Toth and Kellermann 1966, Nature 212, 781
75	18	100	15	0	0	Mills and Little 1970, Ap. L. 6, 197
4,11	18	10	54	0	0	Wardle and Miley 1971, Ap. J. (Lett.) 164, L119
21	19	7	95	4	4	Katgert et al. 1973, A & A 23, 171
73	18.5	100	36	2	6	Murdoch and Crawford MNRAS 1977, 180, 41P
21	18	10	62	8	13	Fanti et al. 1977, A & A 61, 487
4,11	16,2	20	6	1	16	Shaffer and Green 1978, PASP 90, 22
2,6	19	25	240	18	8	Savage and Bolton 1979, MNRAS 188, 599
13	20	14	247	18	7	Sramek and Weedman 1980, A. J. 238, 435
2,6	19	10	122	10	8	Smith and Wright 1980, MNRAS 191, 871
3,6	19	10	70	10	14	Strittmatter et al. 1980, A & A 88, L12
6	17	0.5	22	9	41	Condon et al. 1981, Ap. J. 246, 624
6	16	2	94	27	29	Shaffer et al. 1982 IAU Symp. 97, p. 367
20	18.5	10	30	2	6.5	Mitchell 1983 thesis,
	19.5	10	26	2	8.5	Pennsylvania State University
6	17	0.5	41	7	17	Hutchings and Gower 1985, A. J. 90, 405

Notes to Table I. λ, observing wavelength in cm; $B_{\ell im}$, approximate limiting magnitude of optical sample; S_{min}, approximate limiting flux density in mJy for radio detection; n_{obs}, number of QSO's observed; n_{det}, number of quasars detected above S_{min}; %, percent of QSO's with detected radio emission.

for the wide spread in the observed ratio of radio to optical flux
density. It has also been suggested that quasars may be surrounded by
a cold cloud of absorbing gas, and that only in those cases where the
radio source "breaks through" the absorbing cloud, is there observable
radio emission (Bolton, 1977, Strittmatter et al. 1980). Finally,
there has been the suggestion of Scheuer and Readhead (1980), that due
to relativistic beaming, only the small fraction which have their beams
oriented toward the observer appear radio loud.

The relativistic beaming model has enjoyed great popularity,
because it appears to explain a number of observed phenomena including
superluminal motion, rapid flux density variations, and the absence of
Compton scattered X-rays (e.g., Porcas, 1986). However, the
relativistic beaming model predicts that the number of detected radio
quasars per flux density interval should increase roughly as $S^{-5/2}$.
Although the specific value of the exponent is somewhat model
dependent, the relatively few quasars which are weak radio sources
appear to be inconsistent with simple beaming models (e.g.,
Strittmatter et al. 1980, Smith and Wright 1980).

II. THE VLA OBSERVATIONS OF THE BRIGHT QUASAR SURVEY

In an attempt to understand better the radio emission from
quasars, we have observed all 114 objects in the Schmidt and Green
(1983) Bright Quasar Survey (BQS), using the VLA in the D configuration
with 18 arcsec resolution at 6 cm wavelength. The radio observations
are complete for all BQS quasars down to a flux density limit of
0.25 mJy, and for some objects to 0.15 mJy. All objects detected as
radio sources were reobserved in the A-configuration with 0.5 arcsec
resolution to isolate the flux density contained in the compact core.
A preliminary account of these results has already been reported
(Kellermann et al., 1983) and a more detailed description will be
discussed elsewhere.

The new observations differ from previous studies of optically
selected quasars in several ways.

   (1) It is based on a sample which is optically complete to a
        average limiting B magnitude of 16.16.
   (2) The sample consists of the optically brightest quasars.
   (3) The typical redshift is small, with a median value of only
        $z = 0.19$.
   (4) The limiting radio flux density is well below that of other
        surveys of optically selected quasars.

Thus, for the two-thirds of the objects with $z < 0.3$ the limiting
flux density of 0.25 mJy corresponds to a monochromatic luminosity
about 10^{22} WHz/Hz, whereas most previous radio observations were less
sensitive by several orders of magnitude. Moreover, since the BQS
objects are among the brightest quasars in the sky at optical
wavelengths, the low limiting radio flux density corresponds to a
ratio, R, of radio to optical flux density, S_r/S_o of 0.2 even for

B = 16. The new radio data, together with the previous radio searches
listed in Table 1, help to determine the distribution, $\Psi(R)$, and its
dependence ón optical luminosity and cosmological epoch.

We have been able to detect 73 (79%) of the 92 BQS quasars with
M < −23; 32 (97%) out of the 33 of objects with z < 0.5, but only 13
(59%) of 22 of those with z > 0.8. As suggested by previous observers,
the detection rate apparently depends on optical magnitude. For the 44
quasars with B < 15.5, we detected 39 (89%); whereas for the 48
quasars, faint quasars with B > 15.5, we detected only 28 (63%) above
0.25 mJy.

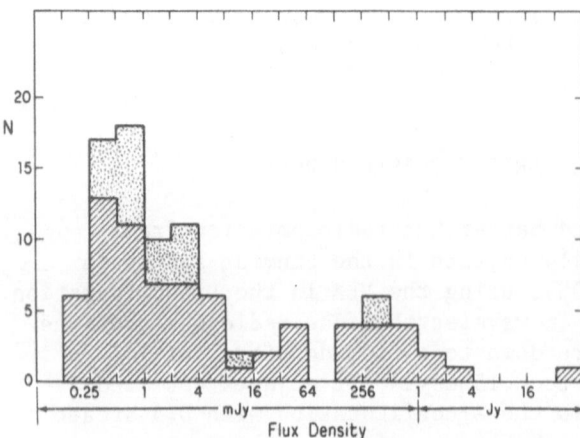

Figure 1. The distribution
of radio flux density of
the 94 detected BQS
objects. The 73 quasars
with M < −23 are shown with
diagonal shading, while
the 21 objects with
M > −23 are shown with
stippled shading.

In Figure 1 we show the distribution of observed flux density for
the detected quasars. The distribution is unusual, in that there is a
small fraction (17%) which are relatively strong with observed flux
density greater than 100 mJy. These are the "radio loud" population
which has been observed in previous radio observations of optically
selected quasars, and which are normally found in radio surveys. In
the range 10 < S < 100 mJy the number of radio detections increases
only slightly, in agreement with previous surveys. But for S < 10 mJy,
the numbers increase rapidly, and 67 (72%) of the 92 quasars have a
total flux density greater than 0.25 mJy. The distribution of radio
flux density is clearly bimodal. The presence of a separate radio loud
population is shown clearly in the distribution of S_r/S_o illustrated in
Figure 2. For the radio loud population the radio flux density is 1000
times greater than at optical wavelengths, but for the "radio quiet"
population the radio and optical flux density is equal to within a
factor of 10. Even for the non-detections, the observed radio limits
are not extreme and do not require that the radio flux density be less
than 10% of the optical flux density.

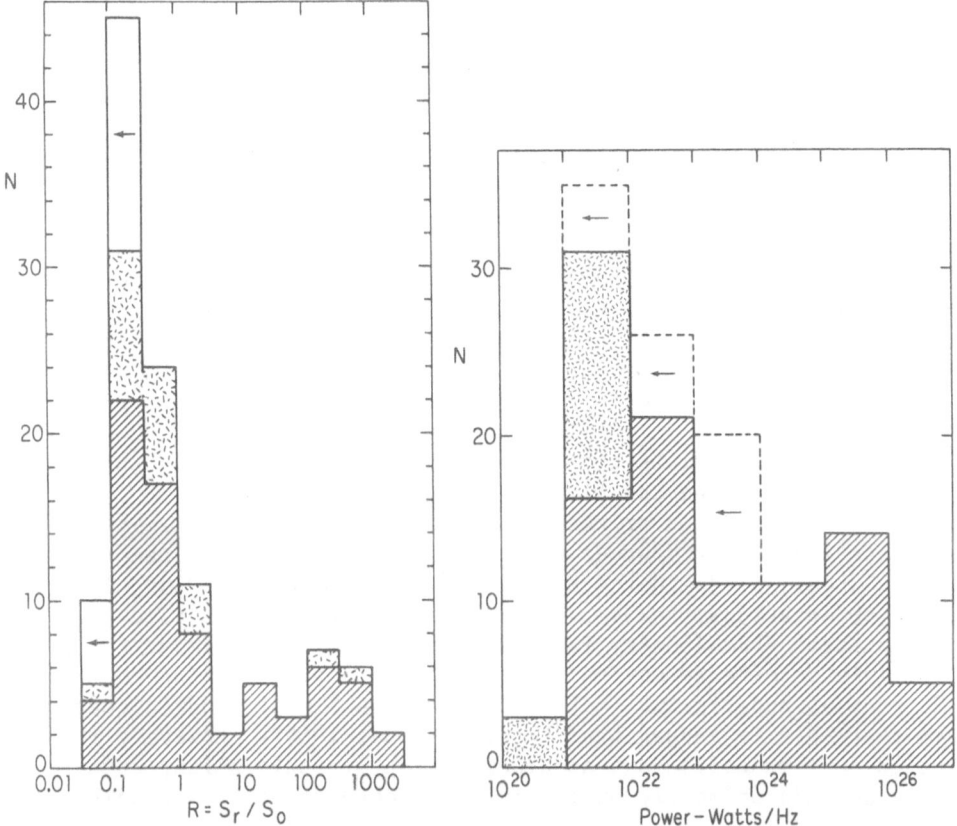

Figure 2. Distribution of S_r/S_0. Shading as in Figure 1.

Figure 3. The distribution of 6 cm absolute monochromatic luminosity and upper limits for the BQS objects. Shading as in Figure 1.

The distribution of absolute radio luminosity for all objects in the BQS is illustrated in Figure 3, which shows that the detected quasars have radio luminosities in the range 10^{20} to 10^{27} WHz, essentially the same range covered by radio galaxies. The non-detections correspond to a typical upper limit between 10^{21} and 10^{24} WHz^{-1}, a value equivalent to weak radio galaxies or Seyfert galaxies. The objects with M > −23, all have radio luminosities at the low end of the distribution between 10^{20} and 10^{22} WHz^{-1}.

III.   SUMMARY AND CONCLUSIONS

The fraction of BQS quasars detected as radio sources depends on apparent magnitude and redshift, but apparently not on absolute

magnitude. The observed radio luminosity has a range roughly
equivalent to that of radio galaxies. A small fraction, between 15 and
20 percent are anomalously strong radio sources, with most of their
radio emission coming from extended regions well removed from the
quasar, and with radio flux density up to 1000 times greater than the
optical flux density.

The bimodal distribution of radio luminosity is clearly not
consistent with the simple interpretation of radio loud quasars being
those with relativistic beams oriented toward the observer. For the
weak source population, however, the numbers do increase in a manner
roughly consistent with the beaming model. If the detection rate were
to continue to increase below 0.25 mJy, in the way calculated from
beaming, then essentially all of the quasars should be detected above
0.15 mJy. However, for the 20 quasars we have not detected, we measure
an average flux density of only 8 ± 17 micro Jy. These may be truly
"radio silent" quasars.

The beam model predicts that the stronger sources, (i.e., those
with beams oriented along the line of sight) should contain a
relatively larger fraction of their flux density in the Doppler boosted
core, whereas the weaker sources with beams oriented more in the plane
of the sky, should be dominated by the extended component. Our high
resolution observations, however, show no dependence of the core
fraction on total flux density for the sample with S < 100 mJy.

Relativistic effects may be important in quasars, and indeed the
preponderence of superluminal motion in nearly all well-studied quasars
suggests that this is indeed true (e.g., Porcas, 1986). However, the
distinction between radio loud and radio quiet or radio silent quasars
apparently depends primarily on intrinsic effects as well as possibly
geometry.

REFERENCES

Bolton, J. G. 1977, IAU Symposium No. 74, D. Jauncey ed., page 85.
Condon, J. J., O'Dell, S. L., Puschell, J. J., and Stein, W. A. 1981,
     Ap. J. 246, 624.
Fanti, C. et al. 1977, Astron. and Astrophys. 61, 487.
Kellermann, K. I., Sramek, R., Shaffer, D., Schmidt, M., and Green, R.
     1983, Proceedings of 25th Liege Symposium, page 81.
Porcas, R. 1986, this volume , p.131.
Scheuer, P.A.G. and Readhead, A.C.S. 1979, Nature 277, 182.
Schmidt, M. and Green, R. F. 1983, Astrophys. J. 269, 352.
Shaffer, D. B., Green, R. F., and Schmidt, M. 1982, IAU Symp. 97,
     page 367.
Smith, M. G. and Wright, A. E. 1980, MNRAS, 191, 871.
Sramek, R. A. and Weedman, D. W. 1980, Astrophys. J. 238, 435.
Strittmatter, P. A. et al. 1980, Astron. and Astrophys. 88, L12.

DISCUSSION

Wampler : Could the gap between the brighter radio sources in the BQS and the fainter radio objects be caused by the apparent insenstivity of the BQS to quasars with z ~ 0.5?

Kellermann : I do not think so. The radio luminosity distribution is so broad (a range of 10^5) that a gap of even a factor of two in the red-shift distribution would be washed out in the flux density distribution.

Kembhavi : If you convolve the distribution of L_R/L_{opt} with optical quasar counts, it should be possible to predict the quasar content of 6 cm source counts at very low fluxes. Has this been done ?

Kellermann : In principal what you suggest is true, but the strong radio quasars, which dominate the quasar component of the radio source counts are only weakly represented in the BQS sample, so the predictions will be subject to large statistical uncertainties. Conversely, while most BQS objects have a 6 cm flux density $\lesssim 1$ mJy, radio source surveys at this flux level contain few, if any, quasars. It will be more productive, I think, to work with both the ratio of L_R/L_{opt} of the BQS sample which is dominated by quasars of relatively low radio luminosity, and the quasar radio number counts to derive the best fitting values of the local luminosity function and its spatial evolution. We plan to do this.

Swarup : What are the typical scale sizes of extended structures seen for quasars with $S_{5GHz} \lesssim 10$ mJy ?

Kellermann : There are a few sources with $1 < S_5 < 10$ mJy with angular sizes in the range $30 < \theta < 60$ arcsec, but for $S < 1$ mJy the overall sizes appear to be less than 20" in most cases.

Cohen : In the bimodal distribution, (a) The weak ones are consistent with a beamed population. What can you say about the strong ones ?
(b) The weak and strong ones have both small-scale and extended emission. Are there systematic morphological differences ?
(c) Very strong ones (3C273, 3C345, etc.) have steep spectrum outer components. Do you get a bimodal distribution if you use only cores (S within 0.̈5) ? Would you get a bimodal distribution with fluxes measured at $\lambda = 1$ cm ?

Kellermann : a) The strong (radio loud) quasars appear to be anomalous. They have a large ration of L_r/L_0 (100 to 1000). Since they are lobe dominated, it is unlikely that they are affected by relativistic beaming.
b)   Not obviously, but we really do not have enough structural information to say much. At low resolution (18") many of the sources are only barely resolved, while the high resolution (0.̈5) data resolves out all of the extended structure.
c)   No. The core fluxes are nearly uniformly distributed in flux density, with a slight excess of weak sources. There is no evidence for

"strong" compact component such as we see for the "extended" components. The ratio L_r/L_o for the cores is about unity-within a factor of 10, so there is no evidence that the compact radio flux is dominated by beaming effects.

Kühr : Among the somewhat stronger radio sources (e.g. S > 100 mJy) we see different distributions in the differential source counts for flat and steep radio spectra objects; is this effect also found in your faint flux density (e.g. S < 100 mJy) sample ?

Kellermann : We do not have spectral information except for the "radio loud" population which is roughly equally divided between "flat" and "steep" spectra. Our structural data on the weaker radio quasars suggest that these are also roughly equally divided between lobe dominated and core dominated sources.

Segal : In the context of a temporarily homogeneous, spatially spherical, space-time model, there is a natural, simple explanation for the discontinuity between the compact flat spectrum sources and the extended steep spectrum sources. Sufficiently long-lived sources should be observable in opposite directions, providing both a primary (short-path) and a secondary (longer-path) image, - which would be expected to have a flatter spectrum (at least in the specific case of the chronometric cosmology). Can your observations invalidate this explanation ?

Kellermann : No.

H.J.Smith : Jim Douglas and Arakel Bozyan are testing for the presence of antipodal objects in Douglas 'Texas Radio Source Catalog (source positions accurate to about ±1 arcsec). So far they find, among the 52,000 sources in the 7.8 sterradians lying in the range ±35° declination, no significant increase of antipodal coincidences over chance expectations (but they also note that sufficiently large random walks in the positions, from light bending in the gravitational fields of intervening galaxies, could preclude reliable detection of antipodal sources).

THE RADIO EMISSION FROM OPTICALLY SELECTED QUASARS

J.A. Peacock and L. Miller
Royal Observatory, Edinburgh.

A.R.G. Mead
Department of Astronomy, University of Edinburgh.

ABSTRACT. We describe models which try to account for the incidence of quasar radio emission in terms of two populations. One is effectively radio quiet, with radio powers $P_{2.7} \lesssim 10^{22}$ WHz^{-1} sr^{-1} characteristic of type I Seyferts. These are presumed to lie in spiral host galaxies. The other population is the classical radio-loud quasars with $P_{2.7} \gtrsim 10^{24}$ WHz^{-1} sr^{-1}. Apparent correlation of radio and optical powers can arise without any real effect being present because the two populations in general have separately-evolving luminosity functions.

1. A DUAL-POPULATION MODEL FOR QUASAR RADIO EMISSION

Most quasars are not radio sources. One of the outstanding problems of quasar research for more than a decade has been to account for this simple fact. The general picture which has emerged is not a simple division into radio-loud and -quiet classes, but of a single population with a very broad distribution of radio-to-optical spectral indices (e.g. Condon et al 1981). The origins of this viewpoint are to be found in a seminal paper by Schmidt (1970) in which an apparent correlation between radio and optical properties of quasars was found. The argument was based on the fact that the redshift distribution of 18 mag quasars selected optically by ultraviolet excess (UVX) was statistically identical to that of 3CR quasars of the same apparent magnitude. If we suppose that quasars of all optical luminosities have a universal distribution function of radio luminosity, then a radio-selected subset of the quasar population should be strongly biased to low redshifts, contrary to observation. Schmidt therefore proposed that there was instead a universal distribution of the radio-to-optical flux ratio R (effectively the spectral index between these wavebands). Astrophysically, this was a very important conclusion since it indicated that the radio properties were primarily determined by the optical properties of the active nucleus.

The applicability of Schmidt's correlation to the most recent quasar data has been re-examined by Peacock, Miller & Longair (1986; PML) and

G. Swarup and V. K. Kapahi (eds.), Quasars, 103–107.
© 1986 by the IAU.

we summarise some of their arguments here. PML made a comparison
between UVX quasar data from three samples (the Palomar Green survey
and the two Braccesi samples) and Parkes radio quasars selected according to the same optical criteria. The Hubble plot of these data is
reproduced in Figure 1.

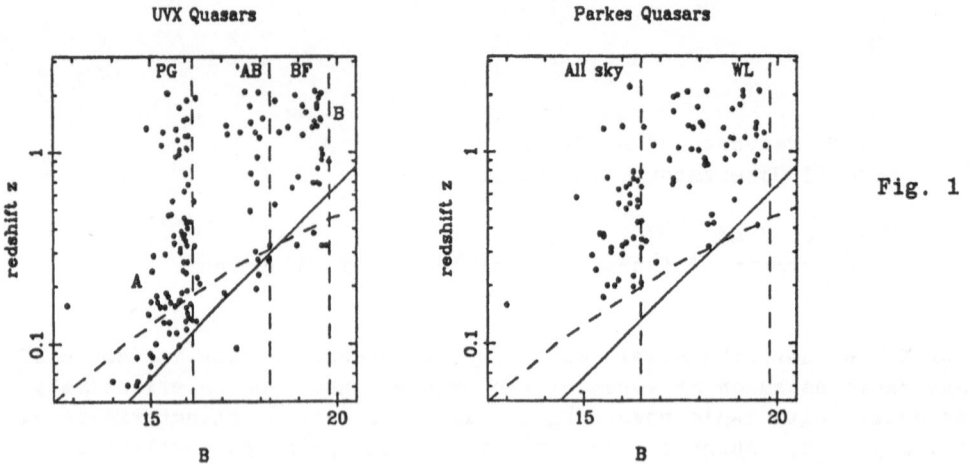

Fig. 1

The solid line marks an absolute magnitude of $M_B = -23$ calculated as
specified by Schmidt and Green (1983) ($H_o = 50$ kms^{-1} Mpc^{-1} ; $q_o = 0.5$;
$F_\nu \propto \nu^{-0.5}$). The dashed line represents the track expected for an
evolving elliptical galaxy (scaled arbitrarily in the horizontal
direction). The main different in the two plots is the lack of radio-
selected quasars at faint M_B. No Parkes quasar at the PG optical level
has $M_B > -24$, whereas the PG survey found many objects with $M_B > -23$.
These were rejected by Schmidt & Green via an arbitrary criterion at
$M_B = -23$; weaker objects, although predominantly stellar in appearance,
turned out to be Seyfert galaxies. The quasar host galaxies (spiral
in the case of Seyferts) are the clue to explaining this behaviour. If
we assume that radio-selected quasars reside in giant ellipticals
(which, for radio galaxies, have $M_B \simeq -22$), it is reasonable that nuclear quasar luminosities of $M_B < -24$ will be needed to outshine the
galaxy completely. At high redshift, the galaxy K-correction causes
appreciable dimming, and hence the lower limit on quasar optical luminosity falls. The Broad-Line Radio Galaxies (e.g. 3C109), which are
found at $z < 0.3$, would be classified as quasars at $z \sim 1$. In short,
the lack of optically weak radio quasars may be understood as an effect
of image classification.

If we assume that quasars with $M_B \gtrsim -24$ have radio emission characteristic of Seyferts, then with $P_{2.7}$ rarely exceeding 10^{22} WHz^{-1} sr^{-1}
(Meurs & Wilson 1984), sensitivities of < 1mJy are required for detection even at $z = 0.1$. In this interpretation, with 'radio-loud' quasars
having $P_{2.7} \gtrsim 10^{24}$ WHz^{-1} sr^{-1} and 'radio-quiet' $P_{2.7} < 10^{22}$ WHz^{-1} sr^{-1} ,
classification simply according to detection will usually yield a
consistent separation of the two classes.

2. VLA OBSERVATIONS OF FAINT QUASARS

It has long been known that faint quasars are harder to detect in the
radio than bright , even at constant redshift, which has been interpreted as evidence for a radio-optical power correlation (e.g. Wills &
Lynds 1978). However, in PML's model, the two classes of quasar can in
principle have different optical luminosity functions: detectability of
two different sets of quasars will then be governed by the relative proportions of the radio-loud and -quiet classes. PML considered this
problem by comparing the detection rates of quasars with $M_B \simeq -25$ at
low and high redshift. Even considering radio limits which corresponded
to the same radio power threshold, the high-redshift (i.e. apparently
faint) quasars were detected less often than their low-redshift counterparts. This is clear evidence for cosmological evolution, but the
existing radio data could not discriminate between the alternatives of
changing proportions of radio loud/quiet quasars or simply some evolution
in the mean level of radio output. To sort this out requires radio
observations deep enough to detect Seyfert-like emission or, more realistically, to set upper limits well into the transitional luminosity range
around $P_{2.7} \sim 10^{24}$ WHz^{-1} sr^{-1}. We shall briefly describe two sets of
observations.

The first (Miller & Hawkins in preparation) consists of 5-GHz VLA snapshots with rms noise $\sim$ 200μJy of 154 quasar candidates (selected via
variability) with B < 21; only 3 were detected. If we assume (via other
smaller samples) a typical $z \sim 1$, this corresponds to a 3σ luminosity
limit of $\sim 10^{23.5}$ WHz^{-1} sr^{-1} - close to the point at which we would
begin to detect Seyferts. This drastic drop in detectability supports
our above hypothesis: optically faint quasars are hard to detect in the
radio, not because of a radio-optical correlation, but because the vast
majority ($\gtrsim$ 98 percent) of quasars with B $\simeq$ 20 are radio-quiet analogues
of Seyferts. Thus, the steeply-rising number counts of optically-
selected quasars down to B $\simeq$ 20 are telling us about strong evolution
of the radio-quiet optical LF only. The task of making models for
the quasar LF (e.g. Marshall 1985) may now be severely complicated by
the need to account for two separately-evolving populations.

To escape some of the complications caused by an unknown degree of
Cosmological evolution, we have taken another set of VLA snapshots of
107 quasars over the (small) redshift range 1.7 < z < 2.5. These
objects were selected by emission lines (Osmer & Smith 1980; Osmer
1980). This allows us to look directly at the variation of radio
properties over 3-4 mag of optical luminosity. In all, 10 quasars were
detected (including 6 bright detections seen by Smith & Wright 1980);
these data are plotted in two ways in Figure 2. First, a histogram of
optical luminosity with the positions of the detections marked: the
detection rate varies from $\sim$ 50 percent at the bright end to $\sim$ 5 percent
at the faint. This corresponds to a difference in slope of 0.8 ± 0.1
for the luminosity functions for the two classes. Second, we plot radio
power versus optical. Note that, for the radio-loud objects, there is
no evidence for any correlation. The distribution of radio powers is

consistent with the form found by PML at low redshift: $F(>P) \propto P^{-1}$ for $P > 10^{25}$ WHz^{-1} sr^{-1}. The radio luminosity distribution flattens for lower powers; we need deeper data to see if there is in fact a lack of quasars with $P \sim 10^{23} -10^{24}$ WHz^{-1} sr^{-1}.

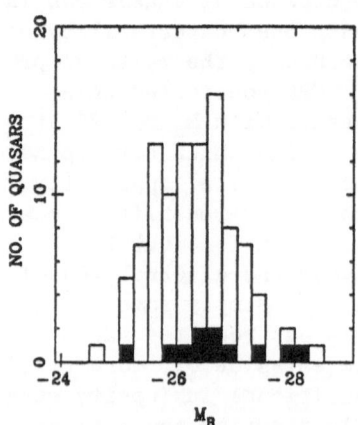 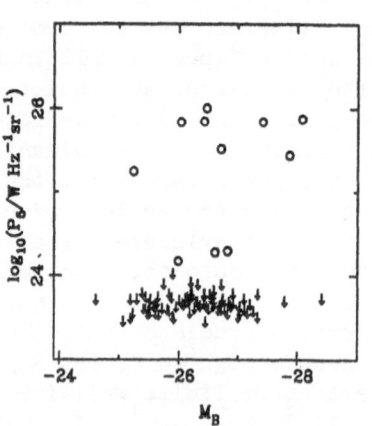

Fig. 2

3. CONCLUSIONS

We have described observations which go some way towards verifying the PML explanation for the radio properties of quasars. In particular, it does seem necessary to account for two radically different populations. Also, radio and optical powers seem essentially uncorrelated: we recommend that use of the parameter radio-optical spectral index be avoided. There remains the possibility of some weak correlation, especially for compact flat-spectrum quasars: this should be looked into. Observationally, the next step is long integrations with many quasars in one VLA beam, to reach the predicted emission at Seyfert levels.

REFERENCES

Condon, J.J., O'Dell, S.L., Puschell, J.J. & Stein, W.A., 1981.
    Astrophys. J., 246, 624.
Marshall, H.L., 1985. Astrophys. J., in press.
Meurs, E.J. & Wilson, A.S., 1984. Astron. Astrophys. 136, 206.
Osmer, P.S., 1980. Astrophys, J. Supply., 42, 523.
Osmer, P.S. & Smith, M.G., 1980. Astrophys. J. Suppl., 42, 333.
Peacock, J.A., Miller, L. & Longair, M.S., 1986. Mon. Not. R. astr.
    Soc., in press.
Schmidt, M., 1970. Astrophys. J., 162 371.
Schmidt, M. & Green, R., 1983. Astrophys. J., 269, 352.
Smith, M.G. & Wright, A.E., 1980. Mon. Not. R. astr. Soc., 181, 871.
Wills, D. & Lynds, R., 1978. Astrophys, J. Suppl. Ser. 36, 317.

DISCUSSION

Petrosian : Your results (and those presented by Kellermann) about two classes are very similar to the conclusion I drew from analysis of the very first optically selected sample. In that paper I suggested that perhaps the class with lower luminosity does not undergo the strong evolution observed at higher luminosities. Is this interpretation consistent with the new and more extensive data ?

Peacock : Your suggestion, I believe, was that the absence of radio-loud quasars at the lowest redshifts (z < 0.4 in the samples available then) was due to evolution of their LF. In the PG sample we see a drop in numbers of radio-loud quasars at $z \simeq 0.1$ which is very abrupt. These two facts make a non-evolutionary explanation in terms of simply stellar/non-stellar image classification more attractve.

Elvis : In comparing the 16 mag and 20 mag samples you go from an 8% to 3% detection rate, scaling appropriately : (1) What are the errors on these number ? (2) How much would you have to change the 20 mag radio limit by to get an 8% detection rate ?

Peacock : The 8% and 3% numbers could be argued to be only marginally inconsistent; however, they are based on published data only. If we use our VLA data to lower both radio limits by a factor 10, the detection rate at 16 mag rises to ~ 20%, whereas that at 20 mag remains about 3%. Our model implies that this will only improve if 20 mag quasars can be observed with sensitivities of ~ 10 µJy.

Burbidge : Is not the radio morphology for giant ellipticals different from that of QSOs ?

Peacock : At a given radio power, the only difference I am aware of is nuclear : radio quasars have relatively brighter central components and more frequently one-sided jets. The outer structurers in extended sources are similar.

Jerzy Machalski and Bina Swarup

A DEPENDENCE BETWEEN OPTICAL AND RADIO LUMINOSITIES OF QUASARS AFTER ALL?

Jerzy Machalski
Astronomical Observatory, Jagellonian University
ul. Orla 171,
PL-30244 Cracow,
Poland

It is still uncertain whether optical and radio emission mechanisms in AGNs are coupled or not, and how? New observational data give more light on this subject. The data used in this analysis are shown in Table I:

Sample	N	Sky area (sr)	S (Jy)	B-(J-) photometry	Red-shift
BDFL	68	4.30	>2.0	41(60%)	68(100%)
GB/GB2	83	0.44/0.091*	0.25÷2.0	16(19%)(UBVRI)	29(35%)
LBDS	25	0.0055	≥0.001	25(100%)(UJFN)	5(25%)
VLA	5	0.0012	≥0.005	-	-

* 0.44 sr for S≥0.55 Jy; 0.091 sr for S<0.55 Jy

Apparent blue (B or J) magnitude of these objects plotted against their 1.4 GHz flux densities (Machalski and Condon 1985) suggested a positive, though weak, $\log F(2500 ℛ) - \log P$ (1400) correlation. Since all samples but one were uncomplete in redshifts, the mean redshifts are estimated for them with two independent statistical approaches employing:
1. the $\Psi(R)$ function of Fanti and Perola (1977) (assumed to be known for $\log R \gtrsim 3.5$, and the optical-luminosity $\{\Phi(F)\}$ and evolution $\{E(z)\}$ functions of Bahcall and Turner (1977);
2. numerical experiments to generate a probable redshift for each QSO candidate, concordant with the observed Hubble diagram for QSOs.
    For each QSO sample complete over A(sr), the number of expected optical quasars (if their $\Phi(F)$ and E(z) functions are known) was compared with the observed number of objects having a specified radio/optical luminosity ratio R. A redshift corresponding to a median distance (redshift) within which a half of the observed objects N(<m,r) should lay is the resultant median redshift. This is shown in Table II

G. Swarup and V. K. Kapahi (eds.), Quasars, 109–110.

along with corresponding optical and radio luminosities for
different samples considered.

Sample	z (median)	logF(2500) (W/Hz)	logP(1400) (W/Hz)	α_{ro}
BDFL	0.86±0.07*	23.24±0.07*	27.44±0.06*	0.71
GB(0.55)	1.18±0.12	23.14±0.14	26.99±0.11	0.65
GB(0.25)	1.22±0.14	23.06±0.16	26.79±0.13	0.63
LBDS	1.48±0.30*	22.91±0.30*	25.85±0.31*	0·50

* observed values (in LBDS sample for five objects only)

The above trend was confirmed with the numerical experiments.
An example is shown in Fig. 1 (left):

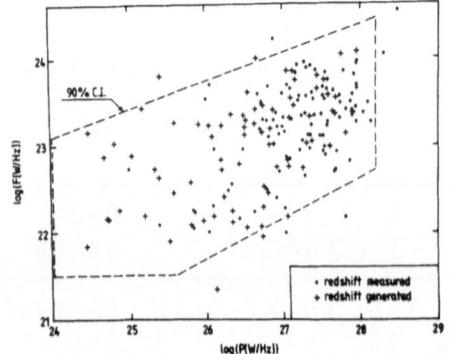 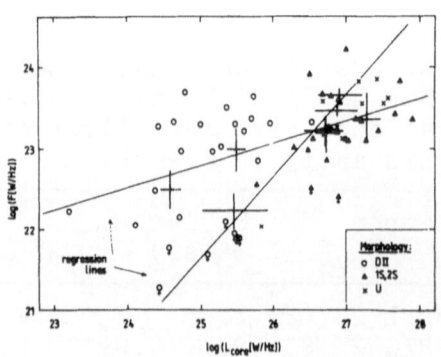

90% confidence intervals (C.I.) for the mean logF in different logP bins bound an area marked in Fig. 1. A logF—logP dependence seen in Table I and Fig. 1 was verified using the
Spearman partial rank correlation test to examine an influence of the logF—z and logP—z correlations, arised mainly from
selection effects. The test showed that the correlation coefficient between logF and logP remains about 0.5. This result
is significant at the 5σ level. A similar correlation was
found between nuclear luminosities in the BDFL and GB/GB2
samples (shown in Fig. 2; right). Since the apparent logF—logP
correlation is mainly caused by faint LBDS+VLA objects, it
suggests that most of these less luminous quasars can be rather compact. If there was true, less luminous radio-selected
QSOs would have smaller intrinsic size than do bright quasars.
It is possible that a negative correlation between radio luminosity and size (cf. Macklin 1982; and references therein)
concerns only very luminous quasars.

REFERENCES
Bahcall, J.N., and Turner, E.L.(1977). IAU Symp. No. 74, p. 295
Fanti, R., and Perola, G.C.(1977). IAU Symp. No. 74, p. 171
Machalski, J., and Condon, J.J.(1985). A.J. (submitted)
Macklin, J.T. (1982). M.N.R.A.S. 199, 1119

A STATISTICAL ANALYSIS OF THE RADIO PROPERTIES OF A LARGE SAMPLE OF 374 OPTICALLY SELECTED QUASARS

Gopal Krishna[1], H.Steppe[2], T.Ghosh[1] and L.Saripalli[1]

[1] T.I.F.R Centre, P.O.Box 1234, Bangalore 560 012 (India)
[2] I.R.A.M., Avd.Divina Pastora 7, Bloque 6/2B, Granada (Spain)

ABSTRACT. Using a large, unbiased sample of 374 optically selected QSOs, of which 54 are radio detected, we have computed for different redshift ranges the probability distribution function $\psi(R)$ of R defined as the ratio of monochromatic luminosities at 15 GHz and 2500 Å in the QSO's rest frame. A significant variation of $\psi(R)$ with redshift is noticed. At small redshifts (z < 0.5), the distinctive feature of $\psi(R)$ is a peak near $R \sim 10^2$, arising due to the QSOs having steep radio spectrum($\alpha_r < -0.5$).

1. THE QSO SAMPLE AND THE PROBABILITY DISTRIBUTION FUNCTION $\psi(R)$

The sample (374 QSOs) consists of 6 sets, each derived from the radio-surveyed portion of a major QSO list, by excluding all objects with $M_B > -23(q_0 = 0, H_0 = 50$ Km s^{-1} Mpc^{-1}) and also those for which QSO classification or tabulated redshift(z), or app.magnitude(m) has been labelled as doubtful/uncertain in the parent list (Table 1). Radio data available for all 54 detections in the sample permit classification into flat or steep spectrum type, based on α_r defined between ~ 5 and ~ 30 GHz and also yield flux density, S_r, at 15 GHz needed for computing 'R' (see Abstract),

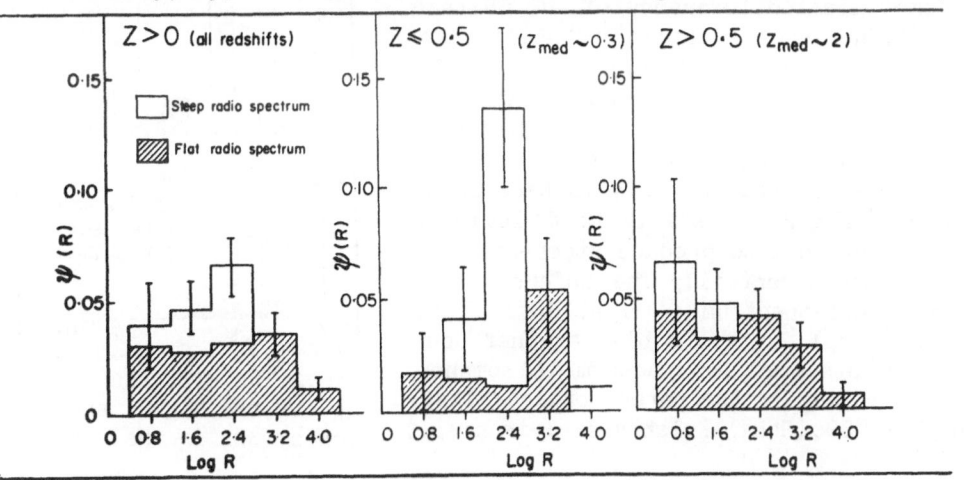

Fig.1: The probability distribution function $\psi(R)$ for different z-ranges.

G. Swarup and V. K. Kapahi (eds.), Quasars, 111–112.

Table 1: The 6 sets constituting our sample of 374 optically-selected QSOs

Set, No. of QSOs (radio detections)	Radio telescope used, frequency*, sensitivity limit	Type+ of the adopted mag. (m) and reference to m & z
1. CTIO[1]: 119(12)	VLA[2], 5 GHZ, 10 mJy	m_{con} at $\lambda_e = 1475$ Å, (1)
2. BQS[3]: 92(23)	Effelsberg[4], 10.7 GHz, 10 mJy	B-magnitude, (3)
3. AAO§: 53(7)	Parkes[5], 5 GHz, 25 mJy	B-magnitude, (5)
4. MCS[6]: 48(7)	Effelsberg[7], 5 GHz, 10 mJy	m_{con} at $\lambda_0 = 4500$Å, (6), (φ)
5. KP[8] : 44(3)	Arecibo[9], 2.4 GHz, ~14 mJy VLA[10], 5 GHz, ~ 1 mJy	m_{con} at $\lambda_0 = 4500$Å, (8,10)
6. BPS**: 18(2)	Westerbork[10], 1.4GHz, ~ 4 mJy	B-magnitude, (11)

References: (1)Osmer and Smith(Ap.J.Suppl.42,333,1980); (2)Condon et al.(Ap.J.244,5,1981); (3)Schmidt and Green(Ap.J.269,352,1983); (4)Steppe et al.(preprint,1985); (5)Savage and Bolton(MNRAS 188,599, 1979); (6)Lewis et al.(Ap.J.233,787,1979); (7)Strittmatter et al.(Astr.Ap.88,L12,1980); (8)Vaucher and Weedman(Ap.J.240,10,1980); (9)Sramek and Weedman(Ap.J.221,468,1978); (10)Sramek and Weedman (Ap.J.238,435,1980); (11)Marshall et al.(Ap.J.269,42,1983); (12)Fanti et al.(Astr.Ap.61,487,1977).

* flux densities at additional frequencies are taken from the references cited here and from the major radio source catalogues.

\+ m_{con} =continuum magnitude, λ_e =emitted wavelength, λ_0=observed wavelength.

§ from the sample given in ref.5. QSOs outside the central 5°x5° are excluded.

φ values of m not given in ref.6 are deduced following ref.(10).

**derived from the 'Bologna-Palomar Schmidt' sample defined in ref.(11).

following Schmidt and Green (Ap.J.269,352,1983). For the (320) QSOs undetected in radio, the quoted limits to flux density were adopted for S_r. Note that all the 3 frequencies refer to the QSO's rest frame. Now, using this database we computed $\psi(R)$ for the entire z-range (0 to 3.3), following ref.(10) (Fig.1). It is found to be consistent with the earlier determinations(refs.10,12), though much better defined here. We next computed $\psi(R)$ dividing the sample into low ($z \leq 0.5$) and high($z > 0.5$) redshift subsamples: a clear variation of $\psi(R)$ with z is noticed (Fig.1). At low-z $\psi(R)$ is marked by a peak arising near $R \sim 10^2$ due to steep-spectrum QSOs.

For the high-z subsample, ($z_{med} \sim 2$), $\psi(R)$ is dominated by flat-spectrum QSOs. This is not likely to be due to an observational bias, since for our major sets of both high-z (CTIO, MCS, AAO) and low-z (BQS) QSOs, the survey frequencies roughly correspond to the same emission frequency of ~15 GHz(see Table 1). The inferred lower probability for steep-spectrum QSOs at high z is also difficult to explain as an artefact of the median optical luminosity of our high-z subsample being higher(by ~3 mag.), compared to our low-z subsample. This is because, as seen from fig.2, optically more luminous QSOs in fact seem to produce steep-spectrum sources, preferrentially. The inferred rarity of steep-spectrum QSOs at high-z may, thus, be real, unless a bias against sources with steep radio spectrum is somehow present already in the parent optical lists of high-z QSOs(Gopal-Krishna etal, in prep.)

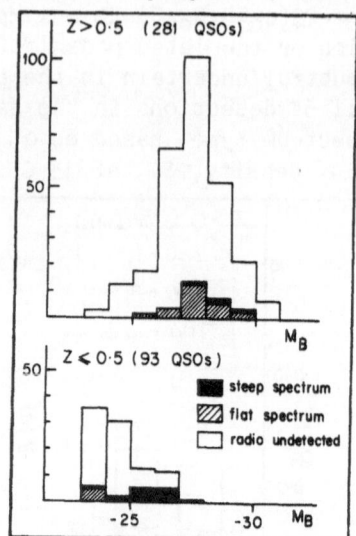

Fig.2: The distribution of M_B for the high-z and low-z QSOs in the sample.

Radio and Optical Observations of "Optically Quiet Quasars"

W. D. Cotton, F. N. Owen
National Radio Astronomy Observatory

M. J. Mahoney
Clark Lake Radio Observatory

1. Introduction

In recent years a number of very steep spectrum, compact radio sources have been discovered (e.g. Cotton 1983, Cotton and Owen 1985, Ulvestad 1985) which have no optical counterpart to the limit of the Palomar Sky Survey. VLBI observations of a number of these have confirmed the very compact (<10 mas) nature of several of these sources. Analysis of the available data in terms of the standard synchrotron model suggest that they contain very weak magnetic fields, large particle densities and may emit detectable infrared and optical emission by inverse Compton scattering in the compact radio source (Cotton 1983). This paper will report on an analysis including new VLBI observations, infrared and optical imaging at KPNO and low frequency radio observations at CLRO of a number of these objects.

2. Analysis

Important physical parameters can be derived for a synchrotron emitting source from observations of the spectrum and size of the source. In particular, the magnetic field strength and relativistic electron densities derived for the sources under consideration (see Marscher 1983) indicate weak magnetic fields and large relativistic electron densities. Under these conditions inverse Compton scattering of the radio photons off the relativistic electrons becomes important. Table 1 shows the observed and derived parameters for several objects; minimum gamma is the minimum bulk relativistic factor for which the model does not violate the data and B corresponds to this model.

Table 1
Observed and derived source parameters

Source	Size mas	Max. freq. MHz	Max flux mJy	Min gamma (z=0)	B mgauss
0752+342	11.	<100	>2000	4.6	<3.3E-3
1015+345	6.	<100	>400	2.9	<4.8E-3
1621+347	17.2	100	2000	2.5	1.1E-2
2147+145	5.8	150	3200	24.	1.2E-3

G. Swarup and V. K. Kapahi (eds.), Quasars, 113–115.
© 1986 by the IAU.

Figure 1 shows the observed spectrum of 2147+145 and several synchrotron model computations intended to reproduce the observed spectrum and size. The dotted line is a model including only synchrotron processes; the dashed line is the same model but including the effects of Compton scattering. This latter model shows that most of the radio photons have been scattered to higher frequencies. The solid line shows a model including relativistic beaming with a bulk relativistic factor of 24 (assuming z=0; 48 if z=1) which is in reasonable agreement with the data. Clearly, the only realistic synchrotron model which adequately reproduces the observations is the relativistic beaming case. There is, however, no direct evidence that the emission is from the synchrotron process.

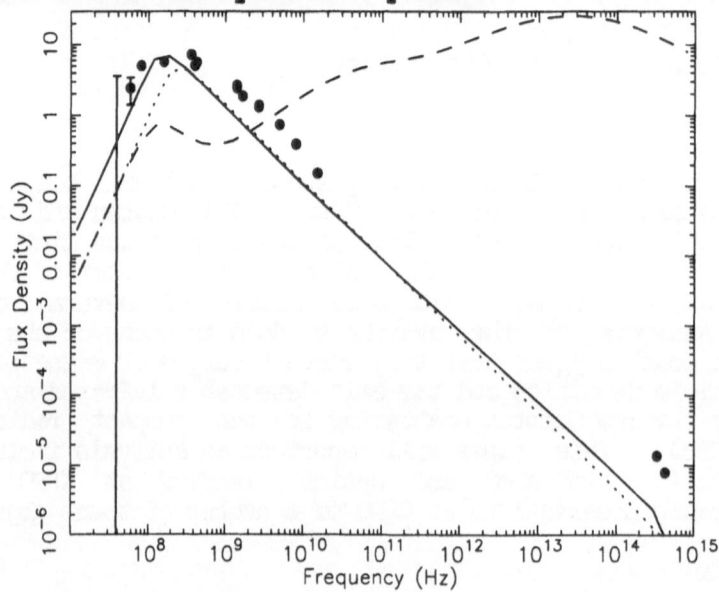

Figure 1 - Observed spectrum and synchrotron models for 2147+145.

3. Conclusions

A number of steep spectrum, compact, optically weak radio sources have been observed which cannot be modeled by synchrotron emission from a non-moving source. The sources can be adequately modeled by a synchrotron emitting source moving towards us at highly relativistic velocities. Another possibility requiring serious attention is that the emission may not be produced by the synchrotron process.

REFERENCES

Cotton, W. D., 1983, Ap. J., 271, 51.
Cotton, W. D. and Owen, F. N., 1985, unpublished results.
Marscher, A. P. 1983, Ap. J., 264, 296.
Ulvestadt, J. M., 1985, Ap. J., 288, 514.

DISCUSSION

Burbidge : Are they galactic stars ?

Cotton : Probably not; several of the objects have only a stellar like component but several also have significant levels of fuzz. The one object with a good VLBI image, 2147+145, has a core - "jet" structure similar to quasars. There is no direct evidence that these are not stars.

Coleman : Could you say what Lorentz factor is required when fitting a synchrotron self-compton model to the low optical flux.

Cotton : If the object is at $z \simeq 0$, $\gamma = 25$ is required, but for $z \simeq 1$ it is $\gamma = 50$.

RESOLVED IMAGING AND SPECTROSCOPY OF QSOs

Paul Hickson
  University of British Columbia
  2219 Main Mall, Vancouver, B.C., Canada
J. B. Hutchings
  Dominion Astrophysical Observatory
  5071 W. Saanich Rd., Victoria, B.C., Canada

ABSTRACT: We have obtained direct images and off-nuclear spectra for
five QSOs having a wide range of radio, optical and X-ray luminosity.
Four objects show absorption features identified with stars in a host
galaxy, and off-nuclear changes in emission line wavelengths. All
objects show off-nuclear changes in continuum coulour, and in emission
line intensities and ratios. The radio loud objects have more luminous
galaxies, strong and extended [OII] and [OIII], and UV-bright nuclei.
They tend to have luminous nuclei, high nuclear to galaxy luminosity
ratio, and blue host galaxies. This paper is a brief summary of results
that will appear in more detail elsewhere.

1. OBSERVATIONS

The objects were observed in June 1985 at the prime focus of the Canada-
France-Hawaii Telescope (CFHT) using FOCAS-II, a holographic grating
spectrograph/imager (Hickson 1986a). The spectroscopic detector was a
photon counting CCD camera (Hickson 1986b) and The CFHT RCA CCD camera
was used as the imaging detector. The spectra have a spatial resolution
of 0.8 arcsec and a spectral resolution of 30Å. Exposures used a long
slit 3 arcsec wide and ranged from 2.5 to 4 hours per object. The direct
images have 0.4 arcsec pixels. They were used to position objects on the
slit, and were later analysed to study the host galaxy morphology, the
luminosities of the galaxies and nuclei, and to determine the seeing
which was uniformly close to 1 arcsec FWHM.

The observational results and data for individual objects are summarized
in Table I.

2. DISCUSSION

Host galaxy spectra are clearly seen in the two objects of lowest
Lnuc/Lgal ratio. The galaxies are marginally seen in the next two and
are not seen at all in the last (which has the highest Lnuc/Lgal of any
so far imaged). The host absorption spectra have the same redshift as
the nuclei. Where there is extended emission, usually several

G. Swarup and V. K. Kapahi (eds.), Quasars, 117–119.

TABLE I

	1512+370	1700+518	1701+610	2130+099	2135-145
z	0.37	0.29	0.16	0.06	0.29
V	15.5	15.2	17.0	14.9	15.4
Lnuc/Lgal	10.0	40.0	0.35	2.6	4.0
Mnuc	-25.0	-24.9	-20.1	-21.5	-23.5
Mgal	-23.0	-21.2	-21.5	-20.6	-22.2
log L6cm (W Hz^{-1})	25.7	22.1	–	<20.0	24.7
log Lx (erg s^{-1})	–	–	43.7	–	45.3
B-V:					
nuc1	(-0.3)	0.0	0.3	0.1	0.6
nuc2	–	–	0.5	–	0.2
off-nuc	(0.0)	-0.1	0.7	0.5	0.1
companion	(0.3)	–	0.1	–	0.2
[OII],[OIII]	Strong	Very weak	Strong	Weak	Strong
[O] extended?	Yes	No	Slightly	No	Yes
Host Abs	(Mg b)	–	Mg b,G,CaII	Mg b,CaII	(Mg b)
Other lines	–	H-beta abs	–	4600Å emis	–
Interacting?	Yes	No	Yes	No	Yes
Other morph	OIII clouds	BAL	Tidal tail	Spiral gal	Twin nuc

Lnuc/Lgal corrected for z, sky,seeing as in Hutchings et al. 1984
H = 100 for all luminosities and absolute magnitudes
Off-nuclear spectra are not corrected for nuclear contamination

components are seen spatially, with velocities up to 1000 km s^{-1}. Some
off-nuclear areas have blue spectra, suggesting hot stellar populations.

As can be seen from the table, the radio-loud objects have more luminous
galaxies, and have strong and extended [OII] and [OIII]. They tend to
have luminous nuclei, high nuclear to galaxy luminosity ratio, and blue
host galaxies. All three objects which are interacting have strong
nuclear (or near nuclear) [OIII]. Two of our objects have been studied
(without spatial resolution) by Boroson and Oke (1984). Our results are
in agreement with theirs, and are consistent with the connections that
they make between radio and optical properties.

3. REFERENCES

Boroson, T. A., and Oke, J. B. 1984, Ap.J. 535, 68.
Hickson, P. 1986a, submitted to Pub. A. S. P.
Hickson, P. 1986b, submitted to Pub. A. S. P.
Hutchings, J. B., Crampton, D., Campbell, B. 1984, Ap.J. 280, 41.

DISCUSSION

Bregman : How are you able to distinguish whether the galaxy underlying a quasar is a spiral or an elliptical galaxy ?

Hutchings : Not by the luminosity-radius relation. The uncertainity in the central intensity of the point spread function allows you to fit exponential or $R^{1/4}$ laws equally well in most cases. Spirals can be recognised by structure, but ellipticals can be disguised by tidal effects or starbursts if an encounter has occurred. While many host galaxies _may_ be ellipticals there are few positive indicators. Some off-nuclear spectra show late type populations which may be the best evidence.

Alighieri : What is the object to the North of the quasar 4C37.43 in your last slide ? Is it a galaxy ?

Hutchings : It is a galaxy. We have a redshift, and it is the same as the quasar (0.37).

Chatterjee : What, if any, are the essential differences between quasars which seem to be associated with interacting galaxies and quasars which seem to be isolated ?

Hutchings : I do not know of any differences in the central QSO. The host galaxies however are frequently tidally distorted and contain hot stars or ionised gas. However, the data are fully consistent with all quasars being ignited by a collision, if their lifetimes are 10^7 or 10^8 years.

Alighieri : Concerning the discussion on whether extended line emission means that galaxies underlying quasars are spirals rather than ellipticals, it is known now that several radio galaxies, classified as ellipticals from broad band imaging have extended ionised gas surrounding them at distances of several tens of kpc.

Hutchings : This tends to support my claim. Ellipticals which have been violently activated bear these other signs. Normal ellipticals do not. When they are very distant and have a quasar core, it is difficult to say definitely that they _are_ ellipticals.

V.Radhakrishnan, Beverley and Derek Wills

ON THE APPEARANCE OF THE HOST GALAXIES OF QUASARS

Smita Shanbhag[1,2] and Ajit Kembhavi[1]
1. Tata Institute of Fundamental Research
   Bombay 400 005
2. Indian Institute of Science
   Bangalore 560 012

ABSTRACT. It is believed that the QSO phenomenon is a form of violent activity in the nuclei of galaxies. This view has found support after the recent detection of nebulosity around a large number of quasars. The nature of the fuzz is not yet clear, but there is some evidence that it is normallly a spiral galaxy hosting a quasar (e.g. Hutchings et al. 1984). However, the experience with radio galaxies suggests that the host of a radio loud quasar ought to be an elliptical galaxy. We have considered the appearance of elliptical galaxies of various redshifts with a view towards determining the nature of ellipticals which could act as hosts for radio quasars at large redshifts. For spirals we have considered the data given by Boroson (1981) and calculated the appearance of different morphological types at large redshifts.

RESULTS

Elliptical galaxies which follow de Vaucouleur's law and Kormendy's relation cannot be visible beyond z = 0.47 in the V band (H_0=50 km/sec/Mpc). Therefore, if ellipticals are observed around quasars at higher z_s they have to have special properties.

Our calculations for ellipticals show that at a given redshift, we can observe only a part of the luminosity function. Lower cut-off of the observed luminosity increases with redshift.

If we consider nearby ellipticals (upto ∿100 Mpc), then the calculations show that the surface brightness of ellipticals decreases for M < -21. This is consistent with observations.

Visibility studies for various morphological types of spiral galaxies show that Sbc - Sc galaxies are most likely candidates for host spiral galaxies of quasars.

CALCULATIONS

We have calculated the surface brightness $\mathcal{M}(r)$ of different types of galaxies as a function of redshift. We assume that a galaxy can be distinguished as such if the 25 mag/(arc sec)^2 isophote has an angular size = 1".

G. Swarup and V. K. Kapahi (eds.), Quasars, 121–122.

Ellipticals

We assume that the surface-brightness profile of ellipticals follows
de Vaucouleur's law. To this we add Kormendy's (1980) relation for
ellipticals:

$$\mu(r_e) = 3.28 \log r_e + 19.45 \quad \text{B mag/(arc sec)}^2.$$

Using these two relations, the number of parameters defining $\mu(r)$ is
reduced to just one: Then, for a given Z, $\mu(r)$ has a minimum and
appears brightest for $r_e = 4.54r$. Choosing this optimal value of r_e ,
and introducing observational constraint as stated above, and solving
the $\mu(\theta = 1")$ equation for Z, we get the limiting value of Z upto which
ellipticals can be seen. Even giant ellipticals like NGC 4073 and CDs
like A2029 disappear before Z = 0.41. For any $Z < z_{lim}$, we get two
limiting values of r_e. The effective radius of an elliptical must be
contained in these limits for it to be visible. The limits on r_e
translate to limits on the absolute magnitude. These limits are shown
in Fig.1. For nearby ellipticals (dist $\sim$100 Mpc) we have r_e (optimum)
= 1.9 kpc which corresponds to M_B = -21. A galaxy brighter than this
would have a lower surface brightness.

Spirals

The parameters describing $\mu(r)$ of spirals are many and we have not been
able to reduce their number as in ellipticals. Preliminary investigations, using various values for disc and bulge parameters as given by
Bōroson have shown that Sb - Sc galaxies are the best candidates for
detection at the redshifts reached by Hutchings et al (Z = 0.6).

References

Hutchings et al. (1984) <u>Ap. J.</u>, **280**, 41
Boroson, T. (1981), <u>Ap. J. Supp.</u> **46**, 177
Kormendy, J. (1984), <u>Morphology and Dynamics of Galaxies</u>, 162.

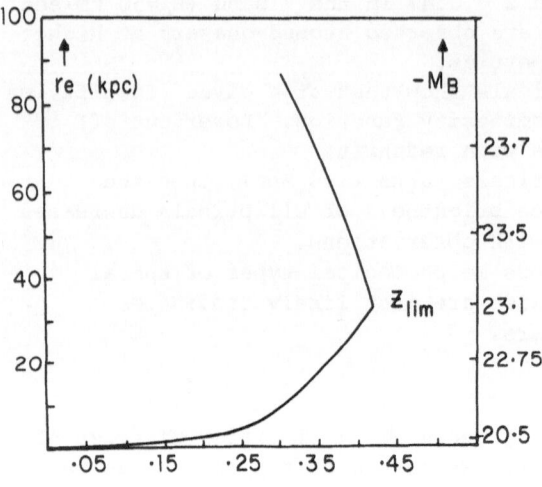

Fig. 1. The bounds on
the effective radius
are shown as a
function of redshift.
Values of the cor-
responding absolute
blue magnitude are
shown on right.

CCD IMAGES OF SOUTHERN RADIO SOURCES

Graeme L. White, Michael J. Batty and David L. Jauncey
Division of Radiophysics, CSIRO
PO Box 76
Epping NSW 2121
Australia

ABSTRACT. We present here the first results of a program to obtain CCD images of a number of southern extragalactic radio objects using the 3.9-m Anglo-Australian telescope.

1. INTRODUCTION

This program is designed to:

(a) Study the optical morphology of a range of southern QSOs and radio galaxies in order to determine the properties of any underlying galaxies and the extent to which the local environment determines the radio properties.

(b) Carry out an accurate comparison of Southern Hemisphere radio and optical reference frames in order to improve the status of the astrometric frames in the south. This is essential to ensure accurate registration of optical and radio maps obtained at high resolution. An accuracy of about 10 milliarcsec will be needed to register images obtained with the Hubble Space Telescope and high-resolution radio maps - e.g. from the VLA and the Australia Telescope.

2. OBSERVATIONS

Images of radio objects are being obtained using a CCD camera at the prime focus of the AAT. Exposures are typically 300 s and taken through an R filter with the doublet corrector. The image scale is 0.49 arcsec per pixel. All images are being obtained through the Service CCD Observing program of the Anglo-Australian Observatory.

PKS 0439-433. The original optical identification for 0439-433, by Peterson and Bolton (1972), is confirmed on the basis of an accurate radio position measured at 2.3 GHz with the Tidbinbilla interferometer (see Jauncey et al., 1983). A redshift of z = 0.594 is reported by Wilkes et al. (1983). Accurate photometric magnitude (V=16.36) and colours (B-V=0.28, U-B=-0.65) have been determined by Adam (1978).

The CCD image (Fig. 1) was obtained in 1.7 arcsec FWHM seeing and clearly shows the radio object to be stellar-like, with an image which overlaps that of a nearby galaxy. The two objects are separated by 4.0 arcsec.

G. Swarup and V. K. Kapahi (eds.), Quasars, 123–124.

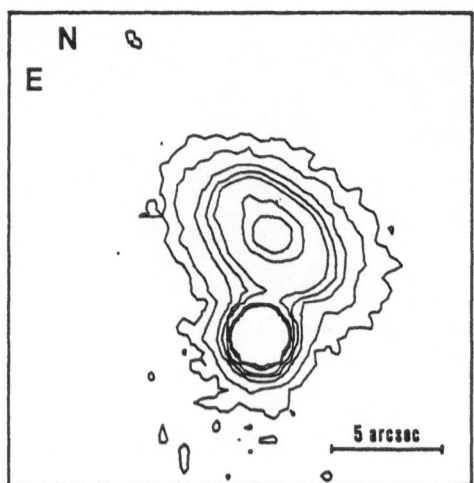

Figure 1. A red (R) filter CCD image of PKS 0439-433 obtained using the AAT. The QSO is the lower stellar-like object.

PKS 2250-412. The original identification as a 20 mag galaxy by Hunstead (1971) is confirmed. The SERC J sky atlas showed the candidate as stellar with a similar object preceding by 6.0 arcsec. The CCD image (Fig. 2), obtained in seeing conditions of 1.3 arcsec, shows three objects, the central one being the optical identification. The preceding object has a soft, galaxy-like appearance while the following stellar-like object is only just visible on the SERC J survey. This latter object is shown on the finding chart by Hunstead. It is either very red in colour, or variable, or both. The image of the identification shows structure extending about 2.5 arcsec at P.A. 90°. This extension occurs in the lower 20% of the image and is probably real, since stellar images in the field do not show similar extension.

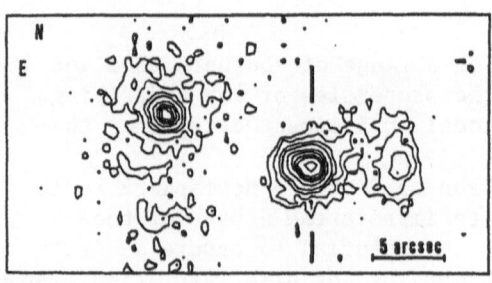

Figure 2. A red (R) CCD image of the radio source PKS 2250-412. The preceding object is a galaxy while the following is either a red star or a variable star (see text). The radio object shows a slight extension at P.A. 90°.

3. CONCLUSIONS

Red CCD images of extragalactic radio source identifications reveal complex morphology for two southern radio QSOs.

4. ACKNOWLEDGEMENTS

We are indebted to the Anglo-Australian Observatory for their Service CCD Observing programs and for software facilities for reducing of these data. GLW acknowledges the receipt of an Australian National Research Fellowship.

REFERENCES

Adam, G. (1978). Astron. Astrophys. Suppl. Ser., 31, 151.
Hunstead, R.W. (1971). Mon. Not. R. Astron. Soc., 152, 277.
Jauncey, D.L., Batty, M.J., Savage, A. and Gulkis, S. (1983). In Quasars and Gravitational Lenses, 24th Liège Astrophysical Colloquium, Institut d'Astrophysique, June 1983, p. 59.
Peterson, B.A. and Bolton, J.G. (1972). Astrophys. Lett., 10, 105.

A CCD SURVEY OF BL LAC OBJECTS

Helmut Kuhr and Josef Fried
Max-Planck-Institut fur Radioastronomie
Heidelberg, F.R.G.

ABSTRACT. A long term project was started at the Max-Planck-Institut
fur Astronomie, Heidelberg, to study a complete sample of 46 northern and
southern BL Lac objects with flux densities exceeding 1 Jy at 5 GHz using
optical spectroscopy, optical polarimetry, and direct deep CCD imaging.

INTRODUCTION

Several studies have been published where authors discuss the underlying
galaxy content of quasars (e.g. Gehren et al. 1984). Little is known,
however, about the nature and morphology of the galaxies BL Lac objects
are possibly imbedded in. The subject of this presentation is an optical direct imaging program concentrating on a complete sample of 46 radio
selected BL Lac objects.

THE SAMPLE

The complete sample contains all objects from the 1 Jy catalogue (Kuhr
et al. 1981) covering the northern and southern sky, which meet the
following criteria :
i) at the time of the radio surveys flux densities exceeded 1 Jy at 5
GHz, ii) on Palomar Observatory Sky Survey plates the optical counterparts are brighter than 20 mag, iii) optical spectroscopy and optical
polarimetry show basically featureless polarized continuous spectra.

OBSERVATIONS

The BL Lac nature of about half of the objects finally selected was previously known in the literature. For the remaining sources various optical observing programs were carried out in order to establish their
membership in the complete sample, which included :
i) spectrophotometric measurements at the Steward Observatory 90 inch
and Multi Mirror telescopes in Tucson (Kuhr, Liebert, Strittmatter to be

125

G. Swarup and V. K. Kapahi (eds.), Quasars, 125–126.

published), and ii) polarization measurements at the Steward Observatory 90 inch telescope (Kuhr, Schmidt, Strittmatter to be published). Subsequent direct deep deep CCD images of high quality and typical integration times of several hours were taken at the two 2.2 m telescopes of the Max-Planck-Institut fur Astronomie on La Silla, Chile, and on Calar Alto, Spain. The standard reduction process used included flat fielding and removal of cosmic ray events, adding and calibrating the various frames, as well as determining the point spread function and subtracting the pointlike sources from the BL Lac objects as described by Gehren et al. (1984).

RESULTS

So far only the four brightest objects of our complete sample were fully reduced and analyzed. Preliminary results indicate that the radial brightness distribution in each case is compatible with that expected for elliptical galaxies.

REFERENCES

Gehren, T., Fried, J., Wehinger, P.A. Wyckoff, S. 1984, Ap.J. <u>278</u>, 11
Kuhr, H., Witzel, A. Pauliny-Toth, I.I.P., Nauber, U. 1981, Astron.
    Astrophys. Suppl. <u>45</u>, 367

CCD OBSERVATIONS OF GALAXIES WITH RADIO JETS

I. Pérez-Fournon[1], L. Colina[2], P. Biermann[3], and J.M. Marcaide[3]

[1] Instituto de Astrofísica de Canarias, Tenerife, Spain
[2] University Observatory, Göttingen, F.R. of Germany
[3] Max-Planck-Institut f. Radioastronomie, Bonn, F.R. of Germany

We present preliminary results of a CCD survey of nearby ($z < 0.1$) northern galaxies which have well defined radio jets. Our main objectives are to: 1) detect - or put constraints to - the presence of optical jets in these galaxies, 2) obtain high dynamic range maps of known optical jets, and 3) make a detailed comparison of our results with those available from the radio maps.

We began this project at the MPIA Calar Alto Observatory, Spain. We used a RCA CCD chip at the Cassegrain focus of the 2.2 m telescope (pixel size = 0.34"). We took at least two 30 minutes exposures in the g and r filters of the Thuan-Gunn photometric system for each object in our survey. We reduced the images in the standard way using flat fields of the twilight diffuse emission. In order to detect asymmetrical features, we subtracted elliptical or spherical models of the stellar light. Two of the observed objects are here briefly discussed.

3C120: This object is a Seyfert galaxy with a strong and variable starlike nucleus. Its associated nebulosity has a unique morphology as firstly noticed by Arp (1975). Wlérick (1981) and Arp (1981) classified some of the optical extensions as jets. The main goal of our observations of 3C120 is to determine which one, if any, of these "jets" is related to the radio jet. This identification is now possible with the help of a good 6 cm VLA map obtained by Walker (1984). The jet in the VLA map bends towards the NW lobe beyond a radio knot at about 4" from the nucleus. Subtracting a spherical galaxy template from the reduced r frame (Fig. 1a) we obtained the residual map (Fig. 1b) which shows an optical extension along the northern edge of the radio jet. The inner part of this optical jet was noticed by Wlérick (1981) who suggested a bending towards the NW optical extension. Comparison of our residual map with the radio one shows that the bending of the radio jet is less than expected for that interpretation. Although the inner optical jet is connected with the NW optical extension, Fig. 1b shows that it continues beyond the radio knot in the same direction as the radio jet.

3C75: This object is associated with the central galaxy of the

G. Swarup and V. K. Kapahi (eds.), Quasars, 127–128.
© 1986 by the IAU.

cluster Abell 400. It is the first case, to our knowledge, having a pair
of twin radio jets (Owen et al., 1985). They originate in the double nucleus of the galaxy. This multiple jet system puts strong constraints to
models of the structure and kinematics of the intergalactic medium in
clusters of galaxies. Our CCD data show an object superimposed on the
eastern part of the northern radio jet at about 24" from the northern
nucleus. This latter object appears slightly extended at a position
where there is no apparent enhancement of the radio emission. Spectroscopic and polarimetric observations are needed to determine its nature.

ACKNOWLEDGEMENTS

I. Pérez-Fournon would like to thank the IAU Symp. 119 Organizing Committee for financial support. L. Colina was supported in part by the German
Science Foundation under grant FR325/21-1.

REFERENCES

Arp, H.: 1975, P.A.S.P. 87, 545
Arp, H.: 1981, in "Optical Jets in Galaxies", ESA SP-162, p.53
Owen, F.N., O´Dea, C.P., Inoue, M., Eilek, J.A.: 1985, Ap. J. Lett. 294,
     L85
Walker, R.C.: 1984, in Proc. of NRAO Workshop No. 9 "Physics of Energy
     Transport in Extragalactic Radio Sources", eds. A.H. Bridle and J.A.
     Eilek, p.20
Wlérick, G.: 1981, in "Optical Jets in Galaxies", ESA SP-162, p.29

(a)   (b)

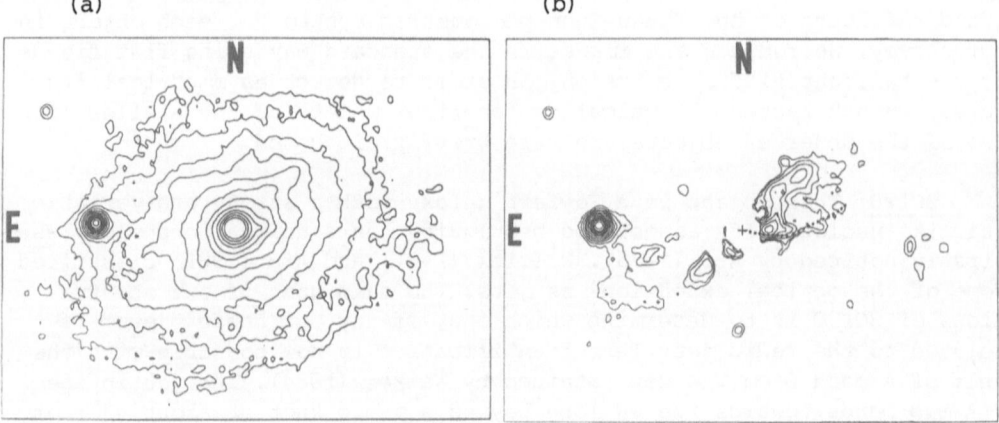

Figure 1. a) Contour map of 3C120 in the r filter. The interval between
contours is 0.5 mag arcsec^{-2} and the minimum level corresponds to a surface brightness of 3% of the local sky background. The box size is 68" x
51".   b) Residual map, excluding a central region of radius 2.4", obtained by subtraction of a spherical galaxy template from the image in
the r filter (Fig. 1a). The interval between contours and box size are
the same as in Fig. 1a. The dotted line shows the position of the radio
jet, assuming that the central radio source coincides with the centroid
of the central optical source.

ENVIRONNEMENT DU NOYAU DE 3C 120

G. Wlérick, A. Soubeyran, B. Servan, L. Renard, D. Horville
Observatoire de Paris
A. Bijaoui, Observatoire de Nice
G. Lelièvre, Société du Télescope CFH
P. Bouchet, European Southern Observatory

ABSTRACT .The region surrounding the nucleus of 3C 120 has been recorded on electronographic plates, in U, B, V and V' colours. It has been possible to substract the contributions of the bright nucleus and the galaxy. Faint condensations appear. Their colours show that they are probably not emitting synchrotron radiation. The brightest condensation A is located 4" West of the nucleus. It does not coïncide exactly with the knot of the radio jet and is very slightly displaced toward the North. It is probably of the same nature as Minkowski's Object : burst of star formation triggered by the collision of the jet with relatively dense extranuclear gas.

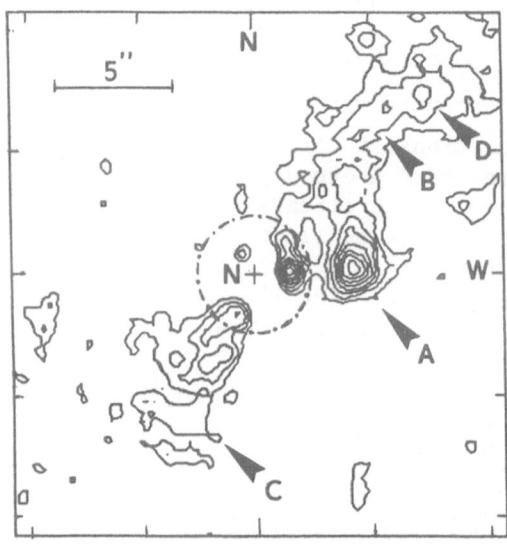

Fig . 1 . 3C 120 .Condensations en couleur U .

Wlérick et al.(1981) ont noté, au voisinage du noyau de 3C120, une structure filiforme émettant dans le continu et dans les raies. Les cartes radio MERLIN (Browne et al.1982) et VLA (Balick et al.1982, Benson et al.1984) font apparaître un jet qui part du noyau vers l'Ouest. Pour comparer les structures optique et radio, nous avons observé 3C 120 avec les télescopes de 3,6 m. ESO et CFH et des récepteurs électronographiques. Les clichés ont été pris dans les bandes U B V et V'. La bande V', 530-650 nm, élimine [OIII] et correspond à l'émission dans le continu. Le noyau de 3C 120 étant brillant, un traitement d'images est nécessaire pour le soustraire. Nous montrons les résultats obtenus avec un logiciel simple:

G. Swarup and V. K. Kapahi (eds.), Quasars, 129–130.
© 1986 by the IAU.

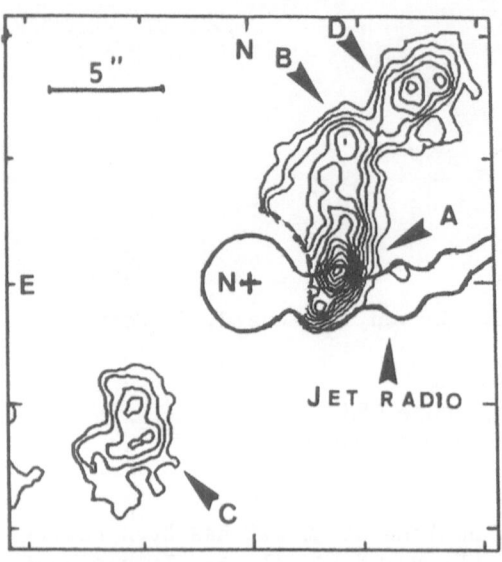

Fig.2 . Condensations optiques
en couleur V' et jet radio .

on prend un fichier centré sur le noyau, on le retourne et on soustrait le fichier retourné du 1er fichier. Le résultat apparaît fig.1 (couleur U). Autour du noyau N, il y a une zône non significative d'environ 2" de rayon. Au-delà de ce cercle, on note 4 condensations; les résultats des mesures de 2 d'entre elles apparaissent tableau 1. Les couleurs des condensations ne semblent pas correspondre à celles d'un rayonnement synchrotron. Les valeurs (U-B) et (B-V) sont voisines des couleurs moyennes trouvées par Lelièvre (1976) et Wlérick et al.(1979) pour la galaxie (noyau exclu). On note la valeur de (V-V') pour A ; il y a donc une forte émission [OIII] coïncidant avec l'émission dans le continu.

Tableau 1. Flux et couleurs des condensations.

Condensation	B	U - B	B - V	V - V'
A	$20,0 \pm 0,15$	$-0,10 \pm 0,20$	$0,95 \pm 0,20$	$-0,40 \pm 0,20$
D	$20,95 \pm 0,15$	$-0,20 \pm 0,20$	$0,75 \pm 0,20$	$0,0 \pm 0,20$

Fig.2 montre simultanément les condensations et l'isophote extérieure du jet radio (Benson 1984).La condensation A est déplacée un peu vers le Nord par rapport au 1er noeud du jet .Cette situation est analogue à eelle de l'objet de Minkowski qui se trouve sur le trajet d'un jet radio (Van Breugel et al 1985) (Brodie et al 1985).Par similitude nous suggérons que A est une zone de formation d'étoile ,ayant pour origine la collision du jet avec une région gazeuse dense .

REFERENCES
Balick et al , 1982 , Astrophys.J. , 254 , 83 .
Benson et al ,1984 , IAU Symposium 110 , 125-126 .
Brodie et al , 1985 ,Astrophys.J. , 293 , L59 .
Browne et al , 1982 , Nature , 299 , 788 .
Lelièvre , 1976 , Astron.Astrophys. , 51 ,347 .
Van Breugel et al , 1985 , Astrophys.J. , 293 , 83 .
Wlérick et al , 1979 ,Astron.Astrophys. , 72 , 277 .
Wlérick et al , 1981 ,Astron.Astrophys. , 102 ,L17 .

COMPACT RADIO STRUCTURE OF QUASARS

R.W. Porcas
Max-Planck-Institut für Radioastronomie
Auf dem Hügel 69
D-5300 Bonn 1
F.R.G.

ABSTRACT. Recent results concerning the compact radio structure of quasars are reviewed. Emphasis is placed on VLBI results from statistical studies, from detailed mapping and from multi-epoch monitoring of variable sources.

1. INTRODUCTION

The description of radio source structures as either "compact" or "extended" is primarily the result of the resolution capabilities of different instruments. Connected element arrays such as the VLA and MERLIN have limiting resolutions of somewhat better than 0.5 arcsecond and hence are ideal for mapping structures with angular extents greater than $\sim$ 1 arcsecond. Frazer Owen describes such extended radio structures of quasars in his review (these proceedings). In this review I will concentrate on the sub-arcsecond structure of quasars, basing my descriptions on the results of VLBI measurements. These range in resolution from $\sim$ 10 mas for the European VLBI network (EVN) down to $\sim$ 0.2 mas achieved on transatlantic baselines at the highest observing frequencies.
    For this review, I have excluded discussions of objects normally classified as Seyfert galaxies, radio galaxies etc., and concentrate only on observational material relating to quasars. The reader is referred to the review of Phinney (1985) for a more theoretical viewpoint, and to Porcas (1985) and the Proceedings of IAU Symposium 110 "VLBI and Compact Radio Sources" (1984) for more extensive treatments of compact radio structure.

2. STATISTICS AND SURVEYS

A division can be made between compact steep spectrum (CSS) sources (spectral index, $\alpha \simeq -0.7$, $S \propto \nu^{\alpha}$) and compact flat spectrum sources. The latter class, which are largely identified with quasars and BL Lac objects, have been extensively observed using VLBI techniques. Zensus et al. (1984) observed a sample of 57 such sources, with flux densities

G. Swarup and V. K. Kapahi (eds.), Quasars, 131–140.

greater than 1 Jy, using a 3-element VLBI array with resolutions down
to 1 mas. Essentially all sources were detected, and typically 50% of
the radio flux comes from components $\lesssim$ 1 mas (this corresponds to
$\lesssim$ 10 pc for most quasars). Barthel et al. (1984) made similar observations on a sample of flat spectrum cores of quasars also exhibiting
extended radio structures. Of 16 objects with core flux > 0.1 Jy,
almost all were detected, again confirming that for flat spectrum
quasars a high fraction of the flux comes from pc-scale components.

A number of surveys of compact radio structure have been made,
notably by Pearson and Readhead (1984) (45 sources > 1.2 Jy mapped) and
Eckart et al. (1985b) (13 sources > 1.0 Jy). Romney et al. (1984) have
mapped 21 sources which exhibit low frequency flux variability, and
Hodges et al. (1984) have observed 9 sources with peaked radio spectra.
A rough classification of structures as "compact", "core-jet", "compact
double" or "complex" can be made (see e.g. Porcas 1985). CSS sources
have been studied with the EVN by Fanti et al. (1985) and with the VLA
at high resolution by van Breugel et al. (1984) and Pearson et al.
(1985). In contrast to radio galaxies, which tend to be "doubles", CSS
quasars are generally core-jets or complex (Fanti et al. 1985).

3. EXAMPLE QUASARS

To illustrate the range of morphologies seen, I have selected 12 quasars
which exemplify the different spectral and structural classifications
which can be made with resolutions > 1 arcsecond. An additional criterion for inclusion of a source was the availability of data on a number
of angular scales. Maps are presented in Fig. 1 which, for each source,
show the radio structures ranging from the most extended structure down
to the most compact known. The maps are arranged so that the compact
core features are aligned across the page. With the exceptions of BL Lac
and 3C273, all the sources have similar redshifts, so that the conversion from angular to linear scales are similar. References to unpublished material are given without date. Where no interferometer is
specified the maps have been obtained using transatlantic VLBI.

3C263 (z = 0.652) and 3C179 (z = 0.846) are both steep spectrum
sources, exhibiting extended, classical double-lobed (D1) structure,
and relatively weak flat spectrum cores. The mas maps show compact cores
and secondary "knot" features, whose position angle (pa) is extremely
well aligned with the extended structure (see also Shone et al. 1985).

3C216 (z = 0.670) and 3C147 (z = 0.545) are CSS sources, with
complex structure in the range 0.1 to 1 arcsecond (a few kpc). Both
sources show compact mas cores (that in 3C216 being more prominent)
with lower brightness extensions; in 3C147 this shows a wiggling
morphology. The intermediate resolution maps show that these jet-like
extensions continue out in roughly the same pa until 1 or 2 kpc from
the core, where they change direction abruptly. At present, the
connection to the arcsecond scale morphology, which has features at
rather different angles, is unclear.

3C454.3 (z = 0.859), 3C345 (z = 0.595) and 3C273 (z = 0.158) belong
to the D2 category, consisting of a dominant, compact, flat spectrum

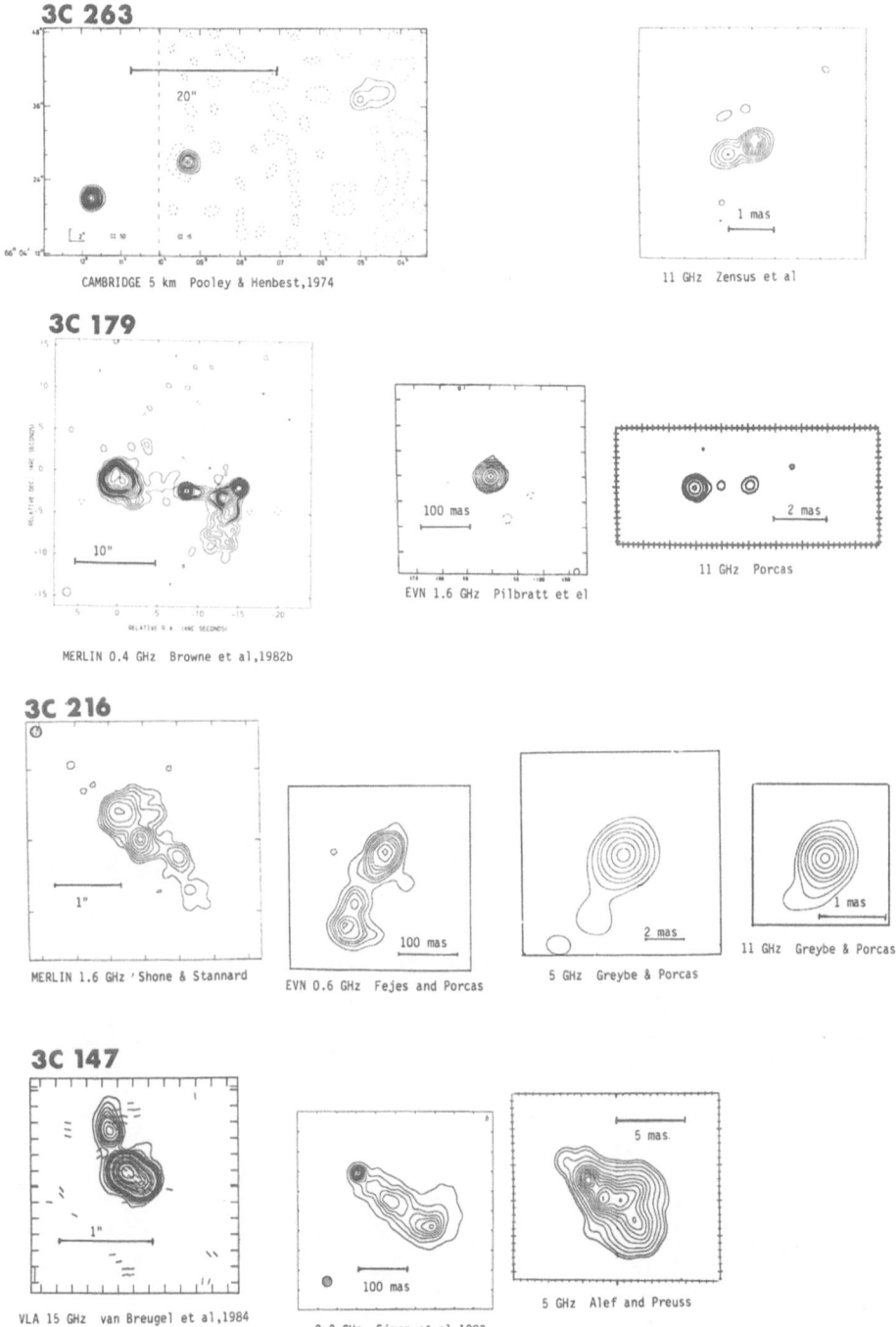

Figure 1: Arcsecond to mas structure of representative quasars.

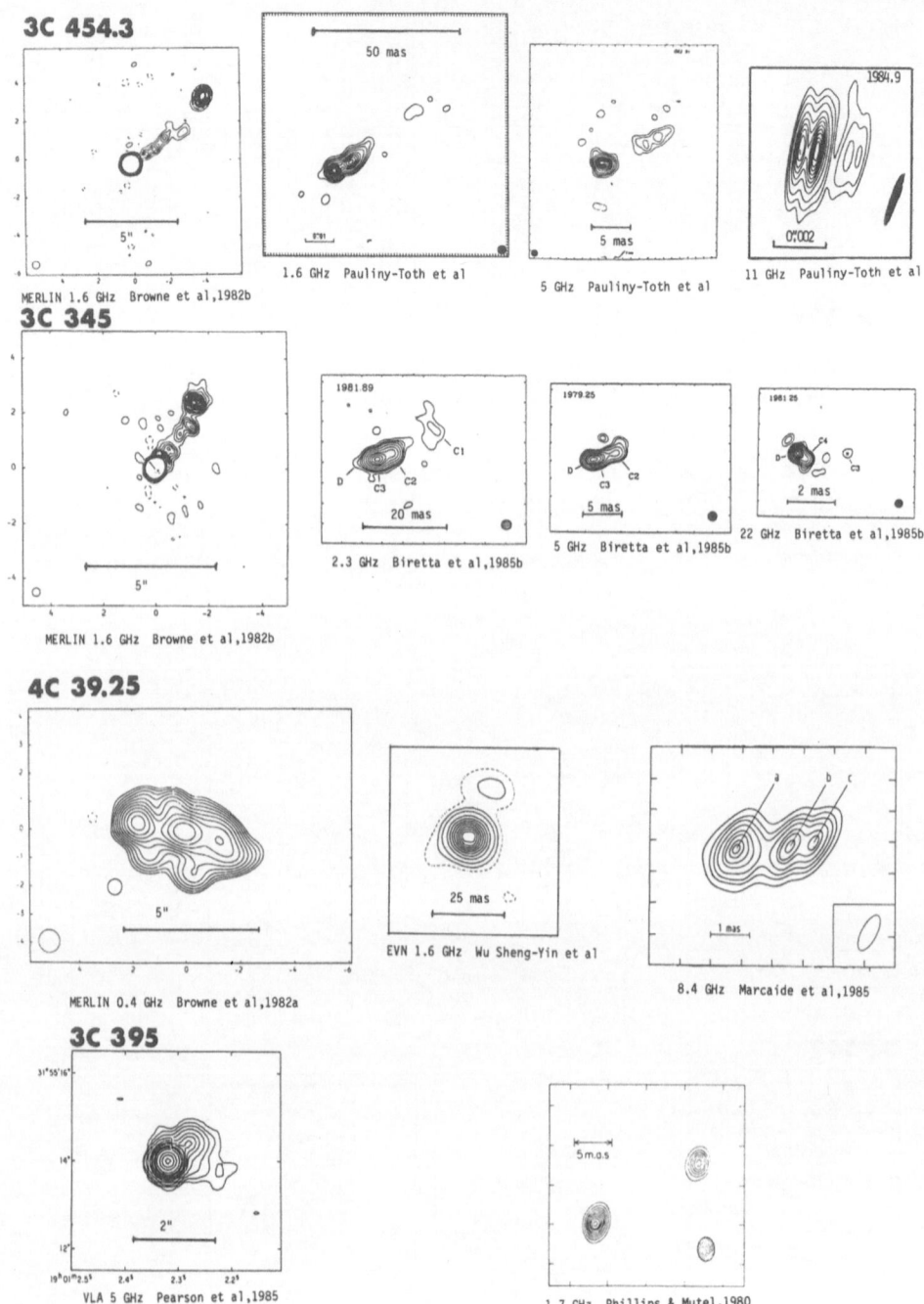

3C 454.3

MERLIN 1.6 GHz Browne et al,1982b

1.6 GHz Pauliny-Toth et al

5 GHz Pauliny-Toth et al

11 GHz Pauliny-Toth et al

3C 345

MERLIN 1.6 GHz Browne et al,1982b

2.3 GHz Biretta et al,1985b

5 GHz Biretta et al,1985b

22 GHz Biretta et al,1985b

4C 39.25

MERLIN 0.4 GHz Browne et al,1982a

EVN 1.6 GHz Wu Sheng-Yin et al

8.4 GHz Marcaide et al,1985

3C 395

VLA 5 GHz Pearson et al,1985

1.7 GHz Phillips & Mutel,1980

Figure 1 (contd.)

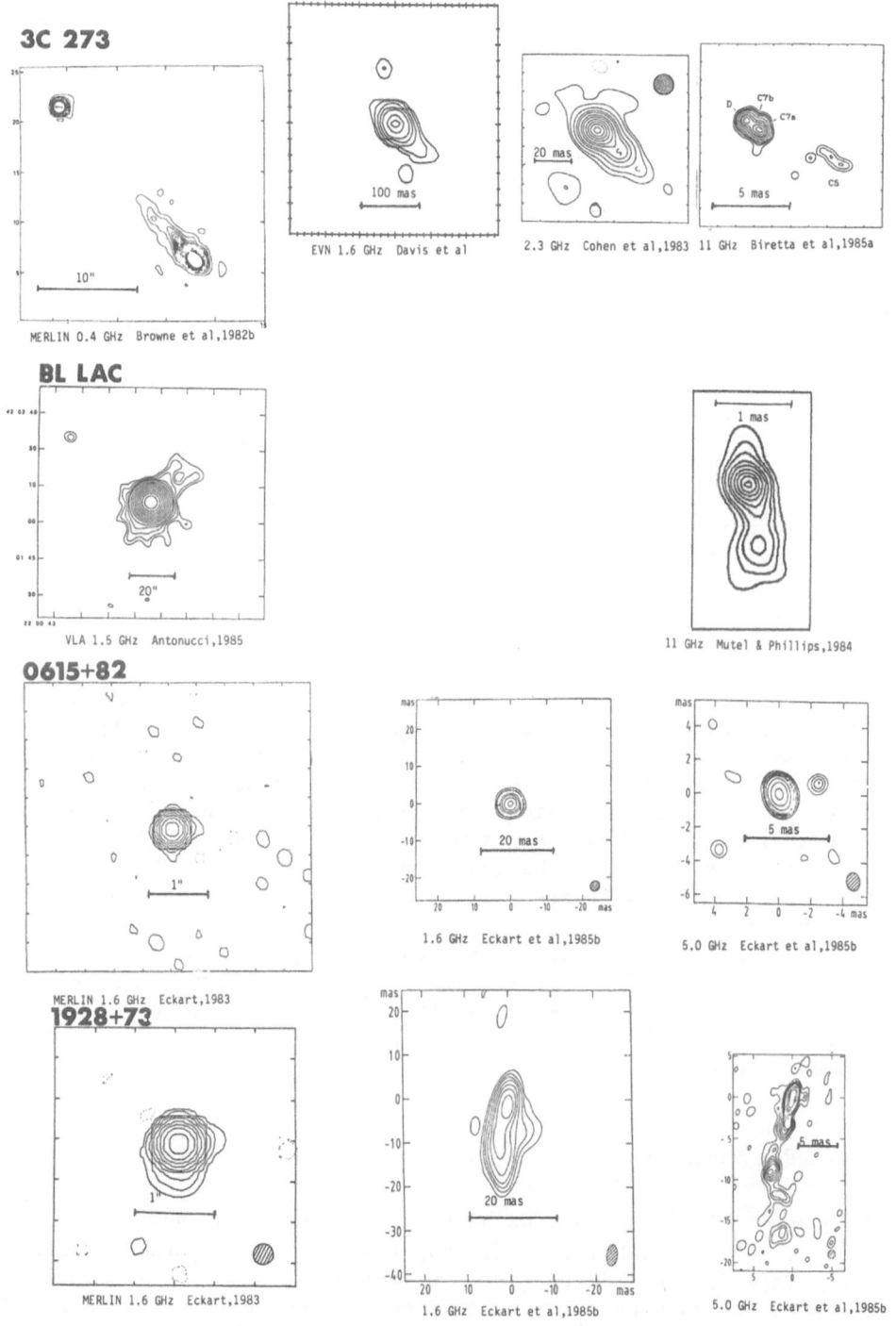

Figure 1 (contd.)

component and an extended 1-sided jet on the arcsecond scale. For these quasars a similar morphology is seen on both intermediate angular scales and in the very highest resolution maps. A clear continuity is traced out by the jet structure, from the innermost core regions out to the arcsecond scale, although in the cases of 3C454.3 and 3C345 there is a large change of pa very near the centre. In these inner regions the jet is composed of a number of knots.

4C39.25 (z = 0.698) and 3C395 (z = 0.635) are also core-dominated in their arcsecond scale structures. 4C39.25 has 2-sided extensions and may be thought of as "D1.5". The mas structure is dominated by 3 compo-nents, the westernmost being a compact core. 3C395 has only 1-sided structure in the VLA map. On the mas scale it consists of 2 widely spaced components, similar to sources in the "compact double" class (Phillips and Mutel 1982). However, the 2 components have different spectra, the western being flatter, apparently the core (Johnston et al. 1983). In both these sources the relationship between mas and arcsecond structures is unclear. It is of interest that the steep spectrum mas component in 3C395 is on the opposite side of the nucleus to the arc-second extension.

BL Lac (z = 0.070), 0615+82 (z = 0.71) and 1928+73 (z = 0.30) are all sources dominated by single, compact flat spectrum components on the arcsecond scale, with hardly any trace of extended radio emission. BL Lac and 1928+73 both show core-jet structures at 1 mas resolution. In 1928+73 the jet is broken up into many knots and extends all the way out to 20 mas. 0615+82 is still core-dominated even at 1 mas resolution and shows little trace of any jet.

The compact structures exhibited by these quasars are fairly typical of compact radio structures in general. In all cases the mas structure is asymmetric even when the arcsecond scale extended structure is symmetric. It consists of a compact, flat spectrum component, iden-tified as the core, and a 1-sided extension which may be broken up into knots. Marcaide (1984) has shown that there is a spectral gradient along the jet extensions in the quasar pair 1038+528A, B. It is perhaps significant that none of these quasars shows the symmetric "compact double" morphology seen in some radio sources identified with less luminous optical objects, the closest case being the (asymmetric) double structure of 3C395. The similarity of the mas structures of these quasars seems to indicate that the structural differences seen on the scales of 100 pc to 100 kpc are due primarily to the galactic environment and not to the central energy sources.

4. STRUCTURAL VARIABILITY

Multiple epoch VLBI observations have now been made for quite a number of quasars, and the number showing "superluminal" structural changes certainly exceeds the $\sim$ dozen listed by Porcas (1985). Not all these sources show the same phenomenon. A number of sources show "classical" superluminal motion, consisting of apparent faster-than-light motion of a compact knot with respect to the core. 3C345 (Biretta et al. 1985b), 3C273 (Unwin et al. 1985), 3C179 (Porcas 1984), BL Lac (Mutel & Phillips

1984) and NRAO140 (Marscher & Broderick 1985) are examples of this. For 3C345 there is also an upper limit of 0.7c for any motion of the core (Bartel et al. 1985). Two sources which show a rather slower motion of ∿ 2c are 3C263 (Zensus et al., in preparation) and 3C279 (Unwin, these proceedings), the latter in contrast to earlier reports of much higher velocities. Other recently discovered examples are the motion of knots in 1928+73 (Eckart et al. 1985a) and 1642+69 (Pearson et al. 1986).

At least two quasars show both moving and stationary knots. Shaffer and Marscher (1985) report that the central component of 4C39.25 (b in Fig. 1) is apparently moving superluminally with respect to the core (c), whilst the easternmost component (a) is stationary. This explains the apparent source contraction reported by Shaffer (1984). Waak et al. (1985) report a similar behaviour for a newly discovered component located between the 2 main mas components of 3C395.

3C454.3 (Pauliny-Toth et al., in preparation) deserves special mention. Following a flux density outburst in 1981, the compact core showed a superluminal brightening (Pauliny-Toth et al. 1984). From 1982 on, an extension to the core appeared, of size of ∿ 1.6 mas in pa∿ -95°, some 35° away from the pa of the jet feature seen at ∿ 6 mas. If this was ejected from the core, it corresponds to motion of $\gtrsim$ 30c (H_0 = 55 km s^{-1} Mpc^{-1}, q_0 = 0.05). However, the extension is composed of a number of knots, and although their relative brightness changes, there is no obvious systematic motion of the knots in subsequent epochs.

Recent mapping of multi-epoch data of the CSS quasar 3C147 (Alef and Preuss, in preparation) has resulted in the identification of knots in the complex mas structure. The knot nearest to the core is apparently moving away at a rate corresponding to ∿ 2c. Pearson et al. (these proceedings) have reported possible superluminal expansion in the other example of a CSS source, 3C216.

It is now clear that superluminal structural changes are very common in quasars, and the phenomenon is by no means confined to the "core dominated" sources. Studies of structural variations in the weak cores of extended quasars (Zensus & Porcas, these proceedings; Barthel et al. 1985) and in CSS sources will hopefully allow us to determine whether the "Unified Schemes" involving relativistic jets can be modified to explain the phenomenon, or whether a radically new model of superluminal motion is needed.

I wish to thank all my colleagues for use of data prior to publication.

REFERENCES

Antonucci, R.R.J.: 1985, NRAO preprint
Bartel, N. et al.: 1985, preprint
Barthel, P.D. et al.: 1984, Astron. Astrophys. 140, 399-404
Barthel, P.D. et al.: 1985, Astron. Astrophys. 151, 131-136
Biretta, J.A. et al.: 1985a, Astrophys. J. 292, L5-L8
Biretta, J.A., Moore, R.L., Cohen, M.H.: 1985b, Caltech preprint No. 20
Browne, I.W.A. et al.: 1982a, Monthly Notices Roy. Astron. Soc. 198, 673-688

Browne, I.W.A. et al.: 1982b, Nature 299, 788-793
Cohen, M.H. et al.: 1983, Astrophys. J. 272, 383-389
Eckart, A.: 1983, Ph.D. Dissertation, Univ. of Münster
Eckart, A. et al.: 1985a, Astrophys. J. 296, L23-L26
Eckart, A. et al.: 1985b (preprint)
Fanti, C. et al.: 1985, Astron. Astrophys. 143, 292-306
Hodges, M.W., Mutel, R.L., Phillips, R.B.: 1984, Astron. J. 89, 1327-
     1331
Johnston, K.J. et al.: 1983, Astrophys. J. 265, L43-L47
Marcaide, J.M., Shapiro, I.I.: 1984, Astrophys. J. 276, 56-59
Marcaide, J.M., et al.: 1985, Nature 314, 424
Marscher, A.P., Broderick, J.J.: 1985, Astrophys. J. 290, 735-741
Mutel, R.L., Phillips, R.B.: 1984, IAU Symp. 110, p. 117-118
Pauliny-Toth, I.I.K. et al.: 1984, IAU Symp. 110, p. 149-152
Pearson, T.J., Readhead, A.C.S.: 1984, IAU Symp. 110, p. 15-24
Pearson, T.J., Perley, R.A., Readhead, A.C.S.: 1985, Astron. J. 90,
     738-755
Pearson, T.J. et al.: 1986, Astrophys. J. (submitted)
Phillips, R.B., Mutel, R.L.: 1980, Astrophys. J. 236, 89-98
Phillips, R.B., Mutel, R.L.: 1982, Astron. Astrophys. 106, 21-24
Phinney, E.S.: 1985, Astrophys. of Active Galaxies and Quasi Stellar
     Objects, ed. Miller, J.S., University Science Books, p. 453-496
Pooley, G.G., Henbest, S.N.: 1974, Monthly Notices Roy. Astron. Soc.
     169, 477-526
Porcas, R.W.: 1984, IAU Symp. 110, p. 157-161
Porcas, R.W.: 1985, Active Galactic Nuclei, ed. Dyson, J.E. Manchester
     Univ. Press, p. 20-49
Romney, J.D. et al.: 1984, Astron. Astrophys. 135, 289-299
Shaffer, D.B.: 1984, IAU Symp. 110, p. 135-136
Shaffer, D.B., Marscher, A.P.: 1985, BAAS 17, 608
Shone, D.L., Porcas, R.W., Zensus, J.A.: 1985, Nature 314, 603-604
Simon, R.S. et al.: 1983, Nature 302, 487-490
Unwin, S.C. et al.: 1985, Astrophys. J. 289, 109-119
van Breugel, W., Miley, G., Heckman, T.: 1984, Astron. J. 89, 5-22
Waak, J.A. et al.: 1985, Astron. J. 90, 1989-1991
Zensus, J.A., Porcas, R.W., Pauliny-Toth, I.I.K.: 1984, Astron. Astro-
     phys. 133, 27-30

DISCUSSION

Mutel : Are there any multi-epoch VLBI observations of D2 Quasars which are <u>not</u> consistent with superluminal motion ?

Porcas : I think this is a case where the dictum "absence of evidence is not evidence of absence" applies. I would not be surprised if all D2 quasars are shown to exhibit superluminal motion. However, as a result of the work which Toni Zensus and I are carrying out, I would not now be surprised if all D1 quasars show superluminal motion !

Hutchings : What is the evidence for optical beaming in the superluminal sources as a group ? Are they line-weak like BL Lacs, do they have characteristic or variable line profiles, polarization or continuum variability ?

Porcas : Many different optical classes are represented among the superluminal sources. 3C 345, 3C 454.3, 3C 216 are OVVs, BL Lac is a BL Lac. These all show high optical variability and polarization. 3C 273, 3C 179, 3C 263 are normal quasars, showing only moderate variability, if any. 3C 120 is a Seyfert galaxy. Whether any of these optical classes show evidence for optical beaming is, of course, a matter of speculation.

Wilkinson : Do you believe there have been <u>any</u> unequivocal detections of components on <u>both</u> sides of the flat-spectrum "nuclei" which you have defined ? i.e. have we seen milliarcsecond counter-jets yet ?

Porcas : I believe Peter Barthel has reported 1 or 2 examples of 2-sided VLBI jets. In general, I believe they are very rare.

Kundt : How many double sources are known, and what are their scales ?

Mutel : Dozen. 10-150 mas.

Swarup : Do you find any increased brightness near the bends seen at around 1 kpc ?

Porcas : Sometimes ! In the CSS source 3C 216 which I showed, there is a strong concentration of radio emission some 120 mas away from the core, at the point where the structure changes direction sharply. On the other hand, in the CSS source 3C 147 which I also showed, there does not seem to be such a knot at the equivalent distance.

Wandel : Do objects like 3C 216, in which the compact structure is not aligned with the larger scale, show spectra different from those of aligned objects ?

Porcas : No. For example the two sources, 3C 216 and 3C 179 both have similar spectra, at low frequencies dominated by steep spectrum emission from optically thin regions, and a slight flattening at high frequencies

(> 5 GHz) as the compact cores start making a significant contribution to the total flux. 3C 179 shows good alignment, 3C 216 not. The difference is in the arcsecond scale structure, the steep spectrum compact sources ($\lesssim$ 1") generally having large misalignments.

Segal : Since you suggest that one must stick with the beaming model since it is the only one known, I should mention that the redshift-distance relation in the chronometic cosmology, $z = \tan^2 \frac{1}{2}$ (distance in radians), reduces all distances to a level that eliminates superluminal motions. Is there any observational reason to reject the very simple explanation that the Friedman distances are wrong ?

Porcas : I did not wish to imply that the relativistic jet aligned at a small angle to the line of sight is the only suggested explanation for superluminal motion. Gravitational lenses and screens, real "tachyonic" motion, non-cosmological redshifts and a much larger Hubble constant all have their advocates and can all explain the bare facts of superluminal motion. The criterion for acceptability is the range of agreement with observations over broad areas in astronomy and physics.

POLARIZATION DISTRIBUTIONS IN COMPACT RADIO SOURCES

David H. Roberts and John F. C. Wardle
Department of Physics
Brandeis University
Waltham, MA 02254 USA

ABSTRACT. We present milliarcsecond-resolution 5 GHz polarization
maps of several active galactic nuclei: one epoch each for the quasar
3C345, the galaxy 3C120, and the BL Lacertae object 0735+178, and two
epochs for the BL Lacertae object OJ287.

1. INTRODUCTION

At the heart of the radio structures of active galactic nuclei is a
compact (angular size mas, physical size pc) radio source coincident
with the optical center of the object. At arcsecond resolution,
these central components usually have flat radio spectra and a rather
low degree of linear polarization. When observed at milliarcsecond
resolution by VLBI, these sources often consist of a few discrete
components with very high brightness temperatures ($\gtrsim 10^{10}$ K). In many
cases the structures mimic those on the arcsecond scale, having an
optically-thick "core" and an optically-thin "jet". The relationship
of the millarcsecond jets to the arcsecond jets is not clear: Do
the milliarcsecond jets "feed" the arcsecond jets, or do they have
different origins? Are the same emission mechanisms and confinement
processes involved? The low integrated polarization of the central
components has been ascribed variously to a tangled magnetic field,
a peaked electron energy distribution, optical thickness, internal
Faraday rotation, or a non-synchrotron emission mechanism. Which, if
any, is correct? The milliarcsecond jets must coexist with the optical
line emitting gas; what is the effect of one upon the other? What are
the kinematic and dynamical origins of the superluminal motions? And
finally, what more can we learn about the "central engines" of active
galactic nuclei by studying the radio structures on the smallest
scales?

In order to further attack these questions, we have developed
techniques of linear polarization measurement for VLBI and applied
them to some of the brightest active galactic nuclei.

G. Swarup and V. K. Kapahi (eds.), Quasars, 141–147.

2. POLARIZATION SYNTHESIS AT MILLIARCSECOND RESOLUTION

All of the data discussed below were taken using the Mark III VLBI
system, recording seven 2 MHz channels of left circular polarization at
four stations (Haystack=K, Green Bank=G, Phased VLA=Y, Owens Valley=O),
and seven channels of right circular polarization at two (G,Y), at
a frequency of 4984 MHz (λ6 cm). The data were correlated for all
parallel- and cross-hand fringes at Haystack Observatory, and were
reduced using our software on the VAX 11/780 and VAX 8600 computers at
Brandeis.

The crucial step in a polarization synthesis is the determination
and removal of the instrumental components of the cross-polarization
response of the interferometer. The procedure for accomplishing this
in the (non phase-coherent) VLBI case has been given by Roberts $\underline{et}$
$\underline{al}$. (1984). The distribution of linearly-polarized radiation may
be described by the complex polarization $P = Q + iU = p \exp(2i\chi) =$
$mI \exp(2i\chi)$, where I is the total intensity, Q and U are the two
Stokes parameters for linear polarization, $p = \sqrt{Q^2+U^2}$ is the polarized
intensity, m is the fractional polarization, and χ is the position
angle of the electric vector. Maps of P may be made from the
decontaminated cross-fringes by several techniques (Cotton $\underline{et}$ $\underline{al}$. 1984;
Roberts $\underline{et}$ $\underline{al}$. 1984; Roberts and Wardle, in preparation). The basic
idea is to make an I-map from the parallel-hand fringes using one of
the hybrid map-making techniques (Pearson and Readhead 1984), use the
results to correct the cross-fringes for the antenna phases, transform
the cross-hand visibilities, and perform a complex CLEAN on the
resulting dirty P-map. The I-map and P-map are thus phase-referenced
to each other, so there is no ambiguity about their registration.

3. THE SUPERLUMINAL QUASAR 3C345

The quasar 3C345 is one of the best-studied examples of the
superluminal phenomenon. It consists of an opaque core denoted
"D", and a series of rapidly-moving knots "C2", "C3", etc., which
form a jet at position angle PA = -75° (Cohen $\underline{et}$ $\underline{al}$. 1983a). Taking
H_0 = 75 km/s/Mpc and q_0 = 0.5, for 3C345 (z = $\overline{0.595}$), 1 mas = 5.1 pc
if we adopt the cosmological interpretation of the redshift.

Figure 1a shows a total intensity map of 3C345 made from limited
data taken in December 1981 (Wardle $\underline{et}$ $\underline{al}$. 1986). The clean beam used
was 3.9 × 2.3 mas with major axis at $\overline{PA}$ = -18°. At this resolution,
the inner (C3) and outer (C2) knots blend together to form the "bulge"
to the west, and are separately visible only in the clean components.
However, Figure 1a is consistent with the map of Unwin $\underline{et}$ $\underline{al}$. (1983),
which was based on observations a few months earlier than ours with
twice the resolution. Figure 1b shows a polarization map, with E-field
vectors superimposed on contours of polarized intensity. The position
of the opaque core D (from Figure 1a) is shown by the cross. Because
two of the VLBI stations detected only one polarization, the u-v
coverage is somewhat poorer than that for the I-map, and the restoring
beam here is 4.7 × 2.5 mas at PA = -25°.

Comparison of Figures 1a and 1b shows that the polarized flux is associated with the jet rather than with the core, and that the two knots C2 and C3 are differently polarized. Summing the clean components of the I- and the P-maps, we find that the core is less than 1 % polarized, C3 is 11 % polarized at χ = 22°, and that C2 is 6 % polarized at χ = 83°. However, the <u>maximum</u> degree of polarization in the source is located on the side of C3 towards the core, and is about 20 %. While this is due in part to dilution by C2 (which has a different χ), it may represent actual polarization structure within C3. The total polarized flux in the map is 260 mJy at χ = 29°, while the integrated polarized flux of 3C345 at this epoch (H. D. Aller, private communication) was about 370 mJy at χ = 33°. Clearly there is some polarized flux associated with scales to which our observations are not sensitive. A lower resolution map made with a slight east-west taper reveals an additional weak component to the west of C2 (Wardle <u>et al.</u> 1986), but because of the paucity of short spacings in our data, we cannot be certain of its reality; it may be associated with the component C1 of Cohen <u>et al.</u> (1983b).

4. THE SUPERLUMINAL GALAXY 3C120

This galaxy provides the nearest (z = 0.033) example of superluminal motion, and is of great interest for the small scales which may be investigated (1 mas = 0.60 pc). An I-map made from data taken in December 1982 is shown in Figure 2a; this is in good agreement with interpolation between the maps presented by Walker <u>et al.</u> (1984). The opaque core and the knots called B, C, and D by these authors are all detected. The P-map in Figure 2b has a peak intensity which is only 2.3 % of that of the I-map; comparison of the two shows that the core is unpolarized, and the knots to have m ~ 11 %, 3 %, and 3 %, respectively. However, the correspondence of the I- and P-structures of the jet is not very tight. The total polarized flux in the map is 24 mJy at χ = -22°, while the integrated flux measured by the VLA was 177 mJy at χ = 171° (corresponding to m = 6.4 %). Thus there is substantial polarized flux in an extended VLBI jet which we cannot map. These results are not unexpected, as Benson <u>et al.</u> (1984) have shown that 3C120 has structure on a continuous range of scales from milliarcseconds to arcminutes.

5. THE BL LACERTAE OBJECT 0735+178

This rather compact source (z > 0.42, > 4.4 pc/mas) has previously been studied at 6 cm by Baath (1984). The I-map in Figure 3a (epoch 1982.9) shows a bright core and two knots at PA = 72°. The polarization structure of 0735+178 is shown in Figure 3b; comparison of the maps shows a complete contrast to 3C345 and 3C120. Here the core is 3.6 % polarized, the inner knot (3 mas from the core) is unpolarized (m < 1 %), and the outer knot (5 mas) has m = 6.5 %. In this source 72 % of the integrated polarized flux appears in the VLBI map.

6. THE BL LACERTAE OBJECT OJ287

This source (z = 0.306, 3.7 pc/mas) has particularly strong and
variable polarization at radio wavelengths. Previous VLBI observations
have shown it to be very compact, and it is often used as a calibrator.
Maps of OJ287 at two epochs a year apart are shown in Figures 4a, 4b,
5a, and 5b. While the total intensity maps show that the source is
only slightly extended (in the south-west direction), the polarization
structure is both striking and variable. Model fitting of the I-maps
at each epoch is consistent with a simple double structure, with a
separation of 1.9 mas and an intensity ratio of 8:1. At both epochs
the P-map also shows a double structure, but with an intensity ratio
much closer to unity. Because of the "interference" exhibited by close
components with electric vectors roughly perpendicular to each other,
the two components in Figure 4b (epoch 1981.9) are actually closer
than they look. Model fitting shows the separation to be ~ 0.9 mas,
with intensities (p = 120 mJy, χ = -19°) and (184 mJy, +83°) for the
NE (core?) and SW (jet?) components, respectively. One year later
the separation had increased to 1.4 mas, and the intensities were
(124 mJy, -79°) and (96 mJy, +73°). Thus the NE component remained
about 4 % polarized but underwent a rotation of position angle of
at least 60 degrees, while the SW component fell a factor of two in
polarized flux but held a roughly-constant position angle. If we
identify the SW components of the I-maps with the SW components of the
P-maps, the fractional polarization of this jet was as high as m = 50 %
in 1981.9, and fell to 25 % a year later. A strict lower limit to the
polarization of the SW component is 15 %.

7. ACKOWLEDGEMENTS

This research has been supported by the NSF and by a grant from NASA
administered by the American Astronomical Society. We thank the staffs
of the VLBI Network observatories and of the Haystack VLBI correlator
for their help, Leslie Brown, Denise Gabuzda, and Bob Potash for their
assistance, and Alan Rogers for his collaboration and expertise.

8. REFERENCES

Baath, L.A. 1984, in VLBI and Compact Radio Sources, IAU Symposium 110,
    ed. R. Fanti et al. (Reidel, Dordrecht), p. 127.
Benson, J.M., Walker, R.C., Seielstad, G.A., and Unwin, S.C. 1984, in
    VLBI and Compact Radio Sources, IAU Symposium 110, ed. R. Fanti et
    al. (Reidel, Dordrecht), p. 125.
Cohen, M.H. et al. 1983a, Ap.J.Letters, 269, L1.
Cohen, M.H. et al. 1983b, Ap.J., 272, 383.
Cotton, W.D. et al. 1984, Ap.J., 286, 503.
Pearson, T.J., and Readhead, A.C.S 1984, Ann.Rev.Astr.Ap., 22, 97.

Roberts, D.H., Potash, R.I., Wardle, J.F.C., Rogers, A.E.E., and Burke, B.F. 1984, in VLBI and Compact Radio Sources, IAU Symposium 110, ed. R. Fanti et al. (Reidel, Dordrecht), p. 35.
Unwin, S.C. et al. 1983, Ap.J., 271, 536.
Walker, R.C., Benson, J.M., Seielstad, G.A., and Unwin, S.C. 1984, in VLBI and Compact Radio Sources, IAU Symposium 110, ed. R. Fanti et al. (Reidel, Dordrecht), p. 121.
Wardle, J.F.C., Roberts, D.H., Potash, R.I., and Rogers, A.E.E. 1986, submitted.

9. FIGURES

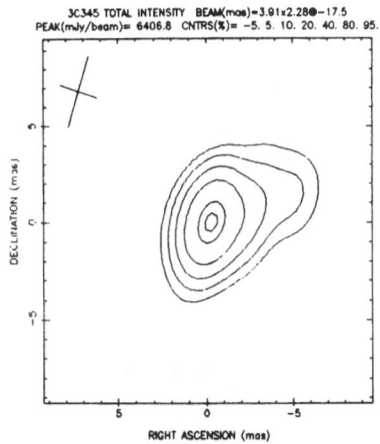 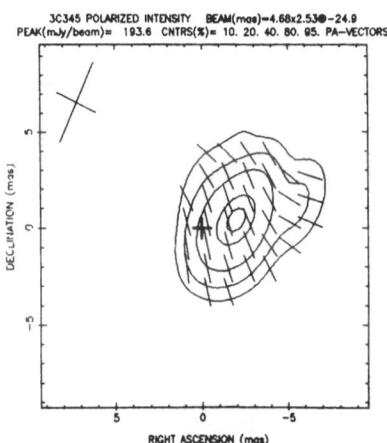

Figures 1a, 1b. I- and P-maps of 3C345, epoch 1981.9.

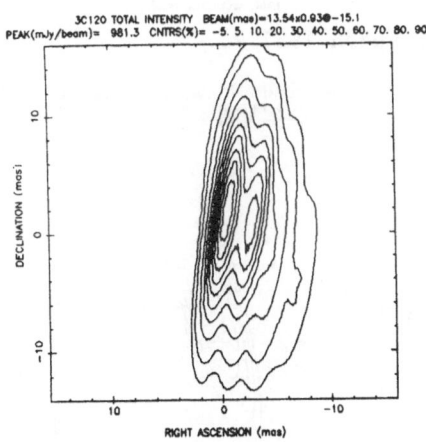 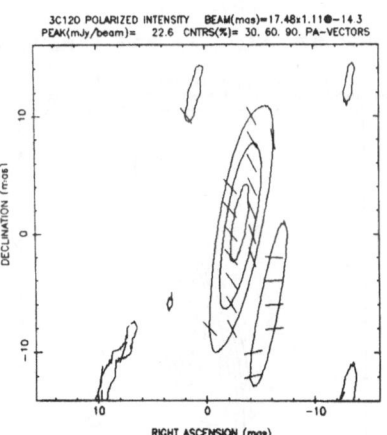

Figures 2a, 2b. I- and P-maps of 3C120, epoch 1982.9.

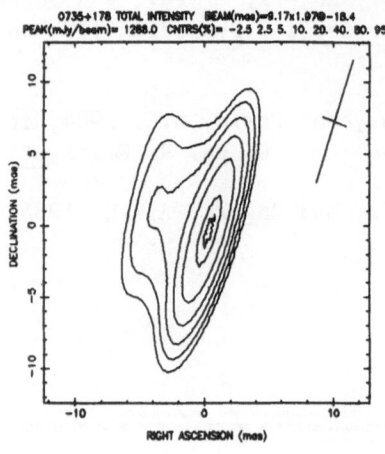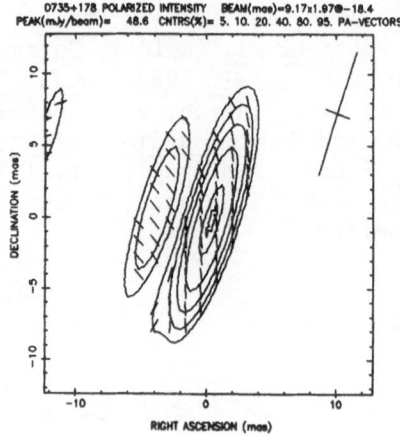

Figures 3a, 3b.   I- and P-maps of 0735+178, epoch 1982.9.

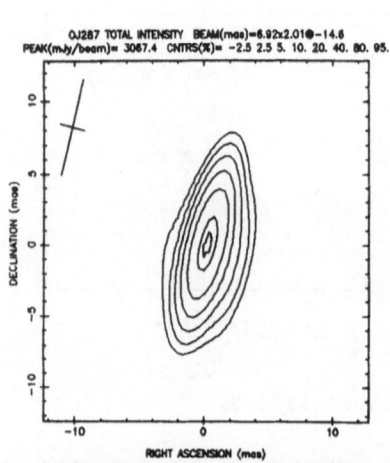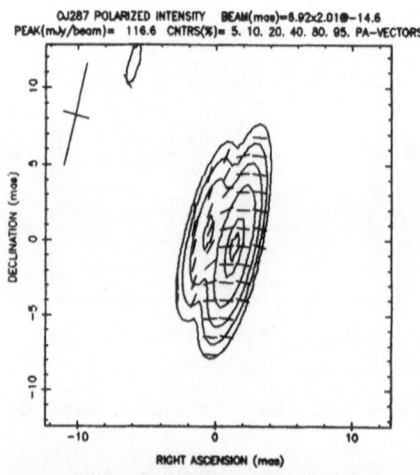

Figures 4a, 4b.   I- and P-maps of OJ287, epoch 1981.9.

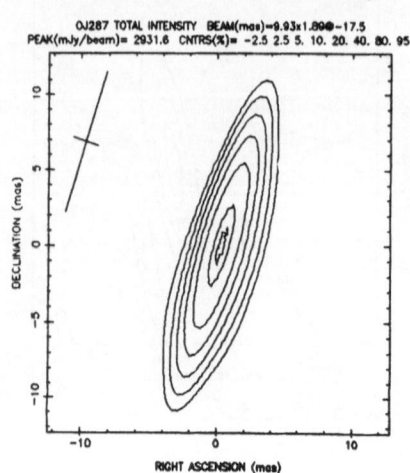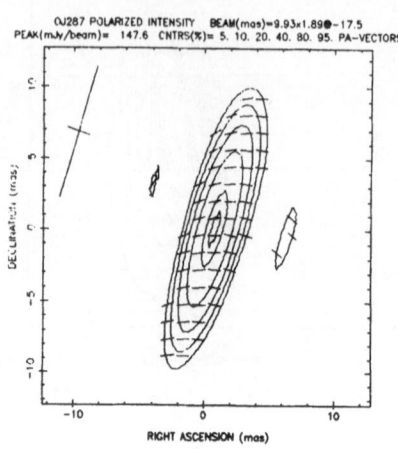

Figures 5a, 5b.   I- and P-maps of OJ287, epoch 1982.9.

DISCUSSION

Cotton : You interpreted the change in position angle as a change in the orientation of the magnetic field. Have you considered synchrotron opacity effects ? Five years ago we did a 13 cm VLBI polarization measurement of 3C454.3 in which the position angle rotated by 90°; we interpreted this as an opacity effect.

Roberts : I agree that a change in the orientation of B along the jet is not only possible explanation of swings in the plane of the electric vector. I would like to have multi-frequency polarization VLBI data in order to put better limits on any opacity effects.

Perez-Fournon : How do your results correlate with the unresolved polarization properties ?

Roberts : In some objects (0735+178, OJ287, and probably 3C345) the VLBI p-map contains most of the polarized flux seen with the VLA (ie, 50-90 percent). However, in 3C120 the milliarcsecond jet apparently has polarised flux on scales we are unable to map with our very limited u-v coverage, and we recover only ~ 14 percent of the arcsecond-scale polarized flux.

" .... BL Lac is a BL Lac."

- Richard Porcas (p.139)

RESULTS FROM A MONITORING PROGRAM OF LOW FREQUENCY VARIABLE SOURCES

L. Padrielli, R. Fanti, A. Ficarra, L. Gregorini, F. Mantovani
Istituto di Radioastronomia,
Via Irnerio 46,
Bologna, Italy

INTRODUCTION

The flux variability of extragalactic radio sources at decimetric wavelengths (Low Frequency Variability LFV) is mostly associated with the nuclei of compact radio sources. But is a not yet well understood phenomenon. The main question still is: where does this phenomenon take place?

Two categories of models have been proposed to account for this phenomenon: a) the intrinsic models, among which a general consensus emerges for the interpretation of the flux variations in terms of synchrotron emission of relativistic electrons beamed in a direction close to the line of sight. In this case the LFV is directly related to the way the 'central engine' produces and transfers relativistic particles. Estimates of the relativistic Lorentz factor (γ) can be derived from the LFV. b) The extrinsic models, that attribute the LFV to propagation effects through the interstellar medium. In this case the LFV could be used for measuring the properties of the interstellar turbulence.

An analysis of multifrequency observations at 0.4, 2.3, 4.8, 8.0 and 14.4 GHz of 51 radiosources selected from the Bologna monitoring program (Fanti et al 1981) over a period longer than 5 years allowed us to identify three classes of sources showing LFV and to evaluate their occurrence in an unbiased sample of variable sources (Padrielli et al, 1986a).

SOURCES DISPLAYING CORRELATED BROAD BAND VARIABILITY

This class contains sources with broad frequency band activity, which appears to be correlated across the whole radio frequency band. The outbursts are either quasi-simultaneous or regularly drifting to lower frequencies, with somewhat reduced amplitude. A light-curve of a source belonging to this class is shown in Fig. 1.

In our sample there are 4 good cases of these objects: 3C 120, 0605-085, 1510-089, BL Lac and some probable ones. The occurrence of this class in our sample of LF Variables is from 10% to 20%.

For the two sources 0605-085 and 1510-089 we obtained two epoch VLBI observations at 18 cm with an array of 8 stations, corresponding to a

G. Swarup and V. K. Kapahi (eds.), Quasars, 149–155.
© *1986 by the IAU.*

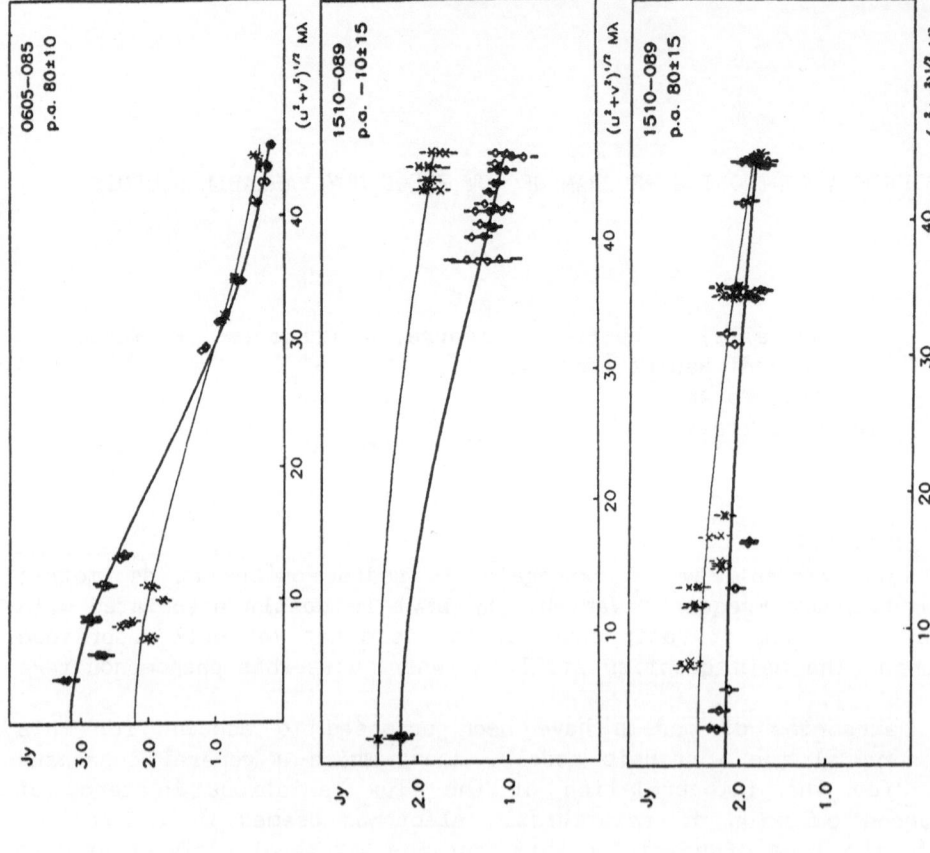

Fig. 2 – Fringe visibility amplitudes vs baseline length in specific directions on the u-v plane. ✗ represents first epoch data (1980.1), ◊ second epoch data (1981.8)

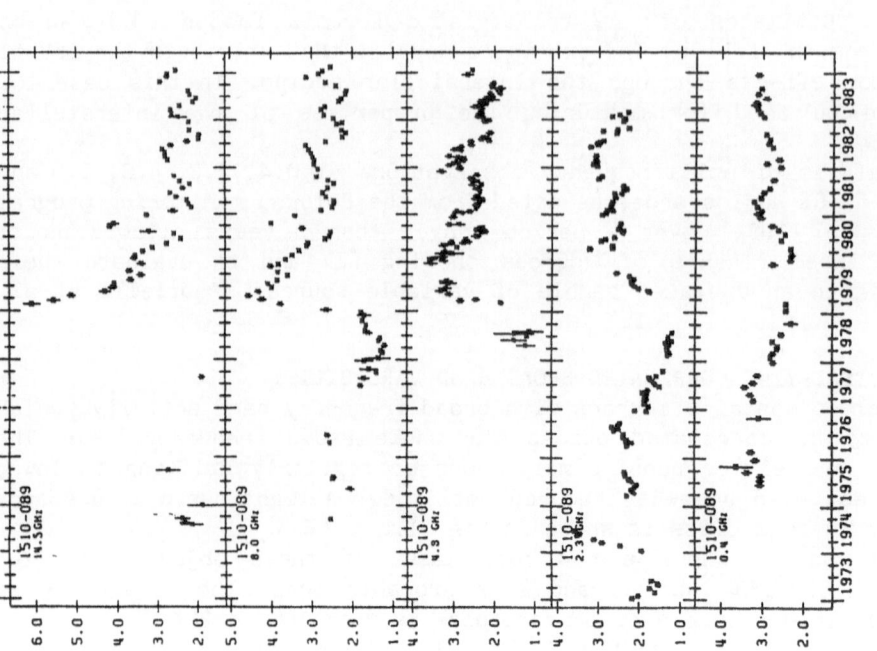

Fig. 1 – Light-curves of the source 1510–089 at different frequencies.

resolving power of 2-3 mas (Romney et al 1984, Padrielli et al. 1986b).

From the time scale of the LFV between the two epochs we can compute the angular diameter of the varying component, its brightness temperature and then the lower limit to the bulk Lorentz factor. For these sources our VLBI observations give evidence of statistically significant structural changes between the two epochs. A careful analysis of the fringe visibilities at the longest baselines (b>30 Mλ) shows that a possible interpretation of the differences is an increase of the angular size of the more compact component. The corresponding expansion rate is in agreement with the Lorentz factors derived from the LFV (with the simple assumption of an angle between the line of sight and the ejection direction of the order of $1/\gamma$). Fringe visibility amplitudes of the two epochs are shown in Fig. 2. 3C 120 and BL Lac are also well known superluminally expanding sources with rates that are in agreement with the γ derived by the LFV.

For the sources of this class, the whole observational scenario is in agreement with models requiring relativistic bulk motions. These models successfully explain an activity extending from the very high frequencies to meter wavelengths.

SOURCES WHOSE VARIABILITY IS CONFINED TO THE LOW FREQUENCY RANGE

This second class contains sources that are strong variables at low frequency ($\nu < 1$ GHz) and weak at high frequency. The occurence of this class of objects is 35% in our sample of LF Variables.

Fig. 3 shows the light-curve of DA 406, source well studied by several authors and prototype of the category (see for references Altschuler et al. 1984).

We have two epoch VLBI measurements at 18 cm only for two objects of this class. 0859-140 did not vary between the two epochs, but 1611+343 (DA 406) had a spectacular variation at 0.4 GHz. If we interpret this variation in terms of relativistic motions, we obtain a lower limit for the Lorentz factor of the order of 12 (H=100 Km s^{-1} Mpc^{-1}, q$_0$=1), which leads to an expected expansion of 0.7 mas (with an angle between the beam and the line of sight of $1/\gamma$). The fringe visibilities do not show evidence for a significant angular increase between the two epochs (Fig. 4).

In this case we do not find a direct evidence of relativistic motions associated with the LFV ant it could be due to a distinct phenomenon.

SOURCES WITH UNCORRELATED BROAD BAND VARIABILITY

This class contains sources which have both high and low frequency activity, that appear to be unrelated, with each other, and display a minimum of activity at the intermediate frequencies ($\nu \sim 1$ GHz). The occurrence of this category of objects in our sample is from 20% to 30%. In Fig. 5 an example of light-curves of this kind of sources is shown.

We have 2 epoch VLBI information for several objects in this variability class, but, due to the lack of correlation of high and low frequency activity any discussion on the structural changes from 18 cm data can only done if we have spectral information, which will allow the determination of the components responsible for the LFV. This is the case of the source 3C 454.3 (Pauliny-Toth et al. 1981). After the first

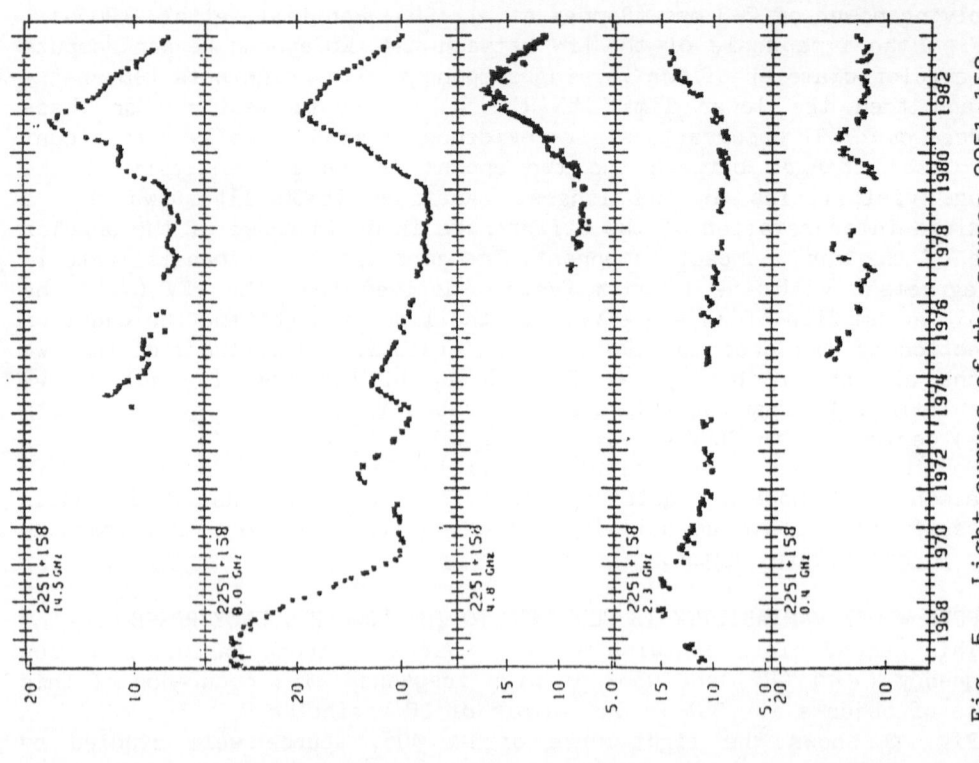

Fig. 5 – Light-curves of the source 2251+158
(3C 454.3) at different frequencies.

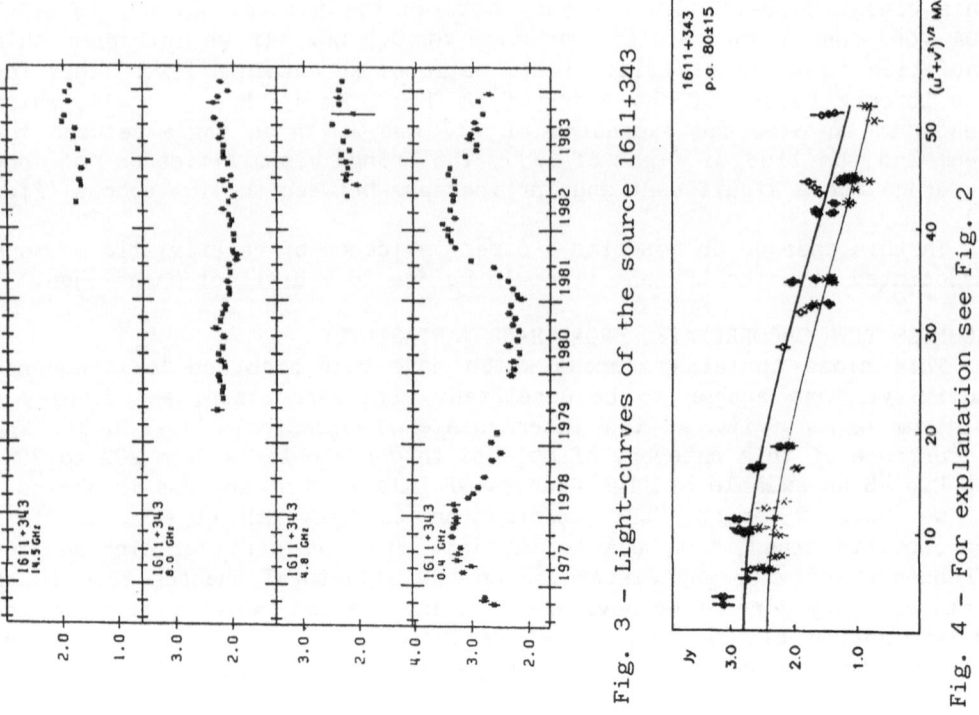

Fig. 3 – Light-curves of the source 1611+343

Fig. 4 – For explanation see Fig. 2

VLBI measurement, the source had a strong burst at 0.4 GHz and the corresponding brightness temperature leads to a γ limit ~11. The two VLBI maps are very similar to each other and the separation between the core and the jet (the region we expect to see at 0.4 GHz) is unchanged (Fig. 6). Other sources in this category do not show the angular expansion that is expected on the basis of 0.4 GHz time variability from relativistic models. The behaviour of the variability of this third class could probably be due to a superposition of the two different processes.

EXTRINSIC INTERPRETATION

We have focused our attention on the objects of the last two classes of LFV, comparing the observed variability amplitudes and time scales at 0.4 GHz with the expectations from effects of the refractive focusing of the large scale irregularities of the galactic medium as it has considered by Rickett et al. (1984).

We define m as the ratio between the flux density r.m.s. and the mean flux density and ΔT as the half power full width of the decorrelation functions of the 0.4 GHz light-curves that span over ~10 years. The theory predicts for the extragalactic sources a dependence of the variability and time scale on the galactic latitude (b) and the angular size of the source itself, for 'relatively extended sources'. Furthermore, the product of m and T should statistically depend only on the galactic latidude.

Fig. 7 gives the plot of $m*\Delta T$ versus sin b. We have divided the sources according to their galactic longitude (l) into two groups, one projected in the direction of the Galactic Centre and the other one in the opposite side. A correlation was not found, but nevertheless there is an interesting difference between the two samples. While the (variability * time scale) product is completely uncorrelated with the galactic latitude for sources in the anticentre direction, the same quantity shows a sort of a lower bound that depends on the galactic latitude for the sample projected in the direction of the galactic centre. In this direction, the source radiation passes through many irregularity patterns, and we expect to find a statistical behaviour in agreement with the theory at least in the lower limits of the scintillation. The slope of the straight line of the limits is about 1.2, in agreement with the expected slope of 1.6.

An independent test on the extrinsic interpretation is obtained by comparing of the B2 and B3 surveys on a time scale of about ten years (Colla et al. 1970, Colla et al. 1973, Ficarra et al. 1985). The surveys consist of 24 hours of observations in the declination range 31°–40°. The covered sky is all in the direction of the galactic anticentre and a comparison between the two sky regions cannot be done.

We have found 42 variables sources over about 2000 sources stronger than 0.4 Jy. Fig. 8 shows the distribution of the percentage of variable sources vs the galactic latitude. The probability of a uniform distribution of this plot is less than 1%.

What can be concluded? The data suggest a certain degree of agreement with the extrinsic explanation for the LFV when it is uncorrelated with the high frequency one. Clearly the problem is not resolved and a better description of the galactic medium should be necessary.

References
Altschuler D.R. et al. 1984, Astron.J. 89,1784
Colla G. et al. 1970, Astron.Astrophys.Suppl.Ser. 1,281
Colla G. et al. 1973, Astron.Astrophys.Suppl.Ser. 11,291
Fanti R. et al. 1981, Astron.Astrophys.Suppl.Ser. 45,61
Ficarra A. et al. 1985, Astron.Astrophys.Suppl.Ser. 59,255
Padrielli L. et al.1986a, Astron.Astrophys. in press
Padrielli L. et al. 1986b, Astron.Astrophys. submitted
Pauliny-Toth I.I.K. et al. 1981, Astron.J. 86,371
Rickett B.J. et al. 1984, Astron.Astrophys. 134,390
Romney J.D. et al. 1984, Astron.Astrophys. 135,289

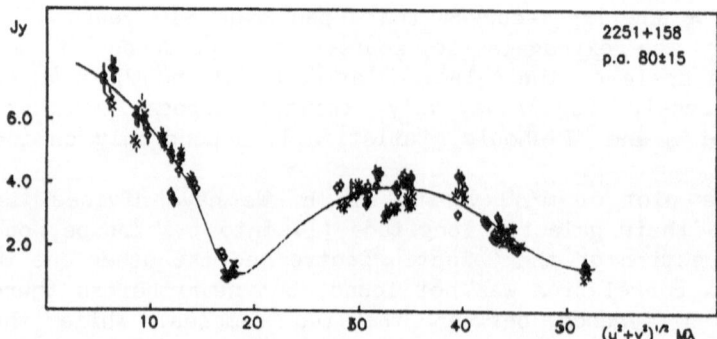

Fig. 6 – For explanation see Fig.2

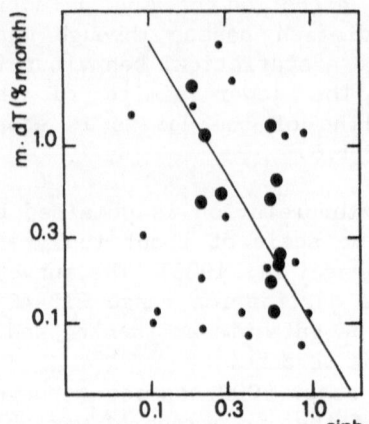

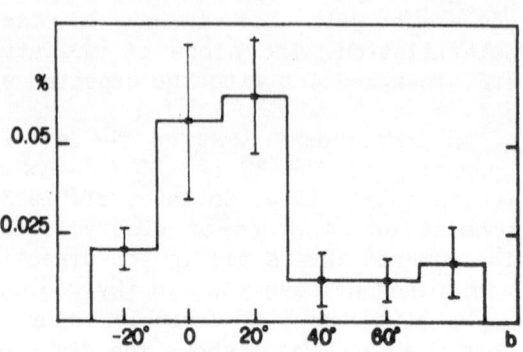

Fig. 7 – m·ΔT plotted vs sin b. ● represents sources with l<90° or l>270°. • represents sources with 90°<l<270°

Fig. 8 – Distribution of the percentage of variable sources vs the galactic latitude.

DISCUSSION

Baldwin : To test the interstellar explanation of variability we have searched for sources which are compact and nonvariable at high frequencies and, from their spectra, are compact also at low frequencies. CTD93 is the best example of these and at 151 MHz it has varied by less than 5% over 18 months.

Burbidge : The important question is : are there some sources which definitely show low frequency variability which is intrinsic ?

Padrielli : There are ~ 20% of low frequency variable sources that show activity correlated across our entire frequency range from 0.4 to 14.5 GHz, with superluminal expanding motions associated to the L.F. variations. I think that this can be interpreted as evidence of intrinsic variability.

Cohen : For the few cases where relativistic motions explain both the low frequency variability and the VLBI changes, what values of Lorentz Factor (and Hubble Constant) do you find ?

Padrielli : The Lorentz factor ranges from 2 to 5 for $H = 100$ km/sec/Mpc and $q_0 = 1$.

Mutel : The large scatter of galactic latitude - variability plot is probably due to known clumpy structure of interstellar turbulence, as deduced for example, from studies of angular broadening of compact sources in the galactic plane.

"Relativistic beaming is such a strong amplifier that we may be "blinded" by a realtively insignificant part of the source coming towards us."

- Roger Blandford (p.221)

EVOLUTION OF THE COMPACT RADIO SOURCE 3C 345

J. Biretta, M. Cohen, and R. Moore
California Institute of Technology
Pasadena, California 91125

The quasar 3C 345 is recognized as a strong, variable, superluminal radio source, an optically violent variable, and a weak X-ray source. We present results of VLBI monitoring between 1979 and 1984 at 2.3, 5.0, 10.7, and 22.2 GHz using antennas of the U.S. and European VLBI Networks. These results are interpreted in terms of a relativistic jet model (Biretta 1985; Biretta *et al.* 1986).

Maps made at 22.2 and 10.7 GHz show development and position angle changes of the new superluminal component C4. The 10.7 and 5.0 GHz maps show superluminal motion of C2 and C3, and they also show curvature of the jet.

Positions of component centroids relative to the "core" (component D) were determined by modelfitting and are shown in Figure 1. The new component C4 (Biretta *et al.* 1983; Moore *et al.* 1983) changes position angle from $-135°$ to $-87°$ as it moves away from the core. Components C3 and C2 are at position angles $-86°$ and $-74°$, respectively, and the 3-arcsecond jet has a position angle of $-30°$. The jet curvature cannot be explained as precession of the central engine; this mechanism cannot account for the position angle change of C4. Furthermore, the curvature between C3 and C2 would require a precession period too short to be explained by known mechanisms. The observed curvature may however be caused by pressure gradients in an external atmosphere if there is pressure balance. Component C5 is a westward extension from the core visible after 1982. Its position angle, $-90°$, indicated that either the jet does not have a fixed path near the core, or more likely, that components initially fill only part of the jet's width.

Component C4 accelerates from $v/c = 1.3$ to 6.5 ($H_0 = 100$). This may reflect a true acceleration, but part or even all of it may be due to bending of the jet away from the line of sight, which can produce an apparent acceleration while the intrinsic velocity remains constant. Component C3 moves with $v/c = 6$ but C2 is faster with $v/c = 9$, which requires a Lorentz factor > 10 and an angle between the jet axis and the line of sight $< 12°$. Both C4 and C3 are systematically farther from the "core" at higher frequencies.

The core and C4 have flat spectra which fall off sharply below 5 GHz, while the spectra of C3 and C2 are steep. The flux decay of C3 is too slow to be explained as adiabatic expansion, and too rapid to be explained by synchrotron radiation losses.

G. Swarup and V. K. Kapahi (eds.), Quasars, 157–158.

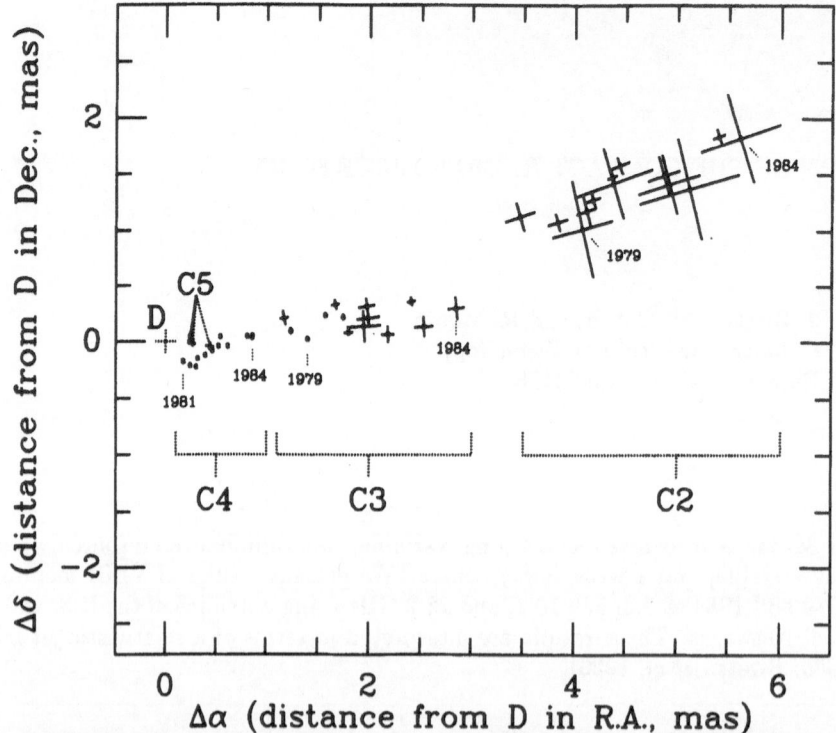

Synchrotron-Compton models for the superluminal components require minimum relativistic Doppler factors of 7, 3, and 5 for C4, C3, and C2, respectively, to explain the weak X-ray flux. The energy density of the electrons is greater than that of the magnetic field, unless the Doppler factor exceeds 15. The observed self-aborption turnovers for C4, C3, and C2 indicate that the magnetic field falls off approximately inversely with distance along the jet.

For component D we have applied a conical jet model with magnetic field $B \propto r^{-1}$ and electron density $n \propto r^{-2}$, where r is distance along the jet. This model requires the jet axis to be within 8° of the line of sight, and gives a Doppler factor of 7. It predicts frequency dependent shifts in the position of D which agree in both sign and magnitude with the shifts observed between D and C4, and D and C3.

REFERENCES

Biretta, J. A.: 1985, *Thesis*, California Institute of Technology.

Biretta, J. A. *et al.*: 1983, *Nature* **306**, 42.

Biretta, J. A., Moore, R. L., and Cohen, M. H.: 1986, *Astrophys. J.*, submitted.

Moore, R. L., Readhead, A. C. S., and Bååth, L. B.: 1983, *Nature* **306**, 44.

OBSERVATIONS OF 3C 273 WITH HIGH NORTH-SOUTH RESOLUTION

M. Cohen[1], J. Biretta[1], R. Schaal[2], G. Comoretto[3],
L. Bååth[4], and G. Nicholson[5]
[1] Caltech, Pasadena, CA 91125, USA
[2] Itapetinga Radio Observatory, INPE, S. J. dos Campos, SP, Brazil
[3] Instituto di Radioastronomia, via Irnerio 46, Bologna, Italy
[4] Onsala Space Observatory, Onsala, Sweden
[5] National Inst. Telecommunications Research, Johannesburg 2000, S.A.

The quasar 3C 273 lies at a declination of $+2°$. Maps made with northern hemisphere VLBI Networks clearly show superluminal motion, but the north-south resolution is poor and details cannot be seen. We have now made two observations at 10.7 GHz with the Itapetinga Radio Observatory in Brazil, and one at 5.0 GHz using the Hartebeesthoek Observatory in South Africa along with the European and U.S. VLBI Networks (Biretta et al. 1985).

A major new result is that the VLBI jet does not have a simple monotonic curvature. The 11-GHz maps show new superluminal components about 1 milliarcsecond from the core at position angles between $-140°$ and $-120°$. Comparison of these positions with those of components farther from the core shows that the VLBI jet first curves north and then south, and ultimately points to the optical jet.

Another result is that components near the core change position angle as they move out. This behavior is similar to that first seen in 3C 345, and occurs at about the same projected distance of 2 pc from the core ($H_0 = 100$ km s^{-1} Mpc^{-1}) (Biretta et al. 1983; Moore et al. 1983).

The observed curvature of the VLBI jet cannot be due to precession of the central engine, since simple models produce neither the non-monotonic curvature nor the position angle changes. The overall curvature to the south is probably caused by external pressures which bend the jet.

REFERENCES

Biretta, J. A. et al.: 1983, Nature **306**, 42.

Biretta, J. A. et al.: 1985, Astrophys. J. Letters **292**, L5.

Moore, R. L., Readhead, A. C. S., and Bååth, L. B.: 1983, Nature **306**, 44.

G. Swarup and V. K. Kapahi (eds.), Quasars, 159–160.

DISCUSSION

Burbidge : Have several groups studied the same superluminal sources and if so, do they agree ? Secondly, what happens when you plot the apparent superluminal velocity against redshift ?

Cohen : 3C 345 has been studied by several groups and the maps are in good agreement. In general, there is an anti-correlation between proper motion and redshift. For example, 3C 120 has the smallest redshift and the largest proper motion. The apparent transverse velocity, calculated for a co-moving observer using a standard cosmology, shows no particular correlation with redshift. In any one source these velocities can differ by a factor 2 or more for different components, and this might be enough to obscure any real redshift dependence.

Bramwell : What are the statistics of the correlation of the epoch of zero separation with known flux outbursts ?

Cohen : In 3C345 there are large outbursts which are correlated with C4 and C3, and perhaps C2. The peak flux density (> 10 GHz) occurs when the component is at a projected angle of 0".3 from the core, not at the extrapolated zero position. There is well-documented acceleration in C4. Other superluminal sources are presumably like 3C 345 and so the general lack of a good correlation between outbursts and extrapolated zero separation is not surprising.

SUPERLUMINAL MOTION IN THE QUASAR 3C 279

S. C. Unwin
Owens Valley Radio Observatory
California Institute of Technology   105-24
Pasadena, California 91125, USA

We present the results of a program of monitoring structure changes in
the radio core of the quasar 3C 279 (z = 0.538) using VLBI.   In all,
nine hybrid maps have been made from observations at frequencies of 5,
11, and 22 GHz, during the period 1981 - 1985. The VLBI structure shows
many similarities with other 'superluminal' radio sources, e.g. 3C 273
(Unwin et al. 1985), such as alignment with a kiloparsec-scale jet, a
one-sided spectral distribution, and superluminal motion.   The proper
motion measured from our observations is v/c $\approx$ 2, only one quarter of
the previously published rate.

The observations consisted of full tracks using at least 5 VLBI
stations, and we made good maps from every session. These maps show the
source to be basically a double, and resolve the core of the
kiloparsec-scale jet seen by the VLA (de Pater and Perley 1983) into two
components B1 and B2 which define an axis in P.A. -130°, within 10° of
the large-scale jet. For one 5-GHz epoch we used 9 stations, including
Hartebeesthoek (South Africa). The long baselines to this southern
telescope greatly improved the north-south resolution, which is usually
very poor for sources at such low declination (-5°). Our map, made with
a 0.9 x 2 milliarcsec beam, shows for the first time that the VLBI
structure is very narrow perpendicular to the jet axis.

Comparisons between maps made at the three frequencies show that
the NE component (B1) is the 'core' of 3C 279, as it has a higher
turnover frequency than the SW component (B2), which is probably a
'knot' moving with the VLBI jet. At 5 GHz, weak emission with a
still-lower turnover is seen beyond B2. As with other superluminal
sources, the spectral distribution defines the sense of the jet to be
toward the outer jet, and it suggests a continuous connection between
the different scales.

Slow proper motion is seen between B1 and B2 at all three
frequencies. To minimize systematic effects, we model-fitted the
visibility data using a pair of circular gaussians to derive the

G. Swarup and V. K. Kapahi (eds.), Quasars, 161–162.

separations. The results are plotted in Figure 1, and the straight line represents an expansion rate of $v/c = 2.3 \, \underline{h}^{-1}$ (for Hubble constant $H_o = 100 \, \underline{h}$ km/s/Mpc), only one quarter of the rate reported by Cotton et al. (1979) for early 1970s data. We believe that both measurements are correct and that the rate has changed. However, this does not imply deceleration, as the 'jet' component followed in the early data is not B2, and is not detected in our observations.

ACKNOWLEDGEMENTS

Many people have contributed to the observation and reduction of data presented here, especially M.H. Cohen, J.A. Biretta, G. Comoretto, T.J. Pearson, and K.R. Lind. The VLBI observations were organized through the US VLBI Network Users Group. VLBI research at Caltech is supported by the National Science Foundation via grant AST 82-10259 to the Owens Valley Radio Observatory.

REFERENCES

Cotton, W.D., et al.: 1979, Astrophys. J. Letters 229, pp. L115-117.
de Pater, I., and Perley, R.A.: 1983, Astrophys. J. 273, pp. 64-69.
Unwin, S.C., et al.: 1985, Astrophys. J. 289, pp. 109-119.

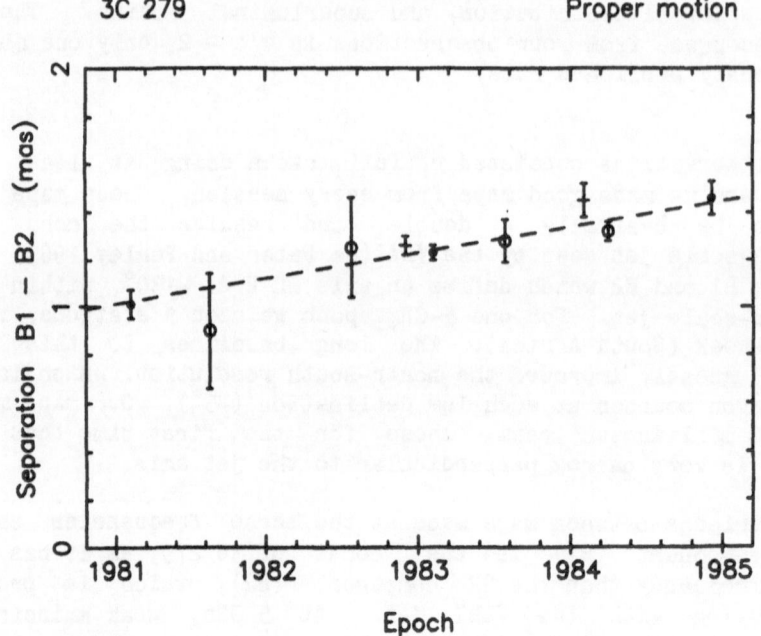

Figure 1. Separation of components B1 and B2 determined by model-fitting, vs. observing epoch. Error bars of $\pm \sigma$ represent the formal uncertainty in the model fit. (Open circles - 5 GHz; filled circles - 11 GHz; crosses - 22 GHz). The fitted straight line has a slope of 0.11 milliarcsec/yr.

NEW SUPERLUMINAL QUASARS

T. J. Pearson, P. D. Barthel, A. C. S. Readhead, and C. R. Lawrence
Owens Valley Radio Observatory
California Institute of Technology
Pasadena, California 91125
USA

Since 1977 we have been engaged on a project to survey the milliarcsecond structure and internal motions of a complete, unbiased sample of 65 radio sources (Pearson and Readhead 1984). One of the goals of the project was to discover more superluminal sources and find out how common they are. The sample contains three established superluminal sources (3C 179, 3C 345, and BL Lacertae); so far we have discovered three new superluminal quasars (0850+581, 1642+690, and 1928+738), and a fourth (3C 216) that may be superluminal. Thus at least six out of the 65 sources are superluminal, and we expect to find more as our observations proceed.

The complete sample was selected from the NRAO–MPIfR 5-GHz Strong Source Surveys S4 and S5 by the following criteria: (a) declination (1950) > 35°, for good u, v-coverage with existing telescopes; (b) galactic latitude $|b| \geq 10°$; (c) total flux density at 5 GHz ≥ 1.3 Jy (at the epoch of the original surveys). Twenty of the objects in the sample are classical double radio sources, while the remaining 45 are compact, core-dominated sources with $\gtrsim 95\%$ of their flux density arising in a region smaller than about 1″. We now have first-epoch VLBI maps of 41 of the core-dominated sources in the sample and second-epoch maps of 20.

We are in the process of obtaining high-quality optical spectra of the entire sample of 65 sources using the double spectrograph on the Palomar 200-inch telescope. These observations have already provided a number of new redshifts (Lawrence et al. 1986), including that of 1642+690.

It is clear from the VLBI observations that not all compact radio sources are superluminal. In particular, there is a class of "compact double" sources that are identified with galaxies or empty fields, show little flux-density variation, and have low polarization (< 1%); none of these has shown large structural changes. On the other hand, at least half the sample consists of unresolved objects and asymmetric "core-jet" sources that are highly polarized and highly variable in both total flux density and structure. This class includes most of the quasars and BL Lac objects. The six known superluminal sources all have a "core-jet" structure, and it is quite possible that all of the objects of this type will turn out to be superluminal when they have been studied sufficiently well.

0850+581, $z = 1.322$: We have maps of 0850+581 at only one frequency, but they show an asymmetric structure that can be described as a "core-jet," with the mil-

G. Swarup and V. K. Kapahi (eds.), Quasars, 163–164.

liarcsecond jet aligned with a curved, large-scale jet, 10″ long. The change in the structure is very suggestive of superluminal motion along the jet, with $v/c = (4.6 \pm 2.3)h^{-1}$, and further observations are in progress to confirm the result.

0906+430 (3C 216), $z = 0.669$: 3C 216 is unlike the other quasars presented here in that the VLBI observations show a milliarcsecond jet aligned roughly perpendicular to the major axis of the large-scale emission; the only comparable source is the quasar 3C 309.1. Both 3C 216 and 3C 309.1 are steep-spectrum compact sources. The change in structure is ambiguous: it could be due to a weak component in the "jet" having brightened in the 4.5-yr period between the two observations; but a plausible explanation is that the change is due to the motion of a component along the jet, with $v/c = (11.5 \pm 2.0)h^{-1}$.

1642+690, $z = 0.751$ (Pearson *et al.* 1986): The radio source 1642+690 has a dominant, compact core with a jet extending 5″ to the south. The VLBI maps of the core show that the milliarcsecond structure is similar to that of the superluminal sources 3C 273 and 3C 345. A milliarcsecond jet is closely aligned with the arcsecond jet, and a "knot" in the jet is moving outwards along the jet at 0.34 mas/yr, an apparent transverse velocity of $v/c = 9.3h^{-1}$.

1928+738, $z = 0.302$ (Eckart *et al.* 1985): The VLBI maps of 1928+738 show a self-absorbed "core" with spectral index $\alpha \approx 0.0$ and a long "jet" extending 17 mas from the core in position angle 165° ($50h^{-1}$ pc). The jet has a spectral index $\alpha \approx -0.5$ and contains at least nine components, five of which were detected at both epochs. The most plausible alignment of the maps from the two epochs indicates that all five of these components have a proper motion relative to the core of ~ 0.6 mas/year, or apparent transverse velocities $v/c \sim 7.5h^{-1}$.

REFERENCES

Eckart, A., Witzel, A., Biermann, P., Pearson, T. J., Readhead, A. C. S., and Johnston, K. J.: 1985, *Astrophys. J. Letters* **296**, L23–L26.

Lawrence, C. R., Pearson, T. J., Readhead, A. C. S., and Unwin, S. C.: 1986, *Astron. J.* **91**, in press.

Pearson, T. J., and Readhead, A. C. S.: 1984, *IAU Symp.* **110**, 15–24.

Pearson, T. J., Barthel, P. D., Lawrence, C. R., and Readhead, A. C. S.: 1986, *Astrophys. J. Letters* **300**, L25–L29.

THE NUCLEAR JETS IN 3C309.1 & 3C380

P.N. Wilkinson   NRAL, Jodrell Bank U.K.
A.J. Kus   T.R.A.O., Torun, Poland
T.J. Pearson   OVRO, Pasadena, U.S.A.
A.C.S.Readhead   OVRO, Pasadena, U.S.A.
T.J. Cornwell   NRAO, Socorro, U.S.A.

ABSTRACT. 8 station VLBI maps of the nuclear jets in 3C309.1 and 3C380 reveal complex, bent, structures which seem hard to reconcile with the concept of ballistic outflow. Together with the observation of "flaring" in the MERLIN/EVN map of the 3C309.1 jet the maps point strongly to the jets being confined fluid flows.

1. THE MAPS

1.1 3C309.1 (z = 0.911)

Fig.1 shows two maps; the top map was made from 8-station European-U.S. VLBI at $\lambda 18$ cm. The restoring beam is 3 mas and the bottom contour is drawn at 0.2% of the peak brightness of 533 mJy/beam. The scale marks 50 mas ($\sim 200h^{-1}$ pc for $H_0 = 100h$ km s^{-1} Mpc^{-1}). The bottom map was made from MERLIN and EVN data and is also at $\lambda 18$ cm. The restoring beam is 50 mas and the bottom contour is drawn at 0.1% of the peak brightness of 2.95 Jy/beam. The scale marks 500 mas ($\sim 2h^{-1}$ kpc).

1.2 3C380 (z = 0.692)

Fig.2 shows an 8 station U.S.-European VLBI map at $\lambda 6$ cm. The restoring beam is 1 mas and the bottom contour is drawn at 0.2% of the peak brightness of 521 mJy/beam. The scale marks 10 mas (~ 40 h^{-1} pc)

2. INTERPRETATION

Both these quasars have overall steep radio spectra but while the over-all structure of 3C309.1 (e.g. Cornwell and Wilkinson, M.N.R.A.S., 196, 1067, 1981) resembles that of many, more extended, sources mapped with the VLA that of 3C380 is very distorted (see Wilkinson et al. Nature, 308, p619 (1984)); nevertheless their nuclear jets appear qualitatively similar. Thus whatever causes the distortions in 3C380 must be situated well away ($\gtrsim$ 100 pc) from the nucleus. The most plausible agent for

G. Swarup and V. K. Kapahi (eds.), Quasars, 165–166.
© 1986 by the IAU.

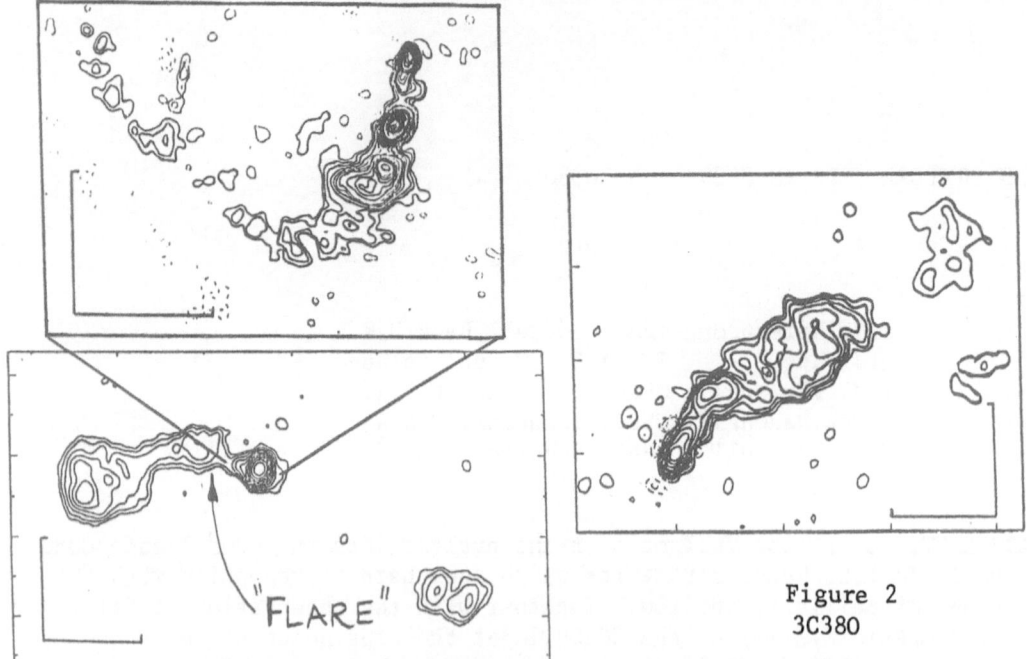

"FLARE"

Figure 2
3C380

Figure 1   3C 309.1

causing the distortions in 3C380 remains dense gas.

   The form of the one-sided nuclear jet in 3C309.1, which is revealed
in its entirety in Fig.1, can not be explained as the locus of ballis-
tically moving "plasmons". The wiggles, the sharp changes in brightness
seen on the 10s of pc scale and the fact that the jet apparently flares
~ 1 h^{-1} kpc (projected) away from the nucleus (bottom map in Fig.1) are
much more likely to be the result of various Kelvin-Helmholtz instabilities in a confined fluid flow coupled with changes in the pressure
exerted by the confining medium.

   The nuclear jets in both cases appear to be very one-sided. In
the case of 3C309.1 we have maps at other frequencies and can unambiguously locate the, flat-spectrum, core as lying at the northernmost
tip. The brightness of any counter-jet more than a few mas away from
this nucleus is $\leqslant 0.04$ of the brightest knot in the visible jet. So far
we have no clear-cut limits on the bulk flow velocity but standard
inverse-Compton/X-ray flux calculations indicate that the Doppler factor
for the strongest knot must be a relatively modest (compared with that
in known superluminals) 3.2. This fits in with motion at $\gamma=2$ (favoured
by Scheuer and Readhead, Nature, 277, 182, 1979, for the nuclei of
classical doubles) at 14° to the line-of-sight. The current evidence,
then, points to 3C309.1 as being a double source whose axis is quite
close to the line of sight. The question as to whether the jet is
intrinsically one-sided or not remains open.

SEARCH FOR SUPERLUMINAL MOTION IN THE WEAK CORES OF EXTENDED QUASARS

J. A. Zensus
California Institute of Technology, Pasadena, U.S.A.

R. W. Porcas
Max-Planck-Institut für Radioastronomie, Bonn, F.R.G.

ABSTRACT. We present a status report on a project to study structural changes in a sample of radio cores associated with quasars with extended double lobes. Of six objects observed, so far two (3C 179 and 3C 263) are superluminal and two more (3C 268.4 and 1951+49) are possible superluminal candidates.

The search for superluminal motion in the weak cores of double radio sources provides a critical test of simple beaming models commonly used to explain the statistics of superluminal sources (Scheuer and Readhead, 1979; Blandford and Königl, 1979; Orr and Browne, 1982). It follows from the requirement of a narrow relativistic jet oriented close to the line-of-sight to the observer that the superluminal effect should be rare in a complete sample of sources selected on the basis of the strength of the extended emission, i.e. without orientation bias due to the central components. Such a sample of sources is provided by the 30 quasars from the Jodrell Bank 966 MHz survey that were mapped with arcsecond resolution by Owen et al. 1978 ($S_{966} \geq 0.7\,$Jy, $m_b \leq 19\,$mag, and angular size $> 10\,$arcsec). We have continued our project to study the milli-arcsecond structures and changes thereof for six cores in this sample with VLBI at $\lambda\,2.8$cm: 3C 179, 3C 204, 3C 263, 3C 268.4, 1732+65, 1951+49 (see Zensus and Porcas, 1984, for a list of source properties). These cores range in flux density between 50 and 300 mJy and are therefore ca. 100 times weaker than the core-dominated superluminal sources. The measurement of superluminal expansion in 3C 179 with a speed of ca. $10c$ ($H_0 = 55\,$km s^{-1} Mpc^{-1} and $q_0 = 0.5$) is well established and reported elsewhere (Porcas, 1981, and Porcas, *these proceedings*). Here we concentrate on the results for the other sources, except 3C 204, which was not detected in our observations. Fig. 1a–c show hybrid maps for 3C 263, 3C 268.4, and 1951+49 (a model fit to the data of 1732+65 shows a compact core, possibly double, extended roughly along the overall source axis). The structural properties of these objects can be summarized as follows:

    a. These cores have simple double structures accounting for large fractions of the total core fluxes and similar to the structures found in core-dominated objects. 3C 179 can be considered a prototype for this class of object.

    b. Despite the lack of spectral information for the components their morphology is suggestive of a "core-jet" structure.

    c. An "orientation memory" is found between features on pc- and kpc-scales. Bending of the structures seems to occur not within the compact regions but rather through interaction of the jets with the external medium.

G. Swarup and V. K. Kapahi (eds.), Quasars, 167–168.
© *1986 by the IAU.*

STRUCTURAL CHANGES: All quasars repeatedly observed show structural changes. From 3 observing epochs for 3C 263 superluminal expansion at a speed of ca. $3c$ was derived (Zensus, Hough and Porcas, 1986). This is the weakest superluminal source found so far ($S_{\lambda 2.8} \sim 150$ mJy). Preliminary results from second epoch data for 3C 268.4 and 1951+49 suggest that both are possible superluminal candidates, too.

CONCLUSION: Our results indicate that superluminal motion is a common phenomenon among the weak cores of extended quasars, in contradiction to expectations based on simple versions of the beaming hypothesis.

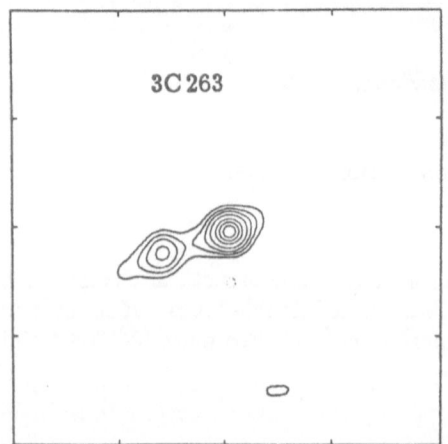

FIGURE 1 :

VLBI-maps ($\lambda 2.8$ cm) of cores in 3C 263, 3C 268.4, and 1951+49 (scale 1 milliarcsecond per tick; size of restoring beam 0.2 x 0.2 m.a.s.).

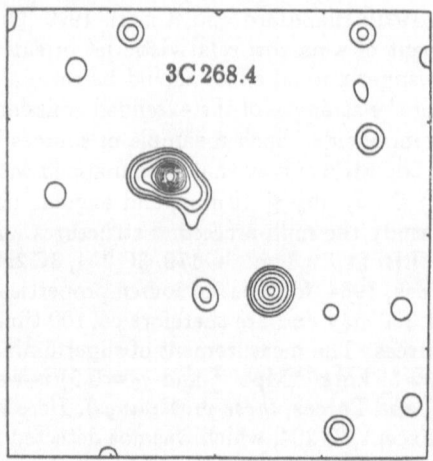

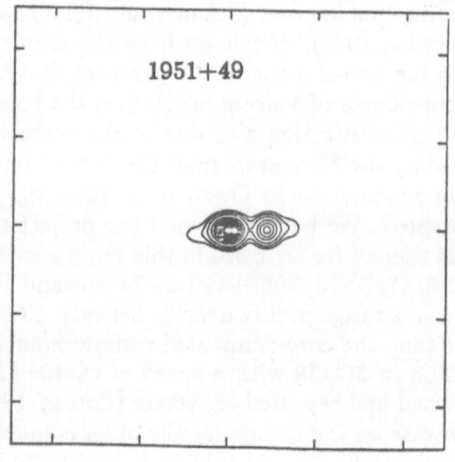

REFERENCES

Blandford, R. D. and Königl, A.: 1979, *Astrophys. J.* **232**, 34
Orr, M. J. L., Browne, I. W. A.: 1982, *Monthly Not. Roy. Astron. Soc.* **200**, 1067
Owen, F. N., Porcas, R. W., Neff, S. G.: 1978, *Astron. J.* **83**, 1009
Porcas, R. W.: 1981, *Nature* **294**, 47
Scheuer, P. A. G. and Readhead, A. C. S.: 1979, *Nature* **277**, 182
Zensus, J. A. and Porcas R. W.: 1984, proc. *IAU Symp.* **110**, 163
Zensus, J. A., Hough, D. H., Porcas, R. W.: 1986, *in preparation*

HIGH RESOLUTION VLBI OBSERVATIONS OF THE NUCLEI OF BRIGHT GALAXIES AND QUASARS AT 327 MHz

S.Ananthakrishnan and V.K. Kulkarni
Radio Astronomy Centre(TIFR), P.O.Box 8,
Uthagamandalam 643001, India

INTRODUCTION

An intercontinental VLBI experiment at 327 MHz involving the Ooty Radio Telescope and telescopes at Jodrell Bank, Westerbork, Torun and Crimea was carried out during December 8-12, 1983. The purpose of the experiment was to (i) establish the feasibility of performing MKII VLBI observations from India and (ii) to study the low frequency structure of nearby galaxies, quasars and some other well known radio sources at metre wavelengths. The sources observed were NGC 262, 315, 1052, 1068, 1265, 1275, 4151, 4486, 7674 and 3C 120; PKS 1055+018, 1148-001 and CTA 102; 3C 237 and SS 433. The bright NGC galaxies chosen were selected from an Interplanetary Scintillation survey of 150 nearby galaxies (Bagri and Ananthakrishnan 1983); they were found to have a flux density greater than about 0.2 Jy in subarcsec components at 327 MHz and have high frequency VLBI observations. The quasars chosen are amongst the most compact sources known at metrewavelengths. The other objects were the well known sources which have been studied extensively at higher frequencies.

OBSERVATIONS AND ANALYSIS

A total of 40 hours of simultaneous observations using the above stations were made. Since this was a pilot survey, each source was observed only for a few hours. The recording of the data was done at each of the stations using a MKII VLBI terminal with a bandwidth of 2.0 MHz (326.0-328.0 MHz). The processing and analysis of the data were done at the MPI fur Radioastronomie, Bonn, FRG, using the MKII VLBI processor.

RESULTS

Table 1 gives the name of the source, correlated flux density for the shortest and longest baselines (corresponding to about 0.5×10^6 and $7\times10^6 \lambda$)and for some of the sources, angular sizes derived from model fitting. It is seen from Table 1 that most of the nearby bright galaxies are found to be resolved on the longest baselines. Based on this pilot survey, another VLBI experiment is planned at 327.0 MHz with a view to making detailed maps of some of the above sources for studying the low frequency structure.

G. Swarup and V. K. Kapahi (eds.), Quasars, 169–170.

Reference

Bagri,D.S. and Ananthakrishnan,S., Proc. of IAU Symp.110 on 'VLBI and Compact Radio Sources', 261p.

Table 1

Source Name	Correlated flux density in Jy		Largest angular size in mas
	Largest baseline	Shortest baseline	
NGC 262	<0.1	0.4	-
315	<0.1	0.3	-
1052	<0.1	0.5	-
1068	~0.1	1.2	-
1265	<0.1	0.15	-
1275	2.0	7.5	30^1
4151	<0.1	0.25	400^2
4486	<0.2	1.3	-
7674	<0.1	0.8	-
3C 120	0.4	2.7	160^3
1055+018	1.5	3.1	15^4
1148-001	1.6	2.9	12^4
CTA 102	2.2	5.5	30^5
3C 237	<0.1	20.0	1200^6
SS 433	<0.1	0.5	860^4

Notes

1 Hybrid map being made.
2 Double; separation 400 mas; each component of size 100 mas; Ratio of Flux densities 2:1.
3 Double; separation 160 mas; component sizes 9 and 90 mas with flux density ratio 1:5.
4 Single component size.
5 Double; separation 30 mas; component sizes 12 and 18 mas with flux density ratio 3:2.
6 Double; separation 1200 mas; each component of size 250 mas; flux density ratio 1:1.

MONITORING OF BL LAC OBJECTS AT 327 MHz

Hari Om Vats and S.S.Degaonkar
Physical Research Laboratory, Ahmedabad-380009, India

P.K.Manoharan and S.Ananthakrishnan
Radio Astronomy Centre (TIFR), Post Box No. 8,
Uthagamandalam-643001, India

ABSTRACT. The variable BL Lac objects OJ 287 and B2 (1308+326) have
been observed using the Ooty Synthesis Radio Telescope (OSRT) at
327 MHz. Preliminary analysis of one of them, namely OJ 287 indicates
that flux variations of ∼ 20% on a time scale of few hours to few
days could be present, but this needs to be confirmed by a larger
data base.

OBSERVATIONS AND RESULTS

The observations were made during 11-16 March 1985. The daily observing session lasted for 4 to 5 hours around local transit of each
source. Every half an hour, the flux density of OJ 287 was calibrated
using 3C 210 (S_{327} = 7.27 Jy). On two days, a control source (0852+12)
of comparable flux density (S_{327} = 1.50 Jy) was also monitored
every half an hour in addition to the calibration source.
    The data were analyzed during standard OSRT analysis package.
Fig. 1 shows the observed flux density of OJ 287 during the observing period of 11-16 March. The dotted line through the data points
is the best eye fit and the bars are ± 1σ variation of the control
source. The flux density variations are prominent on 11, 13 and
14 March. The time scale of the variations on individual days
is 2 - 3 hours. Fig. 2 shows the averaged flux from Fig. 1 plotted
as a function of date. The ± 1σ error bars are derived from the
standard deviation of all the data points on individual days. It is
seen that the flux density of the source on March 11 is significantly above that on all the other days. If this is not excluded
variations in the flux density of OJ 287 of the order of 20% over
time scales of 5 - 10 days seem present in the data. It is clear
that systematic observations over longer periods are required to
confirm these variations.

G. Swarup and V. K. Kapahi (eds.), Quasars, 171–172.

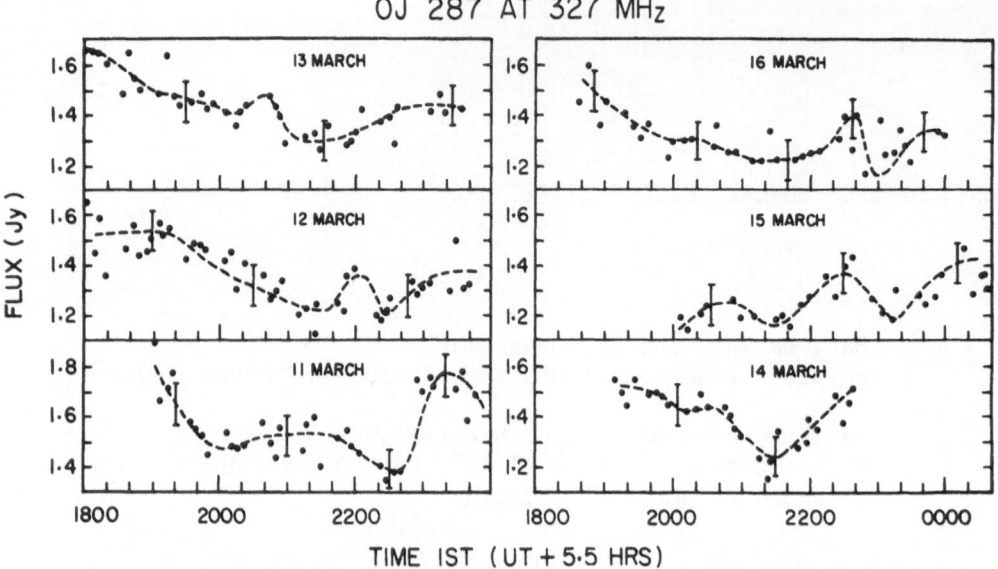

Figure 1. Illustrating the flux variation of OJ 287 with the time for six days (March 11 - 16, 1985)

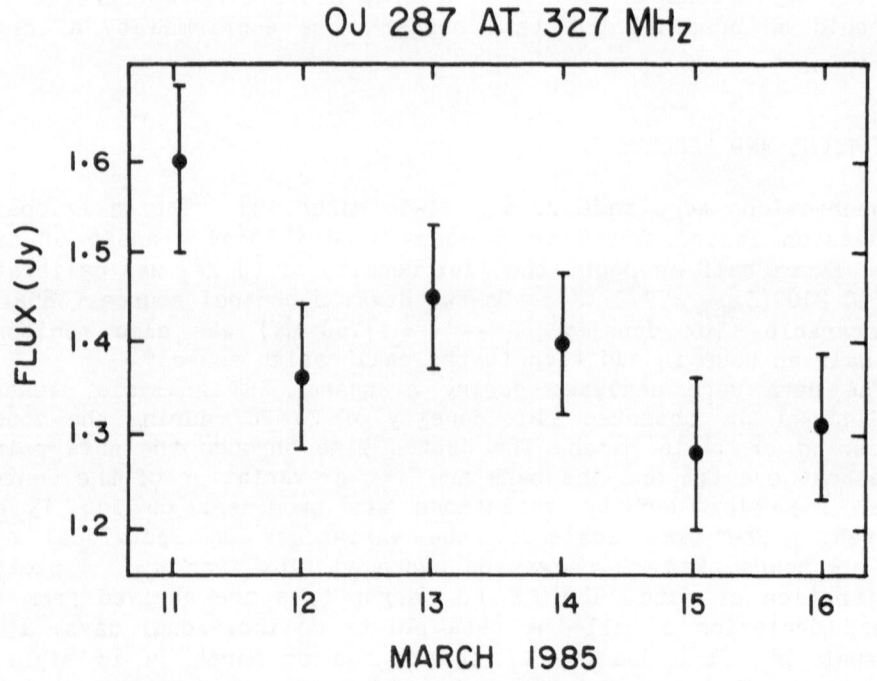

Figure 2. Illustrating the variation of flux (averaged during the daily observing sessions) as a function of day from March 11 - 16, 1985.

THE EXTENDED RADIO STRUCTURE OF QUASARS

Frazer N. Owen
National Radio Astronomy Observatory
P.O. Box O
Socorro, NM 87801
USA

ABSTRACT. Modern radio maps usually allow quasars to be recognized from their radio morphology alone. Most have strong central components, double lobed outer structure and one-sided jets connecting the inner and outer structures. The physics of the sources is poorly understood. The observed bending of the jets, the high minimum pressures observed, and the required energy supply to the lobes are major problems. However, the outstanding problem regarding the extended structure is whether or not this morphology is produced by special relativistic effects or the intrinsic activity level and physics of the radio sources.

1. Introduction, Basic Properties

A decade ago our knowledge of the extended structure of quasars was limited to the fact that they generally resembled radio galaxies. We knew that resolved, high luminosity radio sources, whether identified with galaxies or quasars, most often were double or triple sources centered on the optical identification. Today with a new generation of radio interferometers and new image processing techniques we have a much improved picture and one can, in most cases recognize the difference between a radio galaxy and a quasar from the radio map alone. The two outstanding differences between radio galaxies and quasars with strong radio emission are 1) the strength of the radio jet and 2) the strength of the core component. Starting with the discovery of the radio jet in 4C 32.69 (Potash and Wardle, 1980), it has become increasingly clear that many, probably most or all, radio quasars possess a one-sided radio jet connecting the optical quasar with the extended lobe emission. Often these jets are knotty and show pronounced bends and wiggles, sometimes of 90° or more. Roughly 100 quasar jets have been found to date and none are known which are clearly two-sided. In figure 1 several examples of quasars with the characteristic one-sided structure are shown.

Second, almost all radio quasars with integrated flux densities greater than, say, 1 Jy at 1 GHz, have central components coincident with the optical quasar stronger than 10 mJy (*e.g.* Owen and Puschell, 1984). This statement usually refers to flux densities of the cores at 5 GHz, although most of the central components which have been observed at multiple frequencies appear to have flat radio spectral indices.

G. Swarup and V. K. Kapahi (eds.), Quasars, 173–179.

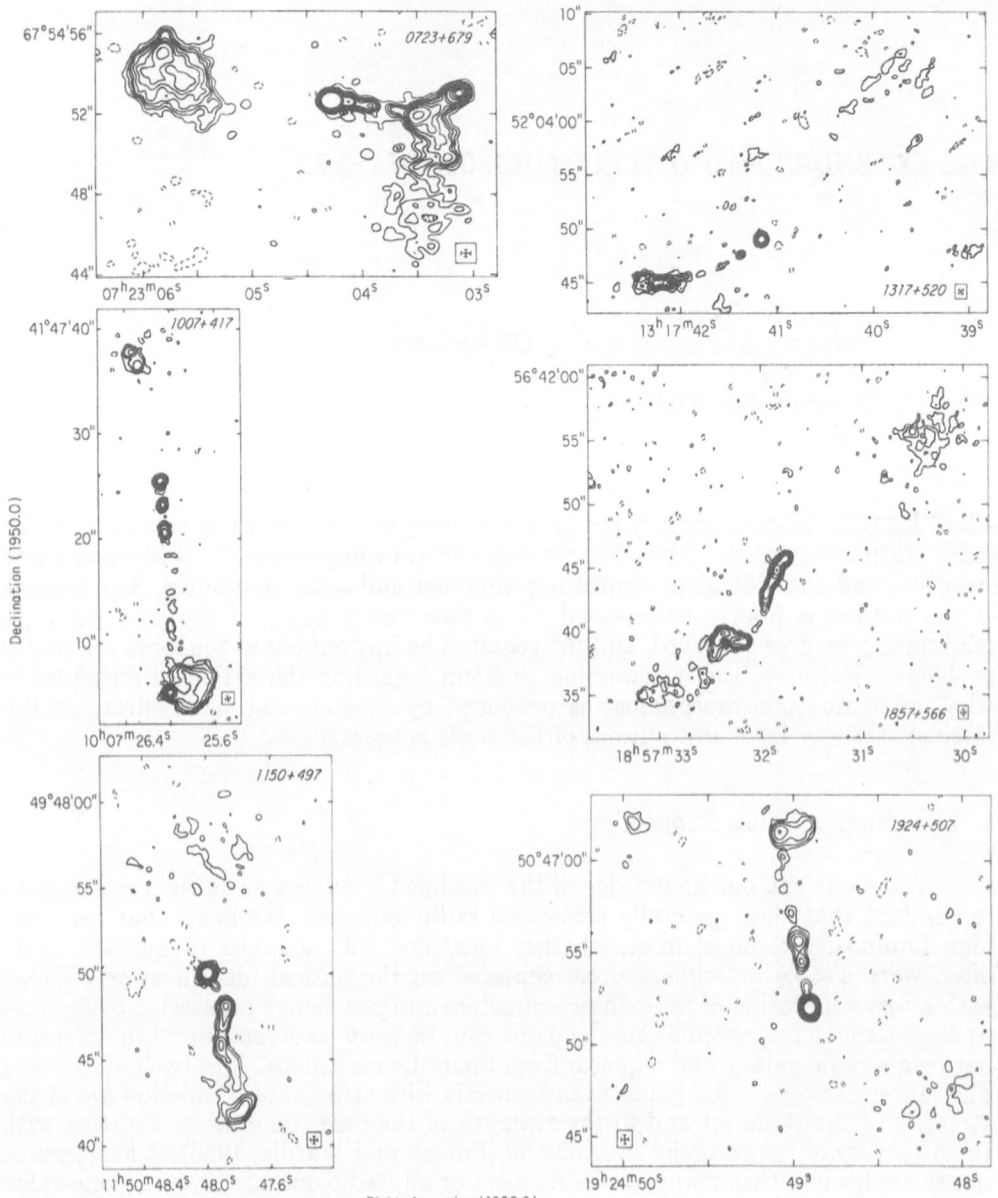

Figure 1. Examples of Radio Quasars with Jets (see Owen and Puschell, 1984).

Powerful radio galaxies also appear to have jets (usually one-sided) and central components; however, their strength, at least relative to the radio lobes, is weaker. Only a few FRII class radio galaxies (Fanaroff and Riley, 1974) have been observed to possess radio jets (*e.g.* Perley *et al.* , 1984; Linfield and Perley, 1984) and the detection of these cases has taken long integrations with the VLA and special image processing techniques. Nonetheless when the observations have been made the jets, also one-sided, have been found. Also the central components are weaker. For high redshift radio galaxies with similar or higher luminosities to the quasars mentioned above the central components are almost always weaker than 10 mJy, and typically for 3CR sources with redshifts greater than 0.3 are about 1 mJy at 5 GHz (Owen *et al.* 1981, Owen and Laing, in preparation).

Thus at the 1 Jy level a powerful radio galaxy can usually be distinguished from a quasar by noting the strength of the central component. A strong, one-sided jet is a further tipoff. The structure of the lobes, of course, is a good indicator of the general range of radio luminosity (Fanaroff and Riley, 1974).

2. Physical Conditions

The understanding of the physics of the extended structures of quasars is dominated at present by the question of whether or not relativistic motion dominates the observed properties of the systems. This question will be discussed in the next section. Otherwise quasars are usually understood within the context of the beam model which can be studied, although imperfectly, in nearerby radio galaxies where we can observe the radio structure at higher linear resolution and have a better understanding of the external conditions. Nonetheless, a number of physical problems will probably, eventually be best attacked in quasars due to their apparently extreme conditions. Two such problems are 1) the combination of the energy supply and with the bending of quasar jets and 2) the confinement of the emitting structures. Potash and Wardle (1980) discussed the bending and energy supply question for 4C32.69 and show the apparent contradiction with the standard assumption of the kinetic energy in the flow being converted into radio luminosity. Barthel (1984) has considered this question for high redshift quasars and considered deflections by dense, cool clouds as the origin of the bending. Others (*e.g.* Gower *et al.* , 1982) have been able to explain some structures by ballistic precession models. However, if the jets act like a fluid as is expected in most modern jet models, a different formalism needs to be used (Hardee ,1984) which has not been applied to many cases. In particular, the quantity, $\omega R/u$ is conserved, where ω is the wave frequency, R is the jet radius, and u is the jet velocity (Hardee, 1986).

The second major problem is that the minimum pressure implied by the brightness of the extended structure (as high as $10^{-6} dynes - cm^{-2}$) appears to exceed any reasonable external pressure (*e.g.* Potash and Wardle, 1980; Owen and Puschell, 1984; Barthel, 1984). This situation also appears to be true in more nearby sources such as M87 (Biretta et al, 1983). However, due to their distance, more possible explanations exist for quasars:

1) The pressure may really be higher than we think near some quasars.

2) The jets could be made up of plasmoids in a channel which are each dynamically confined (*e.g.* Christiansen *et al.* , 1977).

3) The jets could be magnetically confined as has been suggested for M87.

4) The jets could simply have a low filling factor (like the broad line region) with high density regions made up of transients locally out of pressure equilibrium (*e.g.* plasma waves) but with large scale pressure equilibrium maintained. (Borovsky, Taos Jet Meeting 1985). Since $P_{min} \propto \phi^{3/7}$, where ϕ is the filling factor, the overall pressure required goes down.

3. Relativistic Motion, Unified Pictures

The most important question about the extended structure in quasars is whether or not the observational results can be explained by relativistic motion in the jets. Two general possibilities can be considered:

1) Jets look one-sided due to doppler beaming (*e.g.* Orr and Browne, 1982).

2) Quasar Jets are intrinsically one-sided (*e.g.* Wardle and Potash, 1985).

If possibility 1) is considered, at least two subclasses of the model can be considered.

a) All quasars are the same and radio quiet quasars are simply beamed away from us (Scheuer and Readhead, 1979).

b) Radio quasars are radio galaxies with their jets pointed close to our line of sight. The cases with jets near the plane of the sky are called radio galaxies and the nuclei in other wavelength bands (*e.g.* optical) are hidden from us by absorption (Laing, Gull, private communication).

Case 2), intrinsically one-sided jets, also has at least two subclasses.

a) The jets flip-flop back and forth between the two radio lobes (*e.g.* Rudnick and Edgar, 1984).

b) The jet on one side is invisible either because it simply has fewer losses or because its energy is carried by different particles (*e.g.* electrons on one-side, protons on the other).

To attack this problem we have three equations describing the relativistic motion which need to be statisically fit by the data.

For $\beta = v/c$ and $\theta = $ the angle between the jet and the line of sight,

1) The intensity of a continuous jet is enhanced by relativistic motion by

$$S_{obs} = S_0 \times (\gamma(1 - \beta \cos \theta))^{-(2+\alpha)}$$

where $S \propto \nu^{-\alpha}$.

For small θ ,

$$S_{obs} \sim \gamma^{2+\alpha} S_0.$$

2) The value of β from the ratio, R, between the intensity of the approaching and receding sides of a continuous, equal, two-sided jet is

$$\beta = \frac{R^\delta - 1}{\cos\theta(R^\delta + 1)}$$

where $\delta = 1/(2 + \alpha)$.

3) The apparent superluminal motion of a a source is given by

$$v_{app} = (v_j \sin\theta)/(1 - \beta\cos\theta).$$

For $\theta \sim 1/\gamma$, $v_{app} \sim \gamma v_j$.

These three relationships assume isotropic radiation in the frame of the jet and cand be modified somewhat by including a more realistic model (e.g. radiation from shocks, Lind and Blandford, 1985). Relationship 1) must be satisfied statisically given the detailed assumptions of a model. Relation 2) for relativisic motion sets a limit on θ which must also agree statistically with the model. Relation 3) gives the allowed distribution of γ and θ. Sorting out these relationships for any model requires complete data sets and detailed arcsecond and milliacrsecond observations. Such information is not yet available for any large sample. However, some clues are starting to appear as to the answer. (Also see Bridle and Perley for a more lengthy discussion of the observational constraints on this subject.)

Rudnick et al. (1984) have shown that for a few radio quiet (or more correctly radio weak) quasars, the radio emission is extended and resembles observations of nearerby Seyferts. Recent work has also shown that most radio quasars with strong cores possess quite extended radio structure. These sources resemble generally the structure and extended luminosity of radio quasars and galaxies with weaker cores (Browne et al. , 1982; Schilizzi and DeBruyn, 1983; Murphy, Browne and Perley, in preparation). These two pieces of information seems to suggest two classes of parent objects with radio quiet quasars resembling active spiral galaxies and radio loud quasars resembles radio active ellipitical galaxies. Thus case 1a) seems unlikely.

Second, the number of distant radio galaxies and the number of quasars in the 3CR are roughly equal. This implies that under 1b) many of the quasars must be as much as 60 degrees from the line of sight.

Third, some correlations exist for and against the beaming hypothesis. Strong cores correlate with one-sided jet structures (Saikia, 1984; Owen and Puschell, 1984). But exceptions exist such as 1857+566 which has a strong, very distorted jet and a very weak central component (Saikia et al. , 1983; Owen and Puschell, 1984). The orientation of the position angle of the core polarization seems to be generally consistent with the expectations of the amplification of small intrinsic misalignments by projection effects near the line of sight (Saikia and Shastri, 1984). However, the projected linear size distributions for sources with and without jets are not significantly different (Saikia, 1984).

Fourth, many several quasars or radio galaxies with observed superluminal motion have been found to have large, extended radio structures, often approaching the size of the largest known radio quasars and typical large radio galaxies (*e.g.* Schilizzi and deBruyn, 1983; Walker, in preparation for 3C120). Also some of the quasars at the upper end of the largest angular size – redshift diagram are being found to have strong central components and one-sided radio jets (Neff and Brown, 1984; Barthel, in preparation; Hintzen and Owen, in preparation).

With these results and others in this volume it is probably still too early to decide the place of relativistic beaming in radio quasars. More careful statistical studies of quasars currently underway have a good chance of settling the question or at least ruling out some models. The next five years should produce very interesting results.

References

Barthel, P. D. (1984). Ph. D. Thesis, Leiden University

Bridle, A. H. and Perley, R. A. (1984). *Ann. Rev. Astron. Astrophys.*, 22, 319.

Biretta, J. A., Owen, F. N., and Hardee, P. E. (1983). *Ap. J. (Letters)*, 274, L27.

Browne, I. A. W., Clark, R. R., Moore, P. K., Muxlow, T. W. B., Wilkinson, P. N., Cohen, M. H., and Porcas, R. W. (1982). *Nature*, 299, 788.

Christiansen, W. A., Pacholczyk, A. G., and Scott, J. S. (1977). *Nature*, 266, 593.

Fanaroff, B. L., and Riley, J. M., (1974). *M.N.R.A.S.*, 101, 31p.

Gower, A. C., Gregory, P. C., Hutchings, J. B., and Unruh, W. G. (1982). *Ap. J.*, 262, 478.

Hardee, P. E. (1984). *Ap. J.*, 287, 523.

Hardee, P. E. (1986). submitted

Lind, K. R., and Blandford, R. D. (1985). *Ap. J.*, 295, 358.

Linfield, R., and Perley, R. A. (1984). *Ap. J.*, 279, 60.

Neff, S. G., and Brown, R. L. (1984). *A. J.*, 195, 195.

Orr, M. J. L. and Browne, I. W. A. (1982). *M.N.R.A.S.*, 200, 1067

Owen, F. N., and Puschell, J. J. (1984). *A. J.*, 89, 932.

Owen, F. N., Puschell, J. J., and Laing, R. A. (1981). *I.A.U. Symposium No. 97*, 435.

Perley, R. A., Dreher, J. W., and Cowan, J. J. (1984). *Ap. J. (Letters)*, 285, L35.

Potash, R. I., and Wardle, J. F. C. (1980). *Ap. J.*, 239, 42.

Rudnick, L., and Edgar, B. K. (1984). *Ap. J.*, 279, 74.

Rudnick, L., Sitko, M. L., and Stein, W. A. (1984). *A. J.*, 89, 753.

Saikia, D. J. (1984). *M.N.R.A.S.*, 208, 231.

Saikia, D. J. and Shastri, P. (1984). *M.N.R.A.S.*, 211, 47.

Saikia, D. J., Shastri, P., Cornwell, T. J., and Banhatti, D. G. (1983). *M.N.R.A.S.*, 203, 53P.

Scheuer, P. A. G., and Readhead, A. C. S. (1979). *Nature*, 277, 182.

Schilizzi, R. T., and deBruyn, A. G. (1983). *Nature*, 303, 26.

Wardle, J. F. C. and Potash, R. I. (1985). *Physics of Energy Transport in Extragalactic Radio Sources*, ed. Bridle and Eilek, (Green Bank, NRAO), p30.

DISCUSSION

Fabbiano : You mentioned that the difference between core component of quasars and radio galaxies might be explained with beaming or column densities. By looking at FRII 3CR radio galaxies and quasars we find two different classes of quasars. The double lobed 3CR quasars appear to be very similar (but brighter than 3CR galaxies when their core 5 GHz radio luminosities and their X-ray luminosities are compared. They follow the same correlation as the galaxies, suggesting that their cores are only more luminous specimens from radio galaxies cores. A substantial column density in radio galaxies (but not in quasars) would affect this correlation. The radio flat quasars are instead over luminous in radio, for a given X-ray luminosity. Beaming or an extra radio component might be present in these objects.

"Until further notice I offer a free beer for every new QSR having   z > 2.5   and a steep (cm-m) radio spectrum."

- Peter Barthel (p.182)

THE RADIO MORPHOLOGY OF QUASARS AT HIGH REDSHIFT

P. D. Barthel
Owens Valley Radio Observatory
California Institute of Technology
Pasadena, California 91125
USA

The mapping at kpc-scale resolution of the radio sources associated with quasars is fascinating, since it provides us with *morphological* information on a population of objects during $10^{10.3}$ y evolution of the universe, as well as information on a possible epoch dependence of the influence of the ambient medium on the properties of these objects.

In 1980 we started a project to map all Quasi Stellar Radio Sources (QSRs) having $z > 1.5$ and steep or unknown radio spectrum. We extracted the QSRs from the Hewitt & Burbidge (1980) compilation, and used the spectral window 1.4 GHz (or 408 MHz) – 5 GHz with the criterion $\alpha \gtrsim 0.6$ for the steep spectrum sources. The sources originated in the 4C, B2, MC, PKS, Ohio, and Westerbork Catalogues, and a considerable fraction had only one measured spectral point. Using the VLA at 5 GHz, we obtained angular resolutions of $0.4'' - 0.5''$. Although in the following I will only discuss the 5 GHz results, we have in the meantime obtained many complementary data using the VLA at 15 GHz, MERLIN at 1.67 and 5 GHz, and the European VLBI Network at 1.67 GHz (resolutions between $0.03''$ and $0.3''$). The 5 GHz VLA observations resulted in radio maps, polarization data and spectral indices for some 100 QSRs, having z between 1.502 and 3.19.

Although the Fanaroff & Riley (1974) classification would predict edge-brightened double sources for these distant, powerful QSRs, we find a predominance of small, distorted morphologies at high z. It is well known that redshift and intrinsic luminosity are correlated in flux density limited samples (from which these QSRs originate). However, we were able to separate the effects of the redshift (epoch) and the luminosity, and the distortion turns out to be a redshift effect. Two examples are shown in Figure 1. We find that jet curvature/distortion frequently occurs, usually in highly polarized, very localized areas. These distortions occur both on subgalactic (0 – 10 kpc) and larger (> 10 kpc) scale. Summarizing, at high z the radio jets experience difficulties in the formation of extended radio sources.

It is a well known fact that radio spectral index and the presence of extended (kpc scale) emission regions, containing *aging* electrons, are tightly correlated. It is therefore interesting to examine whether the above mentioned high z behaviour is reflected in the spectral index distributions at high redshift. Table 1 shows the numbers of flat spectrum and steep spectrum quasars (nota bene: wide spectral window!) for $z > 1.5$ in consecutive redshift bins. In brackets are the numbers of unresolved or slightly resolved steep spectrum

G. Swarup and V. K. Kapahi (eds.), Quasars, 181–184.
© *1986 by the IAU.*

sources, the so called Steep Spectrum Cores (SSCs). The Hewitt & Burbidge compilation lists 9 QSRs having $z > 3$, whereas the most recent figure is 15.

From Table 1 we find a decrease in the fractions steep/flat and SSC/flat with increasing redshift. There are two possible explanations for this decrease: 1) it is real; and 2) we simply do not detect the very distant QSRs having steep spectra, due to observational selection. There are three arguments supporting the reality of the decrease: 1) the average value of the spectral index for the SSCs decreases with z, indicating evolution within the class of SSCs; 2) although we may miss some of the extended sources at high redshift due to limited surface brightness sensitivity, SSCs at redshifts 4 – 5 should be easily detectable, if they were there; 3) since several (high z) QSR surveys were carried out at high frequency (small beam facilitates identification) a high fraction of flat spectrum sources is to be expected. However, none of the QSRs at $z > 3$ show a prominent increase in radio flux density towards lower frequencies (m-wavelengths), as do the low redshift core dominated sources (e.g., 3C 273, 3C 345). Although the $z > 3$ QSRs tend to be bright at cm-wavelength, they do not appear in low frequency catalogues (e.g., 4C), in agreement with the fact that they do not show extended (1″ scale) structure. This third argument can also be used against the hypothesis of the $z > 3$ QSRs being beamed towards us: where is the larger scale, unbeamed emission?

We conclude that a decrease in the space density of steep spectrum QSRs at high redshift is very likely.

Apart from the apparent general QSO cutoff (see Osmer's contribution in these Proceedings), there are also two arguments for a flat spectrum QSR cutoff at $z \sim 4$: 1) all the $z > 3$ (flat spectrum) QSRs would be easily detectable at $z = 4 - 5$. They are not found, and dust obscuration does not work; 2) Downes *et al.* (Edinburgh preprint 15/85) found a high identification rate for sources in a complete sample of deep PKS selected regions, implying that the weak sources in this sample, although distant, cannot be extremely distant.

Taking these facts together, we propose that the evolution of the quasar phenomenon from an apparent cutoff at $z \sim 4$, to $z \sim 2$ goes together with a decrease in the density of an initially very dense, presumed galactic, surrounding medium. These dense high z environments initially block the propagation of radio jets, and hence prohibit the formation of extended emission regions.

The properties of these high z environments will be spectroscopically investigated in 1986, using the Palomar 200-inch telescope. Speculations on a physical link between these dense high z environments and the general QSO redshift cutoff are left to the reader.

I wish to thank my collaborators George Miley, Richard Schilizzi and Colin Lonsdale. Full account of this work will be given elsewhere. Until further notice I offer a free beer for every new QSR having $z > 2.5$ and a steep (cm – m) radio spectrum.

Table 1. The QSR spectral index distributions in bins of increasing redshift.

spectral index	$1.5 < z < 2.0$	$2.0 < z < 2.5$	$2.5 < z < 3.0$	$z > 3.0$
steep (SSC)	53 (11)	22 (5)	6 (2)	0
flat	59	38	13	9 (15)

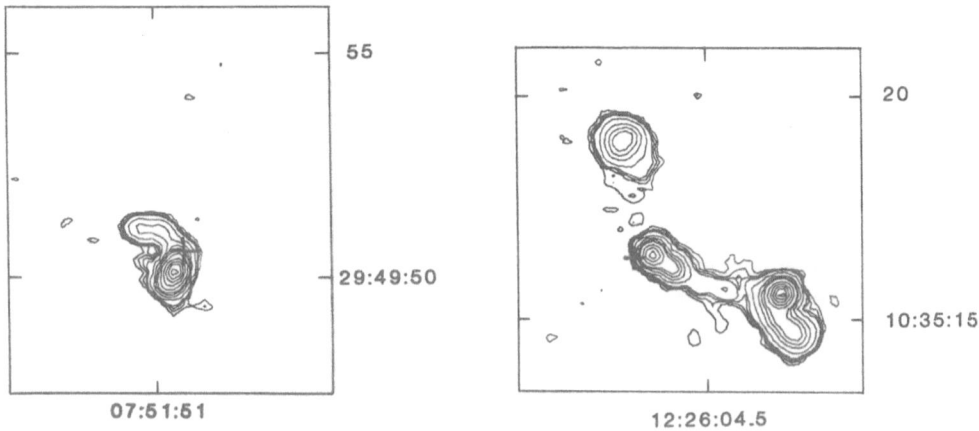

Figure 1. The radio sources associated with the quasars 0751+298 ($z = 2.106$) and 1226+105 ($z = 2.296$).

DISCUSSION

Filippenko : Many of your maps show a substantial offset between the radio core and the optical nucleus. What is your interpretation of this?

Barthel : The scale of the maps, having overall sizes of 5" - 10" is such that the offsets are not larger than 0."5 - 0."7, which is the error in the optical position.

Burke : (1) How dense must the confining galactic medium be in the high-z quasar case ? (2) Is the lack of coincidence between the optical quasar position and a visible radio core caused by uncertainity in the astrometry ?

Barthel : (1) 10^2-10^4cm^{-3}.   (2) Yes.

Murdoch : Many QSOs have steep spectrum at low frequency and flat spectrum at high frequency. Hence for a given observed frequency range, the same source might be observed with a steep spectrum if it occurs at low redshift or with a flat spectrum if it occurs at high redshift due to the large ratio of intrinsic to observed frequency. This could explain at least partially the preponderance of flat spectrum sources at high redshift.

Barthel : I agree. However, none of the $z > 3$ QSRs show a steep increase in flux density towards low frequency (m-wavelength), in agreement with their generally unresolved morphology. On the other hand, Savage & Peterson (Early Evolution of the Universe, IAU Symp., ed. Abell & Chincarini, p. 57) draw attention to the reverse effect.

Kapahi : The lack of very high z quasars in the low frequency surveys could be understood due to the very steep radio luminosity function. If flat spectrum quasars are similar to the normal double quasars, differing only in orientation, one would indeed expect flat spectrum-quasars to extend to higher redshifts because of relativistic beaming.

Barthel : The effect may be important to some extent. My main comment is that we should see a steep increase in flux density towards m-wavelengths, due to the unbeamed component(s), as is the case for e.g. 3C345. This is not seen for the QSRs at $z > 3$.

Kuhr : In your sample you did not find extended radio structure for quasars with $z > 3$; but there exists at least one quasar of redshift $z = 3.41$, the most luminous quasar, for which that was found with the VLA at an angular distance of $0\overset{''}{.}6$ from the nucleus.

Barthel : I was fully aware of that, but considered the overall extent of the source not impressive. Apart from that, also the spectrum of this quasar is not steep. However, the extended component may be the radio jet, finding its way out.

Wilkes : I understood from Fraser Owen's talk that confining and bending of extended structure in low-redshift QSOs cannot be explained in terms of the surrounding medium. Is there any reason why a high density gas should be doing the confining at high-redshift.

Barthel : Frazer Owen was referring to static, thermal confinement of radio plasma in a hot ICM. The cases I showed were subgalactic hotspots with minimum internal pressures 10^{-6}-10^{-5} dyne cm^{-2}. Ram pressure confinement by a very dense (clumpy?) galactic medium seems the only plausible mechanism.

McAdam : Your abstract mentions <u>violent</u> interaction of the radio jets with a medium. What are the indications of violence and do you believe the medium is within or surrounding any galactic (10 kpc) scale ?

Barthel : Calculating the momentum transfer onto a deflecting/confining medium one finds values of about $10^2 M_O$ x 10^3 km s^{-1} per year, which I consider violent. In these cases the size scales are 5-10 kpc from the nucleus ($H_O = 75$, $q_O = 1/2$).

PKS 0812+020: A QUASAR INTERACTING WITH ITS ENVIRONMENT

F. Ghigo, L. Rudnick, U. of Minnesota
P. Wehinger, S. Wyckoff, Arizona State Univ.
H. Spinrad, U. California, Berkeley
K. Johnston, Naval Research Laboratory

ABSTRACT. Slit spectra have shown optical emission line gas in the
vicinity of the radio jet and the north radio lobe of PKS 0812+020.
Velocities up to -1000 km/s with respect to the central quasar are seen
in the north lobe. The [OII] emission strength becomes comparable to
the [OIII] near the north lobe, as would be expected from acceleration
by shocks. We suggest this acceleration occurs by interaction with the
radio-emitting jet.

1. INTRODUCTION

PKS 0812+020 is a z=0.4 quasar situated in a rich cluster of
galaxies. It is the most luminous example of optical emission coincident with a radio hot spot. The optical and radio structure has been
discussed by Wyckoff et al. (1983). Recently, better VLA maps at 20cm,
6cm, and 2cm have been obtained. Also, slit spectra were obtained with
the KPNO CCD Cryogenic Camera on the 4-m telescope.

2. RADIO RESULTS

A straight jet connects the central QSO to the north radio hot
spot. Optical emission (R=22^{m}) is seen in this hot spot on direct
photographs. Near the QSO, the jet width is about 2". Beyond about
4" (40 kpc) from the QSO (the limit of the fuzz) the jet widens to
3".5 to 4".5. We suggest that the jet is confined by the interstellar
medium in the QSO's host galaxy and flares out upon exiting the
galaxy.

The north radio hot spot is embedded in a large diffuse radio
lobe with a sharp gradient at the northern edge, consistent with ram
pressure confinement by an intracluster medium.

South of the QSO, there is a diffuse lobe with a steep brightness
gradient to the SW, and a weak hot spot also coincident with optical
emission. At present, it is unclear whether this is the terminus of
an invisible jet, or the collision of the slowly expanding radio lobe
with a cluster galaxy.

185

G. Swarup and V. K. Kapahi (eds.), Quasars, 185–186.

Figure 1

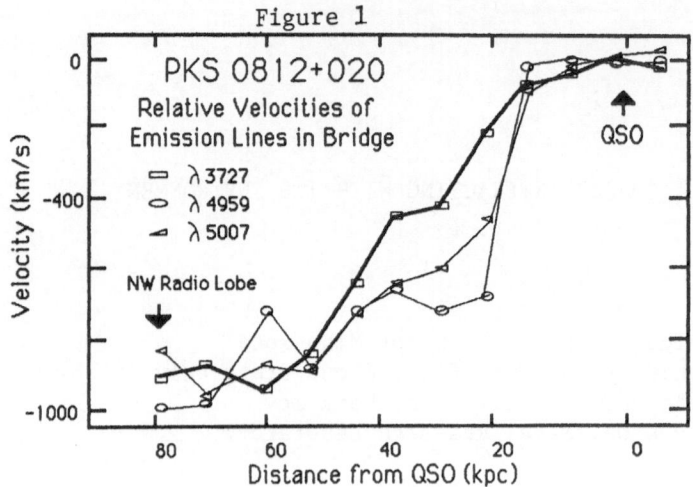

3.   SPECTROSCOPIC RESULTS

Velocities in the emission line gas along the jet are plotted in
Figure 1. Parallel to the radio jet axis, there is a sharp increase in
velocity, reaching almost -1000 km/s with respect to the quasar. This
is followed by an abrupt deceleration at the position of the north hot
spot (as expected for the radio plasma itself). This behavior is
similar to that seen, e.g., in 3C277.3 by van Breugel et al. (1985);
it suggests influence of the radio plasma flow on the emission line
material.

The apparent excitation conditions change with distance from the
nucleus as shown by the [OII] to [OIII] flux ratio changing from about
20 at the QSO to about 1 near the north hot spot.

4.   DISCUSSION

There seems to be an abrupt change in conditions at about 4"-5"
(or 40 kpc) from the nucleus, i.e., at the limits of the quasar's
"host" galaxy. Beyond this interface, the velocities increase, the jet
widens, the extended diffuse radio lobe begins, and the [OII] emission
strengthens with respect to the [OIII]. At present, we are considering
a picture in which most of the emission line material is present
throughout the host galaxy, and photoionized by the core. At the
transition from the interstellar to intracluster medium, the radio
source is perturbed, driving shocks into the thermal material,
accelerating it and increasing the [OII] luminosity.
Supported at Minnesota by NSF grant AST-8315949.

REFERENCES
van Breugel, W., Miley, G., Heckman, T., Butcher, H., Bridle, A., 1985,
    U. Cal. Berkeley, preprint.
Wyckoff, S., Johnston, K., Ghigo, F., Rudnick, L., Wehinger, P.,
    Boksenberg, A., 1983, Ap.J., 265, 43.

THE CORE AND HALO STRUCTURE OF THE QUASAR 4C18.68

Ann C. Gower and J.B.Hutchings
University of Victoria and Dominion Astrophysical Observatory
Victoria, B.C., Canada.

ABSTRACT. We present new VLA observations of this complex low redshift quasar, which was previously modelled as a precessing twin-jet nucleus. A new 2cm map fails to show the predicted curvature near the nucleus. B and C configuration maps have been obtained, to study the halo of the source. These results suggest that the radio source is not very old, and appears to have undergone two major changes of orientation in its $< 10^7$ year history.

RESULTS.

Figure 1 shows the VLA A-configuration 2cm map of the core of the source. This shows a very straight jet from core to bending point and does not conform to the precession model proposed for the 6cm and 20cm maps[1,2]. The recent history of the source seems to involve a constant orientation.

The halo (Figure 2) reveals a third major orientation of the jet in its outer structure. It appears that the source has always been stronger on one side, and that the jet has undergone successive changes of direction at decreasing intervals of time. In each major orientation there are minor wiggles. The polarisation maps and the spectral index maps also suggest a feature along the centre of the long (SW) axis of the halo. Finally, the differential Faraday rotation (Figure 2) suggests a region of variation of the amount of intervening material along this direction.

The spectral index steepens slightly with distance from the core. For the source size, an equipartition field of 10^{-4} g is indicated, for which synchrotron loss timescales are 10^{6-7} years. There is little correspondence with the optical continuum morphology [3] or the [O III] tidal tail to the NE [4]. The scale of the optical tail ($\sim$30 Kpc.) and the distance of the close (interacting?) companions suggest timescales an order of magnitude larger for the formation of the optical structure. However, this does not necessarily rule out a causal connection between the complex optical and radio structures in this quasar.

This work is being published in full elsewhere.

References.
1. Gower A.C. and Hutchings J.B. 1982, *Ap.J.* **253** , L1.
2. Gower A.C., Gregory P.C., Hutchings J.B., Unruh W.G. 1982, *Ap.J.* **262** , 478.
3. Hutchings J.B. *et al* 1984 *Ap.J.Suppl* **55** , 319.
4. Shara M.M., Moffat A.F.J., Albrecht R. 1985 *Ap.J.* **296** , 399.

G. Swarup and V. K. Kapahi (eds.), Quasars, 187–188.

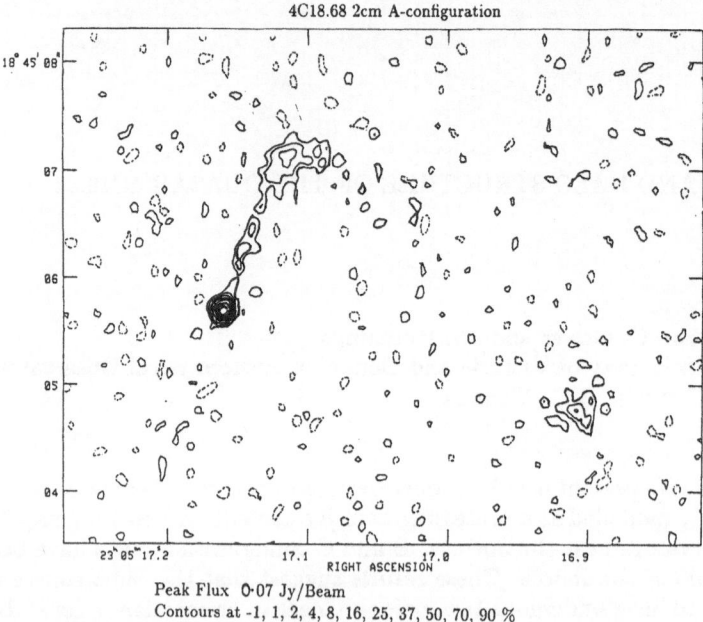

Figure 1. 2cm VLA map of inner structure. Straight jet from core to upper bending point is inconsistent with precession model fitted to longer wavelength maps.

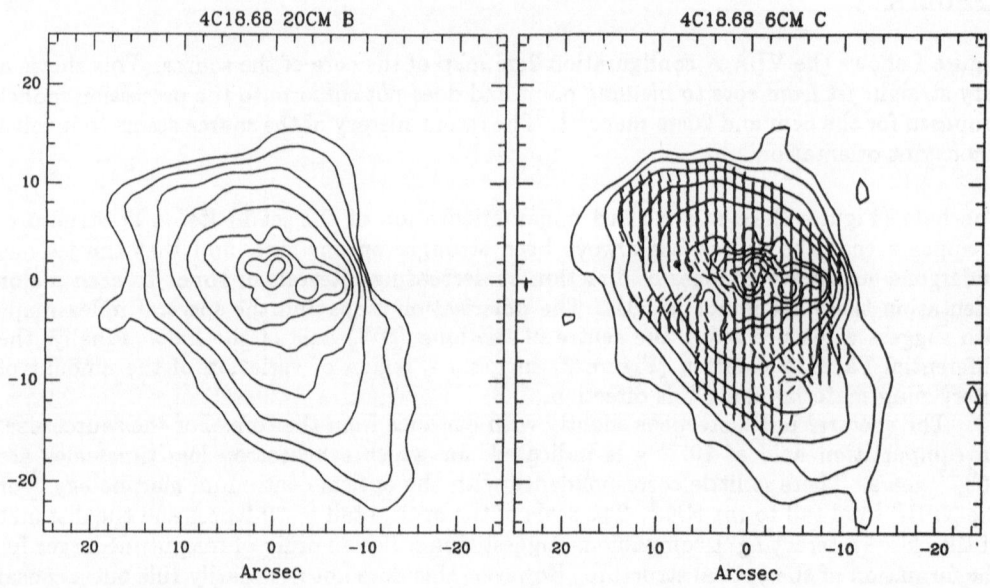

Figure 2. Matched 20cm and 6cm VLA maps, showing halo, with higher resolution inner region maps. Contours lie a factor 2 apart. Note 3 principal directions of ejection: SW, S of W, and NW, from outer to inner structure. Vectors in 20cm map show differential Faraday rotation, which is generally uniform except for SW outer (earlier jet?) direction, inner bending point, and inner SE region. The latter corresponds to the inner part of the optical tail.

THE EXTENDED RADIO STRUCTURES OF QUASARS AND RADIO GALAXIES

J.P. Leahy, P.W. Stephens & T.W.B. Muxlow
University of Manchester,
Nuffield Radio Astronomy Laboratories, Jodrell Bank,
Macclesfield, Cheshire, SK11 9DL
England

ABSTRACT. We have observed 20 classical double radio galaxies and quasars with MERLIN at 151 MHz. We discuss the systematic variation of the bridge structures with radio power and the nature of the identification.

1. INTRODUCTION & OBSERVATIONS

Because spectral index differences between compact 'hotspots' and extended bridge structure are often >1, the dynamic range required to study bridges is more than ten times smaller at 151 MHz than at GHz frequencies. This, together with a 3" resolution giving 10-20 beams across typical bright (3C) sources, and 500:1 dynamic range where not noise-limited, makes MERLIN an excellent tool for studying such bridge emission.

A previous VLA snapshot study (Leahy & Williams 1984, hereafter LW) detected bridges in virtually all sources with P(178 MHz) < 10^{27} W Hz^{-1} sr^{-1}, but for more powerful objects the proportion of detected bridges dropped rapidly. We have therefore observed a number of quasars and very powerful radio galaxies in the hope of finding this "missing" emission at 151 MHz. Examples of the results are shown in Fig. 1.

2. AXIAL RATIOS OF POWERFUL SOURCES

Even at the highest redshifts and powers, we always detect bridges extending back to the host galaxy, at least on one side. It is therefore clear that the large-scale disposition of synchrotron plasma is similar for all classical double sources. Even at 151 MHz the bridges are faint (<10% of the total flux) in the most powerful sources, as predicted by the "compactness" effect of Jenkins & McEllin (1977). The axial ratio (length over width) of such sources tends to increase with increasing power and compactness; this was first quantified by LW. Our new data (Fig. 2) appear to confirm this result at the highest powers; however there is a caveat: only one of our sources (3C68.2) is smaller than 30", whereas the median angular size for 3C sources at z > 1 is about 15". Thus we are sampling only the largest sources at z > 1, but most of the

189

G. Swarup and V. K. Kapahi (eds.), Quasars, 189–190.

population at z < 0.5. A correlation between axial ratio and linear
size could give an apparent correlation with radio power.

3. DO QUASARS LOOK DIFFERENT?

Cores and jets in quasars are usually brighter than in radio galaxies of
similar power. Our data suggest that, at least at lower powers, the
extended structure is also different. Of particular interest is the
tendency for quasars to produce a large, low-surface-brightness lobe at a
large angle to the source axis, <u>on the jet side only</u>. Examples are
3C154, 249.1, 263 and 334. The same structure is known in 3C215 and
3C351 (Laing, private communication) and in 3C179 (Shone et al. 1985).
Too few maps are available to decide how significant this effect may be.

Acknowledgements   PWS and JPL thank S.E.R.C. for support. The data
were reduced on the STARLINK node at Jodrell Bank.

REFERENCES
Jenkins, C.J. & McEllin, M., 1977. Mon.Not.R.astr.Soc., 180, 219.
Leahy, J.P. & Williams, A.G., 1984. Mon.Not.R.astr.Soc., 210, 929.
Shone, D.L., Porcas, R.W. & Zensus, J.A., 1985. Nature, 314, 603.

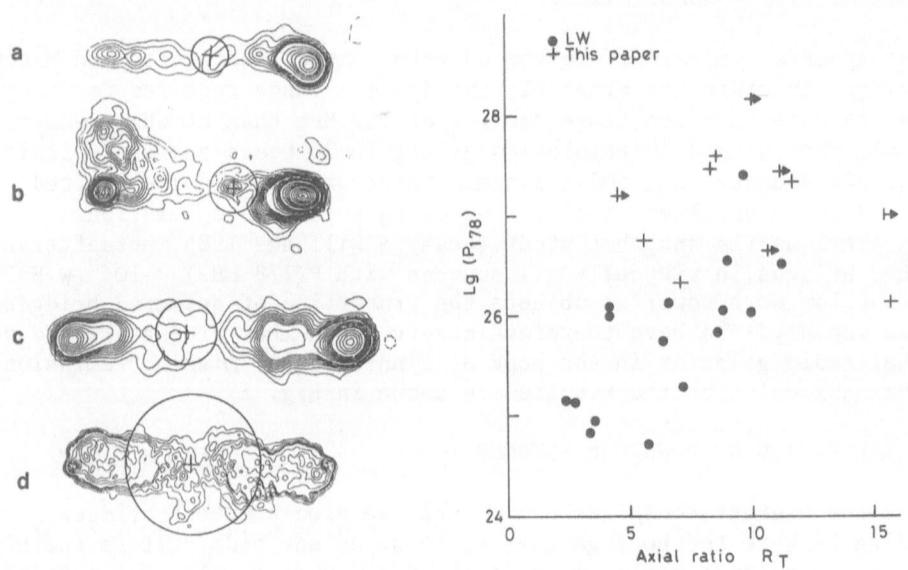

Figure 1. Maps of typical sources. Crosses mark the ID and circles are
a standard 25 kpc radius (H_O=100,q_O=0). a) 3C68.1, QSO, z=1.24; b) 3C154,
QSO, z=0.58; c) 3C268.1, Gal, z=0.97; d) 3C405, Gal, z=0.056.
Figure 2. Plot of Power at 178 MHz against total axial ratio R_T (as
defined by LW) for galaxies in LW and this paper.

THE PECULIAR RADIO STRUCTURE OF QUASAR 1320+299

T.J.Cornwell[1], D.J.Saikia[2], P.Shastri[2], L.Feretti[3],
G.Giovannini[3], P.Parma[3], C.J.Salter[4]

1 NRAO, Socorro, New Mexico, U.S.A.
2 Tata Institute of Fundamental Research, Bangalore, India
3 Istituto di Radioastronomia, Bologna, Italy
4 IRAM, Granada, Spain

The radio source B2 1320+299 is associated with a 20^m QSO. Apart from the core component, it has two outer components on the same side of the QSO; it was therefore classified as of the one-sided ('D2') type. The radio structure is unusual in that the three slightly non-collinear components are apparently unconnected and the projected linear size for any plausible redshift is large for a 'D2' source (Feretti et al. 1982, Astron.Astrophys. 115,423). The radio structure has now been mapped with the VLA (λ20 & 6 cm, B array and λ20, 6 & 2 cm, A array; Figs.1a, b).

Component A : The flat spectrum ($\alpha = 0.2$; $S \sim \nu^{-\alpha}$) component coincident with the QSO has a faint extended lobe on one side (Fig.1b; inset); the core shows no depolarization between λ 20 and 2 cm.

Component B has a steep spectrum ($\alpha = 1.1$). The polarization at the peak is high (19%) at λ6 cm; between λ6 and 20 cm, the depolarization (19 to ~1%) and rotation measure (15 rad m^{-2}) are relatively large. The magnetic field appears to follow the bends in the structure (fig.1b; inset).

Component C also has a steep spectrum ($\alpha = 0.9$) and extended emission towards the north west (fig.1b; inset).

ONE, TWO OR THREE SOURCES ? It is unlikely that the three components of 1320+299 form a single source, in view of a) the absence of any observed radio emission linking them; b) the implied large projected linear size. Component A appears to be an independent source with properties typical of powerful sources with apparent one-sided structure, viz., flat spectrum, a bright compact core, possible optical variability (Feretti, et al. 1982) and a linear size <30 kpc for z < 2(H_0=50 kms^{-1} Mpc^{-1}, q_0=0.5). B and C could well be the edge-brightened lobes of a second source. On the other hand, B taken alone appears to have a structure reminiscent of a "head-tail" source. The PSS prints however, show no optical objects between B and C, or coincident with either of them, nor do they show evidence for a cluster in the region. The quasar redshift and deep optical imaging of the field would help clarify the nature of 1320+299.

Acknowledgement. The National Radio Astronomy Observatory is operated by Associated Universities Inc., under contract with the National Science Foundation.

191

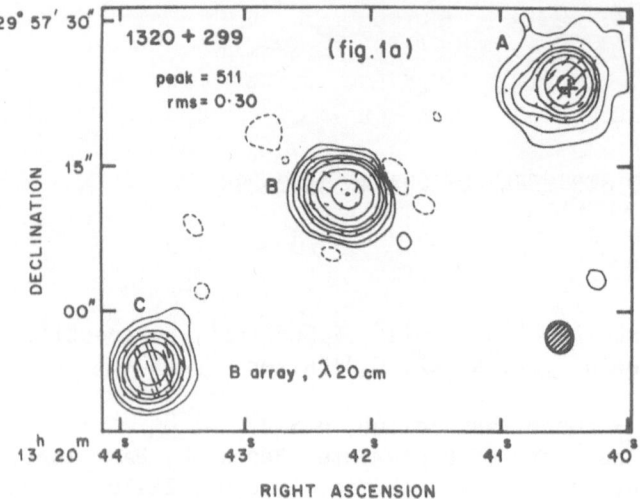

Fig.1a, b. The radio maps. Linear polarization intensity is shown in
the main maps and fractional polarization in the insets. The surface
brightness peak, rms and contour levels are in mJy/beam. Contours :
a) -1,1,3,10,20,40,100,300,500. b) -0.7,0.7,1.4,3,8,20,40,100,200;
insets : -2,-1,1,2,4,8,16,32,64,128.

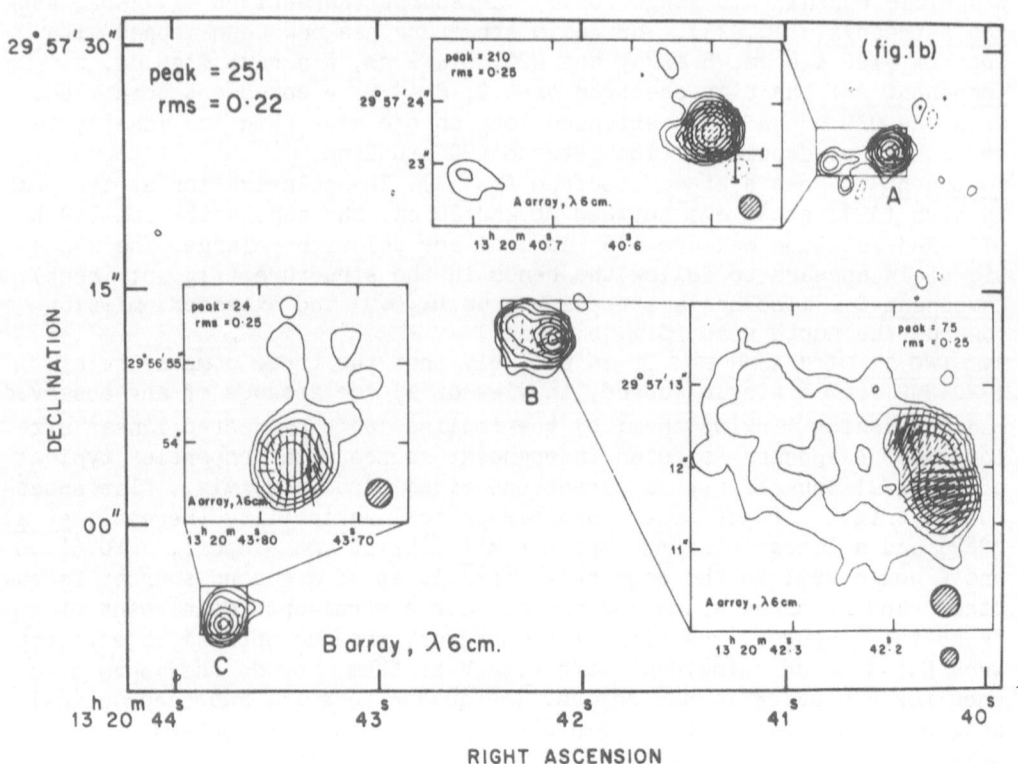

RADIO PROPERTIES OF A 'QUARTER JANSKY SAMPLE' OF EXTRAGALACTIC SOURCES,DEFINED AT 408 MHz

Gopal Krishna[1], L.Saripalli[1], D.J.Saikia[1] and R.A.Sramek[2]

[1] T.I.F.R. Centre, P.O.Box 1234, Bangalore 560 012 (India)
[2] N.R.A.O. (VLA), P.O.Box 0, Socorro, NM 87801 (U.S.A.)

ABSTRACT. For a complete sample of 42 sources, defined over a narrow flux density range around 0.25 Jy at 408 MHz, properties related to size and nuclear radio emission are discussed and compared with the 3CR sample.

1.1 THE QUARTER JANSKY SAMPLE (QJS)

The QJS consists of all 42 sources found in 5C6 and 5C7 surveys,within the range 0.2-0.35Jy at 408 MHz(Pearson and Kus,MNRAS 182,273,1978).The sample was observed with VLA at 1.5 and 5 GHz in A-configuration (and, additionally in C-configuration in case of more diffuse sources) and integrated flux densities were measured at 5GHz with the Effelsberg telescope (D.J.Saikia et al., in preparation).Three of the 42 sources are found to have flat spectra ($\alpha > -0.5$) between 0.4 and 5 GHz.We have established optical counterparts for 13 sources in the QJS, the brightest of them being three 19-mag galaxies.The QJS galaxies are,thus,expected to lie beyond $z{\sim}0.3$ but probably not beyond $z{\sim}2$(R.Windhorst, thesis,1984).Thus,the expected luminosity range for the QJS galaxies is $P{\sim}10^{26}$ -10^{28}W/Hz at 408 MHz.Since 20-25% of sources in both the 10 Jy sample (LRL) (Laing et al.MNRAS 204,151,1983) and the 1 Jy sample (Lilly et al,MNRAS 215, 37,1985) are quasars,we expect${\sim}10$ quasars in the QJS. A majority of these quasars must have steep radio spectra, since the QJS contains only 3 flat spectrum sources.

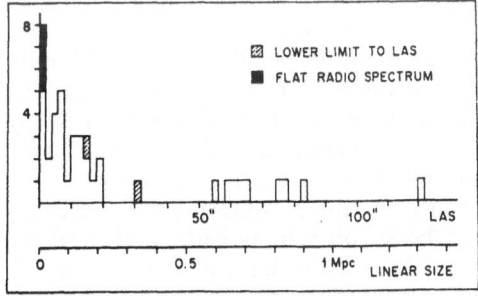

Fig.1:LAS-distribution for the QJS (42 sources). The linear-size scale is computed for $z=1$, $q_o=0$, $H_o=50$ Kms^{-1} Mpc^{-1}.

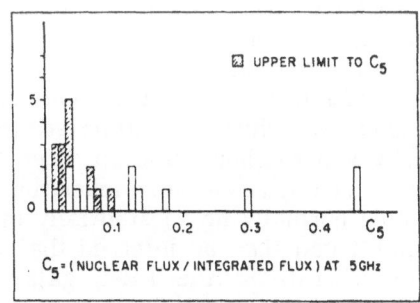

Fig.2: The distribution of core-fraction (C_5) for the QJS-FR II subsample of 23 sources (LAS>6").

193

G. Swarup and V. K. Kapahi (eds.), Quasars, 193–194.

1.2 THE POPULATION OF COMPACT STEEP SPECTRUM SOURCES (CSS)

From Fig.1,only 5 of the 39 steep spectrum sources in the QJS are compact (LAS $\leq$3").This fraction(f=13%) is much smaller than f=14/41=34% reported by Fielden et al.(MNRAS 204,289,1983) for a similar sample defined over 0.1-1 Jy range at 408 MHz.We note that the value of 'f' for the QJS would have been greatly overestimated if LAS had to be derived solely from the 5 GHz(A-array) maps; even moderately extended components seen at 1.4 GHz are often found missing from the 5 GHz maps.We,thus,find no evidence for a substantially higher f for the quarter-Jansky sources as compared to 3CR sources(which are~30 times stronger).Further,spectra of all the 5 CSS sources in the QJS,defined at 0.4,1.4 and 5 GHz,fall with frequency.Down to m_v~22,all 5 of them remain undetected(Perryman,MNRAS,187,223&683,1979).

1.3 EVIDENCE FOR COSMOLOGICAL EVOLUTION OF THE LINEAR SIZE

The expected luminosity range ($\sim 10^{26}$- 10^{28} W/Hz at 408 MHz) corresponds to the Fanaroff-Riley class II(FR-II) morphology for the QJS galaxies whose number and median z are estimated to be about 32 and 1,respectively (Sect.1.1). For them,a median size of 50 to 180Kpc is estimated (Fig.1).From literature, a median size of 350±50 Kpc is estimated for the nearby(z <0.4)FR-II galaxies in the 10-Jy LRL sample,which belong to the same luminosity range as the QJS galaxies(z_{med}~1).A decrease of linear size with z is,thus,indicated(see, also,Kapahi,MNRAS 214,19p,1985).Note that the quoted uncertainty in the median size for the QJS galaxies incorporates plausible deviations of individual galaxies from the assumed z=1(Fig.1,Sect.1.1)and also allows for the choice of up to any 9 steep spectrum QJS sources that have to be dropped from the consideration on account of being potential quasars(additional quasars in the QJS may be present among the 3 flat spectrum sources; Sect.1.1).

1.4 PROMINENCE OF THE RADIO NUCLEI (CORES)

For a subsample containing all 23 confirmed FR-II type sources among the most extended two-third QJS sources(i.e.LAS >6"),values of core-fraction,C_5 (=core flux/integrated flux) could be determined at 5 GHz, (see Fig.2).Further, applying identical selection criteria to the 10 Jy LRL sample, we have derived an equivalent subsample of 89 FR-II sources(with LAS>13")and have,further,est-imated their core-fractions from literature.Whereas only 8%(7/89)sources in this LRL subsample have C_5> 0.1,30%(7/23)sources in the equivalent QJS subsample have such prominent cores(Fig.2).Thus,FR-II type sources with prominent cores are more common in metre-wavelength samples selected at an order-of-magnitude lower flux level,compared to the 10 Jy sample.

Within the QJS subsample of 23 sources, we now confine to the galaxy population which is estimated to be ~ 18, including 4 or 5 sources with C_5> 0.1(Gopal-Krishna et.al.,in prep.).Now,since the cores usually have flatter radio spectra than the lobes,the measured value of C_5 for a distant source would normally be substantially higher than the intrinsic value(radio'K-correct-ion').It can then be inferred that in terms of both radio luminosity and core-fr-action statistics,the FR-II galaxies in the QJS should be intrinsically simi-lar to the nearby FR-II galaxies(z <0.4)in the LRL sample,provided typical redshift for the QJS galaxies were close to 1(Gopal-Krishna et al.,in prep.).

ABSORPTION LINES AND THE RADIO STRUCTURE OF QUASARS

G.Swarup[1], D.J.Saikia[1],M.Beltrametti[2],R.P.Sinha[3] and C.J.Salter[4]
1 Tata Inst. of Fundamental Res.,P.Box 1234,Bangalore,India
2 Max Planck Inst.Phys.Astrophys.,Garching bei Munchen,FRG.
3 Sys. and Applied Sc.Corpn. Hyattesville,Md 20784, U.S.A.
4 Nat. Radio Astr. Obs.,Tuscon,Arizona 85721,U.S.A.

ABSTRACT. We have investigated a possible relationship between the radio structure of quasars and their absorption line systems with velocities, V , in the range 3000-18000 km s^{-1} and find a marginal trend for such quasars to have compact radio structures (Swarup et al. 1986).

From a survey of heavy-element narrow absorption lines in the spectra of 66 quasars of low and intermediate redshifts,Weymann et al.(1979; W^{2}PT) found that the distribution of velocities between the absorption clouds and quasars showed a narrow peak at 0 km s^{-1}, a nearly uniform distribution from 0 to $\sim$ 60000 km s^{-1} and an excess in the range 3000 $\lesssim V \lesssim$ 18000 km s^{-1} .They suggested that the first component is due to absorption by clouds and galaxies in the host cluster, the second is due to 'intervening' galaxies between the quasar and us and the third is caused by material ejected from the quasar. The excess of systems in the range 3000-18000 km s^{-1} was not seen in an independent sample of high redshift quasars by Weymann, Carswell and Smith (1981) and neither this excess nor the peak near 0 km s^{-1} was seen in an unbiased sample by Young,Sargent and Boksenberg (1982). Nevertheless, there is a clear evidence of the presence of intrinsic absorbing material ejected by the central source in Broad Absorption Line quasars and in a class of radio loud quasars that exhibit closely spaced multiple features near the emission redshift (Briggs, Turnshek and Wolfe 1984). It seems interesting, therefore, to investigate radio structure of quasars having sharp absorption lines with 3000 $\lesssim V \lesssim$ 18000 km s^{-1}.

We have compiled available radio and optical observations for the 66 quasars of W^{2}PT, incuding new VLA A-configuration observations obtained by us for 20 sources at 6 cm. Only 45 of the 66 quasars are radio-selected, of which 20 are found to show absorption line features. Of these 20 sources, 7 have absorption systems in the range 3000-18000 km s and 5 out of 7 show compact radio structures. Of the other 13 sources, only 2 are compact.Out of 45 radio-selected quasars, 25 show no absorption lines. Of these, only 2 are compact,

195

G. Swarup and V. K. Kapahi (eds.), Quasars, 195–196.
© *1986 by the IAU.*

20 are extended and radio structure is not known for 3 sources. Only 4 of the 21 optically selected quasars are radio loud, 3 being compact including 1 of them having absorption lines in the above V range.

We have also studied available information about the radio structure for 59 out of 109 radio selected quasars listed by Hewitt and Burbidge(1980) which show absorption line systems. Of the 59 quasars, V lies between 3000 and 18000 km s^{-1} for 23 of them, of which 14 (~60 per cent) are compact. Among the remaining 36 only 9 (25 per cent) are compact. The above trend of quasars with V between 3000 and 18000 km s^{-1} having compact radio structure seems independent of redshift.

The results indicate that either the beams supplying energy to the outer lobes are disrupted in quasars with $3000 \lesssim V \lesssim 18000$ km s^{-1} leading to compact radio structures or the direction of ejection of absorbing clouds as well as the beam axis are inclined at small angles to the line of sight so that the relativistic beaming leads to compact radio structures. However, these results are based on hetrogenous data. In order to confirm the results, it would be desirable to obtain absorption line data of uniform sensitivity and resolution for unbiased samples of compact and extended radio sources in the same redshift range.

REFERENCES

Briggs,F.H., Turnshek,D.A. and Wolfe,A.M. 1984. Ap.J., **287**, 549.
Swarup,G., Saikia,D.J., Beltrametti,M., Sinha,R.P. and Salter,C.J. 1986. Mon. Not. R. astr. Soc., in press.
Weymann,R.J., Carswell,R.F. and Smith,M.G. 1981. Ann. Rev. Astr. Astrophys. **19**, 41.
Weymann,R.J., Williams,R.E., Peterson,B.M. and Turnshek,D.A. 1979. Ap. J., **234**, 33.
Young,P., Sargent,W.L.W. and Boksenberg,A. 1982. Ap.J. Suppl. Ser., **48**, 455.

B2 QUASARS AND RELATIVISTIC MOTION

H.R. de Ruiter[1], A. Rogora[2], L. Padrielli[1]

1 Istituto di Radio Astronomia, Bologna, Italy
2 Dipartimento di Astronomia, University of Bologna, Italy

1. INTRODUCTION

Relativistic motion has been established with a certain degree of confidence only on VLBI scales, but its application to larger scales has encountered severe difficulties (see e.g. De Young, 1984).

Often relativistic effects or more generally Doppler effects are invoked for convenience, in order to explain a particular property of AGN's; we mention the unified scheme proposed by Orr and Browne (1982), which gives a natural explanation for the observed differences between flat- and steep-spectrum radio quasars.

In this note we make use of the B2 sample of radio quasars (Fanti et al., 1979), for which new high resolution (0.3-3 arcsec) maps, made with the VLA, recently have become available (Rogora, Padrielli and de Ruiter, 1985a, 1985b). We have tested two models that both explain some observed properties of quasars as caused by high ejection velocities, thus giving rise to relativistic or, at least, appreciable Doppler effects.

2. THE UNIFIED SCHEME OF ORR AND BROWNE

The first model we tested is the unified scheme, proposed by Orr and Browne (1982). According to them the strength of a radio core relative to the extended emission depends on the orientation of the relativistic beam in the core: a dominant core is beamed towards the observer.

Our data do not sustain the unified scheme; the distribution of R (flux of core/flux of extended regions) does not fit the theoretical distribution based on random orientation of the radio sources. Our values of R are systematically higher by a factor of about five (see figure 1). This could be repaired by increasing the minimum R (R_T in the notation of Orr and Browne, 1982) and increasing the spread in R_T, but then the unified scheme loses its sense - in the extreme the spread in R_T could account for the entire distribution of R. We prefer a simpler explanation: the generally higher values of R reflect an increasing dominance of the radio core over extended emission, when

G. Swarup and V. K. Kapahi (eds.), Quasars, 197-201.

going to lower radio luminosities (as is the case when comparing the B2
sample with the 3C quasars used by Orr and Browne, 1982).

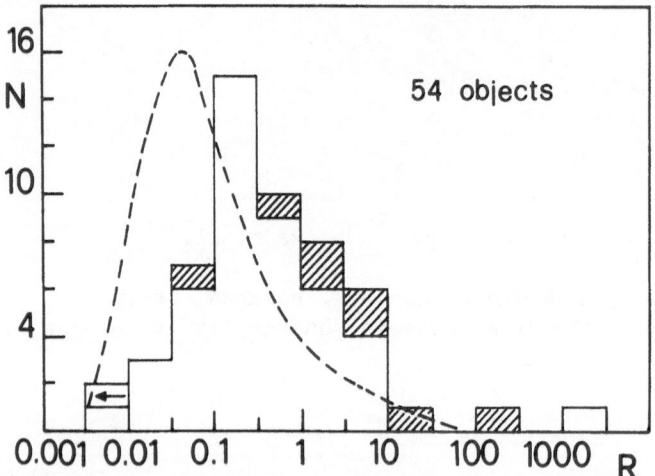

Figure 1. Histogram of R (core flux/flux of extended region) for 54 B2
quasars. Cross-hatched are D2 type sources. The dashed line represents
the distribution predicted by the unified scheme.

3. ARM LENGTH RATIOS

We also tested for Doppler effects that might affect the arm length and
flux ratios of the extended components. As in other samples the best
agreement between observed and theoretical distributions of arm length
ratio ($\underline{1}$) is obtained for velocities of the extended components of the
order of 0.2 to 0.3c (see Katgert-Merkelijn et al., 1980 and Longair
and Riley, 1979). The B2 quasar data are plotted in figure 2. The
agreement between the observed distribution and the one expected from
Doppler effects is only superficial and in particular it does not
explain values of $\underline{1}$ larger than, say, two. Such large ratios, up to
about five, are present in the B2 sample (see figure 2). Moreover, the
expected correlation between $\underline{1}$ and flux ratio of the two extended
components is completely absent. If anything, there is a slight
indication that the highly asymmetric sources have the brightest
component closest to the radio core, opposite to what would be expected
if high velocities were to play a significant role (see also Saikia and
Wiita, 1982, and De Young, 1984).
    We have searched for a different model that does not need Doppler
effects to reproduce the observed distribution of $\underline{1}$.
    The question of arm length ratios has been extensively discussed by
Rudnick and Edgar (1984); none of the models given by them fitted their
data well and the same is true for the B2 sample. However, it is
possible to construct a simple probabilistic model that reproduces the
distribution of $\underline{1}$ satisfactorily, even up to ratios of the order of
five. In this, what might be called free cone model we treat the radio

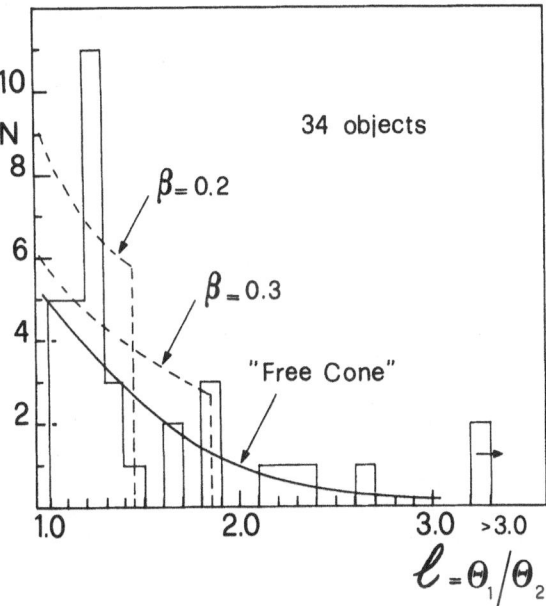

Figure 2. Histogram of arm length ratios (ratio of distance core to farthest and core to closest component), for 34 B2 quasars. Dashed lines give expected distributions due to Doppler effects. Solid line is prediction by "free cone" model (see text).

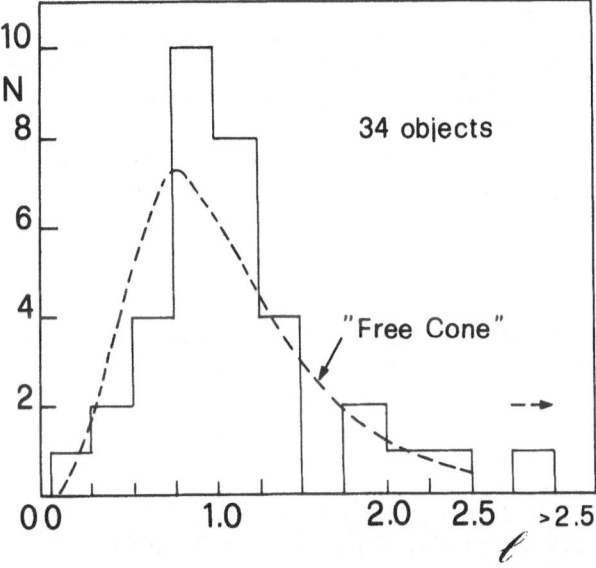

Figure 3. Same data as in figure 2, but arm length ratio is now defined as ratio core-western and core-eastern component. The dashed line is the "free cone" distribution. The mean arm length ratio for the B2 quasars is equal to 1.16 against 1.21 predicted.

source as a cone with a small but finite opening angle. The material that emits the radio radiation moves inside this cone outward with a small, non-relativistic velocity. We then calculate the probability to encounter an obstacle (e.g. cloud) inside the cone and assume that the radio source at that point is stopped. If we always take the ratio western:eastern component $\underline{l}$ will be between zero and infinity. The resulting theoretical distribution has a number of advantages: it has a well defined mean (1.209) and standard deviation (0.978), which often do not exist in this kind of problem. More important, it fits the observed $\underline{l}$ distribution as derived for the B2 sample quite well. In figure 3 the theoretical "free cone" distribution, $f(\underline{l})=3\underline{l}^2/(1+\underline{l}^3)^2$, is shown. Considering the simplicity of the model the agreement with the observed distribution is satisfactory.

It should be stressed that the way we arrived at the theoretical distribution is naive and does not necessarily represent the behaviour of a real radio source. Rather, the free cone model should be considered as an illustration to the fact that the observed arm length ratio distribution can be reproduced without invoking Doppler effects.

4. CONCLUSIONS

We conclude from an analysis of the B2 quasar sample that:
i) no compelling evidence exists for relativistic effects that depend on the orientation of the radio sources. In particular the unified scheme does not seem to work.
ii) arm length ratios can be easily explained by a model that invokes interaction of the radio source (considered as material moving outward in a cone with small but finite opening angle) with randomly located external clouds. In particular no Doppler effects are required to explain the observed arm length ratios.

It follows that the only strong evidence for relativistic effects in AGN's is and remains superluminal motion.

REFERENCES

De Young, D.S.: 1984, in Proceedings of NRAO Workshop No.9, Eds.
    A.H. Bridle and J.A. Eilek, p.100
Fanti, R., Feretti, L., Giovannini, G., Padrielli, L.: 1979,
    Astron.Astrophys. **73**, 40
Katgert-Merkelijn, J., Lari, C., Padrielli, L.:1980, Astron.Astrophys.
    Suppl. **40**, 91
Longair, M.S., Riley, J.M.: 1979, M.N.R.A.S. **188**, 625
Orr, M.J.L., Browne, I.W.A.: 1982, M.N.R.A.S. **200**, 1067
Rogora, A., Padrielli, L., de Ruiter, H.R.: 1985a, submitted
    to Astron.Astrophys.Suppl.
Rogora, A., Padrielli, L., de Ruiter, H.R.: 1985b, in preparation
Rudnick, L., Edgar, B.K.: 1984, Astrophys.J. **279**, 74
Saikia, D.J., Wiita, P.J.: 1982, M.N.R.A.S. **200**, 83

DISCUSSION

Barthel : Two comments : (1) Longair & Riley determined the average hot-spot advance velocity for 3C sources in general - not only quasars. Different beasts ! (2) Did you calculate R at 5 GHz emitted frequency ? Steep spectrum cores may explain some discrepancy. Please note that I am not a fan of beaming, though.

De Ruiter : We did not calculate R at 5 GHz emitted, because not all objects have known redshifts. Since we expect that most quasars in the B2 sample have z of the order of or less than one the discrepancy cannot be resolved this way.

Swarup : Is it not possible that the relativistic beaming model may predict higher values of f_c in the Bologna sample at 408 MHz which is much fainter than the 3CR sample at 178 MHz.

De Ruiter : We have taken the predictions as given by the unified scheme of Orr and Browne. It is true that the Bologna quasars are fainter but not that much. The only way to save the unified scheme is to assume that in the Bologna sample the orientation angle is not random.

Kapahi : Is it possible that some of the B2 quasars you are considering got into the sample because their radio structure was first determined at high frequencies, which led to a confirmation of their identification with B2 radio sources or to the determination of their redshifts ?

De Ruiter : The identifications originally were made using the 408 MHz data, and were based on the uv excess nature of the objects. Only afterwards higher frequency radio data were used to confirm the identifications.

"3C273 contains all the ingredients of a classical double source like Cygnus A, with core, jet, hotspots, wiggle, extended lobe, but on one side only."

- Richard Davis (p.214)

WSRT OBSERVATIONS OF CORE-DOMINATED RADIO SOURCES

A.G. de Bruyn, R.T. Schilizzi
Netherlands Foundation for Radio Astronomy
Postbus 2
7990 AA Dwingeloo
The Netherlands

A few years ago we presented evidence for large scale radio structures around most superluminal radio sources (Schilizzi and de Bruyn, 1983). The extended emission is relatively faint and, in at least two cases (3C120 and 3C345), exhibits a rather complex morphology. Although the sample was small, the conclusion seemed inescapable that the largest angular sizes (LAS) of the superluminals do not permit a large deprojection factor due to foreshortening, without making these sources exceptionally large intrinsically. That is to say, they would be exceptional within the context of the unified scheme (US) which interprets superluminal sources as favourably oriented, strongly beamed, core components of classical double radio sources. To save the US, we concluded that the ejection axes of these sources must undergo a large ($\gtrsim 30°$) change of direction in the period between the creation of the outer and inner emission regions. Superluminality would then be a temporary phase (although possibly recurrent) in the lifespan of a large fraction of the parent population of classical doubles.

In order to extend this result to a larger sample of objects we made observations with the WSRT of 14 sources generally thought to be akin to the superluminal sources. In this sample we included a variety of objects: low frequency variables, optically violent variables and BLLac type objects. The selection criteria for the sample were loosely defined: we chose strongly core-dominated radio sources with a flux density of at least 2 Jy at 6 cm and at accessible declinations ($\delta \gtrsim 15°$) in order to also obtain good resolution in the declination direction (the WSRT is an EW-interferometric array). The sources in the sample were: 0133+47(0C457), 0235+164, 0552+39(DA193), 0735+178, 0836+71, 0851+20(0J287), 0954+55(4C55.17), 1219+28(W Coma), 1308+32, 1358+62, 1611+34(DA406), 2201+42(BL Lac), 2230+11(CTA102), 2251+16(3C454.3).

Observations of 12 hours duration were made of all objects in April and May 1983 with the digital line backend giving a sensitivity of about 0.4 mJy (r.m.s.). Recently, we reobserved three objects (0235+164, 0J287 and BLLac) with the newly constructed broad-band backend yielding a sensitivity of 0.08 mJy after 12 hours integration.

G. Swarup and V. K. Kapahi (eds.), Quasars, 203–206.

The object OJ287 was observed for three 12 hour periods in October and November 1985 to search for rapid intra-day variability. A detailed account of the results on this object will be presented elsewhere (de Bruyn, in preparation).

The reduction of these data was carried out using standard procedures developed for redundancy observations with the WSRT (Noordam and de Bruyn, 1982). A preliminary account of some of the results was presented at IAU Symposium 110 in 1983 (de Bruyn and Schilizzi, 1984).

An important aspect of the reduction of redundancy observations with the WSRT is the production of accurate visibility curve information without recourse to a self-calibration step. The accuracy with which information on sizes and fluxes of resolved emission is derived therefore depends only on the quality of the correlator which is known (empirically) to be very good.

The general result from this work is that all sources show evidence for emission resolved by our $3\overset{''}{.}5 \times 3\overset{''}{.}5 \sin\delta$ beam. In about half of the sources in the sample, this emission is distributed across a region several beams in diameter, at a level never exceeding 2 to 3 times the noise level. In these cases the flux densities and sizes of the extended emission were derived from simple fits to the visibility curves (averaged over 12 hours). Note that this procedure may underestimate the longest dimension if the sources are elongated.

A sample of the results is shown in Figures 1 through 4. A full description of these and the other sources not shown, will be presented elsewhere (de Bruyn and Schilizzi, in preparation). Here we will only summarize the main results:

1)   there are no naked sources when observed with a dynamic range of several thousand to one,

2)   the morphology of the extended emission is generally diffuse, often asymmetric, and in a few cases there are clear suggestions of double structure,

3)   the median LAS is of the order of 100 kpc ($H_o = 50$ kms^{-1} Mps^{-1}, $q_o = 0.5$),

4)   in many cases where this could be determined, the inner (VLBI) axis and the large scale orientation show no relation. Examples in the current sample are OJ287 and 3C454.3, but this result was also found in 3C120 and 3C345 (Schilizzi and de Bruyn, 1983).

We briefly compare these results with the expectations and predictions of the relativistic beaming and unified schemes (Blandford and Königl, 1979; Scheuer and Readhead, 1979; Orr and Browne, 1982). We can make the following inferences:

i) if the intrinsic core/lobe flux ratio is in the range 0.01-0.1, as observed in sources whose extended power is comparable to those in our sample, we need a minimum Lorentz factor between 5 and 10 in order to boost the core power to the observed levels (assuming jet motion within angles of 10°-20° to the line of sight).

ii) it is doubtful that all the sources in our sample are drawn from the same parent population. We have made comparisons with samples of lobe-dominated sources drawn from the 3CR and B2 catalogues, sources which straddle our objects in power and should be the parent population. The observed LAS distributions (expressed in linear size)

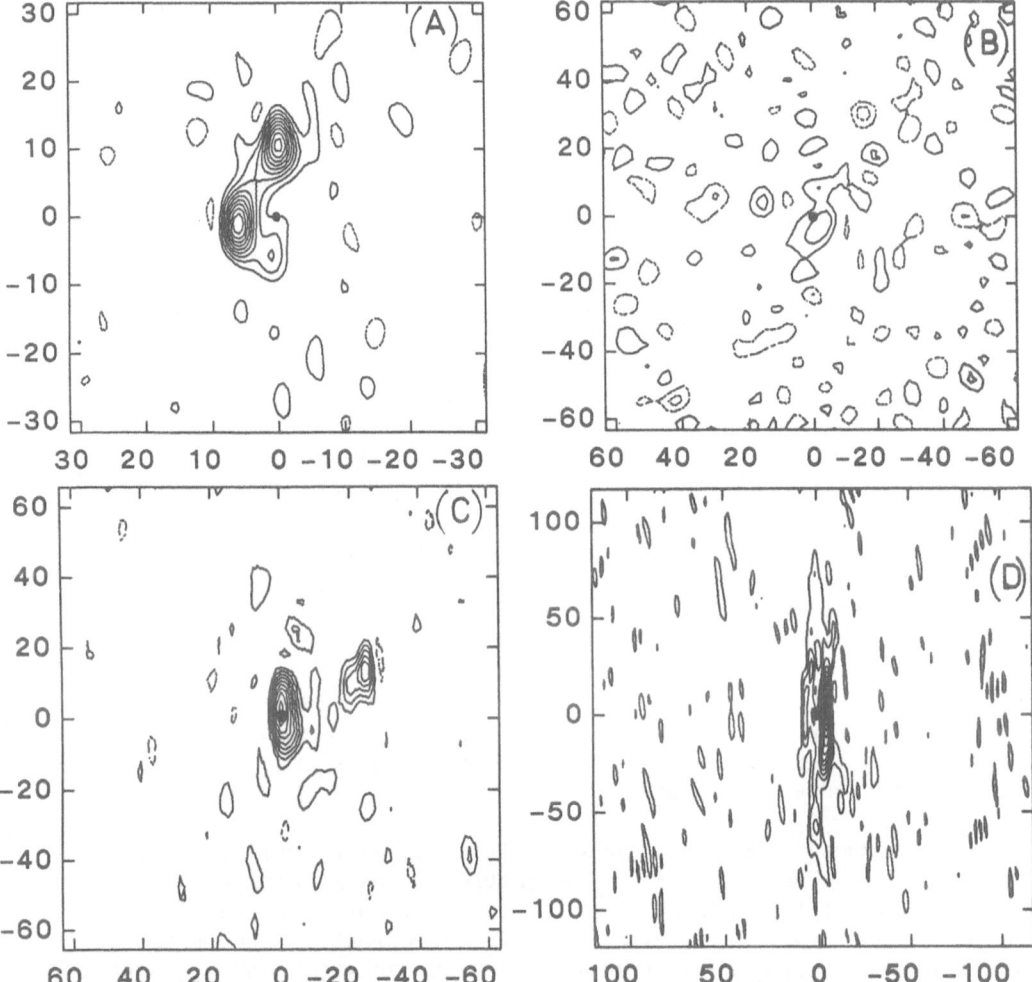

Fig. 1a, b, c, d Maps of the extended emission around four core-dominated radio sources. The axes are labelled in arcsec. In all maps the strong core was subtracted at the centre (shown as a dot). In 0836+71 and 3C454.3 the strong secondary emission within one beam area of the core was also subtracted. The full resolution beam measures 3″5x3″5/sinδ.

a) B2 1308+32: core 1.9Jy; contour levels -1, 1, 2, 3, ..., 9 mJy.
b) 0836+71: map smoothed to 7"x7"/sinδ; core 2.85 Jy; secondary (at Δα=-0″55 Δδ=-1″4) 70 mJy; contour levels -1.6, -0.8, 0.8, 1.6 mJy.
c) 0J287: average of two 12 hour syntheses; core 3.6 Jy; contour levels -0.15, 0.15, 0.30, ..., 1.35 mJy. There is probably continuous emission between the core and the "jet".
d) 3C454.3: core 14 Jy, secondary (at Δα=-3″7 Δδ=+3″6) 260 mJy. contour levels -1.25, 1.25, 2.5, ..., 16.25 mJy. The extensions to the north and south show up more reliably in smoothed maps (and unpublished 21 cm data). The appearance of the structure around the core depends on the amount subtracted at the core position.

of all three sets of objects are very similar. But within the
US/beaming hypothesis, our sources should be foreshortened by some
factor. How large this factor could be is difficult to say because it
depends on the intrinsic morphology. For sources with clear doubles or
distinct high-brightness features the foreshortening correction could
be considerable (up to 5). However, many of the sources in low
luminosity samples of sources are distorted to some degree and have a
transverse size up to half the overall size. Clearly for such sources
the foreshortening correction can be at most a factor of two.

About half of the sources in our sample could fall in this latter
category. For them we can state only that their LAS, corrected for
foreshortening based on a Lorentz factor of 5-10, is still consistent
with them being drawn at random from a sample of randomly oriented
radio galaxies and quasars. This is in agreement with the results of
Antonucci and Ulvestad (1985) who observed a large sample of BLLac
objects. Yet, the maps of OJ287 and 3C454.3 (Figs. 1c and d) reveal
distinct features at a large projected linear distance from the core.
Foreshortening corrections of 3-4 for these sources would make them
exceptionally large, and difficult to fit within the unified scheme.

iii) we observe different position angles for the VLBI and large
scale structures and we observe large projected angular sizes. It is
not possible to reconcile both these observations within the unified
scheme; either there is a large foreshortening (with consequent
problems for the deprojected sizes) and a small intrinsic misalignment
(in which case the observed large misalignment could be attributed to
magnification effects) or there is little or no foreshortening but a
gross (>30°) intrinsic misalignment in these sources in going from the
inner to the outer structure.

If we accept the second possibility (see also Schilizzi and de
Bruyn 1983), we are faced with the problem of explaining why a
disproportionately large fraction of core-dominated sources (compared
to the parent population) has a gross misalignment in structure. The
sources exemplifying this problem are 3C120, 3C345, OJ287 and 3C454.3.
The conclusion is therefore that we must modify, or radically depart
from, the unified scheme. One possibility would be to assume that there
is a causal relationship between the presence of relativistic motion in
core dominated sources and beam wobbling and/or the formation of
extremely large radio structures.

Acknowledgements The Westerbork Synthesis Radio Telescope is operated
by the Netherlands Foundation for Radio Astronomy with the financial
support of the Netherlands Organisation for the Advancement of Pure
Research (Z.W.O.)

References
Antonucci, R.R.J., Ulvestad, J.S., 1985, Astrophys.J. 294, 158.
Blandford, R.D., Königl, A., 1979, Astrophys.J. 232, 34.
de Bruyn, A.G., Schilizzi, R.T., 1984, in "VLBI and Compact Radio
        Sources", IAU Symp. 110, p. 165, Ed. R. Fanti et al.
Noordam, J.E., de Bruyn, A.G., 1982, Nature 299, 597.
Orr, M.J.L., Browne, I.W.A., 1982, Mon. Not. R. astr. Soc. 200, 1067.
Scheuer, P.A.G., Readhead, A.C.S., 1979, Nature, 277, 182.
Schilizzi, R.T., de Bruyn, A.G., 1983, Nature 303, 26.

REDSHIFT DISTRIBUTIONS OF STEEP AND FLAT-SPECTRUM RADIO QUASARS AND RELATIVISTIC BEAMING

V.K.Kapahi
T.I.F.R. Centre, P.B. No. 1234, Bangalore 560012, India

V.K.Kulkarni
Radio Astronomy Centre (TIFR), P.B. No. 8
Ootacamund 643001, India

ABSTRACT. Redshift distributions of almost complete samples of steep as well as flat-spectrum radio quasars are investigated. It is found that flat-spectrum (core-dominated) quasars contain a higher fraction of high redshifts ($z > 1.4$) compared to steep-spectrum ones. The difference can be understood~in the 'unified scheme' in which the two types of quasars differ only in the orientation of their jet axes with respect to the observer.

At high redshifts most of the radio quasars are found to be of the flat-spectrum core-dominated variety (see e.g. Barthel, this Volume, p.181). Although it was suggested by Kraus & Gearhart (1975) and Pacht (1976) that high redshifts are found preferentially in quasars with a 'centimetre excess' spectrum, it has not been possible to confirm this result (e.g. Wills & Lynds 1978) for lack of large and complete samples of steep and flat-spectrum quasars. Recent availability of deep optical identifications and spectroscopy now makes it possible to compare the redshift distributions of almost complete samples of the two types selected from strong and intermediate flux density surveys at high frequencies.

We consider quasars from the following flux limited samples that are near-complete with regard to spectral classification and redshifts, and contain a reasonably large number of sources
1. The SGPS (South Galactic Pole Strong) sample of the 63 sources selected from the Parkes 2.7 GHz surveys that have $\alpha_{5.0}^{2.7} < 0.5$ ($S_\nu \propto \nu^{-\alpha}$); $22^h < RA < 05^h$; $-30° < \delta < -4°$ and $S_{2.7} \geq 0.5$ Jy (Condon et al. 1978; Wright et al. 1983). Identification and spectroscopic data for the corresponding sample of steep-spectrum sources in the region are much less complete. We have therefore compared the z-distribution of the flat-spectrum SGPS sample with that for quasars in the 3CR sample (Laing et al. 1983), which has roughly the same sensitivity limit for steep-spectrum sources ($\alpha \sim 1$).
2. The All-Sky sample of Wall & Peacock (1985;WP) with $S_{2.7} \geq 2$ Jy.
3. The PR sample (Pearson & Readhead 1984) of 65 sources selected from the S4 and S5 NRAO-MPIfR surveys at 5 GHz, and defined by $S_5 \geq 1.3$ Jy; $\delta > 35°$. Some additional redshifts for this sample have been reported recently by C.R.Lawrence et al. (1985, OVRO preprint).

G. Swarup and V. K. Kapahi (eds.), Quasars, 207–210.

Table 1. Numbers of flat-spectrum and steep-spectrum quasars.

Sample	Flat-Spectrum ($\alpha < 0.5$)			Steep-Spectrum ($\alpha > 0.5$)		
	Total	Unknown z	z $\geq$ 1.4	Total	unknown z	z $\geq$ 1.4
SGPS;$S_{2.7} \geq 0.5$ Jy	35	0 (+8 EFs)	17(49%)			
3CR;$S_{178} \geq 10$ Jy				39	2(+6 EFs)	8(20%)
WP-All Sky; $S_{2.7} \geq 2$ Jy	45	7(+1 EF)	13(29%)	26	3(+5 EFs)	2(8%)
PR;$S_5 \geq 1.3$ Jy	20	4(+1 EF)	7(35%)	7	0(+1 EF)	0

The statistics about quasars in these samples are summarized in Table 1 and the z-distributions are shown in Figures 1, 2 and 3. In all the three samples the distributions appear to extend to higher redshifts for the flat-spectrum quasars compared to the steep-spectrum ones. The fraction of quasars with z > 1.4, which is considerably higher for the flat spectrum quasars is indicated in Table 1

The occurrence of higher redshifts among flat-spectrum quasars could partly be responsible for some large scale inhomogeneities in the sky distributions of high-redshift quasars reported in the literature (e g. Shastri & Gopal-Krishna 1983; Arp 1984), because quasar identifications and spectroscopy in some areas of the sky are based largely on surveys at low frequencies and in some regions on surveys at high frequencies (see Kapahi, this volume, p.511)

The different z distributions could of course arise if the two populations are physically quite distinct, with different luminosity functions and/or different evolutionary behaviour with cosmic epoch. It is interesting, however, to see if the distributions are consistent with the predictions of the 'unified scheme' of Orr & Browne (1982), based on the relativistic beaming scenarios of Scheuer & Readhead (1979) and Blandford & Konigl (1979). In this scheme the flat spectrum quasars are normal double quasars seen almost end-on. In order to check this we assume that quasars in a low frequency survey, such as the 3CR, are unbiased with regard to orientation, and have the following average values for the parameters describing relativistic beaming in their cores :

$$R_T \text{ (5 GHz emitted)} = 0.024; \quad \alpha_c = 0; \alpha_e = 0.9 \text{ and } \gamma = 4.7$$

Here R_T is the ratio of the flux density in the core to that in outer lobes (assumed to be moving nonrelativistically) when the jet axis is transverse to the line of sight ; α_c and α_e are the spectral indices of the core and outer lobes respectively and γ is the Lorentz factor in the core. The above values were found by Orr & Browne to provide the best fits to the observed proportions of flat-spectrum quasars in flux limited samples at different frequencies. The only other information required to determine the z-distributions for both steep-spectrum (defined to have $\alpha_{2.7}^5 > 0.5$) and flat-spectrum ($\alpha_{2.7}^5 < 0.5$) quasars in flux limited samples selected at different frequencies is the radio luminosity function (RLF) at different epochs. We have used the local RLF given by Fanti & Perola (1977) for $q_0 = 1$ and simple evolution functions of the type $\rho(z) \propto \rho(z=0) \, e^{m(1-t/t_0)}$, where (t_0-t) is the look back time). An effective cutoff in redshift is also required in order to limit the number of quasars with very high redshifts.

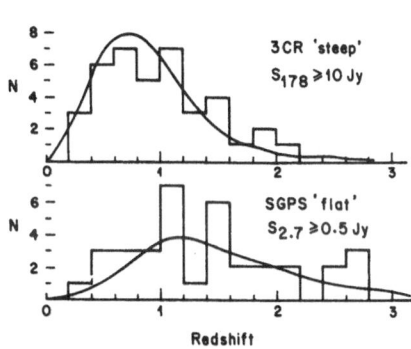

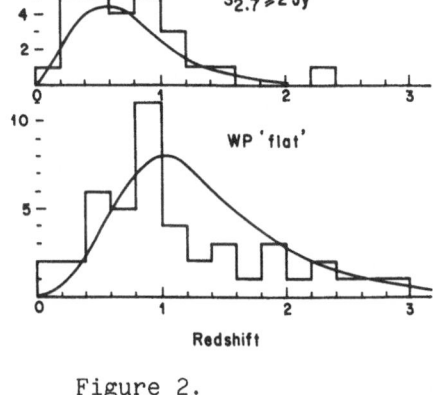

Figure 1.   Figure 2.

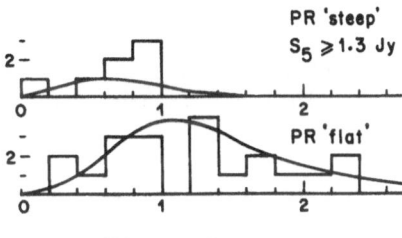

Figs. 1,2 & 3 show distributions
of the observed redshifts of quasars in different samples. Curves
are predictions of the 'unified
scheme'

Figure 3.

A value of m=10 together with an exponential redshift cutoff function of the form $e^{-(z-1)}$ for z > 1 was found to give a reasonable fit
(see figures 1 to 3) not only to the shapes of the redshift distributions but also to the total numbers of quasars observed in the areas of
the sky covered by the different samples. Note that in our calculations
the only normalization of the amplitude of the local RLF has been done
by fitting the observed and predicted number of quasars in the 3CR
sample.

We conclude therefore that the observed redshift distributions of
flat and steep spectrum quasars in high frequency surveys provide additional support to the 'unified scheme'.

References

Arp,H. 1984, Ap.J. 277, L22
Blandford,R.D., Konigl,A. 1979, Ap.J., 232, 34
Condon,J.J., Jauncey,D.L., Wright,A.E. 1978, A.J., 83, 1036
Fanti,R., Perola,G.C. 1977, in 'Radio Astronomy and Cosmology' (ed.)
    D.L.Jauncey, D.Reidel, p 171
Kraus,J.D., Gearhart,M.R. 1975, A.J., 80, 1
Laing,R.A., Riley,J.M., Longair,M.S. 1983, M.N.R.A.S., 189, 433
Orr,M.J.L., Browne,I.W.A. 1982, M.N.R.A.S., 200, 1067
Pacht,E. 1976, A.J., 81, 574

Pearson,T.J., Readhead,A.C.S. 1984, in 'VLBI and compact Radio Sources'
    (eds. R.Fanti, K.Kellermann & G.Setti), D.Reidel, p 15
Shastri,P, Gopal-Krishna 1983, J.Astrophys.Astron., 4, 109
Wall,J.V., Peacock,J.A. 1985, M.N.R.A.S., 216, 173
Wills,D., Lynds, R. 1978, Ap.J.Supp., 36, 317
Wright,A.E., Ables,J.G., Allen,D.A. 1983, M.N.R.A.S., 205, 793
Scheuer,P.A.G., Readhead,A.C.S. 1979, Nature, 277, 182

DISCUSSION

Barthel : Given possible large redshifts in the samples the K corrections can be considerable. You have therefore to be careful in using
spectral indices over 2.7 & 5 GHz. I do agree however that you have
proven that beaming plays a role.

Kapahi : We have for simplicity assumed constant spectral indices α_e and
α_c . The resulting K corrections are of course included in the calculations.

Kuhr : Deep CCD and IR observations of a complete sample of EF sources
with flat spectra from our 1 Jy catalogue indicate that these source
can be identified with radio galaxies, QSOs and BL Lac objects of about
equal numbers. The objects are probably not EFs because of very high
redshifts, but because they all have "cut-off" spectra, i.e., they look
like the rest of the sources between radio and IR but their flux drops
steeply towards the optical.

RADIO JET AND LOBE OF 3C273

R.J. Davis
Nuffield Radio Astronomy Laboratories
Jodrell Bank
Macclesfield
Cheshire
U.K.

ABSTRACT.   The 'superluminal' motion observed in the cores of radio
sources such as 3C273 is now accepted as evidence of relativistic motion
within a few parsecs of the centre, but it is less clear whether such
speeds persist out to kiloparsec scales. The one-sidedness of such
sources is often cited as evidence of relativistic Doppler beaming, but
could equally be intrinsic. New MERLIN maps of 3C273 at 151 MHz and 408
MHz have been made with dynamic range of 4.10^3:1 and 10^4:1 respectively.
These show that (i) there is an extended region or lobe to the south of
the main jet; (ii) the radio emission of the jet is continuous from the
core to beyond the limit of the optical jet; (iii) no counter-component
can be found in the opposite direction to the jet. The ridge-line of
the jet shows a 'wiggle', the wavelength of which decreases by a factor
of 6 along its length. This is interpreted as a deceleration of the
bulk flow along the jet.

1.   INTRODUCTION

3C273 is the archetype of a one-sided source which also shows superluminal motion[1]. This quasar has redshift 0.158 and taking $H_O = 100$
km/s/Mpc 1" corresponds to 1.85 kpc. There is good evidence for relativistic flow at least in the nucleus, and it is tempting to consider
whether the extended one-sided emission is due to Doppler effects. If
relativistic flow is causing Doppler boosting in the nucleus, and if the
kpc jet is moving slowly, then such a source inclined away from the line
of sight (l.o.s.) would appear as a disembodied jet pointing back to a
radio quiet quasar and such sources are not seen. It is thus of great
interest to observe with very high dynamic range at low frequency to
detect any old slow moving components which may be associated with any
invisible counter jet.

2.   OBSERVATIONS AND RESULTS

Measurements of the low frequency spectrum and angular diameters of
3C273 led Foley and Davis[2] to predict the presence of an extended halo

211

G. Swarup and V. K. Kapahi (eds.), Quasars, 211–214.
© 1986 by the IAU.

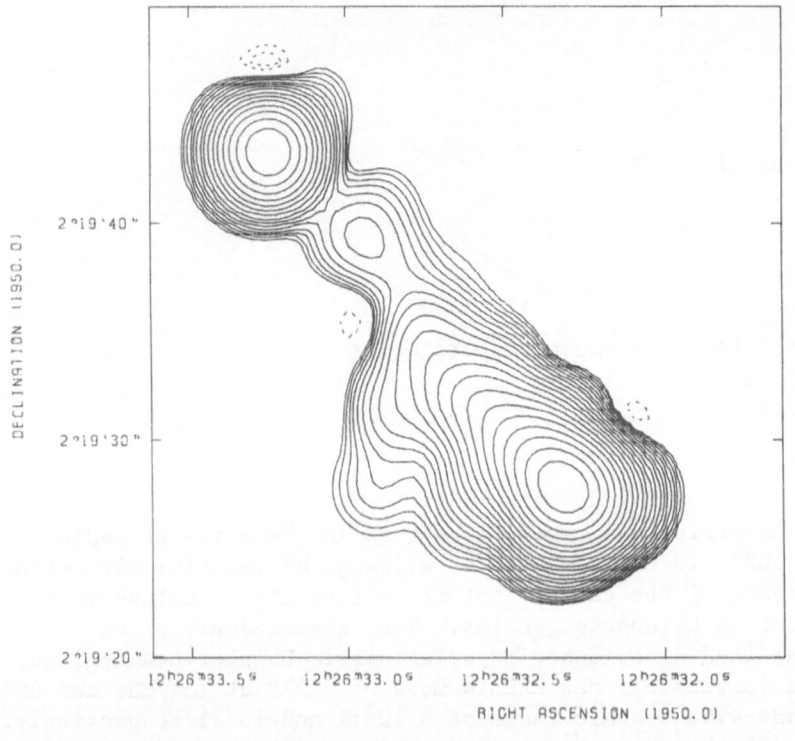

Fig. 1

MERLIN map of
3C273 at 151
MHz

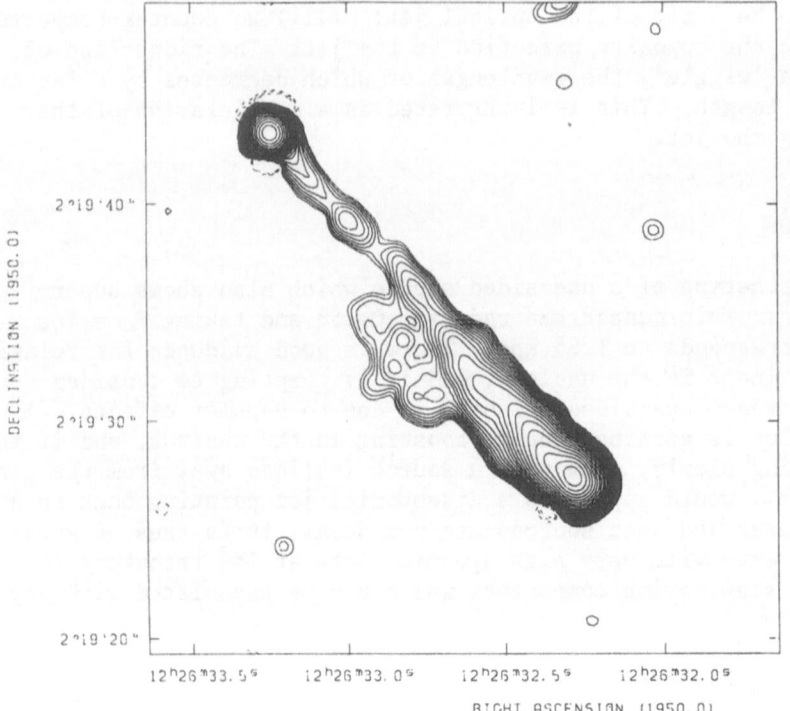

Fig. 2

MERLIN map of
3C273 at 408
MHz

which should be observable at 151 MHz and they expected the halo to
surround the whole source as in 3C345[3] and thus allow the Doppler model
to explain the one-sided nature of the source. Recent advances in the
closure technique[4] have been made at Jodrell Bank enabling maps to be
made down to the thermal noise. Fig. 1 shows a new 3×10^3:1 dynamic range
MERLIN 151 MHz map, and Fig. 2 shows 1981 data at 408 MHz from MERLIN
reprocessed with the new Starlink computer and software giving a 10^4:1
dynamic range map. The new data show a jet continuous from nucleus to
outer hotspot and a steep spectrum ($\alpha = 1.5 \pm .3$) extended region or "lobe"
on the south side of the jet ($\alpha = 0.80 \pm .05$). Nothing above the noise
level is seen on the other side.

3. DISCUSSION AND CONCLUSIONS

3.1. Side to side ratios

A cut made along the jet but in the opposite direction to the jet gives
a peak to peak ratio of 5500:1 between the brightest region of the jet to
the highest noise spike on the other side. Some radiation from the jet
could be Doppler boosted but there are arguments[5] that at least one sixth
is moving slowly and this gives an intrinsic side to side ratio of
>1000:1. The strongest arguments come from the lobe. This is 15 (arc s)^2
on the 408 MHz map and no region of equivalent area on the other side has
one fiftieth of the flux density. Thus peak to peak lobe to counter lobe
ratio is >50:1.

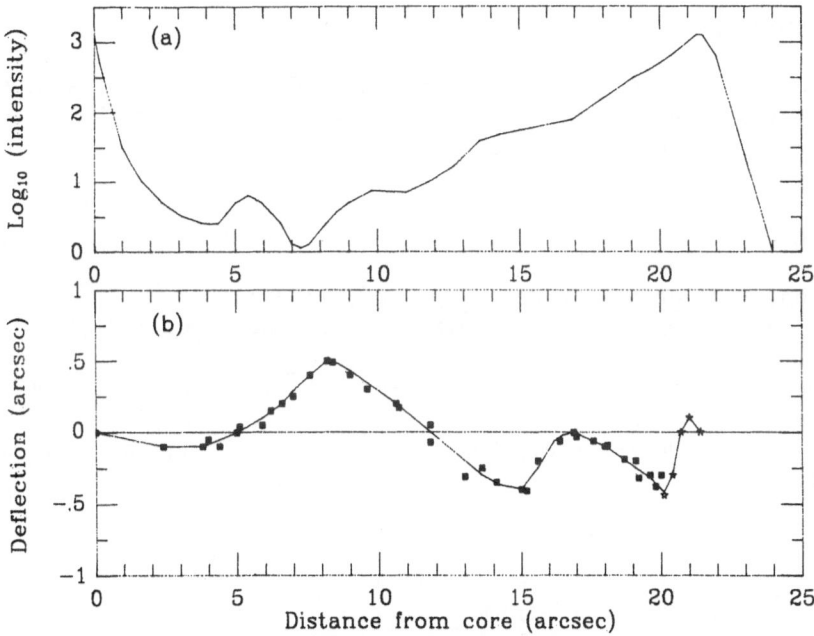

Fig. 3   (a) Plot of intensity along the ridge of the jet.
           (b) Sideways deflection of ridge-line.

3.2. Wiggle

Fig. 3 shows that the wavelength of the wiggle in the ridge line of emission decreases monotonically by a factor of 6 as the radio brightness increases along the jet. This is interpreted as a deceleration of the bulk flow: energy being dumped in the synchrotron reservoir. Strictly this results in a reduction of $\beta_{obs}c$:-

$$\frac{c \beta \sin \theta}{(1 - \beta \cos \theta)}$$

which, if 3C273 is close to the angle to the l.o.s. (θ) for maximum superluminal velocity, reduces to $\beta\gamma c$. One can argue about velocities at each end of the jet a) superluminal flow in nucleus $\gamma\beta \simeq 5$, b) ram pressure arguments concerning the working surface at the head of the jet $\gamma\beta \simeq 0.1^6$, but the new observable is that $\gamma\beta$ or more strictly β_{obs} changes by 6 along the jet. One can then start to build a consistent model for the dynamics of this source.

3.3 Conclusion

It appears that Doppler beaming is insufficient to hide any lobe counterpart and probably it is insufficient to hide any counter head. 3C273 contains all the ingredients of a classical double source like Cygnus A, with core, jet, hotspots, wiggle, extended lobe, but on one side only.

ACKNOWLEDGEMENTS

This work has involved collaboration with R.G. Conway and T.W.B. Muxlow. Thanks must be given to R.G. Noble and Angela Bayley for the software on the new Starlink node at Jodrell Bank.

REFERENCES

1. Pearson, T.J. et al., Nature 290, 365, (1981).
2. Foley, A.R. & Davis, R.J., Mon.Not.R.astr.Soc. 216, 679,(1985).
3. Schilizzi, R.T. & de Bruyn, A.G., Nature 303, 26, (1983).
4. Cornwell, T.J. & Wilkinson, P.N., Mon.Not.R.astr.Soc. 196,1067,(1981).
5. Flatters, C. & Conway, R.G., Nature 314, 425, (1985).
6. Conway, R.G., Davis, R.J., Foley, A.R. & Ray, T.P., Nature 294, 540, (1981).

RADIO STRUCTURE AND THE IR - UV SPECTRA OF QUASARS

Beverley J. Wills and D. Wills
McDonald Observatory and Department of Astronomy
University of Texas, RLM 15.308
Austin, Texas 78712
U.S.A.

Here we present two main results that relate properties of quasar IR-to-UV spectra to the fraction of radio flux density in the compact core.

Wills and Browne (1986) have shown that there is a highly significant inverse correlation between the line widths (FWHM) of broad Hβ and R, the ratio of radio core flux density to that in the extended lobes. An updated version of their data is shown in Figure 1. In the relativistic beaming model for radio sources R is related to the angle, θ, between the radio source axis (the beam direction) and the line-of-sight. The correlation can then be interpreted as motion of the emission line gas confined predominantly to a plane perpendicular to the radio axis - perhaps in a disk. The curve in Fig. 1 shows the expected relation for a specially simple model in which FWHM = $[(4000)^2 + (13000)^2 \sin^2 \theta]^{1/2}$ km/s and the relation between R and θ is as given by Orr and Browne (1982). There must be a range of intrinsic velocity dispersion from one quasar to another.

The simple unified beaming picture requires that aspect invariant optical properties of core- and lobe-dominated sources be the same, but

1. The broad line intensity ratio Fe II (opt)/Hβ is (statistically) smaller in lobe-dominated quasars. This is probably the result of incorrect setting of the continuum level when measuring Fe II. This is because blending of broad lines then causes the Fe II strength to be underestimated more in the broader-lined quasars.

2. Boroson, Persson and Oke (1985, see also Steiner 1981) find that [O III] lines in the fuzz around quasars are significantly stronger for lobe-dominated sources. We suggest that this is still consistent with the beaming hypothesis and is the result of (i) favouring less luminous lobes for core-dominated sources in flux density limited radio surveys, where the number of sources increases markedly for fainter lobes (according to the log N - log S relation), combined with (ii) the known strong correlation between luminosity of [O III] and luminosity of the extended Kpc-scale radio structure. (This relation is strong for the less luminous AGN, and is consistent with quasar data, but as a result of selecting the most luminous objects at larger redshifts, we cannot be sure.)

The second main result is that we find a significant difference between quasars with curved and power law IR (1-5, 10 or 20 μ) spectra in the distributions of both log R and FWHM for broad Hβ (see Fig. 2 where we also give the sense of the IR curvature). The curvature is due to the presence of a flattish ("thermal") component in the optical - UV spectral region, superposed on a steep IR component. These results suggest that R tends to be larger in quasars where the steep IR component dominates. Perhaps the IR component is beamed or otherwise related to the luminosity of the compact radio core.

G. Swarup and V. K. Kapahi (eds.), Quasars, 215–216.
© *1986 by the IAU.*

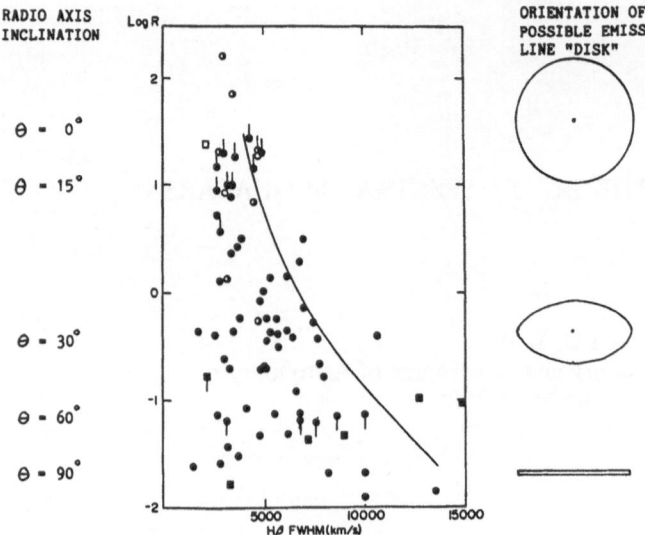

Fig. 1: The correlation between the dominance of the compact radio core, measured by R, and the line width (FWHM).

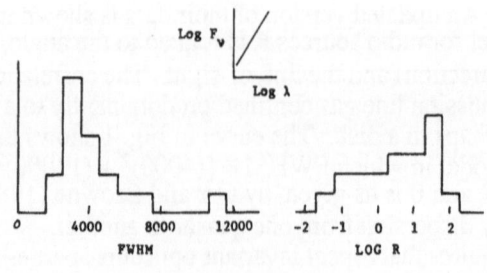

Fig. 2: Histograms of FWHM and Log R for quasars with steep and curved IR spectra.

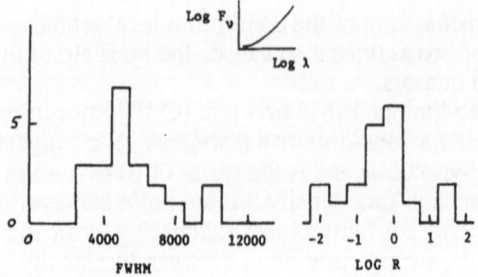

REFERENCES

Boroson, T.A., Persson, S.E., and Oke, J.B. 1985, *Ap.J.*, **293**, 120.
Orr, M.J.L. and Browne, I.W.A. 1982, *M.N.R.A.S.*, **200**, 1067.
Steiner, J.E. 1981, *Ap.J.*, **250**, 469.
Wills, B.J. and Browne, I.W.A. 1986, *Ap.J.*, March 1.

HIGH RESOLUTION VLA OBSERVATIONS OF CORE-DOMINATED QUASARS

C.P.O'Dea and R.Barvainis
NRAO, Edgemont Road, Charlottesville, VA 22903

P.Challis
University of Michigan
953 Physics/Astronomy Building, Ann Arbor, MI 48109

1. THE OBSERVATIONS

We selected 16 core-dominated sources based on two criteria. (1) They have a core-jet VLBI morphology. (2) Previous observations at lower resolution indicated that there was extended structure on scales of $\geq 1"$. The sources were observed in the A configuration of the VLA at 6 and 2 cm with typical resolutions of ~0".45 and ~0".15 respectively. Contour plots of total intensity with E vectors superposed (length proportional to fractional polarization) of four sources with jets are shown in Figure 1.

2. RESULTS

At high resolution two of the sources show no extended structure and seven show only a secondary component. We have discovered one-sided jets in 0605-085, 0735+178, and 1823+568. We confirm the existence of one-sided jets previously reported in 1510-089, 1642+690, 1807+698, and 2037+511. Although there are counter examples, the general trends are as follows. The jets terminate at or near brightness enhancements (hot spots or knots); i.e., they have a morphology which is more consistent with Fanaroff and Riley class II than class I. Note, however, that sensitivity limitations may prevent us from detecting the entire length of the jet. In the jets where we have adequate polarization data the magnetic field tends to be initially parallel to the jet axis. In 0605-085, 1510-089, 1642+690, and 1823+568 the magnetic field flips to become perpendicular to the jet axis at the terminal hot spot.

3. DISCUSSION

The powers of both the core and extended emission in the core-dominated sources studied here are sufficiently large that nearly all of the sources are expected to have one-sided jets, FR II morphologies, and projected magnetic fields which are parallel to the jet axis for roughly the entire length of the jet (Bridle 1984, A.J.,89,979). The observations reported here confirm these expectations. We find that the properties of the extended emission in these core-dominated sources are comparable to those seen in powerful lobe-dominated sources.

G. Swarup and V. K. Kapahi (eds.), Quasars, 217–218.

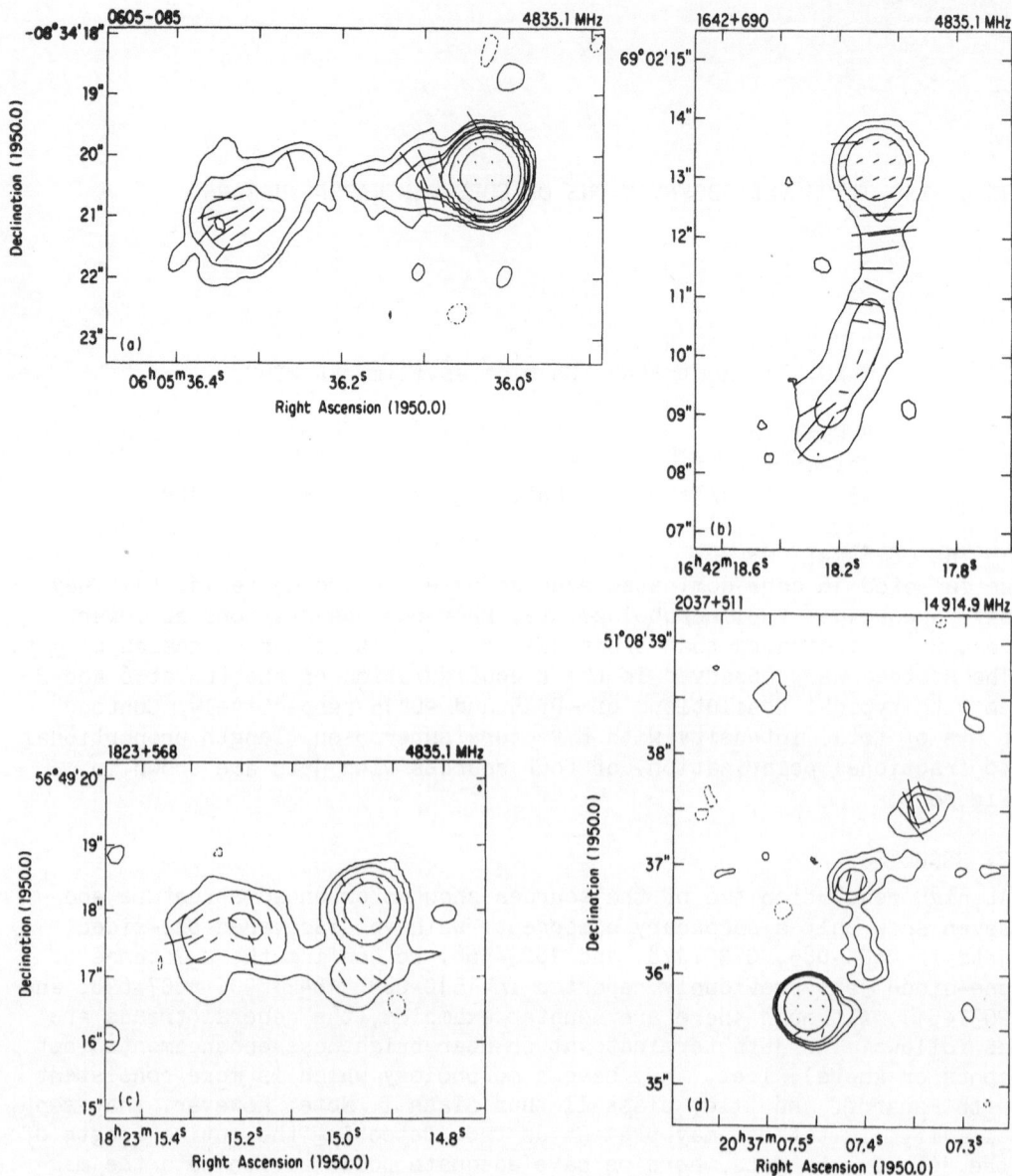

Figure 1. Contour plots of total intensity with E vectors superposed (with length proportional to fractional polarization). Contour levels in multiples of 1.0 mJy per beam and the FWHM of the circular CLEAN beams are as follows. $0605 - 085$: -0.5, 0.5, 1.0, 3.0, 5.0, 10.0, 30.0, and 200.0, $0\rlap{.}''50$; $1642 + 690$: -1.0, 1.0, 7.0, 50.0, $0\rlap{.}''50$; $1823 + 568$: -1.0, 1.0, 5.0, 30.0, and 200.0, $0\rlap{.}''45$; $2037 + 511$: -2.5, 2.5, 5.0, 10.0, 100.0, $0\rlap{.}''2$.

A STUDY OF ONE-SIDED RADIO QUASARS

D.J.Saikia[1], P.Shastri[1], T.J.Cornwell[2], C.J.Salter[3]

1 Tata Institute of Fundamental Research, Bangalore, India
2 NRAO, Socorro, New Mexico, U.S.A.
3 IRAM, Granada, Spain

Although most powerful extended radio quasars have lobes of radio emission on opposite sides of the nuclear component, a significant number appear to have extended emission on only <u>one</u> side. We have observed ~40 such sources (mostly from the compilation of Kapahi 1981) with the VLA, to confirm their classification and study their properties.

The new observations show that many of these sources do not really belong to the one-sided class. In most of these, a component was detected on the other side which was missed earlier either due to its very small separation from the core (e.g., 1509+158, fig.1a), or due to its low surface brightness (e.g., 1012+232, fig.1b).

We find that one-sided sources tend to have more dominant cores and smaller linear sizes. The median value of the fraction of emission from the core at an emitted frequency of 8 GHz, f_c, is ~0.7 for 21 quasars (redshifts known, S(178 MHz)$\gtrsim$2Jy). The median linear size, ℓ, for this sample is ~25 kpc. In comparison, for 131 double-lobed quasars (S(178)$\gtrsim$ 2Jy), the median values of f_c and ℓ are ~0.15 and ~200kpc respectively. The one-sided quasars mapped by Perley (1982) selected at high frequencies have a median of ~0.95. These values are consistent with the possibility that one-sided sources are inclined at small angles to the line of sight (Kapahi 1981).

Two-sided sources tend to have their core polarization E-vector (λ6 cm) normal to the source orientation, with no such trend for one-sided sources, supporting the above interpretation(Saikia & Shastri 1984).

Although most one-sided sources are core-dominant, a few have weak cores, i.e., $f_c \lesssim$ 0.1 e.g., 1729+501,fig.2). Perhaps in these sources we are witnessing the "first flip before the flop" (Rudnick & Edgar 1984).

Acknowledgement. The National Radio Astronomy Observatory is operated by Associated Universities Inc., under contract with the National Science Foundation.

REFERENCES
Kapahi,V.K., 1981. J.Astrophys.Astr. 2, 43.
Perley, R.A., 1982. Astr.J. 87, 859.
Saikia, D.J., Shastri, P., 1984. Mon.Not.R.astr.Soc. 211, 47.
Rudnick, L., Edgar, B.K., 1984. Astrophys.J. 279, 74.

G. Swarup and V. K. Kapahi (eds.), Quasars, 219–220.
© *1986 by the IAU.*

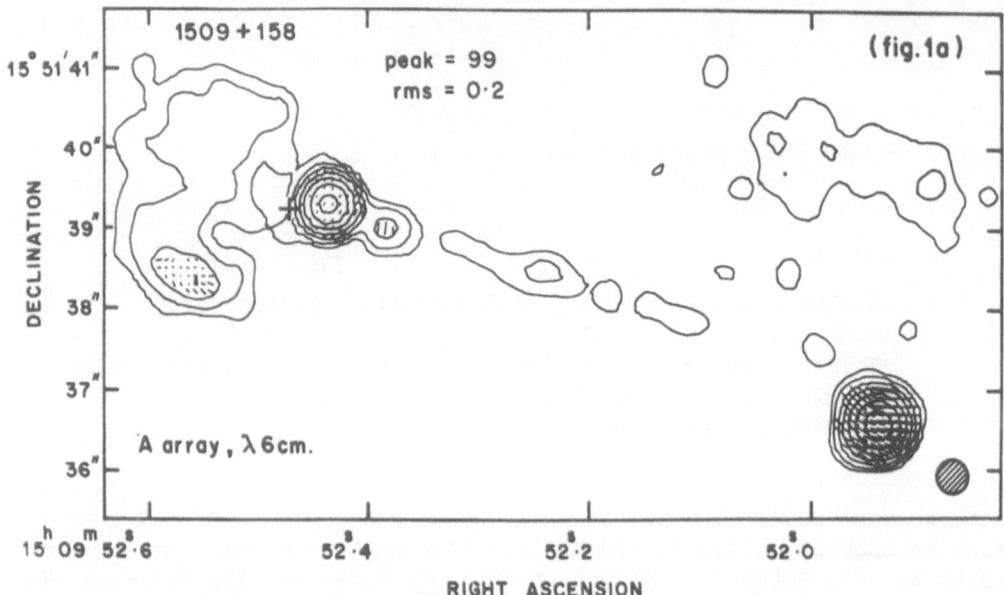

Fig. 1 a,b The radio maps with
fractional linear polarization
superposed. The surface brightness peak, rms and contour levels are in mJy/beam.
Contours : a) -1,1,2,4,8,16,32,
64. b) -4,4,8,16,32,64,128,256,
512.

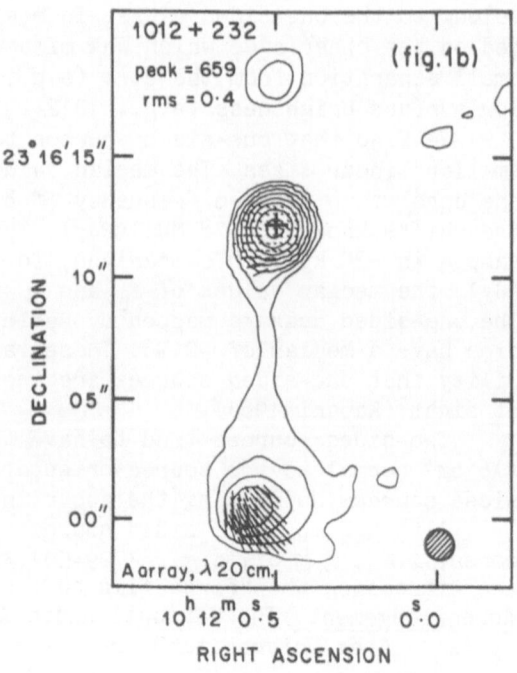

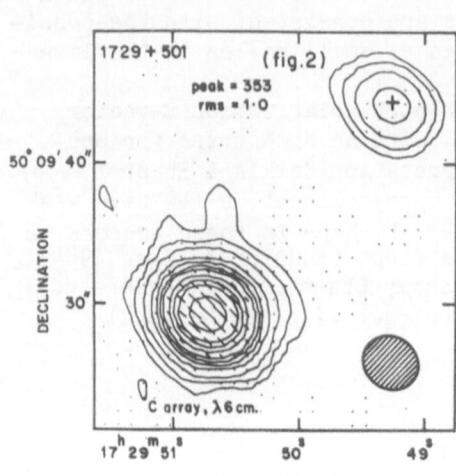

Fig. 2 The radio map with linear
polarization intensity superposed.
Contours : 353x(-0.01,0.01,0.02,0.04,
0.08,0.12,0.16,0.2,0.3,0.4,0.5,0.75).

GENERAL DISCUSSION FOLLOWING PAPERS ON RELATIVISTIC BEAMING

Burbidge : What type of observation would lead you to abandon the relativistic jet hypothesis ?

Blandford : This is a serious question to which I will try to give a serious answer (cf. proceedings at Manchester conference on AGN). My view is not that there are potential observations that absolutely disprove the relativistic beaming hypothesis but instead that there are observations of specific sources that are more naturally explained by alternative models. Three examples include :
i)   Sources in which the flat-spectrum core moves and the steep spectrum jet is stationary (relative to a background point source).
ii) Well defined expansion speeds in excess of 100 c in quasars.
iii) Prevalence of expansion speeds >10 c in steep spectrum radio galaxies and quasars with central jets well aligned with the extended structure.

Kellermann : It seems to me that the issue is not whether relativistic beaming is important, because as we have heard the evidence from superluminal motion and the absence of inverse Compton flux is overwhelmingly in favour. However, some people have tried to explain a wide variety of other phenomena purely on the basis of geometric orientation, and what we have heard today and yesterday suggests that things are not that simple, and that intrinsic effects are also important. But we should not minimise the importance of these attempts at unification, because they have served to focus much of the research that is being discussed here.

Blandford : I wish to make a comment concerning some of the tests of the beaming hypothesis. They are predicated on what I regard as an overly simplistic model of the emission - viz. that there is a stationary ballistic outflow. In such an outflow, the electron losses should be catastrophic unless there is "in situ" acceleration. Now, the most natural way to effect this acceleration is to pass through a shock front. Shock fronts change the velocity of the flow both in magnitude and direction. They therefore change the beaming pattern. (The velocity of the shock front itself, which may appear superluminal, will be different again.) I therefore argue that realistic models of radio emission from compact jets will not give the flux distributions with angle that some of these investigations have been testing. Relativistic beaming is such a strong amplifier that we may be "blinded" by a relatively insignificant part of the source coming towards us.

G. Swarup and V. K. Kapahi (eds.), Quasars, 221.
© *1986 by the IAU.*

Marshall Cohen, G.Srinivasan, Vijay Kapahi and Martin Rees

RESULTS FROM X-RAY SATELLITES

G. Zamorani
Istituto di Radioastronomia
Via Irnerio 46, 40126 Bologna
Italy

ABSTRACT. One of the best studied results of the X-ray observations of
Active Galactic Nuclei is the statistical correlation which has been
observed between the X-ray and optical luminosities in these objects.
In this paper I will review the present situation of the analysis of
such correlation, focusing my attention on the large number of topics
which are more or less directly linked to it.

1. INTRODUCTION

It is not easy to give today a review talk on "Results from X-ray
Satellites" for quasars and Active Galactic Nuclei (AGNs). The main
reason for this is that the X-ray data for these objects (mainly from
the EINSTEIN Observatory) have been extensively discussed and already
presented in many Meetings in the past few years.   Extensive reviews,
which cover many different aspects of the X-ray properties of AGNs, such
as statistical analysis of the correlations observed in various samples
of   quasars (radio quiet vs.   radio loud), X-ray spectra, X-ray
variability, X-ray evolution and luminosity functions, have   been
recently published (see, for example, Zamorani 1984a, 1984b; Avni 1985a,
1985b; Mc Hardy 1985; Elvis and Lawrence 1985). As a consequence, at
least until EXOSAT data are more diffusely presented and discussed in
the literature, it is difficult to say something new on these subjects.
This does not mean that all the possible information has already been
extracted from the EINSTEIN data. On the contrary, very interesting new
results are becoming available only now, as, for example, the results on
the spectral data, presented at this Meeting (Elvis et al. 1986, Wilkes
and Elvis 1986).
    For this reason, and since other papers on a variety of various
topics connected to the X-ray properties of quasars and other AGNs are
presented in this Meeting, this paper will be organized as a sort of
monography. Rather than showing a long list of results, not immediately
related with each other, I will discuss, instead, a single observational
result, the correlation between X-ray and optical luminosities, trying
to show how an in depth analysis of this correlation is directly   linked

G. Swarup and V. K. Kapahi (eds.), Quasars, 223–231.
© *1986 by the IAU.*

to many other both observational and interpretative data, some of which will be addressed in more detail in other papers in these Proceedings.

As an example, some of the questions on which the study of the correlation between X-ray and optical luminosities has some bearing are:

a) evolution, with cosmic time, of the X-ray and optical luminosities;

b) estimate of the X-ray counts and of the contribution of AGNs to the soft X-ray background;

c) variability;

d) low energy absorption in low luminosity AGNs;

e) differences in the X-ray properties of radio quiet and radio loud quasars;

f) comparison with other existing correlations as, for example, the infrared - X-ray correlation.

In the next Sections it will be shown how our knowledge of each of these topics is connected to the study of the correlation between X-ray and optical luminosities.

2. THE CORRELATION BETWEEN X-RAY AND OPTICAL LUMINOSITIES

2.1. The Data

Since the first EINSTEIN observations of radio quiet quasars, the existence of some kind of statistical correlation between X-ray and optical luminosities became evident (Zamorani et al. 1981). On the basis of different samples, this relationship has been described by various authors as a power law ($\log L_x = \beta \log L_o$ + constant), with β in the range 0.6-1.0.

The most recent and complete analysis on this topic is by Avni and Tananbaum (1986; AT) on a large sample (154 objects) of optically selected quasars. Their sample is mainly based on two complete magnitude limit samples, the BQS and the BF quasars with blue magnitude limits 16.16 and 19.8, respectively. The large number of objects, and the wide dynamical range (both in luminosity and redshift) spanned by these objects allowed them to apply a very detailed statistical analysis. Figure 1 shows the X-ray versus optical monochromatic luminosity for the objects analyzed by AT. They fit their data with a simple linear dependence of α_{ox} on cosmological look-back time $\tau(z)$ and on $\log L_o$:

$$\alpha_{ox}(z,L_o) = A_z \, \tau(z) + A_o \log L_o + A + \text{(residual)} \qquad (1)$$

assuming, as initial guess, that the residuals are Gaussian distributed around the best fit line. Determining, with a parametric version of the Detections and Bounds (DB) method (Avni et al. 1980), the allowed region of the parameters A_z and A_o, they find that α_{ox} depends predominantly on L_o, rather than on redshift. In particular, they find that $A_z = 0.0$ is fully consistent with their data. The best fit value of the parameter A_o corresponds to L_x proportional to $L_o^{0.8}$. The dispersion around the best fit line (one sigma) is at least a factor of 3.0. Similar results, from a partially independent sample, have been obtained by Kriss and Canizares (1985).

2.2 Evolution with Cosmic Time of X-ray and Optical Luminosities

As seen in Eq. 1, A_z = 0.0 means that two objects with the same optical
luminosity and different redshift statistically have the same X-ray
luminosity. In other words, this implies that there is no explicit
evolution with cosmic time of the ratio of X-ray to optical luminosity.
On the other hand, within the framework of pure luminosity evolution
models (which have become quite popular in the recent years) one can
interpret the observed correlation between X-ray and optical
luminosities as the path along which single objects move, while dimming
with cosmic time. In this case, a value in the slope different from 1.0
introduces an implicit dependence on redshift: while an object dims,
its ratio of X-ray to optical luminosity is changing. This is an
interesting case in which the same observed correlation has two
different physical interpretations as a function of the assumed
evolutionary model. Theoreticians may tell us which interpretation they
like more.

Moreover, as shown by Avni and Tananbaum (1982), a slope smaller
than one in the correlation implies that, within pure luminosity
evolution models, the cosmological evolution of the X-ray luminosity

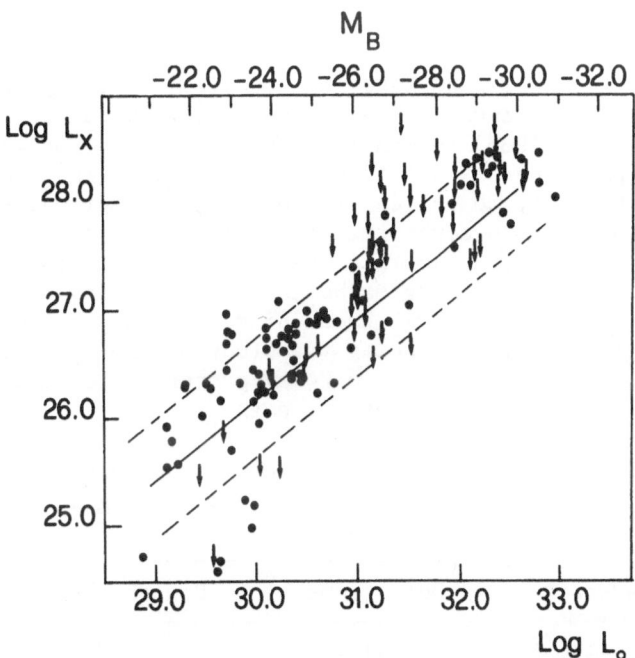

Figure 1. X-ray versus optical monochromatic luminosity for a sample of
154 optically selected quasars (see Avni and Tananbaum 1986). The solid
line represents the best fit (slope 0.8). The two dashed lines
represent the width (one sigma) of the Gaussian distribution of the
residuals.

function should be weaker than the evolution of the optical luminosity
function. The available data seem to be in agreement with this
prediction (Maccacaro, Gioia and Stocke 1984).

2.3 Skewness of the α_{ox} Distribution

In their regression analysis (see Eq. 1), AT found that the
distribution of the residuals around the best fit line is not well
represented by a single gaussian, but requires a significant skewness,
with a longer tail at high α_{ox} (low L_x) and a shorter tail at low α_{ox}
(high L_x). This result is not a purely mathematical exercise, because
the exact shape of this distribution has important numerical
implications in computing, for example, the expected X-ray number
counts. The correlation between L_x and L_o has been used by various
authors to predict the X-ray counts starting from the optical luminosity
and evolution functions (see, for example, Avni and Tananbaum 1986;
Schmidt and Green 1986). All these computations are critically affected
by the detailed shape of the assumed correlation, so that the best
possible analysis of the "observed" data points is crucial for this
purpose. Franceschini, Gioia and Maccacaro (1985) have shown that the
use of the "observed" distribution of residuals around the best fit line
in the L_x-L_o plane leads to significant inconsistencies between
predictions and observations. Most of these discrepancies can be
eliminated if one assumes a narrower "intrinsic" width, which has been
artificially broadened into the observed one due to a variety of
external effects. The most obvious of these effects are variability and
absorption (at both optical and X-ray frequencies) and errors in the
estimated optical and X-ray fluxes as well. A completely independent
analysis led Zamorani (1985) to conclude that long term variability may,
indeed, have produced a significant broadening in the observed
distribution of the X-ray to optical fluxes in the optically selected
sample. While there is no doubt that the net result is a broadening of
any intrinsic correlation, a quantitative estimate of this effect can
only be achieved by obtaining a better understanding of each of the
causes which may determine such broadening. This is an excellent
example of the necessity of linking the study of the L_x-L_o correlation
with the knowledge of other intrinsic X-ray properties of quasars.
    Looking at Fig. 1, we see that while the low luminosity part of
the diagram is reasonably well determined (61 Detections and 19 Bounds
for log L_o < 31.0), the high luminosity region is populated by a large
number of limits (34 Detections and 40 Bounds for log L_o > 31.0). The
situation is even worse in redshift space:

Redshift	Detections	Bounds
0.0-1.0	71	16
>1.0	24	43

This implies that the shape of the distribution of residuals around the
best fit line which results from the DB analysis is largely determined
by low luminosity, small redshift objects. In this respect, it would be

extremely interesting to check whether other high redshift samples (in particular the grism selected sample described in Anderson (1985) and presented at this Meeting by Margon and Anderson (1986)) are consistent with this overall description or require some minor change.

Moreover, the skewness of the α_{ox} distribution, as determined by AT (1986), is largely required by the existence of a few detections and upper limits with a large value of α_{ox}. All these objects are in the low luminosity region of Figure 1, having log L_o < 31.0. Lawrence and Elvis (1982), by comparing hard and soft X-ray luminosities, suggested that a significant fraction of low luminosity AGNs might have their soft X-ray fluxes decreased because of intrinsic absorption. No evidence for a similar effect in high luminosity objects has been found up to now. It is then possible that these high α_{ox} objects are simply those in which low energy absorption is becoming dominant. If this is the case, skewness might not be an intrinsic property of the α_{ox} distribution. A possible way to test this would be to sample, with X-ray detections, the region of low X-ray, high optical luminosity. This can be achieved by obtaining deep X-ray observations of relatively bright (in optical) and high luminosity objects. It is possible that some of these observations are already available in the EINSTEIN data bank; if not, this is something that ROSAT might easily do.

3. COMPARISON WITH OTHER STATISTICAL CORRELATIONS

3.1 Radio Quiet Versus Radio Loud Quasars

It is well known (Ku, Helfand and Lucy 1980; Zamorani et al. 1981; see also Tananbaum et al. 1983) that the X-ray emission of radio loud quasars is, on average, stronger than that of radio quiet quasars. Recently, statistical associations between X-ray luminosity and extended radio luminosity (Feigelson, Isobe and Kembhavi 1984) and between X-ray luminosity and radio core emission (Kembhavi, Feigelson and Singh 1985) have been studied. A review of these results is presented by Kembhavi (1986).

The results of an extensive statistical analysis of the correlations between radio, optical and X-ray luminosities for various subsamples of radio loud quasars are presented by Worrall et al. (1986). They find a significant difference in the L_x-L_o correlation between radio loud with flat spectrum quasars, radio loud with steep spectrum and radio quiet quasars. This difference is such that, for a given optical luminosity, radio loud quasars with a flat radio spectrum are the strongest X-ray emitters, while radio quiet quasars are the faintest ones. Moreover, within each class of radio loud quasars the X-ray luminosity (for a given optical luminosity) tends to increase with increasing radio luminosity. They also find that the fits for the steep, extended radio loud quasars improve somewhat if the radio luminosity from the core alone is used. Their conclusion is that a radio loud quasar can be thought of as a radio quiet quasar with an additional physical component producing the core radio luminosity and an "extra" X-ray emission. This extra X-ray emission would dominate the observed X-ray flux in about 80% of flat radio spectrum objects. If this is true, one might expect to see different X-ray spectral shapes between radio quiet and core dominated radio loud quasars (see Wilkes and Elvis 1986).

3.2 The Correlation Between Infrared and X-ray Luminosities

The existence of a possible correlation between the 3.5 micron nuclear flux and the X-ray emission from Seyfert 1 galaxies was first pointed out by Elvis et al. (1978), and later confirmed by Kriss, Canizares and Ricker (1980). Recently, Malkan (1984) has found that a sample of more than 50 Seyfert 1 galaxies and quasars (both radio quiet and radio loud) shows a very tight correlation between infrared and X-ray luminosities. The best fit between the two luminosities has a slope 1.0, the dispersion around the best fit (one sigma) is only a factor of 2.0 and apparently there is no obvious distinction in this correlation between radio quiet and radio loud objects. Each of these findings is at variance with the results of the analysis of the correlation between optical and X-ray luminosities (see above). The best fit corresponds to an average slope of 1.18 between infrared and X-ray frequencies. This is a confirmation of what Malkan and Sargent (1982) suggested: an extrapolation of the red/infrared power law gives an excellent prediction of the 2 Kev flux, regardless of the ultraviolet or even the radio properties of the AGNs. (See Elvis et al. (1986) and Fabbiano et al. (1986) for more data on this subject).

These results have been taken as an indication that infrared and X-ray emissions are produced, in all these objects, by two related (or even the same) mechanisms, probably non-thermal; at optical and UV

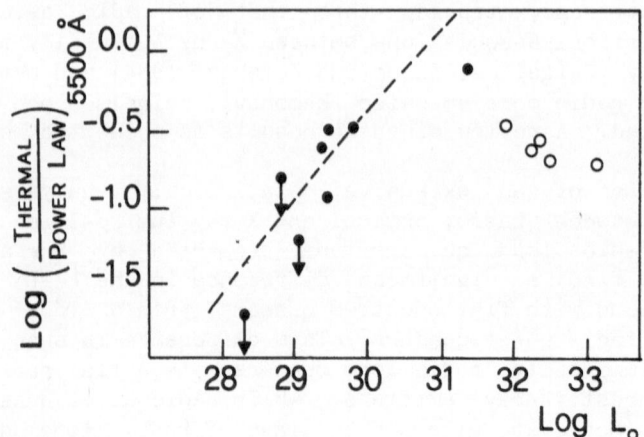

Figure 2. Ratio of thermal to power law components (at 5500 A) versus total optical luminosity. The data are taken from Malkan and Sargent 1982 (filled circles) and Malkan 1983 (open circles). The dashed line represents the trend in this diagram which would be necessary to reconcile with each other the two different slopes observed in the optical - X-ray and the infrared - X-ray correlations.

frequencies, instead, other emission mechanisms (as, for example, Balmer continuum recombination taking place in the broad emission line regions (Richstone and Schmidt 1980; Malkan and Sargent 1982), and thermal emission from an optically thick accretion disk (Malkan 1983)) are dominant, producing the so-called blue bump and hiding the underlying, "universal" power law with a slope of the order of 1.0-1.2. Within this framework, then, the decrease of the ratio of X-ray to optical luminosities with increasing optical luminosity (L_x proportional to $L_o^{0.8}$) can be understood if the ratio of the thermal to the power law component increases with increasing luminosity. This trend was, infact, suggested by the data in Malkan and Sargent (1982) (see Figure 2). Unfortunately, when five higher luminosity quasars were added to the sample (Malkan 1983), the correlation almost completely disappeared. This implies that at least one very elegant and appealing explanation for the observed difference in the L_x-L_o and L_x-L_{ir} correlations is not so clear any more.

At this point, to clarify a little bit the situation, I think that a few obervations would be extremely important:

a) One should try to obtain the spectral decomposition (power law + thermal component) for a larger number of objects. These objects should span a range as wide as possible in luminosity. In view of the difference in the X-ray - optical correlation for radio quiet, radio loud with flat spectrum and radio loud with steep spectrum quasars, all these classes of objects should be well represented in the sample. In costructing such a sample, objects which already have an X-ray measurement should have higher priority.

b) The infrared - X-ray correlation shown and discussed by Malkan (1984) was only for objects up to z = 0.3. It would be extremely important to extend it to higher redshift, by obtaining more infrared observations of high redshift objects. In order not to bias the sample, not only objects which have an X-ray detection, but also objects which have only an X-ray upper limit should be included in the sample. In this respect, it would be particularly interesting to check whether objects that appear to have an exceedingly low X-ray luminosity with respect to the optical are also amomalously infrared faint.

Before these data are available, I think we should be very careful in accepting the conclusion that infrared and X-ray emission mechanisms are strictly related to each other.

REFERENCES

Anderson,S.F. 1985, Ph.D. Thesis, University of Washington.
Avni,Y. 1985a, Proceedings of the XII Texas Symposium on Relativistic Astrophysics, Ann. N.Y. Acad. Sc., in press.
Avni,Y. 1985b, in "Structure and Evolution of Active Galactic Nuclei", G. Giuricin, F. Mardirossian, M. Mezzetti, and M. Ramella eds., (Dordrecht: Reidel), in press.
Avni,Y., Soltan,A., Tananbaum,H., and Zamorani,G. 1980, Ap.J., **238**, 800.
Avni,Y., and Tananbaum,H. 1982, Ap.J.(Letters), **262**, L17.

Avni,Y., and Tananbaum,H. 1986, Ap.J., in press.

Elvis,M.,   Maccacaro,T.,   Wilson,A.S.,   Ward,M.J.,   Penston,M.V.,
    Fosbury,R.A.E., and Perola,G.C. 1978, M.N.R.A.S., **183**, 159.

Elvis,M., and Lawrence,A. 1985, in "Astrophysics of Active Galaxies and
    Quasi-Stellar Objects", J.S. Miller ed., (University Science
    Books), p. 289.

Elvis,M., Czerny,B., Bechtold,J., and Green,R. 1986, This Volume , p.73.

Fabbiano,G., Ward,M.J., Elvis,M., Carleton,N., Willner,S.P., and
    Lawrence,A. 1986, This Volume , p.85.

Feigelson,E.D., Isobe,T., and Kembhavi,A. 1984, Astron.J., **89**, 1464.

Franceschini,A., Gioia,I.M.,and Maccacaro,T. 1985, Ap.J., in press.

Kembhavi,A. 1986, This Volume.

Kembhavi,A., Feigelson,E.D., and Singh,K.P. 1985, preprint.

Kriss,G.A., and Canizares,C.R. 1985, Ap.J., **297**, 177.

Kriss,G.A., Canizares,C., and Ricker,G. 1980, Ap.J., **242**, 492.

Ku,W.H.M., Helfand,D. and Lucy,L.B. 1980, Nature, **288**, 323.

Lawrence,A., and Elvis,M. 1982, Ap.J., **256**, 410.

Maccacaro,T., Gioia,I.M., and Stocke,J.T. 1984, Ap.J., **283**, 486.

Malkan,M. 1983, Ap.J., **268**, 582.

Malkan,M. 1984, in "X-ray and UV Emission from Active Galactic Nuclei",
    W. Brinkmann and J. Trümper eds., MPE Report **184**, p. 164

Malkan,M., and Sargent,W. 1982, Ap.J., **254**, 22.

Margon,B., and Anderson,S.F. 1986, This Volume , p.247.

McHardy,I. 1985, Space Sc. Rev., **40**, 559.

Richstone,D.O., and Schmidt,M. 1980, Ap.J., **235**, 361.

Schmidt,M., and Green,R.F. 1986, Ap.J., in press.

Tananbaum,H., Wardle,J.F.C., Zamorani,G., and Avni,Y. 1983, Ap.J., **268**,
    60.

Wilkes,B., and Elvis,M. 1986, This Volume , p.261

Worrall,D.M., Giommi,P., Tananbaum,H., and Zamorani,G. 1986, This
    Volume , p.263

Zamorani,G. 1985, Ap.J., **299**, in press.

Zamorani,G. 1984a, Proceedings of I.A.U. Symp. No. 110 on "VLBI and
    Compact Radio Sources", R. Fanti, K. Kellermann, and G. Setti
    eds., (Dordrecht: Reidel), p. 85.

Zamorani,G. 1984b, Proceedings of the International Symp. "X-Ray
    Astronomy '84", M. Oda, and R. Giacconi eds., (Institute of Space
    and Astronautical Science - Tokyo), p. 419.

Zamorani,G., Henry,J.P., Maccacaro,T., Tananbaum,H., Soltan,A., Avni,Y.,
    Liebert,J., Stocke,J., Strittmatter,P.A., Weymann,R.J., Smith,M.G.,
    and Condon,J.J. 1981, Ap.J., **245**, 357.

DISCUSSION

Malkan : To update what you said about the correlation of the thermal/
nonthermal ratio with luminosity presented in Malkan & Sargent : though
it is poor for quasars, our new analyses of Seyfert 1's strengthen the
correlation at low luminosities. That is, high and low-luminosity quasars appear to have blackbody/power-law ratios (at 5450 A) of 0.5 to 1,
whereas many Seyfert 1's have ratios of 0.1. So I think it's still possible this correlation may explain the α_{ox}/L trend, and the Baldwin
effect.

Wampler : There are a few (2) broad absorption line (BAL) quasars known
that have M II $\lambda 2800$ in absorption. The M II $\lambda 2800$ absorption line
has a central depth of only about 50% and seems to separate the emission
of a small central source from that of a larger source. At least in the
case of PG 1700+518 the underlying radiation from the small source is a
power law ($f_\nu \propto \nu^{-\alpha}$) with $\alpha = 1.2$. Other BAL quasars may be useful in
separating the various components that contribute to the optical conti-
nuum and aid in understanding of the general connection between the
IR-opt-X ray portions of the continuum.

Segal : An apparently high correlation between the luminosities in two
different frequency bands can easily be the spurious consequence of a
not necessarily valid cosmology in which flux varies rapidly with redshift since the derived luminosities will then have this flux-redshift
function as a common factor. Can you exclude this possibility in the
cases you report.

Zamorani : Yes, I think I can - infact, for the data that I have shown
there is a highly significant correlation also between the observed fluxes, which are obviously independent of any assumption on cosmology.

THE MEDIUM SENSITIVITY SURVEY AND THE X-RAY SELECTED QUASAR SAMPLE

Isabella M. Gioia[1,2], Tommaso Maccacaro[1,2], Rudolph Schild[1],
John Stocke[3], and John Danziger[4]

1) Harvard-Smithsonian Center for Astrophysics, Cambridge, Mass.
2) On leave of absence from Istituto di Radioastronomia, Bologna,
   Italy.
3) Center for Astrophysics and Space Astronomy, Boulder, Colorado.
4) European Southern Observatory, Garching bei Munchen, FRG.

ABSTRACT. We find that the inconsistencies discussed by many authors
between the expected and observed number of X-ray selected quasars
cannot be attributed to incompleteness of the Medium Sensitivity
Survey (MSS). We are presently expanding the survey, this effort will
produce a very large sample of X-ray selected quasars. Preliminary
results are presented.

Quasars were originally discovered through the identification of radio
sources. Radio surveys, however, are not a very efficient way to
discover new quasars. Only a small fraction, about 10%, of the
quasars are in fact radio sources at flux levels detectable in surveys
of a large area of sky (e.g. the Bologna, Parkes, and Cambridge
surveys) and only about 20% to 40% of the radio sources detected in
these surveys (depending on frequency, limiting flux etc.) are
identified with quasars. Optical surveys, taking advantage of the
particular position of quasars in a multicolor diagram, the quasar UV
excess, or the prominence of their emission lines are, by far, the
most efficient method of selecting large quasar samples.

In the last few years, as a consequence of the launch of the Einstein
Observatory, an X-ray telescope capable of detecting sources as faint
as 5×10^{-14} ergs cm^{-2} s^{-1}, X-ray surveys have also become an
efficient way of discovering new quasars. X-ray emission is a common
feature of quasars and the work of Avni and Tananbaum (1986) has shown
that X-ray quiet quasars, if they exist, are limited to a very small
fraction (<8%) of the quasar population. In a soft X-ray survey with
a limiting sensitivity of 10^{-13} ergs cm^{-2} s^{-1}, slightly more than half
of the high galactic latitude sources are Active Galactic Nuclei
(quasars and Seyfert galaxies) and they are detected at a rate of
about 3 per square degree (cf. Gioia et al. 1984).

233

G. Swarup and V. K. Kapahi (eds.), Quasars, 233–238.
© *1986 by the IAU.*

As the samples of X-ray selected quasars grow in number, it is possible to investigate whether they have the same properties of those selected by optical and radio techniques or whether different classes of quasars exist. It is possible, in addition, to test models of quasar evolution derived solely from optical data.

Maccacaro (1984) has pointed out and has briefly discussed the inconsistencies in some of the properties of optically selected and X-ray selected quasars, including their number-counts. More recently, a number of authors have undertaken a quantitative comparison of the properties of optically selected and X-ray selected quasars. We will not summarize here their detailed findings, and we refer the reader to a number of recent publications (Kriss and Canizares, 1985, Franceschini, Gioia and Maccacaro 1986, Schmidt and Green 1986 and Avni and Tananbaum 1986).

The approach followed by these autohrs consists of using the information derived from optical surveys (e.g. a luminosity function plus an evolution model) and combining it with a "description" of the X-ray to optical luminosity ratio to compute expected redshift and luminosity distributions, evolution functions, number counts, etc. for X-ray selected quasars.

The common conclusion is that the current models for the properties of optically selected quasars (luminosity function, evolution, X-ray to optical luminosity ratio) lead to inconsistencies with the observed properties of X-ray selected quasars and, in particular, to the prediction that many more X-ray selected quasars should have been detected than are actually observed.

All the above authors have used the same sample of X-ray selected quasars, the one extracted from the Medium Sensitivity Survey (cf. Maccacaro et al. 1984). This has prompted a detailed examination of possible causes of incompleteness of the MSS quasar sample (Maccacaro and Gioia, 1986).

The published Medium Sensitivity Survey is the result of the analysis of about 350 IPC observations, scattered over the sky, with the restriction that they are at least 20° above or below the galactic plane. These pointing directions are characterized by different values of the hydrogen column density. However, in the survey work, for simplicity, a constant value of 3×10^{20} cm^{-2} has been assumed for the computation of the source X-ray flux and of the image limiting sensitivity. In the soft X-ray band, photoelectric absorption plays an important role and the softer the X-ray band used, the more pronounced are the effects of small changes in the value of the hydrogen column density. The IPC has a significant fraction of its effective area below 0.3 keV. In the MSS however, only the counts recorded in the 0.3 - 3.5 keV band have been used to determine the source parameters. This reduces in part the effects of photoelectric absorption, but does not eliminate them entirely. The rather poor energy resolution of the IPC exacerbates the problem.

Maccacaro and Gioia (1986) have therefore determined, interpolating HI measurements from the recent survey of Stark et al. (1986), the best estimate for the hydrogen column density along the line of sight for the individual IPC images used in the MSS. The distribution of the values of the hydrogen column density has a mean of $\log N_H = 20.5$ or $N_H = 3.16 \times 10^{20}$. This confirms the correctness of the choice of $N_H = 3 \times 10^{20}$ cm^{-2} as a value representative of all the MSS fields. The distribution is slightly asymmetric with respect to the sharp peak, with more fields being characterized by a hydrogen column density smaller than the peak value. The smallest value of N_H seen is 6.1×10^{19}, the largest is 1.4×10^{21}.

The survey has then been divided into two independent subsets, depending on the value of the hydrogen column density pertinent to each observation. The first subset ("HI") is comprised of all the fields for which N_H is larger than 3.37×10^{20} cm^{-2}, and by the X-ray sources detected in these fields. Images of the regions of sky characterized by N_H smaller than 3.37×10^{20} cm^{-2}, and the corresponding sources, define the second subset ("LO"). The value 3.37×10^{20} has been chosen so that the two subsets contain the same number of IPC images.

The possible effects of the neglected non-uniformity of the hydrogen column density on the number of sources detected, has been checked by testing whether X-ray sources are detected in different proportions in the two subsets, and in particular whether the HI subset is characterized by fewer sources than the LO subset.

This analysis has been restricted to the fields for which the neutral hydrogen data are available ($\delta > -40°$). Of the 79 extragalactic sources detected in these images, 41 (52%) belong to the HI subsample and 38 (48%) belong to the LO subsample. These fractions are identical (within the statistical uncertainties due to the limited statistics) to what is expected from the sky coverage of the two subsets. The flux distributions for the sources in the two subsets are also identical.

Maccacaro and Gioia (1986) therefore concluded that there is no evidence of a loss of sources due to the fact that a number (~50%) of IPC images are characterized by values of the hydrogen column density higher than assumed. Consequently, the extragalactic MSS sample, and thus the quasar sample, can be considered statistically complete. Therefore the inconsistency between the expected and the observed number of X-ray selected quasars cannot be attributed to incompleteness of the X-ray selected quasar sample.

Kriss and Canizares (1985) pointed out that some of the discrepancies between the properties of X-ray selected and optically selected quasars must be attributed to the presence of numerous low luminosity, partially reddened and X-ray absorbed objects in the X-ray selected sample. These quasars are not represented in the optically selected samples. These authors have thus restricted their analysis to the MSS quasars characterized by B-V > 0.65. They have shown that when the apparently red objects are excluded the agreement with the optical sample improves considerably. This, however, does exacerbate the discrepancy between the observed and predicted number counts.

Franceschini, Gioia and Maccacaro (1986) have explored a different possibility. These authors have found that the properties of X-ray selected quasars predicted from those of optically selected quasars critically depend on the shape of the distribution of the residuals in the X-ray to optical luminosity correlation. Franceschini, Gioia and Maccacaro have suggested that the observed dispersion σ of the α_{ox} distribution ($\alpha_{ox} = -\log[Lx/Lo]/2.605$) of optically selected quasars suffers from a systematic bias, such that values substantially higher than expected are observed. They interpret this as due to the presence of "noise" which adds to the intrinsic variance of the distribution. When a narrower, "intrinsic" α_{ox} distribution is used, most of the discrepancies, including the number-counts inconsistency, are eliminated.

Clearly there is the need for a larger sample of X-ray selected quasars.

A major expansion of the Medium Survey has begun. As of this writing about 1200 IPC images have been analyzed, yielding a useful area of 600 square degrees. We have detected a total number of about 700 serendipitous sources with fluxes in the range $\sim7 \times 10^{-14}$ to $\sim10^{-11}$ ergs cm^{-2} s^{-1}. The extension of the Medium Sensitivity Survey will be completed early next year. It is expected to contain approximately 900 X-ray sources. Of these, at least 400 are expected to be identified with quasars and Seyfert galaxies (AGN). The present status of the expanded Medium Survey is summarized in Table 1.

Table 1

PRESENT STATUS OF EXPANDED MEDIUM SURVEY

IPC images analyzed	1160	(345)
Total area of sky (sq.deg.)	590	(90)
Sources detected	680	(112)
Area for IPC field (sq.deg.)	0.5	(0.2)
Sources identified	287	(112)
Sources with radio data	260	(102)
Identifications with		
Active Galactic Nuclei:	135	(56)

The numbers in parentheses refer to the published MSS samples (Maccacaro et al. 1982, Stocke et al. 1983, Gioia et al. 1984). A program aimed at the identification of all these sources is in progress. Radio measurements are also being obtained for the AGN in the sample which are easily visible from the VLA. The AGN sample has already grown to 135 objects, more than twice the number in the published sample (cf. Maccacaro, Gioia and Stocke 1984). The basic properties of the X-ray selected quasars are shown in Figures 1 through 3.

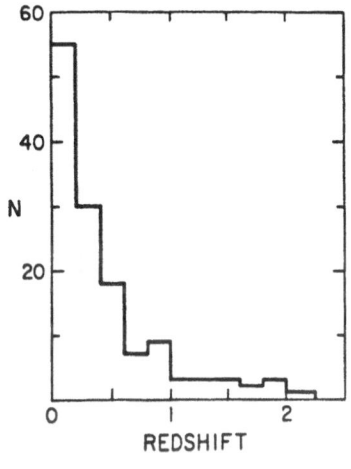

Fig. 1 Redshift distribution.

Fig. 2 Optical luminosity distribution

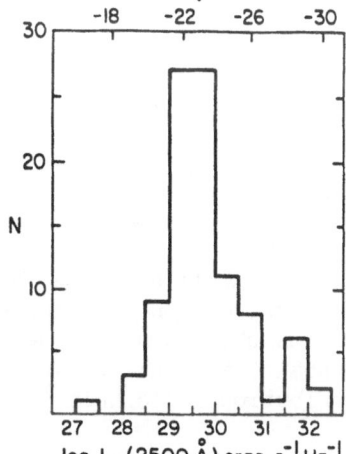

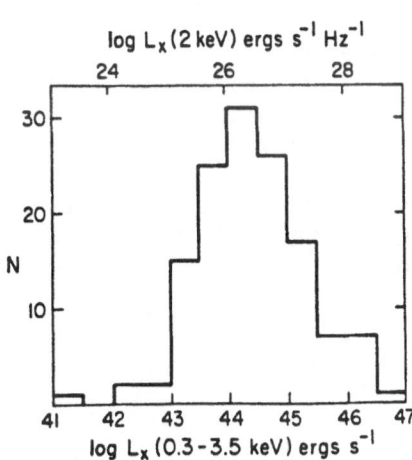

Fig. 3 X-ray luminosity distribution.

Because of the present incompleteness of the optical identification program, the properties of this enlarged AGN sample may be (and probably are) biased in favor of the apparently brighter objects. We therefore caution the reader to keep this bias in mind.

This work has received partial financial support from NASA grant NAS8-30751 and from the Smithsonian Scholarly Studies Program SS48-8-84.

REFERENCES

Avni, Y. and Tananbaum, H. 1986, Ap.J. in press.
Franceschini, A., Gioia, I.M., and Maccacaro, T. 1986, Ap.J.
          in press.
Gioia, I.M., et al. 1984, Ap.J., 283, 495.
Kriss, G. and Canizares, C. 1985, Ap.J., 297, 177.
Maccacaro, T. et al. 1982, Ap.J., 253, 504.
Maccacaro, T. 1984, in "X-ray and UV Emission from Active Galactic
          Nuclei", W. Brinkmann and J. Trumper, eds., page 63.
Maccacaro, T. and Gioia, I.M. 1986, Ap.J. in press.
Maccacaro, T., Gioia, I.M. and Stocke, J. 1984, Ap.J., 283, 486.
Schmidt. M., and Green, R. 1986, Ap.J. in press.
Stark, A.A. et al. 1986, in preparation.
Stocke, J. et al. 1983, Ap.J., 273, 458.

DISCUSSION

Margon : I will describe later in this meeting a program that obtained X-ray information on a large, uniformly selected group of high redshift QSOs. We observe a narrower α_{ox} distribution than reported by Avni and Tananbaum, and this would agree with your speculation that a narrow distribution could help to reconcile the optical and X-ray counts.

Schmidt : Richard Green and I have derived X-ray quasar counts based on Einstein observations of the bright PG quasars. We find agreement with the quasar content of the Medium Sensitivity Survey after introduction of mild X-ray luminosity evolution, approximately $(1+z)^{1.5}$.

Maccacaro : There are several ways to reconcile predicted and observed X-ray number counts. I believe we need larger X-ray selected samples of quasars before reaching definite conclusions.

RADIO INDUCED X-RAY EMISSION IN RADIO QUASARS

Ajit Kembhavi
Tata Institute of Fundamental Research
Homi Bhabha Road
Bombay 400 005
INDIA

ABSTRACT. The relationship between low energy X-ray emission and radio emission in radio quasars is considered. A survey of early results is presented and the current status summarized. Some remarks are made on the simultaneous correlation of X-ray emission with optical and radio emission.

1. INTRODUCTION

Low energy X-ray emission from a few hundred radio, optical and X-ray selected quasars has so far been observed, thanks to the EINSTEIN X-ray observatory. The information is however of a rather limited nature : some data regarding the temporal variability (Zamorani et. al. (1984)) and the low energy X-ray spectrum (see Wilkes and Elvis, these proceedings) is available, but in most cases we have only the X-ray flux, or a 3σ upper limit to it, in the (nominal) 0.5 - 4.5 keV band. In this circumstance it is difficult to understand the nature of the mechanism(s) which leads to the X-ray emission. Some tentative steps towards this may however be taken by considering the statistical correlation of the X-ray luminosity with luminosities in other bands. The pattern of correlations will hopefully lead to the elimination of some mechanisms of X-ray production and a better understanding of the allowed set. We will consider in the following the relationship between the X-ray and radio luminosities of radio quasars.

2. RELATIONSHIP BETWEEN X-RAY AND RADIO EMISSION - EARLY RESULTS

It was established by Ku et. al. (1980) and Zamorani et. al. (1981) that the presence of strong radio emission leads to increased X-ray emission in quasars. This followed from the fact that both optical and radio quasars were found to be powerful ($\sim 10^{44}$ - $\sim 10^{47.5}$ erg/sec) X-ray emitters, with $\langle L_X/L_{op}\rangle$ for radio quasars upto ~ 3 times as high as $\langle L_X/L_{op}\rangle$ for optical quasars. A weak correlation between the optical to X-ray spectral index α_{ox}, and the radio to optical spectral index α_{ro}

239

was also found, which substantiated this conclusion. However there was no direct correlation evident between the radio and X-ray luminosities or fluxes, in contrast to the strong correlation observed between X-ray and optical luminosity, with $L_x \sim L_{op}^{0.7}$.

A very good correlation between the X-ray flux and 90 GHz radio flux was found in a sample of strong millimetre sources by Owen, Helfand and Spangler (1981). These authors argued that such a millimetre source should be a strong X-ray emitter because of the inverse Compton scattering of radio photons to X-ray energies by high energy electrons (Lorentz factor $\gamma \sim 10^3$) in the nuclear region. They observed a set of 25 flat spectrum sources with S (90 GHz) > 1 Jy, of which 12 were quasars, with the IPC detector on the EINSTEIN observatory. Twentyfour of these sources produced a positive X-ray detection, and the 90 GHz to X-ray spectral index α_{mx} was found to be tightly clustered around a mean value $<\alpha_{mx}> = 1.02$ with standard deviation 0.05, which corresponds to a factor of only 2.2 in the ratio of the 90 GHz flux to the X-ray flux. The value $\alpha_{mx} \sim 1$ is consistent with the value expected for the synchrotron self-Compton mechanism. A single synchrotron spectrum from the milli-metre to X-ray wavelengths with slope unity could also explain the observations, but in this case an additional component for the optical and uv continuum would be required, since the flux in these bands lies substantially above the millimetre to X-ray extrapolation.

A sample of optical and radio quasars with known X-ray luminosity was observed by Owen and Puschell (1982) for possible 90 GHz emission. Only 21 of the 51 sources observed were detected at 90 GHz, and α_{mx} was found to have a much wider range than for the strong milli-metre sources. Most optical quasars observed were not detected. It is clear from these observations that the X-ray emission from quasars cannot in general be attributed to the inverse Compton scattering of millimetre photons.

Tananbaum et. al. (1983) have observed a complete sample of 33 3 CR radio quasars in such a way as to produce a positive X-ray detection in every case. Study of the Spearman rank and partial rank correlation coefficients for this sample showed that the X-ray luminosity L_x was tightly correlated with the optical luminosity L_{op}. But L_x showed no correlation either with the total radio luminosity L_t or the redshift z. Tananbaum et. al. argued that the lack of correlation with L_t could be because for the 3 CR sample, which was initially selected at 178 MHz, L_t is dominated by radio lobe emission, which might not be related to the X-ray emission. They therefore considered a restricted sample of 13 "triple" 3 CR quasars with radio core luminosity known from Cambridge 5 km - telescope maps. A rank correlation was found between Log ($L_x/L_{op}^{0.5}$) and α_{ro} at the 98.5 percent confidence level.

3.   RELATIONSHIP BETWEEN X-RAY AND RADIO EMISSION - CURRENT STATUS

There are indications from the X-ray variability data on quasars and high spatial resolution X-ray observations of active galaxies that a significant fraction of the X-ray emission arises in a compact region. X-rays could also arise from the extended radio jets because of the 'in situ' acceleration of electrons in knots, for instance, or the inverse Compton scattering of the cosmic microwave background photons. It is therefore necessary to consider separately the correlation of X-ray emission with core and extended radio emission.

Radio maps at 6 cm with a resolution of a few seconds of arc are now available for a large number of radio quasars with EINSTEIN X-ray observations. In particular, such maps are now available for all the radio quasars from the samples of Ku et. al. (1980) and Zamorani et. al. (1981). The radio data includes VLA observations at 6 and 20 cm of a sample of 35 radio quasars by A. Kembhavi and E. Feigelson, made explicitly to extend the sample available for the study of X-ray/radio correlations.

An exhaustive correlation study of 44 radio quasars with observed extended radio emission and X-ray emission was made by Feigelson, Isobe and Kembhavi (1984). The only significant correlation observed was between the X-ray luminosity L_X and the 178 MHz radio lobe luminosity, with the Spearman rank correlation coefficient $\rho_S = 0.39$, for 39 sources including 4 upper limits for which worst ranks were assumed. This is significant at the 98.5 percent level. No significant correlation was observed between L_X and 5 GHz radio emission, or between L_X and any of the radio spectral or morphological properties like the linear diameter of the radio source, the asymmetry parameter denoting the ratio of distances between the core to the two lobes, or the bending angle between the vectors which connect the core to each lobe. Even the L_X/L_{178} correlation could be partly due to the exclusion of quasars with high L_X but low L_{178}, which could have extended structure below the resolution or dynamic range of the radio maps.

In contrast to the situation with quasars, X-ray emission is tightly correlated with 178 MHz and 5 GHz emission in 3 CR radio galaxies (Feigelson and Berg 1983, Fabbiano et al. 1984). This difference could be due to different production mechanisms for the X-ray emission in the two cases. In the case of quasars the X-ray emission is predominantly of nuclear origin, whereas in the case of radio galaxies it could have a significant component arising in a hot, diffuse interstellar or intracluster medium. Since such a gas would affect the extended radio structure, correlations of the type observed are expected.

The relationship between X-ray emission and radio core emission in radio quasars has been explored in detail by Kembhavi, Feigelson and Singh (1985). The sample here consists of 116 sources, for each one of which either the core radio luminosity at 6 cm is known, or the radio source is unresolved at the scale of a few arcseconds, in which case the whole emission is attributed to the core (see Kembhavi et al. 1985 for the data and details of the analysis). The search for statistical association is based on linear regression analysis in the logarithmic luminosity plane :

$$\text{Log } L_X = b \text{ Log } L_r + \text{constant},$$

where L_X is now the 0.5-4.5 keV X-ray luminosity and L_r the 5 GHz radio core luminosity. Possible spurious correlation between the luminosities, due to the use of redshifts in evaluating them, is examined using the partial rank correlation coefficient, and a correlation accepted as genuine only when this is not significant. X-ray upper limits are ignored in the analysis because they constitute only $\sim$ 10 percent of the sample.

The main results which emerge are the following : (i) There is a very highly significant (confidence level > 99.99 percent) correlation in the whole sample, with $L_x \sim L_r^{0.42\pm.04}$. (ii) The scatter at lower luminosities (see Fig. 1) is much reduced if the sample is split up into subsamples such as unresolved sources, cores of resolved sources, sources with the 6-20 cm core spectral index $\alpha_6^{20} > 0.3$ and those with $\alpha_6^{20} < 0.3$. The regression lines for two subsamples are shown in Fig. 1. (iii) There is a tendency for subsamples with steeper core spectra to have lower $\log L_x / \log L_r$ slopes. The differences in slope are statistically significant. (iv) The steepest slope is obtained for unresolved sources with $\alpha_6^{20} < 0.3$, for which $L_x \sim L_r^{0.71\pm.07}$ (note that 0.3 is used as a dividing line only because a change of slope occurs around this value, one possible reason for which is discussed below. No particular significance is to be attached to the precise value used). (v) The flattest slope, $L_x \sim L_r^{0.35\pm.04}$, is obtained for the cores of resolved quasars.

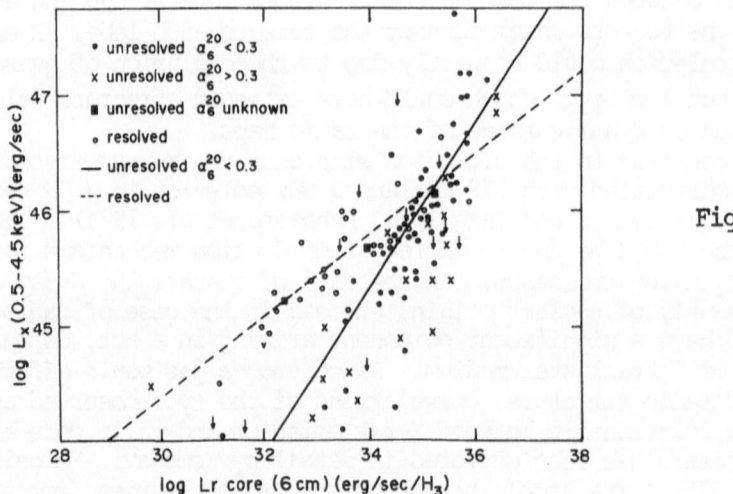

Fig. 1

The flattening of the radio/X-ray slope as one passes to quasars with steeper 6-20 cm radio spectra may be due to the presence, in the presently unresolved radio cores, of extended steep spectrum radio emission which is not related to the X-ray emission. High resolution X-ray observations of Centaurus A (see Feigelson 1982 for a review) show that the radio jet contributes only $\sim$ 1 percent of the X-ray flux. With this in mind we assume that the core radio luminosity L_r may be written as $L_r = L_c + L_e$, where L_c arises in a compact region on the VLBI scale, and is related to the X-ray emission, whereas L_e is steep spectrum extended emission which makes a negligible contribution to the X-ray emission. Since we are observationally limited to considering the correlation between L_X and L_r, the slope of the regression line obtained is

$$b' = \frac{d(\text{Log } L_X)}{d(\text{Log } L_r)} = b \frac{d(\text{Log } L_C)}{d(\text{Log } L_r)} , \qquad (1)$$

where $b = d(\text{Log } L_X)/d(\text{Log } L_C)$ is the slope of the true regression line. If we assume that $L_e = f(L_C)$, it can be shown quite generally (Kembhavi et. al. 1985) that b' may be written as

$$b' = b \times \frac{1 + L_e/L_C}{1 + \gamma L_e/L_C} , \qquad (2)$$

where γ is a constant which enters the functional form of $f : L_e \sim L_C{}^\gamma$. For $\gamma > 1$, $b' < b$, with the value of b' determined by b and the proportion of extended radio emission present in the now unresolved radio core.

To determine b, we note that the mean value of L_e/L_C, at 6 cm, for a given subsample may be expressed as

$$\frac{L_e}{L_C} (6 \text{ cm}) = - \frac{(y^{\alpha_C} - y^{\alpha_r})}{(y^{\alpha_e} - y^{\alpha_r})} , \qquad (3)$$

where $y = 20/6$, α_C is the core spectral index, α_e is the extended emission spectral index, and α_r is the mean observed spectral index α_6^{20} of the presently unresolved radio emission. α_C and α_e have the same values for all subsamples. Using eqn. (3), it may be seen that the right hand side of equation (1) contains the 4 unknowns b, γ, α_C and α_e. Comparing the slopes of 2 subsamples at 6 and 20 cms, 4 equations of the type of equation (1) containing the four unknowns may be obtained. It is possible to solve these equations to obtain values for the 4 unknown parameters. Using the subsamples of unresolved quasars with $\alpha_6^{20} > 0.3$ and $\alpha_6^{20} < 0.3$ gives $b = 0.75$, $\gamma = 1.8$, $\alpha_C = -0.24$ and $\alpha_e = 0.65$. If instead the subsamples $\alpha_6^{20} > 0.3$ and unresolved $\alpha_6^{20} < 0.3$ are used, the values obtained are $b = 0.76$, $\gamma = 1.41$, $\alpha_C = -0.42$ and $\alpha_e = 0.62$. The value of b obtained here may be considered to be the "ultimate" slope of the X-ray/radio regression line for our radio quasar sample.

If the first set of values for the unknown parameters are applied to the subsample of the cores of resolved quasars, it is found that $\langle L_e/L_C \rangle = 3$, $\langle \alpha_r \rangle = 0.5$. The spectral index α_6^{20} is unknown for many of the cores of the resolved quasars. If future observations show that the unknown α_6^{20} are considerably smaller than 0.5, then some additional mechanism must operate in these objects to lower the X-ray/radio regression line slope relative to the slope of the unresolved flat spectrum objects. It can be noticed from Fig. (1) that the cores of resolved quasars have considerably stronger X-ray emission, for a given radio core luminosity, than an unresolved object with the same radio luminosity.

This effect will be further enhanced if the X-ray/radio regression line
slope for these objects is increased by some mechanism of the type described above. This "excess" emission may be traced to the relatively
lower values of α_{ro} compared to other subsamples. Lower α_{ro} values imply
a higher optical luminosity for a given radio luminosity. The optical
contribution to the X-ray emission in these objects is therefore higher,
increasing their total X-ray luninosity to values higher than those for
other subsamples.

The X-ray/radio slope for the flat spectrum quasars, 0.71 ± 0.07, is
identical to the X-ray/optical and optical/radio slopes of the corresponding regression lines, 0.71 ± 0.07 and 0.70 ± 0.09 respectively, for the same
subsample. The identity persists even if the "ultimate" X-ray/radio
slope mentioned above is considered. The reason for this equality of
slopes is not clear. Since a proportionality between the various
luminosities is expected on simple physical grounds, it is possible
that the observed slopes are obtained due to some unknown selection effect.

4. SIMULTANEOUS CORRELATION OF X-RAY LUMINOSITY WITH RADIO AND OPTICAL
    LUMINOSITY

Many lines of evidence suggest that the X-ray emission from a quasar
contains independent contributions from mechanisms associated with the
production of optical and radio continuum radiation. Some of these are:
(1) Optically selected quasars which are relatively radio quiet can be
strong X-ray sources. (2) The presence of strong radio emission
enhances the X-ray emission. (3) "Optically quiet" radio sources are
relatively weaker in their X-ray emission compared to quasars with the
same level of radio emission (Ledden and O'Dell 1983). (4) The X-ray
luminosity shows strong independent statistical correlation with radio
and optical luminosity. It is therefore necessary to consider the
simultaneous fit of L_x to L_{op} and L_r. The condition to be imposed on
any such a fit is that it should reproduce the 2-dimensional correlations
already known.

The 2-dimensional Log/Log correlations described above suggest the
natural extension to 3 dimensions:

$$\text{Log } L_x \; = \; b_{op} \text{ Log } L_{op} + b_r \text{ Log } L_r + \text{constant}. \qquad (4)$$

For radio quiet sources this will produce the observed X-ray/optical
correlation. For a sample of radio quasars, a series of parallel lines
with slope b_{op} will be observed in the X-ray/radio plane, producing the
kind of 2-dimensional fits we have considered above. The slope of the
2-dimensional fit can be somewhat steeper than b_r in equation (2)
because low L_{op} values will dominate for lower L_r and high L_{op} values at
the high radio luminosity end. Tananbaum et. al. (1983) have found, for
the sample of 33 3 CR quasars mentioned in Section 1 that $b_{op} = 0.47 \pm
0.15$ and $b_r = 0.14 \pm 0.12$. The weak radio dependence here is of course
due to the dominant radio lobe emission. For the subsample of unresolved
quasars with $\alpha_6^{20} < 0.3$, we find $b_{op} = 0.44$ and $b_r = 0.39$, with 1σ errors
of $\sim$ 10 percent. Accepting a fit of the type described by equation (4)

would require one to find a physical basis for the relationship

$$L_X \sim (L_{op})^{b_{op}} (L_r)^{b_r} .$$

Another 3-dimensional fit has been suggested by Zamorani(1983). Here one simply adds together the optical and radio related contributions to the X-ray emission:

$$L_X = A_{op} (L_{op})^{b_{op}} + A_r (L_r)^{b_r} , \qquad (5)$$

where A_{op} and A_r are constants. Zamorani (1983) reports the best fit values $b_{op} = 0.75 \pm 0.90$ and $b_r = 0.95 \pm 0.15$ for flat spectrum quasars. It has not been possible to find a least squares minimum for the above fitting function for the subsample of unresolved quasars with $\alpha_6^{20} < 0.3$. However, we will provisionally accept Zamorani's values to see the behaviour of the function in 2 dimensions. Then equation (5) may be written as

$$\text{Log } L_X = \text{Log}(L_{op}^{0.75} + 5 \times 10^{-10} L_r^{0.95}) + 21.9 . \quad (6)$$

The lines described by this function for various values of L_{op} are shown in Fig.2. It is clear that in the region populated by the unresolved, flat spectrum quasars, a behaviour of the form $L_X \sim (L_r)^{b_r}$ may be simulated. The horizontal portions of the $L_{op} = \text{constant}$ lines are however an embarassment, but the situation might improve if the cores of resolved quasars which have a flatter Log / Log slope are included. Much work remains to be done on the 3 dimensional models.

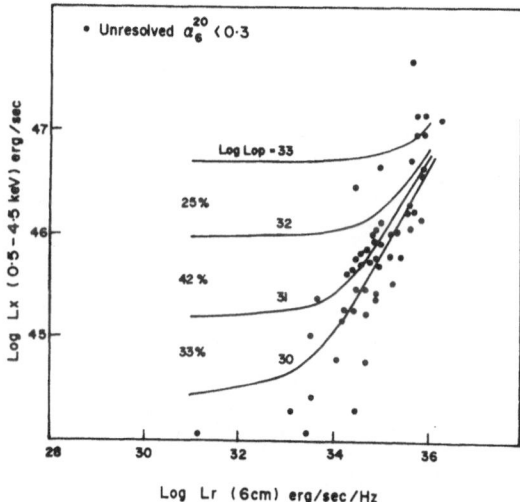

Fig. 2: Lines of constant L_{op} are shown. The percentage of sources in each luminosity interval is indicated.

REFERENCES

Fabbiano, G., Miller, L., Trinchieri, G., Longair, M., and Elvis, M.,
      1983. Ap. J. 277, 115.
Feigelson, E., 1983. In IAU Symposium 97, "Extragalactic Radio
      Sources", eds. D.S. D.S. Heeschen and C.M. Wade, p. 107.
Feigelson, E.D., and Berg, C.J., 1983. Ap. J. 269, 400.
Feigelson, E.D., Isobe, T., and Kembhavi, A.K., 1984. A.J. 89. 1464.
Kembhavi, A., Feigelson, E.D., and Singh, K.P., 1985. MNRAS, in press.
Ku, W.H.-M., Helfand, D.J., and Lucy, L.B., 1980. Nature 288, 323.
Ledden, J.E., and O'Dell, S.L., 1983. Ap. J. 270, 434.
Owen, F.N., Helfand, D.J., and Spangler, S.R., 1981. Ap. J. Letters
      250, L55.
Owen, F.N., Puschell, J.J.   , 1982. A.J. 87, 595.
Tananbaum, H., Wardle, J.F.C., Zamorani, G., and Avni, Y., 1983.
      Ap. J. 268, 60.
Zamorani, G., 1984. In "VLBI and Compact Radio Sources", eds.
      R. Fanti et al.
Zamorani, G., Giommi, P., Maccacaro, T., and Tananbaum, H., 1984.
      Ap. J. 278, 28.
Zamorani, G., et al., 1981. Ap. J. 245, 371.

THE X-RAY PROPERTIES OF HIGH REDSHIFT QUASI-STELLAR OBJECTS

Scott F. Anderson[1,2] and Bruce Margon[1]
Department of Astronomy, FM-20
University of Washington
Seattle, Washington 98195, USA

ABSTRACT. We describe a program aimed at characterizing the X-ray emission of high redshift QSOs. We have obtained slitless spectra of 50 high galactic latitude fields previously imaged at very high levels of sensitivity by the *Einstein Observatory*, generally for original goals unrelated to QSOs. Our survey, covering ~ 17 deg^2 of sky to limiting magnitude $B_{cont} \sim 21$, has yielded ~ 400 previously uncatalogued QSO candidates, each with sensitive new X-ray information available. About 100 of these objects, constituting a "high confidence" set of QSOs, chiefly in the redshift range $1.7 < z < 3$ and thus complementary to previous samples with X-ray data, are used to derive the X-ray properties of high redshift QSOs. Even at these most sensitive available X-ray flux levels, only about 25% of the objects are positively detected in X-rays; thus extensive attention has been given to proper treatment of the upper-limit information. We find a mean optical-to-X-ray slope parameter for the sample of $\alpha_{ox}^{\text{eff}}=1.50$. Our results are combined with those of previous surveys to estimate the fraction of the diffuse X-ray background radiation due to QSOs. QSOs are capable of supplying the majority of the radiation, but the chief contribution comes from an annulus of intermediate redshift, moderate luminosity objects.

1. INTRODUCTION

X-ray observations from the *Einstein Observatory* have increased tremendously our understanding of the X-ray properties of quasi-stellar objects (QSOs), and permitted for the first time reasonably quantitative estimates of the contributions of QSOs to the diffuse X-ray background radiation. Nonetheless controversy remains on a variety of crucial points, *e.g.*, if and how the X-ray properties evolve with redshift and luminosity, and whether QSOs in fact are the dominant contributor to the background. Part of the current uncertainties is doubtless due to the intrinsic complexity of the problem. QSOs exhibit a wide dispersion in their optical and radio properties, so there is little reason to hope that the X-ray emission will be simpler. However, an additional complexity is introduced by the necessary but peculiar selection biases of past X-ray observations of QSOs.

---

[1] Visiting Astronomer, Kitt Peak National Observatory, and Cerro Tololo Interamerican Observatory, National Optical Astronomy Observatories, operated by AURA Inc., under contract with the NSF.

[2] Current address: Steward Observatory, University of Arizona, Tucson, AZ 85721, USA.

G. Swarup and V. K. Kapahi (eds.), Quasars, 247–252.
© *1986 by the IAU.*

Most previous studies have followed one of three approaches. (a). QSOs catalogued previous to the X-ray observations have been studied intensively (e.g., Zamorani *et al.* 1981). These "famous" objects are often atypical of the QSO population, however, as they tend to have abnormally high radio or optical luminosity when selected by these traits. Further, these studies have revealed a remarkably strong correlation between QSO X-ray and radio luminosity, making extrapolations of the rare but well-observed radio objects to the more common but radio-quiet population rather dangerous. Finally, Elvis and Lawrence (1985) have recently demonstrated that radio and optically selected QSOs have very different X-ray spectral slopes, thus calling into question the X-ray intensity calibration of a large fraction of previously published work. (b). X-ray selected QSOs, found as the optical counterparts of serendipitously detected X-ray sources, have been studied by several groups, with particularly large samples available in Gioia *et al.* (1984) and Margon *et al.* (1985). At the risk of stating the obvious, we note that X-ray selected objects must by hypothesis be X-ray bright, and such lists are thus badly biased against QSOs with a small ratio of X-ray to optical luminosity; a population of "X-ray quiet" QSOs would have escaped detection entirely. The very skewed redshift distribution of such X-ray selected samples, which prove to have $< z > \sim 0.4$ even at $B_{lim} = 20$, also stands as a warning against simple application of the X-ray properties of this sample to the entire QSO population. (c). The Braccesi sample of UV-excess selected QSOs has been observed at high sensitivity at X-ray wavelengths by Marshall *et al.* (1984). While this sample is chosen without advance bias towards its X-ray properties, there are other potential complications. The sample size is small, and the nature and completeness of the Braccesi sample has been somewhat controversial (Veron 1983). By selection, the sample has a modest mean redshift of $< z > \sim 1$.

2. OBSERVATIONS

We have recently completed a program aimed at complementing these previous studies through use of an orthogonal selection technique for QSOs with X-ray information. The data base accumulated by the *Einstein Observatory* of course contains information on thousands of still anonymous but reasonably bright QSOs. We know from past studies of the surface density of QSOs, for example, that each 0.5 deg^2 image obtained by *Einstein* included coverage of about 10 QSOs at $V < 19.5$, regardless of the original purpose of that image; this extensive, already extant QSO data base can be exploited if these currently uncatalogued QSOs can be discovered. We have concentrated on 50 X-ray fields obtained with the Imaging Proportional Counter (IPC) at high galactic latitude, $|b| > 30°$, which represent amongst the longest, and thus most sensitive, of all *Einstein* integrations. The original goals of these exposures spanned a large variety astrophysical topics, normally quite unrelated to QSOs. Typical exposure times of our fields were $(1\text{-}5) \times 10^4$ sec, and the data are thus quite competitive in sensitivity with the so-called "Deep Surveys" (Giacconi *et al.* 1979).

At the magnitudes conceivably relevant to these X-ray fluxes, past experience has shown that slitless spectroscopy is a relatively complete and rapid technique for discovery of large numbers of QSOs in a restricted region of sky. Luckily the $38 \times 38'$ area "inside the ribs" of the IPC is very well-matched to the prime focus field of several large telescopes. For our observations we have used a grism at the prime focus of the Kitt Peak and Cerro Tololo 4-meter telescopes, and the "grens" at the Canada-France-Hawaii Telescope prime focus. Typical spectral dispersions employed were 1500 Å mm^{-1}, which is quite sufficient for identification of strong emission line QSOs. The large area of the field, about 100 cm^2, dictates that photographic plates must be used as detectors. For all of our fields we have obtained one or more III-aF emulsions. These plates, baked in forming gas and exposed unfiltered for about 50 minutes, reach $B_{cont} \sim 21.0$ in good seeing, and cover the $\lambda\lambda 3200 - 6900$ Å band. We chose the F in preference to the J emulsion in an attempt to gain wider wavelength coverage, at the expense of the well-known nonuniformity of the emulsion. For many of the fields we further extended the coverage by also obtaining a IV-N plate, hypersensitized with $AgNO_3$. These plates, when exposed for 90 min through a Wr29 filter, can reach about $V_{cont} \sim 21.0$ in good seeing, and cover the $\lambda\lambda 6100 - 8900$ Å band. The total survey covered about 17 deg^2 on about 100 plates. With the goal of avoiding lengthy software preparation, this large but still manageable quantity of spectra (about 10^3 spectra per plate) was reduced manually, with each plate scanned on several independent occasions with a binocular microscope, and underlying grid to systematize the search. A comprehensive description of the observations and reductions is given by Anderson (1985).

Our search resulted in a group of ~ 400 new QSO candidates, chiefly with $17.5 < B < 21.0$, and $1.7 < z < 3$, although the efforts at extended wavelength coverage did result in a significant number of low redshift objects as well. By construction of the sample, each of these objects has available one of the most sensitive possible *Einstein* observations. Without slit spectra to confirm each object, a prohibitive task for this large, faint sample, one must of course worry about sample completeness and contamination. We have extracted from these 400 objects a subgroup of 102 "high confidence" QSOs, objects whose spectra show very prominent and/or multiple emission lines, and often exceptionally blue continua as well. We are confident that the large majority of these objects are in fact bonafide QSOs, and the remainder of the analysis described here will be confined to this sample only. In the small number of objects from this group that we have managed thus far to observe with high spectral resolution, every one has proven to be a QSO.

3. ANALYSIS AND CONCLUSIONS

For ease of analysis we have divided the high confidence sample into two further groups. Our "complete sample" consists of all high confidence objects with $B < 19.5$ and $1.8 < z < 3.0$. Several previous studies have shown that grism selected QSOs are largely ($> 80\%$) complete in this regime. The 40 QSOs in this group, extracted from 12.9 deg^2, yield an internal space density of 3.1 deg^{-2}, which is comparable to that found by previous workers. This complete sample is useful for direct estimates of the contribution of QSOs to the diffuse X-ray background radiation, free of assumptions and/or extrapolations on X-ray evolution, or even external space density estimates. A second useful subset is hereafter referred to as our "narrow sample." This consists of 80 objects in the narrow slice of parameter space confined to $1.8 < z < 2.5$ and $31 < log \, l_{2500Å} < 32$. This restricted sample is useful for testing and extrapolating previous predictions concerning the evolution of X-ray properties with optical luminosity and/or redshift; this group can be treated as an empirical data point, without fear of variation of properties within the sample.

Analysis of the X-ray data at each of the coordinates of our newly discovered QSOs was performed at the Harvard-Smithsonian Center for Astrophysics. We find that despite the high level of sensitivity of the X-ray exposures, only $\sim 25\%$ of our objects yield positive X-ray detections. Proper treatment of the X-ray upper limits is thus essential to maximize extraction of information. We have employed "survival statistics", recently discussed in an astronomical context by Feigelson and Nelson (1985) and Schmitt (1985); the so-called "detection and bounds" method of Avni *et al.* (1980) represents an independent rediscovery of this previously published, more general technique. We have also employed the digital image stacking technique used by Caillault and Helfand (1985) in a slightly different context, and derive equivalent results to the survival analysis. We follow past workers in the field by characterizing our results in terms of the parameter α_{oz}, the slope of a hypothetical power law joining the 2500 Å and X-ray fluxes of each QSO (Tananbaum *et al.* 1979), and also use the mean values, α_{oz}^{eff}, to discuss ensemble properties. The necessary optical photometry has been obtained from image diameter measurements on the *Sky Survey*, and a careful calibration (Anderson 1985) convinces us that our *rms* uncertainties do not exceed a few tenths of a magnitude for $B < 20.2$ (excluding, of course, time variability of the QSO).

We find that the fractional contribution of our "complete" sample, that is, QSOs with $B < 19.5$ and $1.8 < z < 3.0$, to the diffuse X-ray background cannot exceed 3% of the observed background intensity. This is a *direct* measurement, requiring no extrapolation of possible α_{oz} functional dependences, for example. We can combine this result with complementary work by Marshall *et al.* (1984) on the "Braccesi faint" sample, which covers the range $0 < z < 2.2, B < 19.8$. For $z > 3.0$, the space density of QSOs is quite uncertain, but can very conservatively be shown to be a small fraction of all QSOs with $B < 19.5$. We thus conclude that *the total contribution of all QSOs with $B < 19.5$ is less than 20% of the diffuse X-ray background.*

Survival analysis of our "narrow" sample yields an ensemble value of $\alpha_{oz}^{\text{eff}} = 1.50\pm0.03$. This illustrates that we have succeeded in probing a significantly more "X-ray quiet" class than past work; the value for the X-ray selected sample of Margon *et al.* (1985), for example, is 1.26 ± 0.03, and 90% of our objects prove to have smaller ratios of X-ray to optical luminosity than the $\alpha_{oz}^{\text{eff}}=1.42$ inferred by Marshall *et al.* (1984) for their UV-excess selected sample. (There is no conflict between this statement and the fact the the respective α_{oz}^{eff} values for the different works are quite similar, because α_{oz}^{eff} is an extremely "robust" statistic that is dominated by the X-ray loud objects). Avni and Tananbaum (1982) have derived an empirical relation for a dependence of α_{oz} on L_{opt}; our model-independent, directly observed α_{oz} value for the narrow sample is in *excellent* agreement with an extrapolation of their prediction, and adds considerable confidence to their claim that this is in fact a physical dependence rather than a subtle mathematical artefact (Chanan 1983). We can also use this argument in a somewhat different context than normally employed, to state that if α_{oz} does *not* depend on L_{opt}, then, using our derived α_{oz} value and current log N/log S counts, we can conclude that QSOs probably do not dominate the X-ray background radiation; their contribution is $< 50\%$ of the background flux. In short, if one wishes QSOs to make the background, then evolution of α_{oz} must be also accepted.

A significant portion of the background radiation comes from discrete sources unrelated to QSOs, of course; using references cited in Anderson (1985), we estimate 15% of the flux as due to active galaxies, 5% to clusters of galaxies, and 5% to low luminosity active galaxies. If we accept the relationship for α_{oz} functional dependence suggested by Avni and Tananbaum (1982), then the above considerations lead us to conclude that up to 70% of the background flux can be due to QSOs; this is amazingly close to the fraction unaccounted for by the previously mentioned discrete sources, and the agreement is so excellent that it is perhaps fortuitous! Nonetheless we have at least limited, quantitative evidence that QSOs are *capable* of making the dominant contribution to the background flux.

We close by noting that the combination of the form of the α_{ox} evolution, combined with the shape of the QSO log N/log S curve, leads to a specific definition of *which* QSOs dominate the contributions to the background flux. Although there is of course some contribution from a broad range of objects, this distribution distinctly peaks in the range $< z >\sim 1 - 1.5$, $19.5 < V < 20.5$. We thus have the perhaps curious situation that the typical QSO contributing to the background is an intermediate luminosity object in an intermediate redshift annulus of the Universe.

4. References

Anderson, S. F. 1985, Ph.D. thesis, University of Washington.

Avni, Y., Soltan, A., Tananbaum, H., Zamorani, G. 1980, *Ap. J.*, **238**, 800.

Avni, Y., and Tananbaum, H. 1982, *Ap. J. (Letters)*, **262**, L17.

Caillault, J.-P., and Helfand, D. J. 1985, *Ap. J.*, **289**, 279.

Chanan, G. A. 1983, *Ap. J.*, **275**, 45.

Elvis, M., and Lawrence, A., 1985, in "Astrophysics of Active Galaxies and Quasars",
   Proc. 7th Santa Cruz Workshop on Astr. and Ap., J. Miller (ed.).

Feigelson, E. D., and Nelson, P. I. 1985, *Ap. J.*, **293**, 192.

Giacconi, R., *et al.* 1979, *Ap. J. (Letters)*, **234**, L1.

Gioia, I. M., *et al.* 1984, *Ap. J.*, **283**, 495.

Margon, B., Downes R. A., and Chanan, G. A. 1985, *Ap. J. Suppl.*, **59**, 23.

Marshall, H. L., *et al.* 1984, *Ap. J.*, **283**, 50.

Schmitt, J. H. M. M. 1985, *Ap. J.*, **293**, 178.

Tananbaum, H. *et al.* 1979, *Ap. J. (Letters)*, **234**, L9.

Veron, P. 1983, in "Quasars and Gravitational Lenses", Proc. 24th Liège Int. Ap. Colloq.,
   (Université de Liège), p. 210.

Zamorani, G. *et al.* 1981, *Ap. J.*, **245**, 357.

DISCUSSION

Schmidt : On the basis of our X-ray observations of bright quasars and the Medium Sensitivity Survey Quasars, Green and I estimate that only 8-13% of the 2 keV background is contributed by quasars (with $M_B < -23$).

Margon : First, we should note that the MSS and PG QSOs you discuss both have properties quite different from the sample I discuss here. But, more important, let me stress that our paper states that QSOs can make up to 70% of the background. This upper limit depends on many uncertain factors, particularly the mean X-ray spectral slope.

Segal : Isn't your finding that there musst be evolution in the L_x-L_{opt} relation if the XRB is to arise in major part from AGNs (& quasars) restricted to the Friedman cosmology. The chronometric cosmology, which is non-evolutionary, is consistent with this origin for most of the XRB.

Margon : It may be that the X-ray astronomy community's insistence on terming possible functional dependences of the α_{ox} parameter as "evolution" is causing unnecessary semantic confusion. A concise restatement of our finding is that if QSOs are to dominate the X-ray background, it is a necessary (but not sufficient) condition that the ratio of L_x/L_{opt} depend on L_{opt}. What cosmological model one wishes to adopt, in addition to this statement, is quite another issue.

Bregman : In addition to reproducing the X-ray background flux, is the spectral shape also reproduced ?

Margon : Some people feel there is a discrepancy between currently measured X-ray spectra of QSOs and that of the background. I am not alarmed yet because the best measurements of these two quantities are in slightly different energy bands. Further, there are just a handful of good QSO X-ray spectra available now, and those objects are at lower redshift than the shell that dominates the contribution of QSOs to the background.

Elvis : What level of fluctuation in the IPC deep survey background does your 'thin shell' of quasars predict ?

Margon : The shell is not so thin. There is a contribution from a range of objects, with the peak occuring in the $z = 1 - 1.5$ range. For $q_0=0$, we are talking about 1/3 of the volume of the Universe ! I am not certain whether the relatively low count rates available in the IPC could detect the background fluctuations due to this large number of objects.

Burbidge : If the X-ray background comes predominantly from QSOs in a restricted range of z does this not put further constraints on the shapes of their spectra ?

Margon : It certainly does ! QSOs in the $z = 1 - 1.5$ range will have to have spectra considerably more like the background radiation than had been found for the handful of good QSO spectra currently on hand. But so few good spectra are currently available that I consider the issue essentially unconstrained at the moment.

Cowsik : Would the large no $\sim 10^6$ objects needed to explain the isotropy be an argument against QSO's in $z \sim 1 - 1.5$ being the sources of the X-ray background.

Margon : There should be no problem satisfying isotropy measurements, because the QSO log N/log S curve conveniently also peaks near the apparent magnitude range of these X-ray dominant objects : they are very numerous. But a more sensitive imaging X-ray telescope than Einstein will resolve most or all of the background sources, if the QSO contribution proves to be near our upper limit.

X-RAY STUDIES OF SEYFERT GALAXIES AND QUASARS

Claude R. Canizares, Gerard A. Kriss[1], John Kruper and
C. Megan Urry
Department of Physics and Center for Space Research
Massachusetts Institute of Technology
Cambridge, Massachusetts 02139 U.S.A.

ABSTRACT. We present results of studies carried out with the imaging
instruments on the Einstein Observatory. We summarize a statistical
analysis of the X-ray properties of optically selected, radio quiet
quasars including nine new high redshift quasars detected in two deep
X-ray surveys. We find that the X-ray to optical luminosity ratio of
optically selected quasars decreases with increasing optical luminosity.
It depends only weakly, if at all, on redshift. However, the
distribution function does not properly account for the properties of
the X-ray selected Medium Sensitivity Survey sample (MSS). We note that
part of the discrepancy could be due to the presence of red, low
luminosity quasars in the MSS but not in the optically selected samples.
We also summarize some results from a detailed study of the X-ray
properties of 64 Seyfert galaxies. None of the spectral fits performed
for the brightest 20 required unusually steep spectra, although in many
cases the spectral indices were not well constrained. Of the ten
objects with good measurements of the absorbing column density, three
showed excesses above the galactic value while the remaining seven gave
excess columns generally less than 2×10^{20} cm^{-2} and consistent with
zero. Variability studies of the full Seyfert sample showed three
objects to be variable on timescales of a few hours. One of these is
the Seyfert II Mkn 78.

1. X-RAY PROPERTIES OF OPTICALLY SELECTED RADIO QUIET QUASARS

We have recently published the results of a study of optically selected,
radio quiet quasars (Kriss and Canizares 1985). Our data set of 77
quasars and Seyferts and 101 upper limits includes objects from the
literature (through 1984) and new data from our Einstein Observatory
exposures of fields that had been surveyed optically for quasars using
grism techniques (Hoag and Smith 1977, Sramek and Weedman 1978). We
detected 9 of the 23 optically selected quasars; this 40% detection rate
is high compared to that of earlier surveys and merely reflects the long

[1]Present address: Johns Hopkins University

G. Swarup and V. K. Kapahi (eds.), Quasars, 253–259.
© *1986 by the IAU.*

(~ 50 ksec) exposure times. Inclusion of these objects nearly doubles
the number of high redshift quasars in the sample, which is very
important in defining the possible z dependence of their X-ray
properties (see Margon's contribution to this meeting).

Figures 1 and 2 show the X-ray to optical fluxes and luminosities
for the sample. There is a strong correlation in both flux and
luminosity: the ratio of X-ray to optical luminosities scatters by less

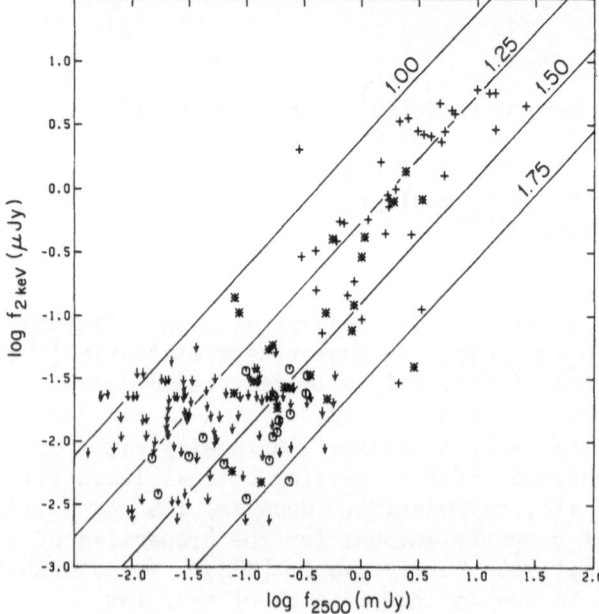

Figure 1. X-ray vs.
optical flux for optically
selected, radio quiet
quasars from Kriss and
Canizares (1985). The
solid lines are the loci
of constant α_{ox} as
indicated. The arrows
indicate upper limits and
the symbols represent
Seyfert galaxies (+), UV
selected objects (*) and
grism selected objects
(circles).

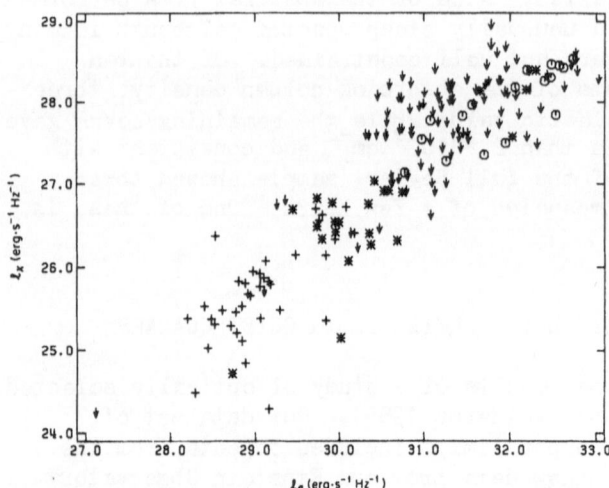

Figure 2. X-ray vs. opti-
cal luminosities for the
objects shown in Fig. 1.

than two orders of magnitude over five orders of magnitude of luminosity
-- from Seyferts to the most luminous quasars. One thing that these
plots show is that the term "X-ray quiet quasar" is simply

inappropriate. Our statistical analysis of the upper limits indicates that they are fully consistent with the detections (i.e. the undetected objects can be drawn from the same parent distribution as the detected objects with high probability). Thus few, if any, quasars lack significant X-ray emission (see also Avni and Tananbaum 1986).

There has been great interest in looking beyond the evident $L_x - L_o$ correlations to see if L_x/L_o, or equivalently $\alpha_{ox} = -[\log L_x/L_o]/2.605$, depends on luminosity or redshift; that is, to see if there is any curvature in Fig. 2 (we take L_x and L_o to be monochromatic luminosities at 2 keV and 2500 A, respectively, in the quasar rest frame). In Kriss and Canizares (1985) we describe the various parametric and non-parametric statistical analyses that we applied to the data (using the methods of Avni et al. 1980). The specific parameters we derive are consistent with those found earlier by Avni and Tananbaum (1982), although our uncertainties are considerably smaller because of our larger sample.

There is now remarkable agreement among all workers in this field about the amount of X-ray emission from optically selected quasars (see Zamorani's and Margon's papers at this meeting and the recent work by Avni and Tananbaum 1986 and Schmidt and Green 1986). All agree that α_{ox} has a strong dependence on L_o corresponding to the relation $L_x \sim L_o^{0.7}$; we are able to rule out independence of the two variables with high significance. On the other hand there is at most only a weak dependence of α_{ox} on redshift, and independence is permitted by our data.

It must be stressed that these relationships are purely phenomenological; their physical meaning is not at all clear. In particular, the recent work on the IR to X-ray spectra of quasars being performed by Elvis and colleagues (see Elvis et al. 1985 and the contributions by Elvis, Fabbiano and Wilkes in this volume) shows that one must be exceedingly cautious in interpreting anything as primitive as a two-point power law index that spans 2.6 decades of frequency. One clear problem is that by choosing to measure L_o at 2500 A we are very sensitive to the relative contributions of both the "big bump" and "little bump" in the optical/UV spectrum. If these change with L_o and/or z it could cause an apparent α_{ox} dependence even if the underlying power law continuum were fixed. We are now trying to obtain better measures of the underlying optical-to-X-ray spectrum in hopes of refining the analysis.

2.   X-RAY SELECTED ACTIVE NUCLEI

A glaring reminder of our lack of understanding about the X-ray properties of quasars is our inability to use the models derived for optically selected objects to account for the observed properties of X-ray selected objects. This was addressed by Kriss and Canizares (1985) as it has been by others (Zamorani and Maccacaro, this meeting, Zamorani 1982, 1985, Maccacaro 1984, Franceschini, Gioia and Maccacaro 1986, Avni and Tananbaum 1986, Schmidt and Green 1986). Briefly, the data of the Einstein Observatory Medium Sensitivity Survey (MSS; Gioia et al. 1984) are not commensurate with the optically selected samples in α_{ox} dependence, z distribution, luminosity function slope and number counts. In Kriss and Canizares (1985), we point out that part of the discrepancy could arise from the presence in the MSS of a large number of low luminosity red objects (with B - V > 0.65) which are not represented in optically selected samples. These tend to lower the redshift distribution of the MSS. Some of the objects may be intrinsically red, but there are also likely to be effects of reddening, X-ray absorption and contamination by galaxy light all of which can distort the α_{ox} distribution. We stress, however, that the red objects cannot explain all the incommensurability of the optical and X-ray samples; the low number counts of the MSS, for example, are exacerbated when the red objects are excluded.

Therefore, we conclude that reconciliation of the optically selected and X-ray selected samples is still a serious problem, and one must approach all discussions of the contribution of radio quiet quasars to the X-ray background with extreme caution. The MSS objects are themselves one small part of the X-ray background. If our models have trouble accounting for this part that is already directly resolved, then we should be exceedingly wary about extrapolations that attempt to explain the part that remains unresolved.

3. X-RAY PROPERTIES OF SEYFERT GALAXIES

We will now switch emphasis entirely and focus on detailed properties of a sample of Seyfert galaxies studied with the Einstein Observatory Imaging Proportional Counter (IPC). This is a continuation of the work of Kriss, Canizares and Ricker (1980). To date we have analyzed data on 50 Seyferts: 37 type I, 10 type II and 3 of ambiguous classification (we have in addition 14 X-ray upper limits).

3.1 X-ray Spectra

We were able to fit spectral models to the data of the twenty brightest objects in the sample. We used a power law modified by absorption in cold material. The mean power law index (defined as $-d[\log f_\nu]/d[\log \nu]$) is 0.86 ± 0.06 (rms), which is slightly larger than the "canonical" value of 0.7 seen in the 2-10 keV range for brighter Seyferts (e.g. see Mushotzky 1984). Although many of our indices are not well constrained, none of our fits requires the very steep spectra (spectral indices as large as 2) found by Elvis et al. (1985) for several quasars.

One interesting result involves the very low absorbing columns that we are finding for some Seyferts. Figure 3 shows the 90% confidence limits on neutral hydrogen column density along the line of sight after subtraction of the Galactic column taken from the 21 cm survey of Stark et al. (1985). Of the ten objects with good N_H determinations, all of which are Seyfert I's, three show significant excesses over the Galactic values. But the remaining seven objects have extremely low excess columns, 2×10^{20} cm^{-2} (which would correspond, for example, to a density of less than 0.06 cm^{-3} over 1 kpc). This shows that we have

very clear lines-of-sight to the central engines of at least some
Seyfert galaxies, which is encouraging for future studies of their
intrinsic properties.

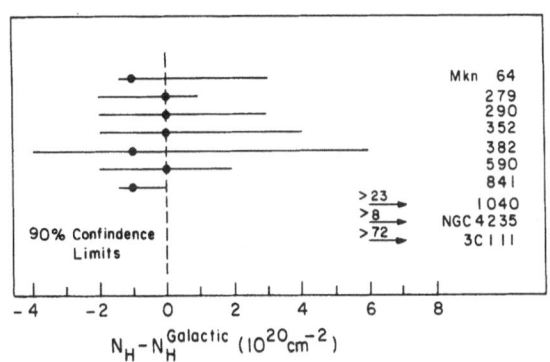

Figure 3. Residual column density after subtraction of the Galactic column for the ten Seyfert I galaxies listed.

3.2 Short Timescale Variability

The full sample of Seyferts was used to search for variability on
timescales comparable to the observation time, which was typically about
an hour. We used both chi squared and Kolmogorov-Smirnov (KS) tests for
departures from constant flux. For all but three objects, the tests
were negative, indicating limits of between 7 and 60% on the amplitude
of any short timescale variability. There are three Seyferts for which
the data are not consistent with a constant source intensity. These are
listed in Table 1. The variability timescale is a few hours, which
corresponds to a size scale of $\sim 10^{14}$ cm.

Table 1

Variable Seyfert Galaxies

Name	Type	Probability Source is Constant	Timescale (hrs)
IIIZw2	I	$\ll 10^{-3}$	~ 10
Mkn 766	I	10^{-5}	~ 2
Mkn 78	II	$< 10^{-3}$	~ 2

This brings the number of Seyferts with observed short timescale
variability to about a half dozen (Pounds 1985). Two of these are
Seyfert type II's (Mkn 78 reported here, and NGC 5506) which strengthens
the conclusion that at least some members of this heterogeneous class
have compact active nuclei similar to those in Seyfert I's and quasars
(see Elvis and Lawrence 1985). Our data fit into the rough correlation
between variability timescale and luminosity presented by Barr at this
meeting. Barr and Mushotzky (1986) use this correlation to suggest that
the emitting regions of all Seyferts and quasars are marginally in the
regime in which electron-positron pair production begins to dominate the
energetics.

ACKNOWLEDGEMENTS. We thank Mike Rohan for assistance in carrying out
the analysis of the short timescale variability of Seyfert galaxies and
Elaine Aufiero for preparation of the manuscript. This work was
supported in part by NASA grant NAG 8-494.

REFERENCES

Avni, Y., Soltan, A., Tananbaum, H., and Zamorani, G. 1980, Ap. J.,
     238, 800.
Avni, Y. and Tananbaum, H. 1982, Ap. J. (Letters), 262, L117.
Avni, Y. and Tananbaum, H. 1986, Ap. J. (in press).
Barr, P. and Mushotzky, R. 1985 (preprint).
Elvis, M., Green, R., Bechtold, J., Schmidt, M., Neugebauer, G.,
     Soifer, B., Mathews, K., and Fabbiano, G. 1986, Ap. J. (submitted).
Elvis, M. and Lawrence, A. 1985, in Astrophysics of Active Galaxies and
     Quasi-Stellar Objects, Miller, J. (ed.), (Mill Valley CA: Univ.
     Sci. Books), 289.
Franceschini, A., Gioia, I., Maccacaro, J. 1986, Ap. J. (in press).
Gioia, I. M., Maccacaro, T., Schild, R. E., Stocke, J. T.,
     Liebert, J. W., Danziger, I. J., Kunth, D., and Lub, J. 1984,
     Ap. J., 283, 495.
Hoag, A. A. and Smith, M. G. 1977, Ap. J., 217, 362.
Kriss, G. A. and Canizares, C. R. 1985, Ap. J.
Kriss, G. A. and Canizares, C. R. and Ricker, G. R. 1980, Ap. J., 242,
     292.
Maccacaro, T. 1984, in X-ray and UV Emission from Active Galactic
     Nuclei, Brinkman, A. and Trumper, J. (eds.), MPE report 184, 63.
Mushotzky, R. 1984, in X-ray Astronomy, M. Oda and R. Giacconi (eds),
     (Tokyo: Institute of Space and Astronautical Sciences).
Pounds, K. A. 1985, in Galactic and Extragalactic Compact X-ray Sources,
     Y. Tanaka and W. H. G. Lewin (eds), (Tokyo: Institute of Space and
     Astronautical Sciences), 261.
Schmidt, M., and Green, R. 1986, Ap. J. (in press).
Stark, A. A., Heiles, C., Bally, J., Linke, R. 1985 (in preparation).
Sramek, R. A., and Weedman, D. W. 1978, Ap. J., 221, 468.
Zamorani, G. 1982, Ap. J. (Letters), 260, L31,
_____, 1986, Ap. J. (in press).

DISCUSSION

Veron : Do Seyfert 2 galaxies have on average a larger column density than the Seyfert 1 galaxies ?

Canizares : None of the twenty cases we have studied is a Seyfert 2. This is because Seyfert 2's are generally lower luminosity and have insufficient flux to allow us to measure their spectra. Also, I should note that by selecting objects with sufficient signal we are also biasing ourselves against heavily absorbed objects, because these will have few counts in the IPC band.

Malkan : Your tight upper limits on the internal absorption column densities in Seyfert 1's are very nice. It's gratifying that they also imply rather strict upper limits on the reddening of the continuum (say $E_{B-V} < 0.07$ mag), which we are also deducing from the optical/UV observations of many AGN's.

Margon : With regard to your comment that the fractional contribution of QSOs to the X-ray background is highly uncertain, I want to stress that the 70% fraction quoted in our own work is most certainly an upper limit. There are very significant uncertainties, in particular the effect stressed at this meeting by Elvis and Wilkes: the tremendous uncertainty in the appropriate X-ray spectral index implies that even the measured X-ray fluxes from individual QSOs may be highly uncertain !

Canizares : I agree.

Belinda Wilkes, Rosalinda and Richard Puetter and Ranu Kundu

THE DIVERSE SOFT X-RAY SLOPES OF QSOS

Belinda Wilkes and Martin Elvis
Harvard-Smithsonian Center for Astrophysics
60 Garden Street
Cambridge, Massachusetts 02138 USA

ABSTRACT. We present _Einstein_ spectra for 30 QSOs which indicate a large diversity in their soft x-ray slopes, unlike the uniform slopes found for AGN higher x-ray energies. The slope is related to radio properties, x-ray luminosity or possibly redshift, but we cannot definitively distinguish these possibilities.

The recent calibration of the _Einstein_ Imaging Proportional Counter (IPC) (Harnden and Fabricant 1985) allows the investigation of faint x-ray source spectra. This provides a unique sample with which to study the soft (0.2-3.5 keV) x-ray spectra of QSOs. The sample now includes 30 QSOs with well constrained x-ray slopes ($\Delta\alpha \lesssim \pm 0.5$).

A histogram of the best fit power-law slopes α ($F_\nu \propto \nu^{-\alpha}$) for these 30 QSOs clearly shows a large range from 0.4 to 2.2 (Figure 1). This is in sharp contrast to previous results for active galaxies which showed a strikingly uniform slope, of α=0.7 (Mushotsky 1984). We believe (Elvis et al. 1986) that this contrast is mainly due to the IPC sensitivity which allows it to detect QSO's selected by uv-excess or radio emission rather than only the brightest 30 hard x-ray sources in the sky.

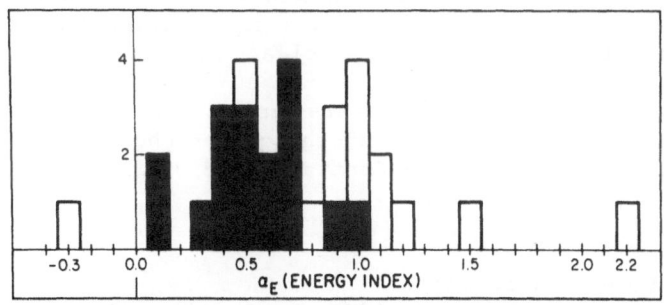

Figure 1:
The distribution of α_e in our sample. RL QSOs are shaded RQ open.

Figure 1 divides the sample into radio-loud (RL) and radio-quiet (RQ) QSOs on the basis of their radio to optical slope (α_{ro}>0.35 is RL, Zamorani et al. 1981). Although the errors (typically ±0.2) introduce some scatter. There is clearly a strong tendency for RL objects to have flatter slopes than RQ. The 17 RL QSOs have a mean α = 0.54±0.06, whereas the 13 RQ have α = 1.00±0.16. A KS test shows this difference to be 99% significant. This result is readily

G. Swarup and V. K. Kapahi (eds.), Quasars, 261–262.

interpretable if RL QSOs have a strong Inverse Compton component
missing in RQ QSOs.

The RL QSOs in our sample, however, are more x-ray luminous than
the RQ with little overlap in L_x, so the RL/RQ distinction may instead
be a luminosity effect. A division of the sample at $L_x = 10^{44}$ erg s^{-1}
yields a marginally significant (95%) difference in x-ray slope. Also
AGN x-ray spectra may steepen at low energies. (This could explain
part of the difference between our result and earlier, hard x-ray,
measurements.) This steepening would be more easily observed in low
redshift QSOs, since in them we observe a lower intrinsic energy
range. Division of the sample at $z = 0.2$ yields a steeper mean slope
for low-redshift QSOs at 95% confidence. Since z and L_x are highly
correlated in our sample we cannot distinguish these two dependencies.
From this data alone it is not immediately clear which parameter
determines the x-ray slope. The study of Worrall et al. (this
meeting) suggests that radio emission may be the key factor.

Spectral fits to the x-ray data allow us to determine the total
line of sight absorbing column to each QSO. We find (Figure 2) that
the column density of cold material is, on average, smaller than that
due to our Galaxy as based on 21 cm data (Stark et al. 1985).

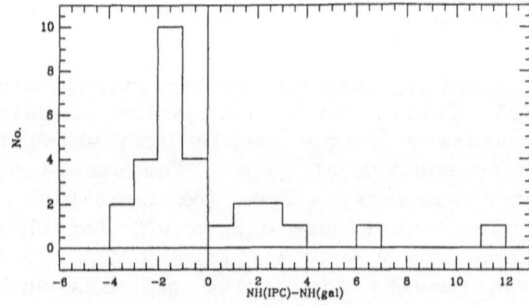

Figure 2:
The differences in N_H
measured by the IPC and
radio (21 cm).

A two-component x-ray spectrum which steepened at low energy, could
produce this discrepancy. However ΔN_H does not correlate with
redshift, $\alpha_\bullet$ or radio properties so this is unlikely. It would also
be surprising if the ΔN_H values produced this way were always just a
few $\times 10^{20}$ atom cm^{-2}. Possibly, the radio measurements of the Galactic
column (which are based on a large beam survey) may systematically
overestimate the column although it is not clear how this could
happen.

This first survey shows that x-ray spectra are clearly a rich
subject with great potential for understanding the quasar phenomenon.

This work was supported in part by NASA Contract NAS8-30751.

REFERENCES

Bechtold, J., et al. 1986, Ap.J., submitted.
Elvis, M., Wilkes, B.J., and Tananbaum, H. 1985, Ap.J., 292, 357.
Elvis, M., et al. 1986, Ap.J., submitted.
Harnden, F.R., Jr. and Fabricant, D. 1985, in preparation.
Mushotsky, R.F. 1984, Advances in Space Research, 3, no. 10-13, 157.
Stark, A.A., et al. 1985, in preparation.
Zamorani, G., et al. 1981, Ap.J., 245, 357.

A STATISTICAL STUDY OF THE RELATIONSHIP BETWEEN X-RAY, OPTICAL AND RADIO LUMINOSITY FOR A SAMPLE OF QSOs

D. M. Worrall[1], P. Giommi[2], H. Tananbaum[1], and G. Zamorani[3]
1. Harvard-Smithsonian Center for Astrophysics, Cambridge, MA, U.S.A.
2. EXOSAT Observatory, ESA-ESOC, Darmstadt, F.R.G.
3. Istituto di Radioastronomia, Bologna, Italy.

ABSTRACT. Our statistical study finds that the X-ray luminosity, l_x, for an average QSO of given optical luminosity, l_o, is an increasing function of radio luminosity, l_r. An average QSO of given l_o and l_r has higher l_x if the radio spectrum is flat than if it is steep. The correlation between l_x, l_o and l_r appears to be improved if l_r refers only to core emission, l_{rcore}. Assuming an average radio-loud QSO is a radio-quiet QSO with a separate component which produces l_{rcore} and l_{xcore}, l_{xcore}/l_{rcore} is larger if the core is that of a QSO with extended radio structure than if that of a compact flat-spectrum QSO.

Various authors have found that samples of QSOs exhibit a dependence of X-ray luminosity, l_x, on optical luminosity, l_o, of the form

$$\log l_x = \log(A l_o^\alpha) \qquad (1)$$

(The logarithm of the function is given to illustrate that the regression analysis is for $\log l_x$ rather than l_x). Different QSO samples have given values of α between ≈ 0.5 and ≈ 0.9 (e.g., Avni and Tananbaum 1986, and references therein). However, it has also been pointed out that the X-ray luminosity of QSOs is influenced by their radio luminosity, l_r (e.g., Kembhavi, Feigelson and Singh 1985; and references therein). We have studied this dependence by applying eq. [1] to a sample of 227 QSOs divided into 3 subsamples in bands of $\log(l_r/l_o)$. Our fits use the Detections and Bounds method of Avni et al. (1980). Luminosities refer to monochromatic source-frame values at 5 GHz, 2500 Å and 2 keV for l_r, l_o and l_x, respectively. The X-ray measurements were made with the Einstein Observatory Imaging Proportional Counter. The radio and optical data were taken from many published sources. We find that all 3 subsamples can be fit with the same value of α, but the normalization, A, increases with increasing $\log(l_r/l_o)$. Furthermore, we find that, for the 114 QSOs in our 2 subsamples of highest $\log(l_r/l_o)$, i.e., those QSOs we define as radio-loud, A is greater for objects with flat radio spectra (FRS) than for objects with steep radio spectra (SRS).

For investigative purposes, we have chosen a model with a simple dependence of l_x on both l_o and l_r,

$$\log l_x = \log(K l_o^\alpha l_r^\beta) \qquad (2)$$

G. Swarup and V. K. Kapahi (eds.), Quasars, 263–264.

For our FRS QSO sample, we find that the sigma of the Gaussian distribution of the residuals for a fit to eq. [2] is significantly smaller than that for a fit to eq. [1], with $\alpha \approx \beta \approx 0.45$. This suggests a connection between the radio luminosity and at least some of the X-ray luminosity for these objects. FRS QSOs are core compact, and the report of $\approx 75\%$ VLBI visibility for objects of this type (Zensus, Porcas and Pauliny-Toth 1984) suggests their radio cores are typically ≤ 20 pc. For our SRS QSO sample, $\alpha \approx 0.45$, whereas β is smaller but still greater than zero. We have divided the 46 objects in our SRS sample on the basis of radio morphology. Our SRS-Compact sample is populated by 16 objects. These are an interesting group for study because their radio structure is typically complex, with structure on all scales between ≈ 10 pc and ≈ 10 kpc (e.g., Pearson and Readhead 1984). Unfortunately, our 16 objects have an insufficient spread in l_r for conclusions to be drawn. Our SRS-Extended sample is populated by 24 objects. For these objects, when l_r refers only to the luminosity within ≈ 30 kpc ($H_o = 50$ km s^{-1} Mpc^{-1}; $q_o = 0$) of the core, l_{rcore}, the fit to eq. [2] appears to be improved, although statistical uncertainties prevent a strong statement.

In order to investigate physical mechanisms, we have fit the samples to a functional form which separates the dependence on optical and radio luminoisty,

$$\log l_x = \log(A l_o^\alpha + B l_r^\beta) \tag{3}$$

We find that the term in l_o for the FRS sample is consistent with that for the radio-quiet sample (for the radio-quiet sample, the term in l_r is assumed to be negligible). This implies that we may consider a radio-loud QSO as a radio-quiet QSO with a separate component which produces l_{rcore} and a contribution to l_x, which we can call l_{xcore}. This applies also to the SRS-Extended objects, when l_r refers only to l_{rcore}. A value of $\beta \approx 0.88$ fits both the FRS and SRS-Extended (core radio) samples. We then find that the ratio of B for the SRS-Extended (core radio) to FRS samples is 5.5 ± 3.6 (90% errors). This means that the ratio l_{xcore}/l_{rcore} is larger if the core is that of a QSO with extended radio structure than if that of a compact flat-spectrum QSO. This could be interpreted as due to relativistic beaming depressing the X-ray relative to core-radio luminosity in FRS but not SRS QSOs. This is in qualitative agreement with the prediction of the unified scheme of Orr and Browne (1982), although our results do not imply a sufficiently large effect to be in quantitative agreement.

REFERENCES

Avni, Y., Soltan, A., Tananbaum, H. and Zamorani, G. 1980, Ap.J., 238, 800.

Avni, Y. and Tananbaum, H. 1986, Ap.J., in press.

Kembhavi, A, Feigelson, E.D. and Singh, K.P. 1985, MNRAS, in press.

Orr, M.J.L. and Browne, I.W.A. 1982, MNRAS, 200, 1067.

Pearson, T.J. and Readhead, A.C.S. 1984, VLBI and Compact Radio Sources, eds. R.Fanti et al. (Reidel), p15.

Zensus, J.A., Porcas, R.W., and Pauliny-Toth, I.I.K. 1984, Astr. Ap., 133, 27.

X-RAY SELECTED BL LAC OBJECTS: TIME VARIABILITY AND X-RAY SPECTRUM

D. Maccagni[1], P. Barr[2], B. Garilli[1], I.M. Gioia[3], P. Giommi[2],
T. Maccacaro[3], R. Schild[3]

[1]*Istituto di Fisica Cosmica, CNR, Milano, Italy*
[2]*EXOSAT Observatory, ESOC-ESA, Darmstadt, FRG*
[3]*Center for Astrophysics, Cambridge, Mass., U.S.A.*

The *Einstein* Observatory Medium Sensitivity Survey contains four sources
identified with BL Lac objects. The radio and optical properties of
these objects have been described in Stocke *et al.* (1985).

The X-ray properties of these X-ray selected BL Lac objects have been
probed by means of repeated observations with the *EXOSAT* Observatory.
The X-ray spectrum has been measured only for two BL Lacs, namely 1E
1402.3+0416 when its intensity was highest and 1E 0317.0+1835. In both
cases the *EXOSAT* Medium Energy experiment data were fitted with a power
law in the 2 to 6 keV energy range: the amount of absorption due to the
galactic hydrogen column densities and energy slopes respectively of
1.6 (+0.3, -0.1) and 1.5 (+0.6, -0.5) were obtained. Within the *EXOSAT*
observations, only 1E 1402.3+0416 has shown fast time variability (a
factor 2 decrease in about 6 hours; see Giommi *et al.* 1986). Long term
variability can be checked by comparing *Einstein* and *EXOSAT* fluxes. The
comparison however is strongly dependent on the spectral slope and on
the amount of absorption. In order to compare the *Einstein* and *EXOSAT*
fluxes of our sources, we either used the measured spectral parameters
or we assumed a spectral slope of 1.5 with an uncertainty of ±0.5. Long
term variability is present only in 1E 1402.3+0416 (Giommi *et al.* 1986)
and possibly in 1E 1235.4+6315, provided its spectral slope is steeper
than 1. It is remarkable that in the case of 1E 1207.9+3945, for which
we have the best monitoring (10 observations spread over 15 months with
EXOSAT, plus 2 *Einstein* observations; see Figure 1), no variability has
been detected.

Many models (e.g., Bonometto and Rees 1971) predict that spectral
slopes greater than 1, as we observed in two objects and as it seems to
be common for many BL Lacs, are most easily obtained for compactness
parameters of the sources $> 10^{30}$ ergs cm^{-1} s^{-1}. Since these X-ray
selected BL Lacs have luminosities of the order of 10^{45} ergs s^{-1}, the
corresponding radial dimensions are 10^{15} cm and therefore their typical
variability time scale should be 10^{4} seconds. Our observations have not
shown this type of variability but in the case of 1E 1402.3+0416. This

G. Swarup and V. K. Kapahi (eds.), Quasars, 265–266.

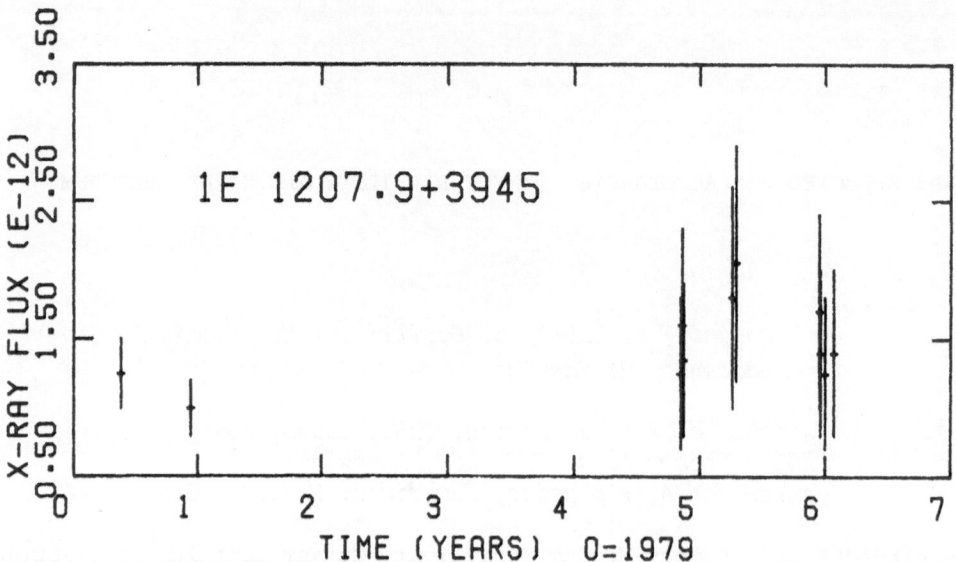

Figure 1: The X-ray light curve of 1E 1207.9+3945. The X-ray flux is in the 0.5-3.5 keV energy range and EXOSAT fluxes have been converted assuming a spectral index of 1.5±0.5.

could hint to emission mechanisms where pair production is not important and the resulting X-ray spectrum simply reflects the electron injection spectrum.

References
Stocke, J.T., Liebert, J., Schmidt, G.D., Gioia, I.M., Maccacaro, T., Schild, R.T., Maccagni, D., and Arp, H.C. 1985, Ap. J. *298*, 619.
Giommi, P., Barr, P., Gioia, I.M., Maccacaro, T., Schild, R., Garilli, B., and Maccagni, D. 1986, Ap. J. (in press).
Bonometto, G.T., and Rees, M.J. 1971, M.N.R.A.S. *152*, 21.

X-RAY BURSTING ACTIVITY IN THE BL LACERTAE OBJECT PKS 2155-304

D. Maccagni[1], L. Chiappetti[1], L. Maraschi[1,2], D. Molteni[3],
M. Morini[4], E.G. Tanzi[1], A. Treves[1,2], A. Wolter[2].

1 Istituto di Fisica Cosmica, CNR, Milano, Italy
2 Dipartimento di Fisica, Università di Milano, Italy
3 Istituto di Fisica, Università di Palermo, Italy
4 IFCAI, CNR, Palermo, Italy

1. INTRODUCTION

The bright BL Lac object PKS 2155-304 (m_V=13; z=0.117) was observed
with EXOSAT at five epochs (1983 Oct. 31, Nov. 30, 1984 Nov. 6, 7 and
11), for a total of about 30 hours of exposure time. Here we present
data and results obtained with the Low Energy (LE) telescopes, in the
band 0.05 - 2. keV, and with the Medium Energy (ME) argon proportional
counters in the range 1. - 6. keV.

2. OBSERVATIONS

The band chosen for the ME data is the one in which the source to
background ratio is higher; the LE data were obtained with the Channel
Multiplier Array in conjunction with the Lexan filter. In both experiments the source exhibited different intensity levels at various epochs:
the lowest intensity on 1983 October 31, the maximum at peak intensity

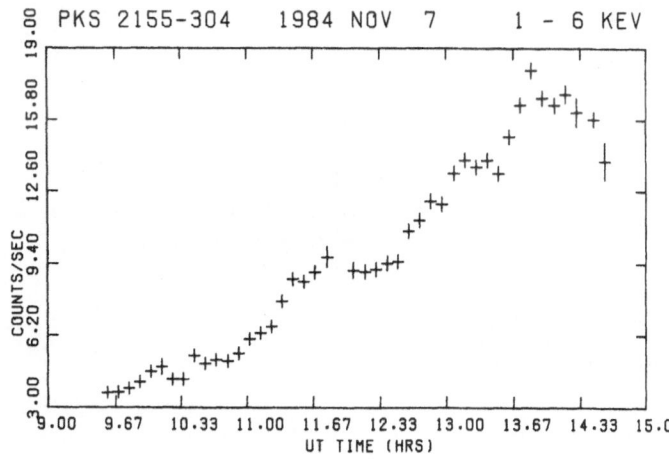

Fig. 1: Argon counters
ME might curve recor-
ded on 1984 Nov. 7.

G. Swarup and V. K. Kapahi (eds.), Quasars, 267–268.

on 1984 November 7. This is the highest dynamic range of intentsity variation recorded for this source with a single instrument. From the light curves it is apparent that variability over time scales of months and days is common for this source, superposed on short-term variability, particularly in the observations of 1984 November 6 and 7, in which the 1-6 keV flux varied by a factor 2.5 in 5 hours and by a factor 5 in 4.5 hours respectively. In Fig. 1 the ME light curve for November 7 is shown: a variation time scale ($\tau = F(\Delta F/\Delta t)^{-1}$) as short as 1 hour (the minimum observed) is apparent around 13:30 UT. The overall behaviour of the LE count rate mimics the ME one, but with smaller dynamic range. During periods of activity there are indications of hardening of the spectrum (see Fig. 2).

The energy spectrum between 1 - 6 keV is well fitted by a power law with a low-energy cutoff due to photoelectric absorption. The photon spectral index ranges between 2.51+0.27 and 2.92+0.12. The values for the absorbing column density derived for the combined LE+ME fits are lower than those given by by the ME data alone: this, together with the different range of variability in the two energy bands, suggests either the presence of an additional soft spectral component with a distinct temporal behaviour, or the hypothesis that absorption present above 1 keV does not affect lower energies (warm absorber).

3. DISCUSSION

The maximum luminosity, recordered on 1984 Nov. 7 (assuming isotropic emission and H_o=100 Km s^{-1} Mpc^{-1}) is L=10^{46} erg s^{-1} in the range 0.1 - 6 keV, comparable with the UV luminosity (L=3×10^{45} erg s^{-1}) and greatly exceeding that in other bands. The maximum luminosity variation observed in the 1 - 6 keV range yields (dL/dt)=10^{42} erg s^{-2}. This observed values of L and dL/dt imply severe constraints on the source structure. Under the hypothesis of Eddington limit, the luminosity yields a lower value of the mass M>10^8 M$_\odot$, which corresponds to a gravitational radius R_g= 3×10^{13} cm. The observed minimum time scale of variability, therefore, corresponds to an emission region of only $3R_g$, a rather restrictive value.

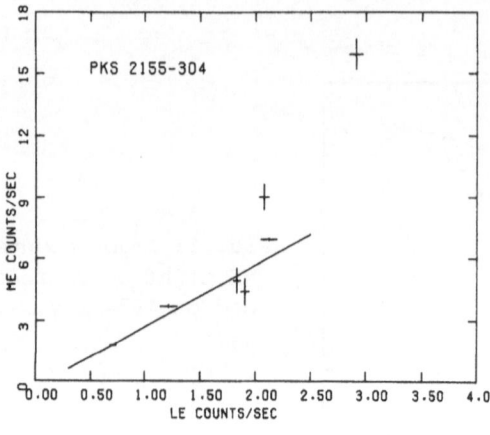

Fig.2: ME counts versus LE counts for all obser- vations. (The peak inten- sity of 1984 Nov. 6 and 7 are not included in the linear regression shown).

RAPID X-RAY VARIABILITY IN ACTIVE GALACTIC NUCLEI

P. Barr   EXOSAT Observatory, Astrophys. Div., ESA
R. Mushotzky   Lab. for High Energy Astrophys. GSFC, NASA
P. Giommi   EXOSAT Observatory, Astrophys. Div. ESA
J. Clavel   IUE Observatory, Astrophys. Div., ESA
W. Wamsteker   IUE Observatory, Astrophys. Div., ESA

SUMMARY

Recent EXOSAT observations of active galactic nuclei are presented. Unlike earlier X-ray satellites (all of which flew in low earth orbit), the deep orbit of EXOSAT allows long continuous observations of celestial X-ray sources, uninterrupted by earth occultation etc. We present the results of EXOSAT observations of several AGN which have been seen to vary rapidly (timescale 0.2-6 hours). We also consider the implications of rapid variability in AGN. For Seyfert galaxies and quasars, we find a highly significant correlation between the timescale of variability and their X-ray luminosity. They are not, howwever, bounded either by the (classical) Eddington limit nor by efficiency arguments. We sugest, rather, that the emitting plasma is dominated by electron-positron pairs.

Several AGN have been seen to vary on short timescales. The nearby spiral M81 was found to contain a powerful X-ray source (L_x - 3×10^{40} ergs/s) which showed repeated variations by a factor two in ten minutes or so. This is the lowest luminosity for any identified AGN. In three out of five EXOSAT observations (each lasting between five and ten hours), NGC 4593 was seen to vary by a factor two, always on a timescale of two hours or so. The more luminous Seyfert I NGC 7469 varied by a factor two in six hours during one observation (out of three). A six hour observation of the Seyfert I galaxy Fairall-9 in July 1985 revealed repeated rapid fluctuations ba a factor two in 15 minutes or so. The implied efficiency of conversion of matter to energy for this object (Fabian & Rees 1978) is $\eta \geqslant 8\%$, as extreme as for the rapid variability found by Tennant et al. (1981) for the (much lower luminosity object) NGC 6814.

What can be learnt from observations of rapid X-ray variability in AGN? We have compiled from the literature an extensive sample of AGN whose reported X-ray variability is reliably established, time resolved and the fastest reported for that particular object. In Fig. 1 we plot the logarithm of the rest-frame 2-10 keV luminosity, $\log L_x$, against the time for the source intensity to double, Δt. The BL Lac objects are considered separately from the Seyferts and

G. Swarup and V. K. Kapahi (eds.), Quasars, 269–271.
© *1986 by the IAU.*

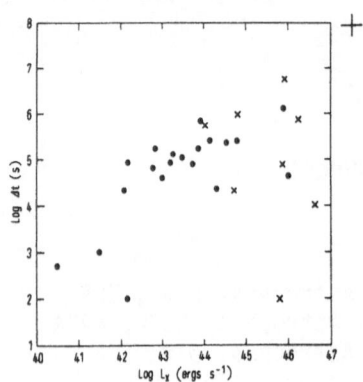

FIGURE 1.
The two-folding timescale of variability, log Δt, versus the mean 2-10 keV rest-frame luminosity, log L_x, for Seyferts and quasars (circles) and for BL Lac objects (crosses). A typical error bar is also shown.

quasars. For the Seyferts and quasars we find a significant trend, writing

$$\log L_x = A \log \Delta t + b \text{ ergs s}^{-1}$$

we find a highly significant correlation with a = 0.9±0.2, b = 39±1 and linear correlation coefficient r_c = 0.7. There is no such trend for the BL Lac objects, r_c = 0.2 Consider the Seyferts and quasars. If their hard X-ray spectra extend to 1 MeV (Bassini & Dean 1983) with spectral index of 0.7 (Rothschild et al. 1983), a bolometric correction of 1.0 to log L_x is indicated. Within the errors the correlation between L_x and   t is linear. Thus the X-ray luminosity for this sample can be written

$$L_x/ \Delta t = 10^{40} \text{ ergs s}^{-2}$$

and, since the source size R $\leq$ c Δt, L/R $\geq 10^{29.6}$ ergs s^{-1} cm^{-1}

What constrains the variability in Seyferts and quasars to this relationship? The implied efficiency of conversion of matter to energy is not large, $\eta \gtrsim$ 0.5% nor is the Eddington limit a problem. Since L/R $<$ L_{Edd}/R_s where L_{Edd} is the classical Eddington limit and R_s the Schwarzschild radius, the observations indicate only L $<$ 0.001 L_{Edd}. One possibility is that the plasma is marginally pair dominated. 2-photon annihilation will lead to the formation of electron-positron pairs. Many workers have studied this phenomenon; a review with particular relevance to AGN is given by Svensson (1984). Copious pair production will occur if the compactness parameter ℓ = L σ_T/Rm_cc^3 > 4π , where L is the γ-ray luminosity. Since the spectra of Seyferts and quasars are so hard, most of the luminosity is radiated as γ-rays. These observations indicate ℓ > 10 and it is highly likely that copious pair production is occurring.

REFERENCES
Barr, P. & Mushotzky, R.F. submitted to Nature
Bassini, L. & Dean, A. 1983. Ast.Ap., 122, 83
Fabian, A.C. & Rees, M. 1978. Proc. XXI Cospar Meeting, p281.
Rothschild, R. et al. 1983. Ap.J., 269, 423.
Tennant, A.F. et al. 1981. Ap.J., 251, 15.
Svensson, R. 1984. Proc. Munich AGN Conference p138.

DISCUSSION

Bregman : Could the few BL Lac sources that appears to radiate in the super Eddington regime have incorrect estimates for L_X because the redshift is unknown ?

Barr : This may be a problem for 1402+04 since the quoted redshift is a lower limit based on the lack of a host galaxy on deep CCD frames (see Maccagni et. al. - this symposium). However, the redshift of 0323+023 seems well established. In addition we have seen two new results presented at this symposium-PKS0548-322 (Agrawal et.al.) both of which have equally high implied efficiencies.

Filippenko : In response to the previous question (where did the redshift of H0323+022 come from ?) : The redshift of H0323+022 came from me - it is 0.1471 ± 0.0005

Barr : Good.

"Beaming is often a panacea for quasar problems, but the absence of corresponding optical and radio features has dissuaded us from suggesting ad hoc *X-ray beaming."*

- Dan Harris (p.276)

X-RAY INTENSITY VARIATIONS IN BL LAC SOURCES H2155-304 AND PKS 0548-322

P.C.Agrawal[1], K.P.Singh[1] and G.R.Riegler[2]

[1] Tata Institute of Fundamental Research
Colaba, Bombay 400005, India

[2] Jet Propulsion Laboratory, Caltech, Pasadena, USA

ABSTRACT: The X-ray Observations of two BL Lac Objects H2155-304 and PKS 0548-322 made with HRI and MPC on the Einstein Observatory show intensity variations on time scale of hours in both the sources. X-ray spectra of the two BL Lacs are derived. Limits on the mass of the accreting compact objects are obtained from time scale and magnitude of variations. Implications of the results are briefly discussed.

In this paper we report variations in the X-ray intensity and spectra of two BL Lac objects H2155-304 and PKS 0548-322. The results are based on observations made with the High Resolution Imager (HRI) for about 10^4 S for each source and Monitor Proportional Counter (MPC) on the Einstein Observatory. The HRI field of view was centered on the two BL Lac objects and the MPC data were obtained simultaneously.

The positions of the X-ray sources obtained from the HRI images agree well with the radio and optical positions of the two BL Lac sources to within 0.6", thus confirming their identification. From the observed radial surface brightness distribution of the X-ray sources, it is inferred that the extent of the sources is $\leq 1"$ which implies a linear extent of ≤ 2.7 kpc for H2155-304 and ≤ 1.8 kpc for PKS 0548-322 (Hubble Constant = 50 km s^{-1} Mpc^{-1}).

X-ray light curves in the soft energy band (0.1 - 4.5 KeV) of HRI show X-ray flux variation by about 20% over time scale of 8 hours for H2155-304 whereas it is nearly constant for PKS 0548-322 for the entire duration of observations. Correlated but much larger intensity variations are detected in 1.2 - 3.5 KeV (by ~ 40%) and 3.5 - 10 KeV (by ~ 100%) bands of MPC for H2155-304. In case of PKS 0548-322 no significant intensity variation is detected in 1.2 - 3.5 KeV band. But its intensity changed by a factor of two in about an hour in 3.5 - 10 KeV interval.

X-ray spectra of the two BL Lac were derived for different time intervals from the MPC data in 1.2 - 10 KeV range. No reasonable fits were obtained for any set of observed spectra with either thermal bremsstrahlung or exponential or black-body models. The spectra are however well fitted with power law models for some of the epochs. The best fit energy spectral

273

index (α) is found to vary between 1.9 and 2.6 for different epoche for H2155-304. However the significance of these variations is small due to poor spectral fits and the spectral fits are consistent with a constant value of α = 2.16. The spectral index for PKS 0548-322 is observed to vary largely between 1.0 and 1.6 with the exception of one epoch where the best fit value of α is as high as 0.5. The average value of α for this source is 1.14. There is a slight indication that the source spectrum is flatter when the flux is higher.

Comparing the flux values and spectra derived from present observations with those obtained earlier by several researchers (for a summary see Urry 1984), it is concluded that there exist long term intensity and spectral changes in both the BL Lac sources. A large change in the average intensity of H2155-304 is observed over the years 1978 to 1979.

The results presented here confirm the earlier reports of variability over time scale of less than a day by Agrawal and Riegler (1979) and Synder et al (1980) in H2155-304. The X-ray spectra of the two BL Lacs are very much softer than those of the Seyfert galaxies. From the rapid variability constraints can be placed on the nature and origin of apparently non-thermal X-ray emission from the two BL Lacs. Assuming that the sources are powered by accretion onto a massive black hole, an upper limit on the black hole mass is obtained by requiring that the fluctuation time be longer than the light travel time across a radius 10 times the Schwarzschild radius. Lower limit on the mass of the accreting object is obtained from the requirement that the luminosity should not exceed the Eddington Limit. Using the observed time scale of intensity variations and magnitude of luminosity changes the following mass limits are obtained : $1.6 \times 10^7 \, M_\odot \le M \le 6 \times 10^8 M_\odot$ for H2155-304 and $6 \times 10^6 \, M_\odot \le M \le 8 \times 10^7 \, M_\odot$ for PKS 0548-322. These variations are further consistent with $\sim$ 10% efficiency for conversion of matter into radiation at the origin of X-rays from a central source in a BL Lac.

The very soft, power law (α > 1), low energy (< 10 keV) spectra and indication of the presence of a variable flat high energy extension for the BL Lacs are similar to those found previously for some other BL Lacs, like Mrk 421 and Mrk 501 (Mushotzky et al. 1979, Worrall et al. 1981). A plausible explanation is that the low-energy X-ray component is the high frequency tail of a synchrotron spectrum which extends from the radio region, while the higher energy X-rays are produced by first-order compton scattering of the synchrotron photons (SSC models). In these non-thermal emission models, an increase in intensity should result from an injection of energetic particles or change in the acceleration rate, and the entire spectrum from radio to X-rays should rise uniformly.

REFERENCES :
Agrawal, P.C. and Riegler, G.R. 1979, Ap.J. (Letters), 231, L25.
Snyder, W.A. et al. 1980, Ap.J. (Letters), 237, L11.
Urry, C.M. 1984, PhD. thesis (NASA Technical Memorandum 86103).
Mushotzky, R.F. et al. 1979, Ap.J. (Letters), 232, L17.
Worrall, D.M. et al. 1981, Ap.J., 243, 53.

THE X-RAY JET IN 3C 273

D.E. Harris and C.P. Stern
Harvard-Smithsonian Center for Astrophysics
60 Garden St.
Cambridge, MA 02138, USA

ABSTRACT. A subtraction technique is applied to the X-ray image of 3C 273 in order to determine the location and other properties of emission associated with the jet.

I. INTRODUCTION

Willingale (1981) demonstrated the existence of X-ray emission from the jet of 3C 273 by applying the maximum entropy method to the observation made with the high resolution image (HRI) of the Einstein Observatory. In order to determine the location and intensity by a different method, we have reanalyzed the HRI image by attempting to subtract the core emission with a Point Response Function (PRF), thereby ridding the residual array of the scattering wings of the PRF. In the following sections we summarise the method and results, and then compare them to new radio and optical maps. Details will be published elsewhere (Harris and Stern, 1986).

II METHOD

The first step is to create an image by accepting only those time intervals during which the star trackers were "locked" onto the pre-assigned guide stars. From experience with many HRI images, we have found that the "locked" images of unresolved sources have a narrower and more circular response than the usual images produced with standard processing.

The chief problem of our method is the difficulty of finding a PRF which matches that of the observation. Imperfect aspect solutions and spatial corrections necessary for rectifying the image compromise the integrity of the PRF in minor ways which are different for each observation. After examining several exposures of bright sources which were unresolved, we chose Cyg X-2 to define the PRF even though the core of Cyg X-2 was slightly broader than that of 3C 273 in one dimension. The image of Cyg X-2 was rolled to match the detector coordinates of 3C 273 and then shifted in RA and DEC to align the cores to ± 0.02 arcsec. Normalization was based on the relative counts in the core of the PRF. The sub-

G. Swarup and V. K. Kapahi (eds.), Quasars, 275–276.

traction was imperfect, leaving a residual positive bar with negative pits on each side (caused by the wider core width of Cyg X-2).

III. RESULTS

The residual map clearly shows a feature located 15.8" from the core in PA = 222°2, well above the local fluctuations. This feature contains 0.7% of the total counts of 3C 273 and the derived flux for a power law energy spectrum with exponent 0.8 is 5E-13 erg $cm^{-2}s^{-1}$ (0.5 - 3.0 keV). The corresponding luminosity (Ho = 50 km s^{-1} Mpc^{-1}) is 8E43 erg s^{-1} (0.5 - 4.5 keV). There is no evidence that the feature is extended along the jet. Because of the uncertainties in the subtraction, it is difficult to ascertain if the distribution is wider than the 6" resolution which results from the 3.5" PRF and the Gaussian smoothing function applied after subtraction.

Our X-ray position is located on the radio and optical jet approximately in the center of the optical "plateau" discussed by Roser and Meisenheimer (1985), which corresponds to the rising part of the radio jet (see Davis et al. 1985 for recent radio maps). Thus we find a curious progression: moving out from the core, the brightest peak in the X-ray occurs before that of the optical, which is followed in turn by the brightest radio feature at the end of the jet.

Estimates of parameters for various emission mechanisms have not provided encouraging results. For a stationary source (no relativistic beaming) a synchrotron model (radio/optical/X-ray) would require high energy electrons with a half-life of only 8 days ($\nu = 10^{18}$HZ). For thermal bremsstrahlung, a mass of order of 10^{10} $M_\odot$ is required, and although inverse Compton models are possible, there is no obvious photon population of high energy density.

Beaming is often a panacea for quasar problems, but the absence of corresponding optical and radio features has dissuaded us from suggesting ad hoc X-ray beaming.

ACKNOWLEDGEMENTS

We thank T. Muxlow and H.J. Roser for sending us their results in advance of publication. This work was supported by a student aid grant from Harvard University and by NASA contract NAS8-30751.

REFERENCES

Davis, R.J., Muxlowe, T.W.B., Conway, R.G. 1985, Nature (in press).
Harris, D.E., Stern, C.P. 1986 (to be submitted, Ap.J.Letts).
Roser, H.J., Meisenheimer, K. 1985 (preprint).
Willingale, R. 1981, Mon. Not. R. astr. Soc. 194, 359.

III

SPECTRAL LINE STUDIES

"It is of course not unusual to conclude that all of these problems need further observational (and theoretical) investigation ..... "

— Beverley Wills (p.284)

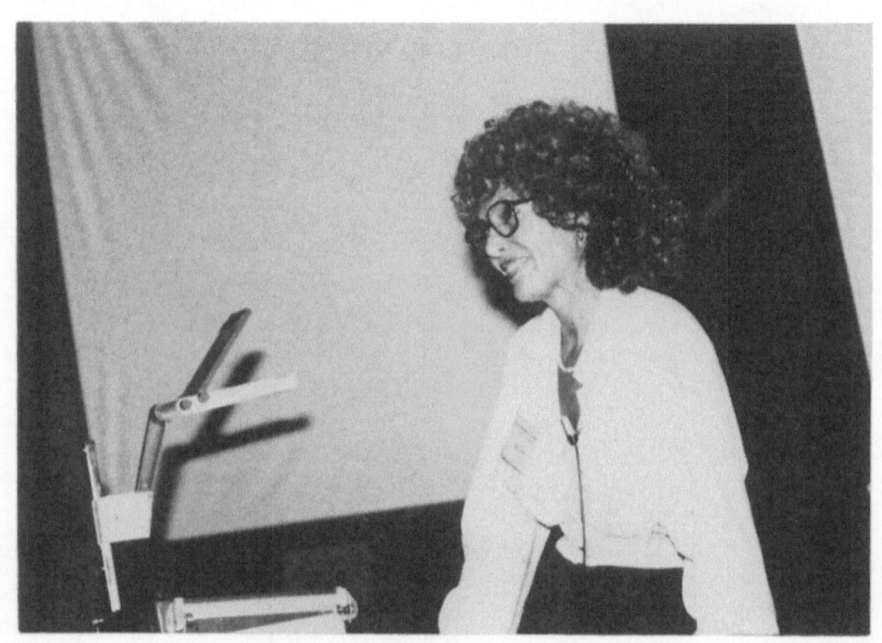

Suzy Collin-Souffrin

A SURVEY OF QSO EMISSION LINES

Beverley J. Wills
McDonald Observatory and Department of Astronomy
University of Texas, RLM 15.308
Austin, Texas 78712
U.S.A.

ABSTRACT. "Standard" photoionization models of the broad line region (BLR) consist of numerous small optically thick clouds of identical ionization, density, temperature and optical depth, moving under the influence of the gravity and radiation field of a 10^{7-8} $M_\odot$ black hole and confined by a high temperature gas. Such models have provided a good description of observed line strengths and widths with only very minor modification. Although photoionization remains the most important heating mechanism in the BLR, new observations point to a wider range of physical conditions and to clues about the geometric and dynamic arrangement of the emitting gas. I want to highlight the observations that have led us to this view.

1. INTRODUCTION

I shall relate the properties of various classes of luminous QSO-like objects and present a global description of the emission line region. This is a dangerous (however convenient!) approach because there are important differences between high luminosity compact and extended radio QSOs, radio quiet QSOs and the lower luminosity active nuclei of broad line radio galaxies, Seyfert 1 and other galaxies, and it remains to be seen whether or not these differences are only of degree and can be encompassed in some "unified scheme." The volume "Astrophysics of Active Galaxies and Quasi-Stellar Objects" (Miller 1985) contains excellent reviews of the situation at the time of the 7th Santa Cruz Workshop on Astrophysics, held in July 1984. I will therefore emphasize the most recent work. In the following, AGN will refer to those lower luminosity active nuclei where the evidence for a surrounding galaxy is quite strong. QSO will refer to the higher luminosity objects, and a radio-loud QSO is a quasar.

1.1. The Standard Model

Excitation of narrow and broad lines is by photo-ionization. (For the most recent evidence for this, from the strong correlation between strengths of lines and non-thermal continuum, see de Robertis [1985] and Kinney et al. [1985]). Line ratios are primarily determined by the ionization parameter which is the ratio of ionizing photon density to electron density at the surface of the cloud or filament facing the central continuum radiation source. The highly ionized H+ region produces the high ionization lines He II

G. Swarup and V. K. Kapahi (eds.), Quasars, 279–288 .

λ1640, CIV λ1549, and some Lyα. The depths of the BLR clouds, which can be very extended if X-rays and γ-radiation are strong, produce lower ionization species: Mg II λ2800, H I and Fe II lines. Most CIII] λ1909 is produced at intermediate depths. Abundances are solar or slightly higher (Gaskell *et al.* 1981, Uomoto 1983). For the history and details of the calculations see Davidson and Netzer (1979) and Netzer and Ferland (1984). Table I gives typical parameters of a "Standard Model".

TABLE I. Characteristics of the "Standard Model"

	Broad Line Region (BLR)	Narrow Line Region (NLR)
Overall dimensions	0.1 - 1 pc	> 1 Kpc
Ensemble FWHM	5000 km/s	500 km/s
Filling factor	10^{-8}	small
Covering factor	0.2	small
Individual clouds:		
Optical depth, τ(Lyc)	10^5	> 1
Density	$10^{9.5}$ cm^{-3}	10^5 cm^{-3}
Ionization parameter, U	10^{-2}	$10^{-2.5}$
Temperature	10^4 K	10^4 K
Size	10^{12} cm	> 10^{12} cm

It has been recognized almost from the beginning that a higher ionization component is needed to explain the strengths of O VI and N V lines and that a hot (10^8 K) intercloud gas is needed to pressure-confine the BLR clouds. Below we summarize recent observations indicating other ways in which the standard model must be modified.

2. THE NARROW LINE REGION (NLR)

- For radio galaxies, with radio structure ~ 100 Kpc, the rotation axis of the narrow line gas is aligned with the direction of the radio "jet", but in general not with the stellar rotation axis (Heckman *et al.* 1985, Simkin 1979). There are also more complex motions. Narrow band and continuum imaging and spectroscopy of steep spectrum radio galaxies and QSOs suggest an external origin for the gas, perhaps disk gas captured during mergers of galaxies (e.g., Stockton and MacKenty 1984, Heckman *et al.* 1985, Spinrad and Djorgovski 1984).
- Narrow line profiles give information about the NLR closer to the nucleus: for Seyfert galaxies asymmetry is observed in the [OIII] λ5007 line (Whittle 1985a, Heckman *et al.* 1984, Vrtilek and Carleton 1985). The line peaks are at the galaxy systemic velocity, and asymmetry increases toward the base of the line, in the sense of there being more emission to the blue of the peak. This implies radial motions together with obscuration, either (i) infall with dust at the back of the narrow line clouds (away from the ionizing source) or (ii) outflow with pervasive obscuration or a central region hiding clouds on the far side. Unlike compact quasars and radio-quiet AGN, limited data for Kpc-scale radio emitting objects do not show systematic [OIII] asymmetry (Heckman *et al.* 1984).

These observations suggest that the rotating outer NLR knows about the central engine, and that radial motions and obscuration are important in the inner NLR. How does this relate to the BLR?

- The [OIII] line widths in AGN correlate with broad Hß widths and are independent of physical conditions in NLR, and optical non-thermal luminosity (Shuder 1984, Heckman *et al.* 1984, Whittle 1985b). This suggests an indirect link between NLR and BLR kinematics, such as related special geometries with a resulting similar dependence of projected velocities on inclination to the line of sight, e.g., disk geometry (see below). The relation may hold for QSOs but the scatter is large.
- [OIII] luminosity declines markedly with increasing broad line contribution, along the sequence Seyfert 2 - Seyfert 1.8, 1.5 - Seyfert 1, almost independent of physical conditions in the NLR. This suggests that the BLR is shielding the NLR from the central continuum, and if, as is believed, the BLR covering factor is small (0.1-0.2), the covering factor for the NLR must also be small. A flattened, coplanar system for both the BLR and NLR, e.g., a disk, is then a possible configuration (Cohen 1983).
- Osterbrock and colleagues have noted that high ionization narrow lines, present in broad line AGN, e.g., [Fe X], tend to be broader and blue-shifted compared with low ionization narrow lines. More detailed studies show, for some Liners and Seyfert 1s, correlations of increasing ionization potential and/or critical density with increasing narrow line width (e.g., Filippenko, this symposium, Filippenko and Halpern 1985, de Robertis and Osterbrock 1984, Whittle 1985c). These suggest higher ionization and densities with increasing velocity probably as the central ionizing source is approached. But the NLR is more complex than that: at least in some cases there is a range of densities for a given ionization level. In one well-studied case, NGC 3712, velocities (FWHM) range from a few hundred to 2000 km/s, with a corresponding range in density from 10^2 to $10^{7.5}$ cm^{-3} - approaching the densities and velocities of the BLR.

All this suggests co-planar disk-like structures and a physically merging (but distinct) NLR and BLR.

3. THE BROAD LINE REGION

Many new observations suggest a wider range of optical depths, densities and ionization parameter in the BLR, as a function of velocity and/or distance from the ionizing continuum.

3.1. Line Strengths

- Fe II emission is much stronger than previously thought (e.g., Wills *et al.* 1985). Typically, the intensity ratio Fe II/Lyα ~ 1-2. In two extreme Fe II emitters, Fe II/Lyα >30 (Wampler 1985, 1986)! Very high densities, $> 10^{11}$ cm^{-3}, large optical depths (10^7) and/or low U (10^{-3}) can produce strong low ionization lines. This requires a separate optically thin emission region of high U to produce the high ionization lines. The intensity ratio Fe II/Mg II λ2798 ~ 8, compared with 1.5 - 4 from models. In Wampler's two strong Fe II emitters the ratio may be 50 to 100. Mechanisms other than photoionization have been suggested (see Joly, this symposium, Collin-Souffrin, this symposium). Is iron overabundant?
- Hydrogen Lines: Balmer, Paschen and Brackett line ratios require more than reddening, and high optical depths and densities are needed (Miller 1985). Balmer

continuum and high order Balmer lines are too strong, and extreme X-ray or γ-ray fluxes may be required (Wills *et al.* 1985). The intensity ratio Lyß/Lyα suggests that Lyα may come from an optically thin region (Wilkes 1984, 1985, Green *et al.* 1980). So perhaps there is more than one kind of BLR cloud. Wills *et al.* (1985) suggest that the Balmer continuum may come from a hot accretion disk.

- Broad [OIII] is suspected in some Seyfert 1 nuclei and QSOs (van Groningen and de Bruyn 1985, Meyers and Peterson 1985, Wills *et al.* 1985). This suggests the existence of BLR densities near the [OIII] critical density of ~ 10^6 cm^{-3}. So BLR densities overlap with those of the NLR.

- A feature at 1909 Å may not be all CIII]. If confirmed this provides further evidence that in some QSOs CIII] λ1909 may have been collisionally suppressed and there could be densities of $10^{10.5}$ cm^{-3} and higher (Wampler 1986, Hartig and Baldwin 1985, see also Gaskell *et al.* 1981).

All these observations indicate a range of physical conditions in clouds giving rise to the broad lines, perhaps more than one kind of heating mechanism and even weird abundances.

3.2. Broad Line Profiles

- Quasars do not have "narrow" broad lines. The FWHM always exceeds 2000 km/s (Wills and Browne 1986).

- FWHM (Hß) is inversely correlated with R, the ratio of radio flux in the compact core to that in the extended lobes. In the relativistic beaming model, R measures inclination, so this correlation may be explained by a disk-like BLR with its axis parallel to the radio axis (Wills and Browne 1986, see also Wills and Wills, this symposium).

- High ionization lines are (statistically) blue-shifted with respect to low ionization lines, and with respect to the narrow lines, e.g., Lyα and CIV λ1549 are blue-shifted ~1000 km/s with respect to OI λ1304 and CII λ1335, and Mg II is blue-shifted by ~600 km/s with respect to the NLR (Gaskell 1982, Uomoto 1984, Wilkes 1985).

More detailed profile studies show:

- He I λ5876 and Hß are broader than Hα in Seyfert 1s (Osterbrock and Shuder 1982) suggesting that U and electron density increase with increasing velocity (inwards). But this effect is smaller in high luminosity AGN and not detected in QSOs (Shuder 1984). In QSOs Balmer lines are probably narrower than He II λ4686 (Wills *et al.* 1985). OI and CII may be narrower than CIV and Lα (Wilkes and Carswell 1982) and Mg II and Hß narrower than CIV (in Seyferts and QSOs, Mathews and Wampler 1985, Joly *et al.* 1985). That is, lower ionization lines are narrower.

- Broad Hß is quite symmetric, very similar in shape (but not width) from one object to another and also similar in shape to the narrow lines in high-ionization Seyfert 1s. So a unique acceleration mechanism (or geometry, or both) may operate in all NLR and BLR (de Robertis 1985).

- Lyα and C IV are very symmetric in most cases, and profiles are logarithmic. The symmetry of Lyα is consistent with radial motions only if the emitting region is optically thin (e.g. Wilkes 1984, Wilkes and Carswell 1982). But in the standard model asymmetry is expected; Lyα is from an optically thick region and escapes only from the side of the cloud facing the continuum source. An optically thin Lyα region may also be indicated by Lyß/Lyα ~0.006 (Wilkes 1984).

As in the case of the line strengths, the profile results suggest a range of physical conditions and in addition show that clouds of different densities, ionization parameter, optical depth etc. have different kinematic properties.

3.3 Broad Line Variability

Line variability is probably ubiquitous among Seyfert 1 nuclei, and recently more data have become available on profile variability and the response of various UV and optical lines to changes in the continuum, with some surprising results. Probably the most extreme are for the well-studied cases of NGC 4151 (Ulrich *et al.* 1984 and Antonucci and Cohen 1983) and Akn 120 (Peterson *et al.* 1985), where BLR radii of a few light days are deduced. As an example of recent trends, in Table II I reproduce results from a re-analysis of earlier data by Gaskell and Sparke (1985).

TABLE II. Results from the Variability of NGC 4151

Line	Radius of BLR (light-days) (approximate)
He II	2 - close to ionizing continuum?
CIV, Hα, Hβ	6
Mg II	>12 perhaps to > 100

The smallest time scales imply high densities of 10^{10} to > 10^{12} cm^{-3}, and similar results are derived for other Seyfert 1 nuclei. For QSOs the time scales are probably longer, but the trends may be similar (Gondhalekar, this symposium) i.e., emission line regions lie closer to the continuum source and are of higher density than in the standard model. Of note is the case of 3C 446: the strength of CIII] varies in proportion to continuum strength with a lag < 30 days and CIV does not vary (Stephens and Miller 1985)! Lyα varies with the continuum with a lag less than a few months (Bregman *et al.* 1985). These results certainly cannot be explained on the standard model!

3.4 Line Polarization

Results mainly from the work of Antonucci (1984) and Rudy *et al.* (1983) show some clear trends:

- In AGN, position angles of optical continuum polarization are often perpendicular to the Kpc scale radio structure, but nearly always parallel in higher luminosity BLRGs. (Compare with the earlier results of Stockman, Angel and Miley (1979) who found that continuum linear polarization for low polarization quasars tends to be aligned with extended radio structure.)
- In at least some AGN and QSOs the polarization appears to be wavelength independent.
- The position angle of broad line polarization is the same as for the continuum (tentatively for QSOs too).

The mechanism for line polarization must be electron or dust scattering. With respect to the distant observer the regions emitting the lines and continuum see the same scattering geometry. The standard model does not account for this. The geometry of the BLR, continuum, and/or scatterers must know about the central engine. (This also resurrects the idea that electron scattering could contribute to line broadening). It will be important for models, to compare the polarization for low and high ionization broad lines, since these may arise predominantly at somewhat different distances from the ionizing continuum.

4. SUMMARY - A POSSIBLE PICTURE FOR THE EMISSION LINE REGIONS

The NLR and at least the low ionization BLR are in a flattened disk-like configuration with the axis aligned with the axis of the central engine. The evidence comes from rotation curves for the outer NLR, apparent shielding of the NLR by the BLR, correlation of broad and narrow line widths, correlation of broad line width with apparent inclination of the radio axis, and the alignment of line and continuum polarization with the radio axis. This picture is also supported by the similarity of NLR and BLR line shapes and other evidence for the merging of the NLR and BLR.

There is accumulating evidence for different clouds with a range of properties in both the NLR and BLR, which also points to a merging of the NLR and BLR. Densities in the NLR and BLR range from a few hundred to $> 10^{11}$ cm^{-3}; velocities (FWHM) from a few hundred to 10^4 km/s. Probably density, ionization and velocity increase toward the central ionizing source. But the situation is more complicated than that. Some emitting regions of high ionization in the BLR are optically thin (as also previously suggested for N V etc.), and some of low ionization in the NLR are optically thick. Higher ionization lines are broader and blue-shifted relative to low ionization lines in both the NLR and BLR, which points to a stratified emission line region (as does the existence of a distinct NLR and BLR in the standard model) with radial motions and some obscuration, even in the higher luminosity objects. Seyfert 1 line variability results lend further support to a stratified model and suggest distances of the BLR from the continuum in AGN ranging from a few light days for high ionization material to hundreds of light days for low ionization line-emitting regions, perhaps even merging into an accretion disk.

This picture is far too simple and there are so many missing pieces in the puzzle that we have had to make big assumptions about the basic similarity of different classes of object. It is of course not unusual to conclude that all of these problems need further observational (and theoretical) investigation, but I believe that a comparison of different classes of AGN and QSOs is especially important.

I gratefully acknowledge useful discussions with H. Netzer, C.M. Gaskell and D. Wills, and thank them and others, especially D. Alloin, M. Joly and A.L. Kinney, for giving me preprints.

REFERENCES

Antonucci, R.R.J. 1984, *Ap.J.*, **278**, 499.
Antonucci, R.R.J., and Cohen, R. 1983, *Ap.J.*, **271**, 564.
Boroson, T.A., Persson, S.E., and Oke, J.B. 1985, *Ap.J.*, **293**, 120.
Bregman, J.N., Glassgold, A.E., Huggins, P.J., and Kinney, A.L. 1985, *Ap.J.*,
    submitted.
Cohen, R. 1983, *Ap.J.*, **273**, 489.
Davidson, K., and Netzer, H. 1979, *Rev. Mod. Phys.*, **51**, 715.
de Robertis, M. 1985, *Ap.J.*, **289**, 67.
de Robertis, M.M., and Osterbrock, D.E., 1984, *Ap.J.*, **286**, 171.
Filippenko, A.V., and Halpern, J.P. 1984, *Ap.J.*, **285**, 458.
Gaskell, C.M. 1982, *Ap.J.*, **263**, 79.
Gaskell, C.M., Shields, G.A., and Wampler, E.J. 1981, *Ap.J.*, **249**, 443.
Gaskell, C.M., and Sparke, L.S. 1986, preprint.
Green, R.F., Pier, J.R., Schmidt, M., Estabrook, F.B., Lane, F.B., Wahlquist, H.D.
    1980, *Ap.J.*, **239**, 483.
Hartig, G.F., and Baldwin, J.A. 1986, *Ap.J.*, in press.
Heckman, T.M., Illingworth, G.D., Miley, G.K., and van Breugel, W.J.M. 1985,
    Ap.J., **299**, 41.
Heckman, T.M., Miley, G.K., and Green, R.F. 1984, *Ap.J.*, **281**, 525.
Joly, M., Collin Souffrin, S., Masnou, J.L., and Nottale, L. 1985, *Astr.Ap.*, in press.
Kinney, A.L., Huggins, P.J., Bregman, J.N., and Glassgold, A.E. 1985, *Ap.J.*, **291**,
    128.
Lawrence, A., and Elvis, M. 1982, *Ap.J.*, **256**, 410.
Mathews, W.G., and Wampler, E.J. 1985, *Publ.A.S.P.*, in press.
Meyers, K.A., and Peterson, B.M. 1985, *Ap.J.*, in press.
Miller, J.S., ed. 1985, Astrophysics of Active Galaxies and Quasi-Stellar Objects (based
    on the 1984 Santa Cruz Astrophysics Workshop held in honor of D.E.
    Osterbrock) (Mill Valley, California: University Science Books).
Netzer, H. 1985, *Ap.J.*, **289**, 451.
Netzer, H., and Ferland, G.J. 1984, *Publ.A.S.P.*, **96**, 593.
Osterbrock, D.E., and Shuder, J.M. 1982, *Ap.J.Suppl.Ser.*, **49**, 149.
Pettini, M., and Boksenberg, A. 1985, *Ap.J.(Letters)*, 294, L73.
Rudy, R.J., Schmidt, G.D., Stockman, H.S., and Moore, R.L. 1983, *Ap.J.*, **271**, 59.
Shuder, J.M. 1982, *Ap.J.*, **259**, 48.
Shuder, J.M. 1984, *Ap.J.*, **280**, 491.
Simkin, S.M. 1979, *Ap.J.*, **234**, 56.
Spinrad, H., and Djorgovski, S. 1984, *Ap.J. (Letters)*, **280**, L9.
Stephens, S.A., and Miller, J.S. 1985, *Bull.A.A.S.*, **17**, 1007.
Stockman, H.S., Angel, J.P.R., and Miley, G.K. 1979, *Ap.J.*, **227**, 55.
Stockton, A. and MacKenty, J.W. 1983, *Nature*, **305**, 678.
Ulrich, M.H., *et al.* 1984, *M.N.R.A.S.*, **206**, 221.
Uomoto, A.K. 1984, *Ap.J.*, **284**, 497.
van Groningen, E., and de Bruyn, A.G. 1985, *Astr.Ap.*, in preparation.
Vrtilek, J.M., and Carleton, N.P. 1985, *Ap.J.*, **294**, 106.
Wampler, E.J. 1985, *Ap.J.*, **296**, 416.
Wampler, E.J. 1986, preprint.
Whittle, M. 1985a, *M.N.R.A.S.*, **213**, 1.
Whittle, M. 1985b, *M.N.R.A.S.*, **213**, 33.
Whittle, M. 1985c, *M.N.R.A.S.*, **216**, 817.

Wilkes, B.J. 1984, *M.N.R.A.S.*, **207**, 73.
Wilkes, B.J. 1985, *M.N.R.A.S.*, in press.
Wilkes, B.J., and Carswell, R.F. 1982, *M.N.R.A.S.*, **201**, 645.
Wills, B.J., and Browne, I.W.A. 1986, *Ap.J.*, March 1.
Wills, B.J., Netzer, H., and Wills, D. 1985, *Ap.J.*, **288**, 94.

DISCUSSION

Hutchings : You mentioned velocity differences between high and low
ionisation lines. Is this due to asymmetry or a shift of the entire
profile ?

Wills : Velocity shift is just a first order measure of profile difference, and most broad line data cannot stand more detailed analysis. The
highest signal to noise data suitable for comparison of high and low
ionization is that obtained by Wilkes and Carswell (1982) and Wilkes
(1985), but the low ionization OI and CII lines are weak.
    For narrow lines, at least [OIII], the shift does appear to be due
to increasing asymmetry from the peak to the baseline, the peak being
at the systemic velocity. (See also the paper by Filippenko in this
proceeding).

Hutchings : Could you comment on the very asymmetrical or double-peaked
broad lines that Gaskell has picked out in the past.

Wills : Gaskell has advocated the gravitational influence of binary
black holes to explain some apparent double-peaked and complex line
profiles (Nature, 1985, <u>315</u>, 386). This fits in nicely with ideas on
the formation of activity in nuclei through galaxy mergers.

Bregman : I and coworkers (Huggins, Glassgold, and Kinney) have seen
Lα line variation in 3C 446 (z=1.404) in 2 months. If the lines are
symmetric about the central source, then $n \geq 10^{14}$ cm^{-3} is implied.
Instead, we argue that the density is lower, which implies that the
clouds are distributed anisotropically. A significant amount of ionizing gas probably lies near our line of sight and at several light years
from the ionizing source.

Wills : I think this is a possible explanation but would feel uneasy
about applying it if the same phenomenon were to be observed in many
more QSOs. I think other more generally applicable geometries may be
possible, but whatever explanation is preferred the CIV λ1549 emission
is unlikely to be emitted in predominantly the same region as the CIII]
1909 or Lα .

Gondhalekar : Comment : We have studied QSOs up to $z \sim 1$ with IUE and
we find variability of emission lines in most objects.
Question : In how many objects do you see symmetric and irregular Lα

profiles ? In UV spectra we find a large incidence of irregular Lα
profile.

Wills : Only the highest signal-to-nosie observations where one can
subtract the influence of blended lines and possible absorption, seem
to be suitable for Lα profile studies (e.g. Wilkes and Carswell 1982).
Some QSOs had symmetric lines and not others, but the sample was very
small.

Filippenko : It is true that many Lα profiles appear to be much more
symmetrical than one would expect from a model in which optically thick
clouds are moving radially. However, Kallman and Krolik have recently
proposed a model in which the optical depth to electron scattering in
the hot confining medium is approximately 0.1; in this case, the back
sides of clouds are ionized to a fairly great extent, and the line
profiles become more symmetrical. This optical depth to electron scattering is not high enough to wipe out continuum variability.

Wills : This model seems to require a fairly specific set of conditions,
and I think it remains to be seen whether all this is theoretically
reasonable.

De Bruyn : You mentioned the presence of broad [OIII] emission in some
Seyfert 1's and said this was evidence for densities less than $10^6 cm^{-3}$
in the BLR. Ernst van Groningen and I have strong evidence for broad
[OIII] emission in many Seyferts (there is a late poster on these res-
ults) and conclude that densities between 10^6 and 10^7 are called for.
We believe this emission to come from a region <u>distinct</u> from the proper
NLR and BLR rather than the BLR itself. Could you comment on this ?

Wills : I think this supports the idea that the whole emission line
region is more complex than we thought-different regions and/or clouds
or filaments of different physical properties.

Khachikian : What is your opinion about   variations of [OIII] lines
in NGC 1275 during some days ?

Wills : I have not thought much about that. Perhaps it is possible in
lower luminosity objects to have some [OIII] in a very small region ?
There is another example in a paper by Antonucci (Ap.J.) where the narrow line appears to disapper completely.

Malkan : You mentioned that some AGN's (e.g. Mrk 486) have the same
polarization position angle in their continua and broad emission lines.
These all turn out to be reddened, dusty AGN's, so the dust-scattering
explanation of polarization looks fine. However, there are many less
dusty AGN's (e.g. NGC 4151, 5548, MKn509) which have polarized continua,
but essentially unpolarized Balmer emission lines. It's possible that
many AGN's may have this sort of polarization, and it may require a
different explanation (i.e. not based on dust scattering).

Wills : I was really referring to the higher luminosity broad line radio galaxies and possibly QSOs where Antonucci favours electron scattering and Rudy and colleagues favour dust. One must explain the observations where line polarization <u>is</u> observed. Where it is not observed it
may not exist or may be below the sensitivity of the observations.

Kundt : Talking of split emission lines is in low-luminosity objects :
didn't Cyril Hazard report on split emission lines in (high-luminosity)
high-redshift objects ?

Turnshek : The object was 1011+09, also studied by Foltz et.al. (1983,
P.A.S.P.). This object is a BAL QSO and Foltz et.al showed that the
reported splitting is probably due to absorption.

STUDIES OF NARROW EMISSION LINES IN AGNs

Alexei V. Filippenko
Department of Astronomy
University of California
Berkeley, CA 94720
USA

ABSTRACT. Optical spectra having moderately high resolution ($\sim 2 - 5$ Å) are being used to study the profiles of narrow emission lines in active galactic nuclei (AGNs). It is often found that forbidden lines associated with high critical densities for collisional deexcitation are the broadest. A good example is [O III] $\lambda4363$ [$n_e(\text{crit}) \approx 3 \times 10^7$ cm^{-3}], whose width can be more than twice that of [O III] $\lambda5007$ [$n_e(\text{crit}) \approx 8 \times 10^5$ cm^{-3}]. The tight correlation between line width and $n_e(\text{crit})$ implies that a much larger *range* of densities ($\sim 10^2 - 10^7$ cm^{-3}) must be present among clouds in the narrow-line region than was previously believed. At times there almost appears to be a continuity between the narrow- and broad-line regions. In some objects the dense, high-velocity clouds are optically thick to ionizing radiation, since they emit [O I] $\lambda6300$ as well as species of much higher ionization (such as [Ne V] $\lambda3426$). These results help eliminate several difficulties in photoionization models of LINERs. It may also be possible to use the observed line widths as probes of the gravitational potential in AGNs.

1. INTRODUCTION

In the standard picture of quasars and Seyfert galaxies, radiation produced by a "central engine" is thought to photoionize gas located in two rather distinct volumes (see, e.g., Weedman 1977). Clouds in the broad-line region (BLR) live $\sim 0.1 - 1$ pc from the nucleus, are quite dense ($n_e \approx 10^8 - 10^{10}$ cm^{-3}), and have characteristic velocities of $\sim 3000 - 10000$ km s^{-1}; the corresponding parameters for the narrow-line region (NLR) are $\sim 10 - 500$ pc, $\sim 10^3 - 10^4$ cm^{-3}, and $\sim 300 - 1000$ km s^{-1}. The temperature of the ionized gas, determined primarily by the equilibrium between photoionization heating and radiative cooling, is $\sim 10000 - 20000$ K in both regions.

The emission lines in AGNs often have nearly symmetrical profiles. When observed at spectral resolutions of ~ 10 Å, those from the NLR are well described by single Gaussians, and in a given object the widths of different lines are similar (Koski 1978). Hence, photoionization models have generally dealt only with the total intensities of individual lines (e.g., Davidson and Netzer 1979). The results have been encouraging, but important difficulties remain.

G. Swarup and V. K. Kapahi (eds.), Quasars, 289–294 .
© *1986 by the IAU.*

It has recently become clear that this simple picture has serious short-comings, and that any quantitative model based on it can provide at best a rough physical description of AGNs. Osterbrock (1979), for example, has stressed that in certain objects there is a range in density among the narrow-line clouds ($n_e \approx 10^3 - 10^6$ cm^{-3}). There is increasing evidence that several distinct families of clouds exist in the BLR, and that $n_e \gtrsim 10^{11}$ cm^{-3} in at least some of the gas (Puetter 1986). Moreover, many broad emission lines exhibit complex profiles. In this paper I will show that the *profiles of narrow emission lines* can also be used to impose constraints on realistic models.

2. A CLOSE LOOK AT THE NLR

Optical spectra having moderately high resolution ($\sim 2 - 5$ Å) were obtained with an Intensified Reticon on the 2.5-m du Pont telescope at Las Campanas Observatory (Chile) in order to study the narrow emission lines in several AGNs. These are illustrated in Filippenko and Halpern (1984; FH84) and in Filippenko (1985; F85).

One object of particular interest is MR 2251–178, the first QSO to initially be identified as a result of X-ray observations. Its spectrum is characterized by very broad permitted lines and considerably narrower forbidden lines (and permitted-line cores) superposed on a strong nonstellar continuum. The lines are undoubtedly produced by photoionized gas. On the other hand, the intensity ratio (R) of [O III] $\lambda\lambda4959{+}5007$ to [O III] $\lambda4363$ is only ~ 10, far lower than normal in AGNs. If $n_e \lesssim 10^4$ cm^{-3}, this implies that $T_e \gtrsim 10^5$ K, which is much too high for the O^{++} region. Osterbrock, Koski, and Phillips (1976) recognized this problem in several broad-line radio galaxies, and suggested that $n_e \approx 10^6$ cm^{-3} in clouds which produce the [O III] emission.

Direct evidence that high densities do indeed resolve the problem in MR 2251–178 is found from a comparison of the line profiles. As shown in Figure 9 of F85, there is a wide range of line widths and profile shapes, indicating that clouds having different physical properties are present in the NLR. In particular, lines associated with high critical densities [n_e(crit)] are *broader* than those with low values of n_e(crit), so they must be formed in the denser, more rapidly-moving clouds. [O III] $\lambda4363$ [n_e(crit) $\approx 3 \times 10^7$ cm^{-3}], for example, is much broader than [O III] $\lambda5007$ [n_e(crit) $\approx 8 \times 10^5$ cm^{-3}]; line width is therefore not correlated with ionization potential (χ), unlike the case in some AGNs (De Robertis and Osterbrock 1984). Note that a large range of densities would not necessarily be excluded if the correlation between line width and χ were better than that between width and n_e(crit), but an unambiguous conclusion would then be difficult to reach without a considerably more detailed analysis.

Since the [O III] lines in MR 2251–178 have such different profiles, it is of interest to examine the ratio R as a function of velocity (i.e., position in the line profile). If the gas is photoionized by ultraviolet radiation from the nucleus, then $T_e \approx 15000$ K in the O^{++} zone, and the changing value of R directly demonstrates that high-velocity clouds ($v \gtrsim 1000$ km s^{-1}) have $n_e \gtrsim 5 \times 10^6$ cm^{-3} (see Fig. 10 in F85). This suggests that an "intermediate" region exists between the NLR and BLR.

The same results are found in Pictor A, a broad-line radio galaxy with strong forbidden lines spanning a wide range of ionization states. As shown in

Figure 1, there is an excellent correlation between line width and n_e(crit). Since line emission per unit mass of gas increases linearly with n_e below n_e(crit), and remains constant above it, different forbidden lines act as effective tracers of n_e when there is a sufficiently large range of densities among the clouds. This is precisely what must be happening in the NLR of Pictor A. In fact, it is known from the [S II] $\lambda\lambda 6716, 6731$ doublet that clouds having $n_e \lesssim 500$ cm^{-3} are present in the NLR, yet the [O III] ratio R indicates that densities in excess of 10^6 or even 10^7 cm^{-3} must also exist (F85). Once again, the highest densities are only an order of magnitude or two below those in the BLR, and the lines produced in the "intermediate" region are quite broad. Similar conclusions have been drawn by Carswell *et al.* (1984).

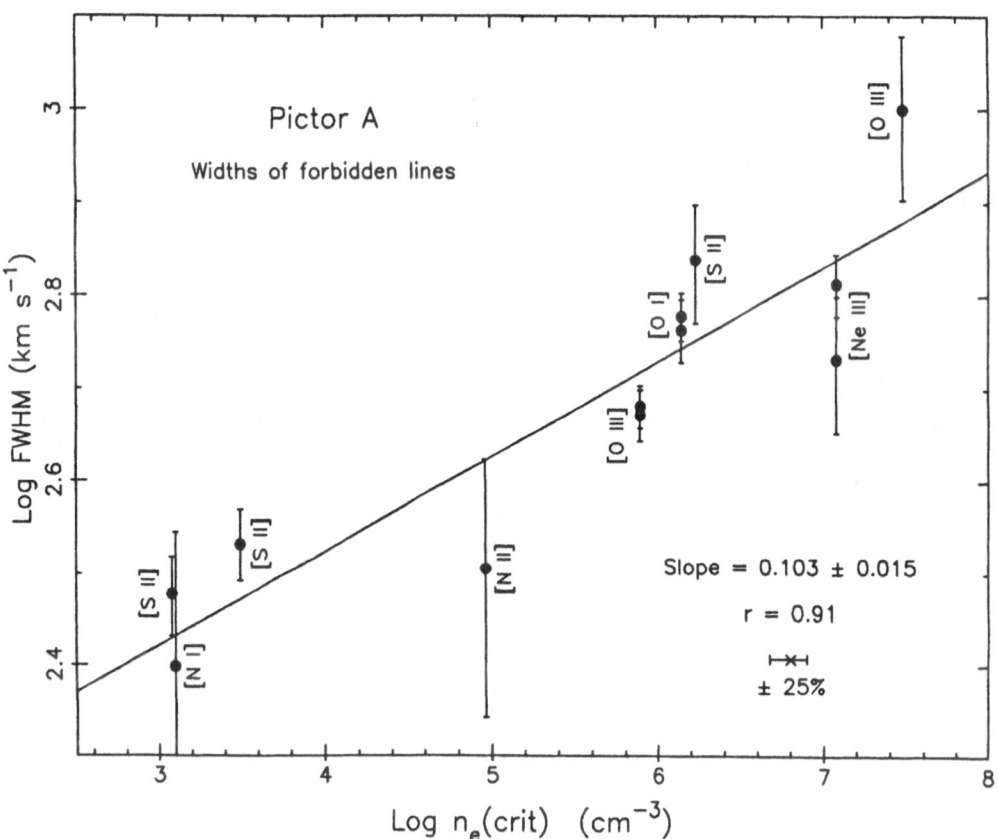

Figure 1: Log FWHM is plotted against Log n_e(crit) for 12 forbidden lines in Pictor A. An error bar of $\pm 25\%$ indicates probable uncertainties in theoretical calculations of n_e(crit). A linear least-squares fit to the points is shown; the correlation coefficient is 0.91, and the probability that the two variables are uncorrelated is virtually zero. Clouds in the NLR must therefore span a very large range of densities.

The correlation between line width and n_e(crit) may be useful as a probe of the gravitational potential in AGNs (FH84; Filippenko and Sargent 1986). If clouds in the NLR move around the nucleus in orbits of the form $v \propto r^\beta$, and if the ionizing flux from the central engine is constant over a dynamical time scale, then $v \propto (nU)^{-\beta/2}$, where U is the ionization parameter ($U \propto Ln^{-1}r^{-2}$). Under the assumption of Keplerian orbits, and if U is independent of distance from the nucleus, this becomes $v \propto n^{0.25}$, which is close to the observed relation ($v \propto n^{0.2}$) in the Seyfert 1 galaxy NGC 7213 (FH84). If the potential is logarithmic, on the other hand, no differences in line width are expected. Many other conditions are also possible, of course.

3. THE IONIZATION MECHANISM IN LINERs

Filippenko and Sargent (1985) and F85 have shown that the range of densities found in the NLR of MR 2251–178, NGC 7213, and Pictor A is also present in many "low-ionization nuclear emission-line regions" (LINERs; Heckman 1980). The strengths of transauroral and auroral lines (e.g., [O III] λ4363) therefore tend to be enhanced relative to the corresponding nebular lines ([O III] $\lambda\lambda$4959, 5007), and the emitted spectra can strongly resemble those produced by shock-heated gas. In fact, heating by shocks was originally thought to be the ionization mechanism in LINERs (e.g., Koski and Osterbrock 1976; Fosbury et al. 1978), but modern photoionization models (Ferland and Netzer 1983; Halpern and Steiner 1983) are more successful at reproducing the observed spectra, especially when the great range of densities is properly taken into account (Péquignot 1984).

One way in which the models have been modified is that large optical depths are now included in the NLR. The presence of [O I] λ6300 with width comparable to that of [O III] λ5007 in Pictor A and other objects (FH84; F85) demonstrates that the densest clouds ($n_e \gtrsim 10^6$ cm^{-3}) must in some cases be optically thick to ionizing radiation. Soft X-rays, however, are able to penetrate deep into these clouds, ionizing roughly 10% of the hydrogen atoms in vast regions (see, e.g., Halpern 1982). The great strength of [O I] λ6300 and certain other lines is then easily explained, because they are best produced in partially-ionized gas of just this type.

ACKNOWLEDGMENTS

I would like to thank the night and day crews at Las Campanas Observatory for their help. The figure in this paper is reproduced courtesy of *The Astrophysical Journal*. Much financial assistance was provided by the Miller Institute for Basic Research in Science (U.C.B.), as well as by the organizing committee of IAU Symposium 119.

REFERENCES

Carswell, R. F., Baldwin, J. A., Atwood, B., and Phillips, M. M. 1984, *Ap. J.*, **286**, 464.
Davidson, K., and Netzer, H. 1979, *Rev. Mod. Phys.*, **51**, 715.

De Robertis, M. M., and Osterbrock, D. E. 1984, *Ap. J.*, **286**, 171.

Filippenko, A. V. 1985, *Ap. J.*, **289**, 475 (F85).

Filippenko, A. V., and Halpern, J. P. 1984, *Ap. J.*, **285**, 458 (FH84).

Filippenko, A. V., and Sargent, W. L. W. 1985, *Ap. J. Suppl.*, **57**, 503.

Filippenko, A. V., and Sargent, W. L. W. 1986, in *Structure and Evolution of Active Galactic Nuclei*, ed. G. Giuricin *et al.* (Dordrecht: Reidel).

Fosbury, R. A. E., Mebold, U., Goss, W. M., and Dopita, M. A. 1978, *M. N. R. A. S.*, **183**, 549.

Halpern, J. P. 1982, Ph.D. thesis, Harvard University.

Halpern, J. P., and Steiner, J. E. 1983, *Ap. J. (Letters)*, **269**, L37.

Heckman, T. M. 1980, *Astr. Ap.*, **87**, 152.

Koski, A. T. 1978, *Ap. J.*, **223**, 56.

Koski, A. T., and Osterbrock, D. E. 1976, *Ap. J. (Letters)*, **203**, L49.

Osterbrock, D. E. 1979, in *Active Galactic Nuclei*, ed. C. Hazard and S. Mitton (Cambridge: Cambridge University Press), p. 25.

Osterbrock, D. E., Koski, A. T., and Phillips, M. M. 1976, *Ap. J.*, **206**, 898.

Péquignot, D. 1984, *Astr. Ap.*, **131**, 159.

Puetter, R. C. 1986, these proceedings.

Weedman, D. W. 1977, *Ann. Rev. Astr. Ap.*, **15**, 69.

DISCUSSION

Burbidge : Why do you suppose that the clouds are bound ?

Filippenko : The clouds themselves are almost certainly not self-gravitating bound systems, if that is what you are asking. It may be true, however, that the clouds are bound to the nucleus; if so, one can study the gravitational potential by examining the line widths. Of course, many other effects e.g. radiative acceleration may contribute to the line widths, and in this case my model is too simplistic. It was just presented as an example.

Wandel : The cloud kinematics in the [OIII] emission region may be dominated by the gravity of the galactic nucleus, as suggested by the correlation between [OIII] line width and stellar velocity dispersion (Heckman and Wilson 1984), while the (narrow) lines of higher critical density, which are supposedly emitted from smaller radii, may have profiles due to different kinematics. Is there a qualitative difference in the profiles of those line (except from just being broader), which could suggest a different kinematic setting ?

Filippenko : This is difficult to address observationally, since different portions of a line profile are produced by different regions, but one cannot easily determine the relative contributions. [OI] 6300 is a good example of this. Moreover, the signal-to-noise ratios in the lines of highest critical density (e.g., [Ne V]) are not sufficiently high. I will look into this question more thoroughly in future studies.

Petrosian : It appears that the range of densities in narrow line regions and broad line regions, are beginning to overlap. Is it, therefore, still useful to think about two distinct regions or should one consider a continuum for cloud densities and velocities for the whole line emitting region.

Filippenko : I believe the answer depends on the types of questions that are being addressed, and on the particular objects under consideration. In general, the classification into a BLR and a NLR is still a useful one, especially since some active galaxies show no evidence for a BLR at all. Moreover, in certain objects there appears to be much less emission from the intermediate-density region than from the low-density and high-density regions. There is increasing evidence, however, that a real continuity in density and velocity exists between the NLR and BLR in many AGNs, and in these one must take the data seriously when constructing models.

Dultzin-Hacyan : With respect to the separation between BLR and NLR, is it not a bit artificial in view of what has been said, and also taking into account the reports by Drs. van Groningen and K. Meyers on observed broad [OIII] lines ?

Filippenko : Let me once again emphasize that a main conclusion from my studies is that in many objects there is indeed a continuity in density and line width, as opposed to discrete BLRs and NLRs. The work you refer to supports this view.

McAdam : The presence of dense clouds in the narrow line region is necessary for any intrinsic absorption mechanism for low frequency variability in quasars - what range of collision frequencies and covering factors do you think are present in your sources ?

Filippenko : The covering factors are probably in the range 0.01-0.1. Collision frequencies of particles in the clouds are roughly 10^{-3} to 1, depending on the density.

Alloin : In the NLR of NGC 3783, we found this relationship between the critical density and the line width, by studying, through gaussian decompositions, profiles in details for lines of different excitations from [OII] up to [FeX]. Indeed, when one goes to lines of higher excitation and larger critical density, one requires the addition of a larger blue-shifted component. This can be understood in terms of an increase towards the centre, both of the density in the NLR filaments and of the mean "turbulent" velocity which governs the line-width, as well as of a radial velocity component.

    I wish to emphasize the fact that it might be time, now, to perform a more detailed confrontation between photoionized models and line intensities in the NLR, considering the line ratios within a given slab of material in the velocity space, rather than the integrated ones.

Filippenko : I agree whole heartedly.

THEORETICAL STUDIES OF THE LINE EMISSION SPECTRUM

S. Collin-Souffrin
DAF, Observatoire de Meudon, 92195 Meudon Cédex
and Research associate at Institut d'Astrophysique, Paris
France

ABSTRACT. Our knowledge of the physical conditions and of the structure
of the broad line emissive region -the BLR- is reviewed. First we derive
the model independent constraints on the different zones emitting the
broad lines. Then we discuss photoionized models. In a first step the BLR
is assumed to be made only of one type of cloud emitting all the broad
lines. We show that this model is unable to explain the observed spectrum.
We thus assume that the high and low ionization lines are emitted by different clouds, still photoionized. This assumption is also in contradiction with the observations and we are led to the idea of a large variety
of emitting clouds. Finally the hypothesis of a purely radiative heating
mechanism should also be questioned.

In this review, I will consider only the broad emission line region (BLR),
the main reason being that the BLR lies very near the centre, so any information concerning its structure, dynamics and kinematics, abundances,
mechanism of heating, is important for our understanding of the central
engine. Nevertheless one must not forget that the physics and kinematics
of the narrow line region may be linked in some sense with that of the
BLR. Due to the lack of space, I will concentrate on the methods which
are used in the computation of line intensities (dropping the studies of
line profiles).

In spite of considerable progress in the last decade, many problems concerning the BLR remain to be solved, and the areas of controversies are
still numerous. So my aim will be to summarize which conclusions can be
taken for granted, and try to show that some of the most widespread ideas
concerning the BLR are in fact not firmly established. The plan is as follows : section 1 gives an outline of the overall properties of the BLR,
i.e. its structure and physical properties as they can be deduced almost
directly from the observations. Section 2 presents a review of photoionized
models, of line excitation mechanisms, and a discussion of some related
problems. Some conclusions are given in the third section.

I. General properties of the BLR (non model-dependent)

In the following, I shall refer to "high ionization lines", HIL, for $L\alpha$

295

G. Swarup and V. K. Kapahi (eds.), Quasars, 295–306 .
© *1986 by the IAU.*

and CIII], CIV, HeI, HeII, NV, OVI lines and to "low ionization lines",
LIL, for Balmer, MgII, CII, FeII lines. A fundamental idea of my talk
will be to relax the generally accepted assumption of a unique emitting
region for these two kinds of lines. Indeed, B. WILLS has shown in her
talk, on the basis of kinematical arguments, that the LIL and HIL are
most probably emitted by different regions. In the following I will show
that similar conclusions can be reached using physical arguments derived
from photoionized models : in fact, there is probably a large range of
physical conditions in the regions producing the different lines.

I.a. electron density, n_e

The absence of a broad component in the forbidden lines is interpreted,
as due to collisional de-excitation so that n_e must be greater than 10^7
cm^{-3}. On the other hand, the presence of the broad semi-forbidden CIII]
1909 line (which starts being de-excited at $n_e > 10^9$) implies $n_e < 10^{10}$
cm^{-3} in the HIL region. More refined arguments based in particular on HeI
line intensities suggest a value in the range 10^9-10^{10} cm^{-3} (cf the review
of DAVIDSON and NETZER 1979, for instance). However, if the assumption
that all the HIL including the CIII] line are emitted by the same region
is relaxed, then the upper limit on the density is higher : the critical
density of a given line depends on the optical thickness. For instance
it is $\simeq 10^{12}$ cm^{-3} for L_α if it is produced in clouds which are optically
thick to the Lyman continuum and it can even be higher if L_α is emitted
by optically thin clouds.

I.b. temperature, T_e

There is no direct way to know T_e in the BLR contrary to the NLR, except
when both the lines CIII]1909 and CIII]977 are observed. In this case the
ratio of these lines always indicates that $T_e < 30000$ K. This is not enough
to produce C^{++} by collisional ionization so that ionization is definitively radiative in the HIL region.

I.c. overall dimension, R

An upper limit to the size of the BLR can be set from the variation time
scale of the emission lines ; this method is obviously restricted to dimensions of a few light years at most, which is the case of Seyfert galaxies : .01 < R < .1 pc.

I.d. ionization parameter, U

This parameter is defined as the ratio of the ionizing photon number density to the electron density : if the emitting region with an electron
density n_e is located at a distance R from a continuum source of lumino-
sity L_ν at the Lyman edge, U is equal to $L_\nu/(\alpha 4\Pi R^2 hcn_e)$, α being the spectral index of the Lyman continuum (notice that different definitions of
U are used in the literature). In the case of a radiative ionizing mechanism -i.e. in the HIL region- it is this parameter which determines the
degree of ionization of the elements. The computation of detailed "photoionized models" can provide the line intensity ratios as functions of U.
In particular, the line ratios L_α/CIV 1549/CIII]1909 indicate U $\sim 10^{-2}$
in most objects (cf. DAVIDSON and NETZER, 1979). Although this result is
not obtained directly from the observations it can be considered as firmly established, if all HIL are produced by the same region, since it de-

pends only weakly on future improvements possibly introduced in photoionized models and on the yet unknown parameters -column density, spectral
shape - (cf. below). An immediate deduction from the observed or extrapolated value of L_V is the size of the emitting region, R, $\simeq$.1 to 1 pc
for Seyfert galaxies,$\simeq$1 to 10 pc for quasars. This value is larger in
Seyfert galaxies than the value deduced from the variation time scales,
especially for high ionization lines (CIV, HeII): it means that in fact
all HIL are not produced by the same region and that, at least in these
objects, the highest ionized lines are emitted closer to the central
source.
The remarquable constancy of the ionization parameter has impelled an
important discussion. It can be well explained in the framework of the
2-phase model of KROLIK, McKEE and TARTER, 1981: these authors assume
that the BLR is made of small dense clouds confined by the external
pressure of a hot dilute medium. KROLIK et al have shown that, owing to
a thermal instability, the two phases can coexist only for a small range of
radiation-to-gas pressure ratio corresponding to the "measured" value of U.
From the comparison of the intensity ratios L_α/CIV/CIII] observed in a
number of quasars and AGN with those computed in a grid of photoionized
models, MUSHOTZKY and FERLAND, 1984, have shown that the ionization parameter decreases as the luminosity, $U \simeq L^{-0.25}$. This decrease would be the
cause of the well-known BALDWIN effect. (Notice that recently NETZER,
1985a, has suggested another interpretation of the BALDWIN effect which
could be due to the different contribution of the continuum emission of
an accretion disk, when viewed at various angles of sight). Finally
FERLAND and ELITZUR, 1984, have shown that the value of U depends,
through the line radiation pressure, on the density and on the column
density of the emitting gas: clearly there is a strong link between the
dynamical and physical state of the emitting gas which is not yet understood and deserves further studies.

I.e. covering factor, $\Omega/4\Pi$

This parameter represents the fraction of the continuum source covered
by the BLR. From direct observations of Lyman discontinuity in high redshift quasars as well as from detailed photoionized models, it is generally concluded that the covering factor is small, $\simeq$.03 to .3. However
it could be of the order of unity in NGC4151 (cf FERLAND and MUSHOTZKY,
1982) and more generally in low luminosity objects.

I.f. abundances and dust

Photoionized models are consistent with heavy element abundances in
quasars being equal to the cosmic ones (cf. for instance GASKELL, SHIELDS
and WAMPLER, 1981). However a factor $\simeq$3 discrepancy is not excluded.
These last authors conclude that dust exists in large amounts. In any case,
if dust were present inside the BLR, it would be heated by L_α and X-rays
to a temperature larger than 500K and would radiate in the near IR range
(RUDY and PUETTER, 1982). On the other hand, the presence of a large
amount of dust surrounding the BLR is a subject of controversy on which
I will come back in the next section.

I.g. geometrical structure

It is widely admitted that the BLR is made of a large number of high

velocity clouds which fill only a small fraction, f, of the emitting
volume (f≈10^{-6}, from our previous estimation of R, and using the observed luminosity of Lα, which is about equal to a pure recombination case
B). The reasons of this consensus are: the great widths of the lines, due
to Doppler motions (although a broadening due to electron scattering is
not excluded in the wings), the absence of broad absorption with P-Cygni
profiles due to an emitting-absorbing continuous medium, like in BAL
quasars, (contrarywise, the narrow absorption components observed in the
line wings of some Seyfert nuclei indicate that the velocity dispersion
or the "turbulent" velocity inside a given cloud is small, as it can
also be inferred from dynamical arguments), the overall size of the BLR,
as compared with the effective emitting volume, and the irregularities
of the profiles in some objects (which actually can also be due to temporal variations of the ionization rate in a continuous medium).
One can get around any of these arguments, by assuming for instance a
disk shaped structure whose emitting region is confined to a thin outer
shell. In any case a most important parameter of these models is the
column density, N, of the emitting region in the direction of lowest size.
If one assumes that all the lines are emitted by the same clouds detailed
photoionization computations (KWAN and KROLIK, 1981) show that N is limited to the range 10^{22}-10^{23}cm^{-2}.

II. Detailed computations: photoionized models and line excitation mechanisms

II.a. photoionized models

Let us compute different time scales implied by the physical conditions
of the BLR:

1-ionization time scale: (flux of ionizing photons x absorption cross-
section)$^{-1}$≈10^{-2}s

2-cooling time scale: n_e x kT/(rate of cooling ≈ $10^{-23}n_e^2$) ≈ 10^2s

3-recombination time scale: (n_e x rec. coef.)$^{-1}$ ≈ 10^4s

4-dynamical time scale: thickness of the cloud/sound velocity ≈ 10^6s

5-life-time: it is not known in general but in several dynamical models
proposed for the BLR it is of the order of R/(macroscopic velocity)≈10^8s.
From the comparison of these time scales, it is clear that the BLR
clouds are in thermal and ionization equilibrium. Pressure equilibrium
should also be achieved unless variations of the ionizing flux take place
in a time smaller than about one week. Therefore the BLR can be described by clouds or shells in thermal, ionization and pressure equilibrium,
photoionized by an external source of UV and X-ray continuum: we shall
call hereafter the computation of such a model a "photoionized model"
(sometimes constant density is assumed instead of pressure equilibrium).
The parameters needed to build a photoionized model are: the ionization
parameter U, the density n, the column density N, the spectral shape of
the continuum, the chemical abundances, the velocity dispersion.
The first models (cf the extensive review of DAVIDSON and NETZER, 1979)
were mainly similar to models of planetary nebulae, except that they had
a harder ionizing spectrum, which induces a mixing of many stages of ionization, and a higher density. They produced hydrogen line ratios very
near to the pure recombination values case B, except for a small increase

of the Balmer decrement due to the influence of collisional excitations
when the density was large enough ($\simeq 10^9$). In particular the ratio
Lα/Hβ was $\simeq 40$
Then BALDWIN (1977) discovered that the observed Lα/Hβ is $\simeq 5$ in quasars
and this was found later on, with IUE observations, to be also the case
in Seyfert nuclei. Corrected for galactic extinction Lα/Hβ is $\simeq 10$.
Actually since 1975, several works aiming at computing the HI spectrum
in conditions of high density and large optical thickness, have drawn
the attention to the fact that line ratios are very different from case
B: in particular, Lα photons are destroyed by collisions during the many
diffusions they undergo (we shall say that Lα is thermalized), while
Balmer photons are actively produced by collisional excitations from the
second level (which is then in thermodynamical equilibrium with the
first level). These computations were made for an homogeneous shell with
given temperature and density, using a "mean escape probability", p_e,
which is the probability that a photon created at the center of a shell
escapes outside. Finally, FERLAND and NETZER, 1979, and KWAN and KROLIK,
1979, using p_e as a local quantity and taking into account the ionization
from excited HI-levels, produced the first photoionization models really
adapted to the BLR (the previous high density models can be however considered as being good for the HIL).
The difficulty of the problem lies in the particular structure of BLR
clouds. Each cloud consists of 3 different regions: facing the continuum
source is a highly ionized region -the HII zone-emitting Lα and the OVI,
NV, CIV, CIII, HeII, HeI lines; its column density is about 10^{23}U. The
far side possibly consists of a completely neutral and cold region. In
between there is a partially neutral gas (the degree of ionization
running from .5 to .01), relatively warm (5000 to 8000 K), which we call
the excited HI zone - the HI* zone-emitting the MgII, FeII, CII lines
and the bulk of the Balmer lines: in this zone, which is heated by the
X-ray continuum, trapped Lα photons insure a high population of the
second level of HI, and the gas is ionized by the incident Balmer contin-
uumn,by collisions due to thermal electrons (from the excited HI levels)
and/or due to high energy non thermal electrons (produced by inner shell
photo-ionizations of heavy elements, as stressed first by WEISHEIT,
SHIELDS ans TARTER, 1981).
One should realize that they are many unknown parameters in this study,
so the uniqueness of any solution is controversial.
First we have to ask what is the real intrinsic broad line spectrum ?
Are there good reddening indicators ? We have seen that HI line ratios
are strongly model-dependent. The narrow lines cannot be used since the
reddening of the BLR and NLR are not necessarily the same. The intensity
ratio OI 1302/8446 which is generally taken as a constant, because these
lines are issued from the same level, is modified by fluorescence processes in a thick shell (cf GRANDI, 1983). The only potential reddening
indicator is the intensity ratio HeII 4686/1640 (cf NETZER et al, 1985)
but there are only very few observations of this ratio. On the other
hand, the 2200A feature which is also a potential reddening indicator,
is always weak in quasar spectra.
KWAN and KROLIK, 1981, have shown that a model called by them "Standard
Model" fits roughly an average quasar spectrum. In this model the HII

and HI* zones have respectively a density of $4\ 10^9$ and of $2\ 10^{10}\ cm^{-3}$, and a thickness of 10^{12} and $5\ 10^{12}$ cm. It has largely been adopted in the literature to represent a standard BLR cloud. They were followed by different computations aiming at finding the best parameters accounting for the BLR spectrum of a given object (FERLAND and MUSTHOTZKY, 1982, for NGC 4151, NETZER, WILLS and WILLS, 1982, for 3C351). BOLDT and LETTER, 1984, computed models for flat UV and X-ray spectra assumed to represent young quasars, MARTIN and FERLAND, 1980, studied the influence of dust, NETZER, 1980, WILLS, NETZER and WILLS, 1985, COLLIN-SOUFFRIN et al, 1985, produced models trying to fit the observed FeII spectrum, Mac Alpine, 1981, and NETZER, ELITZUR and FERLAND, 1985, discussed the HeII problem. Kwan, 1984, has produced the most extensive grid of models and given some useful spectral diagnostics for physical parameters. However his results should be taken with some caution because he restricted himself to a peculiar spectral shape which overestimates the X-ray flux with respect to the optical and UV flux, and also because his results for large column densities are uncertain.

II.b:the transfer problem
There has been a great debate these last years concerning the use of the escape probality formalism in conditions of finite but very high optical thickness (the optical thickness of Lα reaches 10^{10} and that of Hα 10^5 ; even the Balmer continuum may have an optical thickness larger than one). CANFIELD and PUETTER, 1981, CANFIELD, PUETTER and RICCHIAZZI, 1981a and 1981b, PUETTER et al, 1982, HUBBARD and PUETTER, 1985, have extensively discussed the ability of escape probability to account for non local effects in radiative transfer and they have proposed a more elaborate probabilistic method. Unfortunately for now their method has been applied only to a pure hydrogen cloud, so temperature gradients are not consistently computed. On the other hand, COLLIN-SOUFFRIN et al, 1982, and COLLIN-SOUFFRIN and DUMONT, 1985, have compared the escape probability formalism to the exact transfer treatment in the case of BLR clouds and shown that it gives good results (within 50% for most of the line intensities) provided that it is used correctly : in particular if the Balmer continuum has a non negligible thickness, a method developed by ELITZUR and NETZER, 1985, for line fluorescence, and extended to the case of the overlapping of lines and continuum by NETZER, ELITZUR and FERLAND, 1985, is very appropriate . (Notice that line fluorescence processes are common in the formation of broad line spectrum : OI 8446 is formed through Lα fluorescence (cf NETZER and PENSTON, 1976, and GRANDI, 1981 and 1983), OIII Bowen lines through HeII fluorescence (cf NETZER et al, 1985), and highly excited FeII lines through resonance FeII lines (cf NETZER and WILLS, 1983)). However a large incertainty (by a factor 2 or 3) remain on the intensities of resonance lines emitted by the H I* zone, such as MgII 2800, until realistic generating functions are proposed for these photons from exact transfer calculations. Finally the transfer of the diffuse Balmer continuum should be solved exactly in order to compute correctly UV line intensities. Note that the constraint found by KWAN and KROLIK on the column density from the ratio CII 2326/Lα, $N < 10^{23}$ is not valid because the line 2326 is reabsorbed in the HI* zone by ionizing excited HI atoms (cf COLLIN-SOUFFRIN et al, 1985).

Another problem tighly linked with transfer is the anisotropy of the emitted radiation, stressed by FERLAND, NETZER and SHIELDS, 1979 : due to their different depths of formation, the line intensities depends on the angle between the surface and the direction of emission, like in a stellar atmosphere. COLLIN-SOUFFRIN et al, 1985, have shown that the anisotropy could be very large for some lines, namely that Lα, FeII UV multiplets, MgII 2800, are emitted preferentially towards the ionizing source, but that unfortunately this effect is strongly model-dependent. It will clearly be a very large factor of uncertainty in the computation of line spectra or of line profiles in a kinematic model.

II.c:comparison of computed and observed line ratios

As a matter of fact, all the models produced so far fail to account for at least some features (which might be of fundamental importance). Let us for instance consider the Standard Model, which fits best the overall features of an average quasar :
- the ratio CIII]1909/CIV1549 is too weak. It means that the density of the HII zone is too large.
- the ratio Hα/Hβ (which is actually slightly larger than computed by KWAN and KROLIK, owing to their escape probability approximation) is too large. It means that the density of the HI* zone is too small, (cf the discussion of COLLIN-SOUFFRIN, DUMONT and TULLY, 1982).
In other words, HIL and LIL should be emitted by different clouds, HIL by clouds having $n_e < 2 \times 10^9$ and small column densities $(<10^{23}U)$ and LIL by clouds having $n_e > 10^{10}$ and much larger column densities.
- the lines of highest ionization, OVI, NV, HeII, are too weak ; the problem of the OVI and NV lines cannot be solved in the framework of a one component model for HIL - i.e. only one type of clouds -, as was already noted by DAVIDSON in 1977 : clouds with a larger ionization parameter are required (more dilute or closer to the source of continuum) or clouds with a small column density $(<< 10^{23}U)$. The problem of the HeII line spectrum has been debated (Mac ALPINE, 1981) but it seems now well understood with the improvements in the computation of the ionization and thermal structure of the He^{++} zone (NETZER et al, 1985).
- the FeII lines are too weak. The "FeII" problem has raised up a strong controversy. The FeII spectrum consists of numerous UV and optical blends which carry about 1/4 of the whole broad line luminosity. PHILLIPS,1978, proposed an excitation by a continuum fluorescence but COLLIN-SOUFFRIN et al, 1979, 1980, and GRANDI, 1981, showed this mechanism to be insufficient and suggested a collisional excitation in a relatively hot (T ≃ 10000K) gas. NETZER, 1980, included FeII for the first time in a photoionized model. JOLY, 1981, using a sophisticated atom (14 levels) showed that the FeII excitation requires a temperature > 10000K and a high electron density ≃ 10^{11} cm : such values seemed difficult to reach in the HI* zone of a photoionized model. COLLIN-SOUFFRIN et al, 1982, considering that the intensity ratio (FeII lines/HI lines) was too weak in existing photoionized models suggested as well as later CLAVEL et al, 1983, that the BLR contains a non radiatively heated dense region producing mainly FeII lines. NETZER and WILLS, 1983, and WILLS et al, 1985, considered a very elaborate FeII atom, including more than 3000 lines, taking into account the fluorescence processes between all the lines and introduced this atom in a photoionization computation. Then, by

synthetizing a whole broad line spectrum and comparing it to the observations, they put into evidence what they called a"Lα/FeII problem",
namely that the computed FeII line intensities were too weak by a factor
3 with respect to Lα in any photoionized model. COLLIN-SOUFFRIN et al,
1985, tried an ultimate experience : assuming particularly favorable
conditions for the emission of FeII lines (a large flux in hard X-rays,
very large column density and density) they computed several photoionized
models, none of which gives a sufficient energy in FeII lines emitted in
the optical range compared to Hβ, still by about a factor 3 (notice for
the further discussion on the energy budget that this conclusion is in-
denpendent of the assumed reddening). A last possibility in the framework
of a radiative heating is a Compton heated cloud if there is a very intense flux of γ rays : this possibility has not yet been explored.
FERLAND and MUSHOTSKY, 1984, have also proposed cosmic ray heating.
Other possibilities, such as the emission of a disk atmosphere, could
also be explored.

II.d:the energy budget problem

NETZER, 1985b, has pointed out an important problem concerning the
energy budget of the BLR : the total observed luminosity of the BLR
amounts to 8 Lα (taking into account the bound-free and free-free emission). Assuming a column density $N = 10^{23}$ cm^{-2} and a power law spectrum
in the range 200 eV to 40 keV having the average observed slope of .7,
he computed the ratio of the total luminosity to Lα, as a function of
the spectral index α in the range 13.6 to 200 eV and he found that the
observed ratio corresponds to α ≃ .4, in clear contradiction with the
observations which give α ≃ 2. To reduce the discrepancy, NETZER proposed : i. that the BLR spectrum is reddened so the intrinsic ratio is
lower than the observed one; for instance E(B-V) = .2 gives an emitted
ratio of 3.4 which corresponds to α = 1.; besides, ii. that the ionizing
continuum itself is also reddened and/or intrinsically more brighter
than the observed one owing to a special geometry (if it is due to the
surface emission of a disk for instance).
However I would like to propose an alternative solution. Actually the
total observed luminosity of the BLR is shared between the HIL (2.7 Lα)
and the LIL (5.2 Lα), the latter consisting of FeII emission (1.7 Lα)
and the other lines and continua (3.5 Lα). Assuming a small amount of
reddening (E(B-V) = .1) it is easy to account for the HIL luminosity
with a relatively steep index α = 2. For the LIL we have two possibilities. We have seen that the medium emitting the FeII lines should be
heated by a non radiative mechanism. So it would be a natural solution
to assume that all the LIL or at least a non negligible part of them are
emitted by this non radiatively heated medium. It is also possible that
only the FeII lines are emitted by such a medium. In this last case the
other LIL could be produced in clouds of very large column density,
$N \gg 10^{23}$ cm^{-2} which are able to absorb high energy X-rays, hν ≃ 10 keV.
However in order to be efficient enough without producing hard X-ray
absorption in the observed continuum, these "clouds" should have a special geometry, for instance a disk illuminated from above.

III. Conclusions

It is clear from the previous discussion that no firm conclusion concerning the structure and even the physical parameters of the BLR can be drawn from the computation of photoionized models alone : we can only ascertain that there is a large range of physical parameters in the BLR, from low density and/or column density clouds emitting lines of the most ionized species up to large density and column density clouds emitting the LIL, and there are good reasons to think that the LIL or at least the FeII lines are emitted by a non radiatively heated region. To better understand the situation, other constraints should be introduced :
i. from dynamical arguments concerning for instance the structure and the evolution of the clouds. This will be the subject of the talk of J. PERRY;
ii. from kinematical models built to fit the line profiles.
A considerable amount of work which cannot be reviewed here has been recently devoted to the computation of line profiles. They deal with different kinematics - rotation, radial outflow or inflow - and use different descriptions of the line emission which is generally parametrized as a function of the distance to the center. An important problem is the anisotropy of the emission, mentioned previously. Unfortunately no clear conclusion is yet reached since every model is able, according to the choice of some arbitrary parameters, to give results in agreement with the observations (logarithmic profiles, asymmetries...). It seems nevertheless that the most satisfactory results are obtained by assuming that rotation (or bound orbits) and radial outflow are both present. A complete self consistent description would require a dynamical model able to relate the velocity field to the matter distribution and to the physical parameters of the clouds and to compute simultaneously the - anisotropic — line emission of each cloud, a real Titan work ! Actually a most powerful potential tool which can avoid such an effort would be to reduce the free parameters by detailed and very accurate observations of line profile variations.

Acknowledgments : I gratefully aknowledge G. Stasinska for a careful reading and critical discussion of the manuscript.

REFERENCES

BALDWIN, J.A., 1977, Mon. Not. R. Astron. Soc. 178, 67.
BOLDT, E., LETTER, D., 1984, Astrophys. J. 276, 427.
CANFIELD, R.C., PUETTER, R.C., 1981, Astrophys. J. 243, 381.
CANFIELD, R.C., PUETTER, R.C., RICCHIAZZI, P.J., 1981a, Astrophys. J. 248, 82.
CANFIELD, R.C., PUETTER, R.C., RICCHIAZZI, P.J., 1981b, Astrophys. J. 249, 383.
CLAVEL, J., JOLY, M., COLLIN-SOUFFRIN, S., BERGERON, J., PENSTON, M.V., 1983, Mon. Not. R. Astr. Soc. 202, 85.
COLLIN-SOUFFRIN, S., DUMONT, S., HEIDMANN, N., JOLY, M., 1979, Astron. Astrophys. 72, 293.
COLLIN-SOUFFRIN, S., DUMONT, S., HEIDMANN, N., JOLY, M., 1980, Astron. Astrophys. 83, 190.

COLLIN-SOUFFRIN, S., DELACHE, Ph., DUMONT, S., FRISCH, H., 1981, Astron. Astrophys. 104, 264.

COLLIN-SOUFFRIN, S., DUMONT, S., TULLY, J., 1982, Astron. Astrophys. 106, 362.

COLLIN-SOUFFRIN, S., DUMONT, S., 1985, Submitted to Astron. Astrophys.

COLLIN-SOUFFRIN, S., JOLY, M., DUMONT, S., PEQUIGNOT, D., 1985, Submitted to Astron. Astrophys.

DAVIDSON, K., 1977, Astrophys. J. 218, 20.

DAVIDSON, K., NETZER, H., 1979, Rev. of Modern Physics 51, 715.

ELITZUR, M., NETZER, H., 1985, Astrophys. J. 291, 464.

FERLAND, G.J., NETZER, H., 1979, Astrophys. J., 229, 274.

FERLAND, G.J., NETZER, H., SHIELDS, G.A., 1979, Astrophys. J., 232, 282.

FERLAND, G.J., ELITZUR, M., 1984, Astrophys. J. Letters 285, L11.

FERLAND, G.J., MUSHOTZKY, R.F., 1982, Astrophys. J. 262, 564.

FERLAND, G.J., MUSHOTZKY, R.F., 1984, Astrophys. J. 286, 42.

GASKELL, C.M., SHIELDS, G.A., WAMPLER, E.J., 1981, Astrophys. J., 249, 443.

GRANDI, S.A., 1980, Astrophys. J. 238, 10.

GRANDI, S.A., 1981, Astrophys. J. 251, 451.

GRANDI, S.A., 1983, Astrophys. J. 268, 591.

HUBBARD, E.N., PUETTER, R.C., 1985, Astrophys. J., 290, 394.

JOLY, M., 1981, Astron. Astrophys. 102, 321.

KROLIK, J.H., McKEE, C.F., TARTER, C.B., 1981, Astrophys. J. 249, 422.

KWAN, J., KROLIK, J.H., 1979, Astrophys. J. 233, L91.

KWAN, J., KROLIK, J.H., 1981, Astrophys. J. 250, 478.

KWAN, J., 1984, Astrophys. J., 283, 70.

Mac ALPINE, G.M., 1981, Astrophys. J. 251, 465.

MARTIN, P.G., FERLAND, G.J., 1980, Astrophys. J. 235, L125.

MUSHOTZKY, R.F., FERLAND, G.J., 1984, Astrophys. J. 278, 558.

NETZER, H., PENSTON, M.V., 1976, Mon. Not. R. Astr. Soc. 174, 319.

NETZER, H., 1980, Astrophys. J. 236, 406.

NETZER, H., WILLS, B.J., WILLS, D., 1982, Astrophys. J. 254, 489.

NETZER, H., WILLS, B.J., 1983, Astrophys. J., 275, 445.

NETZER, H., 1985a, Mon. Not. R. Astr. Soc. 216, 63.

NETZER, H., 1985b, Astrophys. J. 289, 451.

NETZER, H., ELITZUR, M., FERLAND, G.J., 1985, Preprint.

PHILLIPS, M.N., 1978, Astrophys. J., 226, 736.

PUETTER, R.C., HUBBARD, E.N., RICCHIAZZI, P.J., CANFIELD, R.C., 1982, Astrophys. J. 258, 46.

RUDY, R.J., PUETTER, R.C., 1982, Astrophys. J. 263, 43.

WEISHEIT, J.C., SHIELDS, G.A., TARTER, C.B., 1981, Astrophys. J. 245, 406.

WILLS, B.J., NETZER, H., WILLS, D., 1985, Astrophys. J. 288, 94.

DISCUSSION

Saslaw : Recently Mark Whittle and I have been calculating emission line profiles for the broad line region of AGN. We have been particularly interested in the effects on line profiles of including distributions of cloud properties (such as radii, velocities and masses) in any region of the nucleus. For any given model, these distributions are derived by injecting an initial distribution of clouds and following their thermal and mechanical evolution along their resulting trajectories. We have developed an analytical technique which provides insight into simple models, and have computed more complicated models numerically. The results show that the dispersion of cloud properties in any region can alter the line profiles substantially compared to the case where all clouds in the region are represented by their mean properties. The ratio of flux in the core to that in the wings is especially sensitive to the cloud distribution. The moral seems to be that it will require a much better detailed understanding of gas dynamical processes as well as observations of many line profiles with high spatial resolution to constrain models usefully. It is usually easy to fit one or two profiles of an AGN reasonably well, but much more difficult to fit the whole gamut with physically self-consistent cloud distributions.

Malkan : You mentioned the observation of Netzer and Wills, that FeII UV lines seem to make a surprisingly high fraction of the BLR output, and the total energy is so high, it could require a rather flat FUV continuum. I would just like to warn that measuring the flux of these low-contrast FeII lines is very tough.When we draw in a curved (thermal) UV continuum we measure FeII to be much weaker (a factor of 2 or more). So it may still be too soon to be sure that the "FeII problem" is as serious as they have suggested.

Wills : One can reduce the measured strength of Fe II + Bac lines may be by 20% (or even 30% in a few cases) if the continuum underneath the FeII is like that from an accretion disk with a narrow range of temperature. Such a (relatively) narrow peaked disk distribution seems unlikely, although more theoretical work on the expected emergent continuum is needed. More likely there is a very broad peak to any disk component, which will not alter the measured FeII + Bac significantly. The observed continua may not even require a disk component at all, but just a steepening component in the IR and a flatter component in the optical-UV region.

Kembhavi : (1). It has been reported that the soft X-ray spectra of radio quiet quasars are harder than those of radio-loud quasars. Could this fact be used to explain some of the differences in the emission line spectra of these objects ?

(2). It has been suggested that most quasars are radio-quiet because the radio emission is absorbed by clouds very close to the continuum emitting region. Could such a model be constrained using emission line observations ?

Collin-Souffrin : (1). Up to now, there have been no suggestions made
to attribute the differences between emission line spectra of radio quiet and radio loud quasars to their different X-ray spectra. It should
be looked at in the future. To my knowledge, the only explanation, concerning the weakness of FeII lines in radio loud objects, has been proposed by Ferland and Mushotsky (1984): they suggest that an extra heating due to cosmic rays - related to the presence of a radio source -
quenches the FeII lines.

(2). On the otherhand, Krolik, Tarter and McKea (1981) have suggested that the extra heating due to the absorption of radio emission
could lead to the suppression of the thermal instability which is at the
origin of the two phase-medium in some objects containing very compact
radio sources (BLLac objects for instance). This could cause the dispa-
rition of the BLR.

GAS DYNAMICS AND FORMATION OF THE BROAD LINE REGION OF QSOs

Judith J. Perry
Institute of Astronomy
University of Cambridge
Cambridge CB3 OHA
England

ABSTRACT   Observational and theoretical constraints on self-consistent gas dynamical models of the line emitting regions of QSOs and AGNs are reviewed.   General equations governing the global flows are analysed and models for the formation and acceleration of the line emitting "clouds" are discussed.

INTRODUCTION   Currently, both theoretical and observational studies place QSOs at the centres of young galaxies. Yet a striking feature of QSO spectra is the appearance of broad emission lines from heavy elements whose abundances do not appear to differ significantly from solar.   The emitting gas must have been first processed in stars and later released into the local interstellar medium (ISM).   The observed nonthermal continuum is equivalent to several solar masses of rest mass energy per year.   If, as it is usually assumed, this radiation comes from the conversion of the gravitational energy of gas accreted by a massive black hole then mass fluxes through the emission line region must be at least, roughly, the rest-mass equivalent of the radiation. Under the quite reasonable assumption that the accretion is not spherically symmetric (there is reason to assume that most QSOs, like Seyferts, are in spiral galaxies (Malkin et.al., 1984; Hutchings et.al., 1984; MacKenty & Stockton, 1984)) the continuum radiation, which is intense enough to be important dynamically, can drive a powerful wind over a large solid angle.   Outflow is directly observed in a significant fraction of QSOs, those which have broad absorption lines (BAL) in their spectra (Turnsek, 1984).   Whatever the global flow pattern in any particular object, whether accretion, winds or some complex combination of the two, it must be optically thin to electron scattering above the photosphere of the non-thermal emission region.   This requirement severely limits the allowed gas density in the flow; the interaction of the non-thermal radiation with this low density ambient gas determines the electron temperature, which is estimated to be $\sim 10^8$K. Yet immersed in this hot gas are relatively cool, very dense clouds or condensations which emit the observed broad lines. The widths of the lines imply Mach numbers of the line emitting gas with respect to the hot ISM, $M > 4$.

The presence of a QSO in the nucleus of a galaxy must have profound

G. Swarup and V. K. Kapahi (eds.), Quasars, 307–316 .

consequences for the structure and evolution of the host galaxy.   Both
their intense radiation fields and their high energy gas flows will
significantly alter the ISM (Begelman, 1985; Dyson & Perry, 1985).   The
broad emission line region (BLR) can be viewed as a tracer of such
activity, and cannot be treated out of the context of the QSO-galaxy
interaction. For example, the apparently normal abundances restrict the
source of the line emitting gas. Accretion of primordial intergalactic
gas is ruled out if it is assumed that density perturbations in such
flows give rise to the line emission.

Theoretical explanations of QSO spectra require an understanding of
the structure and evolution of the parent galaxy. Sadly, we have no
direct measurements of the central stellar density, its initial mass
function (IMF), or the mass of the purported central black hole; nor is
there any reason to assume, a priori, that the IMF is the same as that
in normal galaxies. Yet, as we shall see, the stellar population may
well be the most important contributor to the phenomenon. The challenge
facing theorists is that of constructing a self-consistent dynamical
model which can account, simultaneously, for the central fuel supply and
for the appearance of high speed cool clouds containing heavy elements
abundant enough to produce the observed broad emission lines. Hopefully,
such models will contribute to our understanding of the host galaxies.

RADIATION FIELD   Because of the paucity of x-ray and γ-ray observations
estimates of the bolometric luminosities of QSOs and AGNs are uncertain.
On the basis of available spectra it is reasonable to use a 'canonical'
value of $L_{bol} = 10^{47} L_{47}$ erg/s. (then $0.1 < L_{47} < 100$ for QSOs and 10^{-5}
$< L_{47} < 10^{-2}$ for AGNs.) The equivalent mass consumption rate, at a
conversion efficiency of ϵ, is $\dot{M}_a = 1.75(L_{47}/\epsilon) M_\odot$ per year.

Inverse Compton cooling and electron-scattering radiation driving
depend on only the bolometric luminosity. Uncertainties in the dynamics
arise because all other radiation - gas interactions depend on both the
spatial and the spectral distribution of the radiation neither of which
is fully known. It is conventional to use extrapolations to fill in the
unobserved parts of the spectrum, subject to constraints from observations of e.g. the x-ray and γ-ray background. The canonical spectra of
Krolik, McKee and Tarter (1981, hereafter KMT) are used here.

The physical state of optically thin gas in equilibrium with these
radiation fields is most conveniently discussed through its ionization
parameter (KMT), defined as $\Xi = F_{ion}/cnkT = 2.3P_{rad}/P_{gas}$. F_{ion} is the
ionizing flux, n the number density of the gas, T the gas temperature
and c and k the speed of light and Boltzman's constant. P_{rad} is the
radiation pressure and P_{gas} the thermal gas pressure.

At high Ξ (> 10) the gas temperature is determined by the balance
between x-ray Compton heating and inverse Compton cooling and is roughly
10^8 K.   At low Ξ (< 0.3), the gas temperature is determined by the
balance between photoionization heating and line cooling and is about
10^4 K. KMT demonstrated that a two-phase pressure equilibrium can exist
only in the relatively narrow range $0.3 < \Xi < 10$.

If a QSO 'turns on' in a 'normal' galaxy, the ISM undergoes strong heating towards the Compton temperature (Begelman, 1985). There is a concommittant increase in pressure which will have strong gas dynamical effects. Unless input of cool material from stars can balance this the radiation field can e.g. excavate out the cool ISM to ~ 10 kpc.

In addition to its role in determining the thermal state of the ISM, the radiation field plays an important gas dynamical role through the force it exerts on the gas (Mushotzky, et. al. 1971, Kippenhahn et. al., 1974, 1975; Mathews, 1974,1982; Mestel et.al., 1976). We return to this effect later.

OBSERVED PROPERTIES OF THE LINE EMITTING AND INTERCLOUD GAS   The observed excitation of the line emitting gas is compatible with $T \sim 10^4$ K for the high ionisation lines and with somewhat cooler gas for the low ionisation lines. Although detailed interpretation of the line spectra is difficult, there appears to be agreement that the gas is in clouds or filaments of mean density $n \sim 10^9$ (but, in some cases, of 10^{11} cm^{-3}, Collin, 1985). Photoionisation models based on the 'canonical' spectrum provide the best agreement with observations and restrict Ξ to the range $0.3 < \Xi < 2$ (KMT). However, before dynamical models are worked out in detail, the extent of the admixture of shock heating and collisional ionisation will remain uncertain. Based on existing models, the characteristic distance of the emitting gas is determined from the ionisation parameter of the gas. This is determined (in parsecs) to be $r_{pc}^2/L_{47} \sim 1$. The clouds or filaments occupy at most 10% of the volume available, and are often characterised by volume filling factors as low as 0.1%. The total amount of line emitting gas present is not well determined, and is very model dependent. Estimates of the mass (in solar masses) generally yield a rough limit of $M_c < 200 \, L_{47} M_\odot$.

The extremely small size inferred for the clouds implies that their sound crossing or expansion time scale (~ 1 yr) is very much shorter than the dynamic crossing time of the BLR itself (~ 150 yr). If clouds are created continuously at a rate equal to that of their destruction no difficulty exists. If, however, it is assumed that they are injected into the region and survive to cross it, an external confinement must operate throughout their journey. The thermal pressure in the line emitting gas is inferred directly from observation; there is no serious question as to its lower limit. No direct observations of the intercloud medium itself exist. It is, however, constrained to be optically thin above the effective photosphere of the central source. Unless current extrapolations of the QSO continuum are seriously in error, its temperature must be ~ 10^8 K. These two conditions, when combined, place severe upper limits on the intercloud thermal pressure. These limits immediately rule out models of pressure confined clouds in equilibrium with intercloud gas whose density varies as r^{-2} (as in e.g. a wind) since these flows are either optically thick or their thermal pressure at the BLR is insufficient to confine the observed filaments.

CLOUD KINEMATICS   Line widths are well defined. These exceed 3000 km/sec, are characteristically about 5000 km/sec, and in extreme cases are known to extend to over 10,000 km/sec. It has not been possible to demonstrate that all the observations are compatible with only a single general flow pattern. Significantly, this suggests that, in fact, there is no universal flow pattern. The only spectra whose interpetation demands a particular flow pattern are those in which BALs are present. The BAL gas appears to have a higher ionisation, lower density, higher velocity and to be further away from the nucleus than is the BLR gas. Interpretation of the BAL spectra demands outflow, although no fully successful model exists to date. Hartig and Baldwin (1985) find possible kinematic differences between the BLR in QSOs with and without a BALs.

Many QSOs, and most AGNs, have a narrow emission line spectrum superimposed upon the ubiquitous and defining broad emission line spectra. These lines arise in gas whose density is orders of magnitude less than that of the BLR. Typically situated well outside the BLR it is characterised by velocity widths compatible with Keplerian motion. Narrow emission lines appear to be absent in spectra which show broad absorption lines. This lends support to the supposition that gas flows in QSOs vary markedly between objects. Thus whatever mechanism is responsible for the BLR must be operative in a wide variety of flows.

We use units of 0.01c (=3000 km/sec) throughout, and designate speeds in these units by ω. The "clouds" then typically have $\omega = 5/3$; ω ranges from > 1 to ~ 4. Gas at 10^8 K has a sound speed $\omega_c \sim 0.4\sqrt{T_8}$, so the Mach number of the line emitting gas with respect to the hot gas is $M = 2.5 \omega T_8^{-1/2}$. M is thus > 4. Whether it is assumed that the line emitting gas is carried along in a global flow, or is moving through an extended atmosphere, it is clear that supersonic phenomena are involved.

If a central black hole (of $10^8 M_8 M_\odot$) dominates the gravitational mass in the core the Keplerian velocity at r_{pc} is $\omega_k \sim 0.22\sqrt{(M_8/r_{pc})}$. To account for the observed line widths in terms of Keplerian velocities requires $M_8 > 20$.

INTERCLOUD MEDIUM It is not possible to discuss cloud dynamics without invoking an intercloud medium, either to provide thermal or ram pressure confinement in the case of long lived clouds, or to provide a source for the clouds in transient cloud models.

As a measure of mass-flux rates in the core of the QSO-galaxy we use units of the mass-flux equivalent of the bolometric luminosity, $\dot{M}_a$, defined earlier. The total mass of gas in the ISM interior to the observed BLR at r_{pc} is $M_T \sim 0.1 \Gamma nr_{pc}^3 M_\odot$, where n is the number density of the ISM at the BLR. Γ is a geometrical factor which depends on the radial dependance of n ($\sim r^{-\alpha}$). For flat density profiles ($\alpha=0$) $\Gamma = 1$; for $\alpha = 2$ profiles $\Gamma = 3$. Flat density profiles arise naturally in mass loaded flows, (Perry & Dyson, 1985b), see below. The gas must leave the region in, roughly, the mean dynamical crossing time, $t_d \sim 475 \ r_{pc}/\omega$ yr.

(A more complete discussion of mass flux rates in a variety of flows is given in Perry & Dyson, 1985a, hereafter PD. For example, in a thermal wind $\omega \sim 1.2$, whereas in a flow whose velocity is the same as that of the clouds $\omega \sim 1.7$.) Thus the gas must be replaced, or injected, at a rate $\dot{M} \sim 2 \ 10^{-4} \Gamma \ n \omega r_{pc}^2 M_\odot/yr$. We follow current theory and adopt $\epsilon > 0.1$. This gives $\dot{M}/\dot{M}_a \sim 1.2 \ 10^{-5} \Gamma \ n\omega(r_{pc}^2/L_{47})$, and $n < (8/\Gamma)10^4(L_{47}/r_{pc}^2 \ \omega)(\dot{M}/\dot{M}_a)$.

The optical depth can now be determined, if it is assumed that the density profile steepens outside the BLR (i.e. $\alpha \sim 2$). In that case, $\tau \sim 4 \ 10^{-6} \ \Gamma' n r_{pc}$, where Γ' is a geometric factor dependant on α, which for $\alpha = 0$ is $\Gamma' = 1$ and for $\alpha = 2$ is $\Gamma' \sim 50$. Replacing n through $\dot{M}/\dot{M}_a$ and setting $L_{47}/r_{pc}^2 \sim 1$ then yields, for $\alpha = 0$, $\tau = (0.32/\omega)\dot{M}/\dot{M}_a$, whereas for $\alpha = 2$, $\tau = (5.3/\omega)\dot{M}/\dot{M}_a$. The requirement that $\tau < 1$ then, in turn, limits $\dot{M}/\dot{M}_a$: if $\alpha = 0$ $\dot{M}/\dot{M}_a < 3 \ \omega$; if $\alpha = 2$ $\dot{M}/\dot{M}_a < 0.2 \ \omega$.

The thermal pressure in the ISM in equilibrium with the 'canonical' QSO spectrum corresponds to $\Xi = 4.6(\omega\Gamma/T_8)(\dot{M}_a/\dot{M})$. Cloud confinement may be effected, in a transient manner, by dynamical pressure. The stagnation pressure in any flow is greater than the thermal pressure by a factor $\sim 6.4 M^2$. This translates into an ionisation parameter which is reduced by a factor $(T_8/4\omega^2)$ over that in the undisturbed flow (PD). In a region which experiences the stagnation pressure therefore (ie. behind a shock) the ionisation parameter is $\Xi^* \sim (1.2\Gamma/\omega)(\dot{M}_a/\dot{M})$. Imposing the condition that the flow be optically thin, as derived above, then gives, for $\alpha = 0$, $\Xi \sim 1.5/\tau$, $\Xi^* \sim 0.4/\tau\omega^2$; for $\alpha = 2$, $\Xi \sim 74/\tau$, $\Xi^* \sim 19/\tau\omega^2$. A striking 'coincidence' is immediately apparent: the dynamic pressure ionisation parameter in flat profile flows is precisely equal to, or slightly less than, that which is observed in the BLR, implying that dynamic pressure confinement is operating in the BLR. The thermal pressure, in contrast, is less than that required - unless the ISM is optically thick and $\dot{M}$ exceeds $\dot{M}_a$ by a large factor. Another consequence of the difference in pressure confinement mechanisms is in the total mass of the ISM which must be present in the BLR. Let us contrast two ISM whose velocity structure and temperature are the same. If a thermal pressure confined BLR is to exist in the first and a dynamic pressure confined BLR in the second, the density everywhere, and thus the total mass, in the first must be M^2 times that of the second.

Before turning to the gas dynamical details of global flows and of dynamic creation of the BLR we pause to review some of the most important insights which have been gained by study of individual aspects of the problem, in particular in the study of long lived clouds.

CLOUD KINEMATICAL MODELS 'Permanent' cloud models have been extensively studied. Mathews (1982) and his co-workers studied radiatively accelerated cloud models. They showed that in order to achieve the observed velocities, the clouds must be pressure confined by a medium which exerts little drag, i.e. a relativistic plasma. Such ISM simply expand away from the BLR and would have to be continuously regenerated. Furthermore, the clouds are subject to serious disruptive instabilities. Kwan and Carrol (1982) considered low angular momentum clouds which fall into centre of the galaxy from regions 1 kpc distant. The clouds are confined by a combination of ram and thermal pressure exerted by a hot ISM. Extremely massive ($\gg 10^9 \ M_\odot$) black holes are required to produce the observed orbital velocities. Since ram pressure on the leading edge of the cloud is not balanced to the rear, the clouds will expand and dissipate. It is unclear if they can survive for long enough to explain the observations. The disruptive effects produced by motion through the ISM could be avoided if the clouds were carried along by a fast moving wind. Beltrametti (1981) considered the creation of clouds by thermal instabilities in a wind. However, Compton cooling stabilises the wind unless $L_{47} < 0.1$, which effectively rules out most QSOs. All these models have the inherent difficulty that they require a high density ISM to provide the thermal pressure confinement which, as outlined above, leads to high optical depths and large mass fluxes.

The difficulties encountered in producing viable permanent cloud models has led to the consideration of transient cloud models. KMT, in their excellent and wide ranging paper, noted that there should be continuous mass interchange between the hot and cool phases on a timescale less than the dynamical crossing time of the region. The gas velocities result from the expansion of the region as a whole. However, since thermal pressure confinement is required, such two phase models suffer from the same optical depth and mass flux difficulties as the permanent cloud models. Transient clouds confined by dynamical pressure as a way out of these difficulties was first proposed by Dyson and Perry (1982). Such transient models depend on mass injection throughout the region; such mass injection gives rise to flows whose properties are quite different from those where all sources of mass lie outside of the boundaries of the region of the flow. We now turn to these models.

GLOBAL FLOW MODELS   We assume that QSOs occur in the nuclei of young galaxies (Rees, 1985). The stellar population must be young; young stars "experience vigorous episodes of mass loss" (Lada, 1985). The extent of this early outflow stage of stellar evolution, and the enormous energies associated with it, are only now becoming apparent, and we refer the reader to Lada's (1985) comprehensive review. Vigorous mass loss in an environment of high stellar density, as is presumed to be present in the nucleus of QSO galaxies, can be expected to create a rich, turbulent and very energetic ISM. Even if external sources such as accretion from the distant reaches of the galaxy, or winds driven off the surface of a central accretion disk, are present, it is necessary to understand what we term mass-loaded flows.

Mass loading has several important consequences, which we consider in turn. Firstly, and most importantly for BLR models, the density profile is dramatically altered from that in external source flows. The mass conservation equation, $\nabla \cdot \rho v = \dot{\rho}$, can be solved (spherical symmetry is used here for simplicity) for $\rho(r)$ in terms of $\dot{\rho}(r)$, the local mass injection rate: $\rho(r) = [\int_{*}^{r} \dot{\rho}(\zeta) \zeta^2 d\zeta]/r^2 V(r)$. For any particular velocity profile, the density profile is flatter than in the absence of mass loading. This has immediate and profound consequences resulting from the restrictions on the optical depth, as we have seen. Optically thin, flat density-profile flows, for example, have densities at the BLR orders of magnitude greater than r^{-2} flows of the same optical depth.

Secondly, mass loading, because of its dynamic effects, alters the the topology of the solutions of the wind equation fundamentally. In order to examine these effects we now consider the equation of momentum conservation. The effective acceleration due to the pressure gradient, $g_p = -\nabla P/\rho$, can be rewritten by replacing P through the equations of state and of mass conservation to yield an effective pressure term, g_p:

$$g_P = v_c \, d(\ln M)/dr \; + \; g_D \; - \; (v_c/M)(\dot{\rho}/\rho) \; - \; R\nabla\,(T/\mu)\;.$$

g_D is the geometric term which arises because of the divergence of the volume element; absent in plane parallel flows, in spherically symmetic flows it $= 2v_c{}^2/r$. The third term is the deceleration due to the local absorption of momentum by the injected mass (we assume the injection is isotropic and contributes no net momentum to the flow). The last term, due to temperature gradients, acts in the direction of decreasing T.

The equation of momentum conservation is

$$d(\rho v)/dt = \rho(g_P + g_R + g_0 - g_G)$$

where the radiative acceleration $g_R = \int \varkappa\,(\nu)F(\nu)d\nu/c$; $F(\nu)$ is the local radiation flux. The acceleration due to the gravitational field of the central hole and of the stellar core population is g_G; g_0 is any "other" acceleration such as that due to relativistic particles. We replace the velocity through the Mach number, $M = v/v_c$. Inserting the explicit form for g_P in the momentum equation (and noting that now $d(\rho v)/dt = \rho\dot{v} + v\dot{\rho}$) yields the general flow equation:

$$v_c^2 \, \frac{M^2 - 1}{M} \, \frac{dM}{dr} \; = \; (g_D + g_R + g_0 + g_T) - (g_G + g_{\dot{\rho}}) \; = \; g+ - g-$$

where $g_{\dot{\rho}} = (M^2 + 1)(v_c/M)(\dot{\rho}/\rho)$ and $g_T = -R\nabla\,(T/\mu)$. The mass injection deceleration, $g_{\dot{\rho}}$, for the general density profile given above becomes $g_{\dot{\rho}} = (M^2+1)v_c^2 \, r^2\dot{\rho}(r)\Big/\int \dot{\rho}\,(\zeta)\zeta^2 d\zeta$. We expect that $\dot{\rho}(r)$ is related to the local IMF, stellar mass density and to local dynamics, such as tidal and collisional disruption. All these effects lead us to suppose that $\dot{\rho}(r)$ may be flat in the core, but most probably falls with r, steepening as r increases; if $\dot{\rho}(r)$ in the core is given by a power law ($\dot{\rho} \sim r^{-\beta}$) then $g_{\dot{\rho}}(r) = [(3 - \beta)(M^2 + 1)]v_c{}^2/r$. This combined with g_D gives an inwardly directed effective acceleration, $g_e = [M^2(3 - \beta)+(1 - \beta)]v_c{}^2/r$.

Regular sonic points (M = 1) exist only if g+ = g- at the sonic point. Analysis of permitted flow topologies then follows from analysis of the relative radial dependence of g+ and g- (Perry & Dyson, 1985b). For example, thermal pressure driven winds have supersonic branches only if $g_D > 0$, i.e. in divergent geometries. Radiatively accelerated supersonic winds occur only if g_R is flatter than g_G, as e.g. in essentially one-dimensional stellar and disk winds where the radiative force does not fall off above the surface as quickly as gravity. Optically thin, radiatively accelerated winds do not exist above point sources because the radial dependence of the radiative and gravitational accelerations are the same; the radiation force contributes only to a reduction in the effective gravity everywhere, reducing the overall density in the flow. This geometrical conspiracy prevents radiative forces from driving point source winds to highly supersonic velocities - even near QSOs. In order to achieve highly supersonic winds the geometrical tie between g_R and g_G must be broken. This can be achieved by QSOs if, for example, the continuum emission is geometrically extended, if the winds originate above extended accretion disks, or if the opacity varies in the flow.

The sonic point condition in mass-loaded flows (when g_o = 0) is g_R = g_G + $2(2 - \beta)v_c^2/r$. For all β < 2, this gives g_R > g_G at the sonic point. If both terms go as r^{-2}, radiative forces everywhere dominate gravity if they do so at the sonic point. However, the slower decay in $g_{\dot\rho}$ guarantees that this term dominates; hence, regular sonic points do not exist because if M > 1 for r > r_c then dM/dr < 1; the situation is precisely reversed for inflows. [Several aspects of irregular sonic points are discussed by Beltrametti (1981), and references given there. Galactic winds such as those discussed by Mathews and Baker (1971) have regular sonic points in the presence of cooling, when g_T plays the crucial role.] If, in addition to mass loading, geometrical effects conspire to break the tie between g_R and g_G, very highly supersonic winds with flat density profiles can result. The only analytic solutions which currently exist are those of Beltrametti and Perry (1980). Those winds have uniform mass and luminosity generation throughout the core. The velocity increases as r in the core, resulting in a wind density which is independent of r! These winds are radiatively accelerated in their supersonic branches and achieve subrelativistic speeds. Although their models were oversimplified, most of the important physical effects are clearly visible in their solutions. More detailed investigation of the effects of mass loading in more realistic geometries throws up a variety of complex flows (Perry and Dyson, in preparation).

It is certainly unrealistic to assume a strict power law for $\dot\rho(r)$; further complications arise due to the expected sporatic and local nature of the injection. Furthermore, the injection is expected to be time dependent and to depend on the evolutionary history of the galactic core. We will return to these questions elsewhere (Perry and Dyson, in preparation); we now turn to the more local effects of mass injection – in particular to how it creates the broad emission line region in QSOs.

CREATION OF THE BROAD LINE REGION Mass injection into supersonic flows can only occur when mediated by stand off shocks. Furthermore, in the cores of QSO-galaxies, where we expect energetic mass loss by stars and through stellar collisions and tidal disruption, the mass injection is itself supersonic. The collision of two supersonic gas streams results in a classic two shock pattern familiar from interstellar gas dynamics. The shocked wind gas is heated to temperatures well in excess of the equilibrium temperature of the ISM. Its pressure is the stagnation pressure of the flow. Because the gas is superheated it tries to cool. Inverse Compton cooling by the QSO continuum is very effective near the QSO; with increasing distance from the source the diminution of the flux reduces its effectiveness. In the high pressure shocked region the gas cools isobarically by Compton cooling followed by unstable radiative line cooling. We (PD) find that in e.g. shocks around supernovae within about 1 pc of the central source, the trapped gas cools and fragments before it flows out of the high pressure, shocked region. The fragments have densities, temperatures and ionisation parameters in agreement with BLR observations. They are accelerated by the pressure gradients within the shock to speeds characteristic of the local flow, eventually leave the shocked region and reexpand into the flow. (Details of the energy balance, cooling times, flow patterns, are given in PD).

A simple dynamic picture of the BLR thus emerges. Mass injection creates or augments the ISM in the core of a QSO galaxy. In the process stand-off shocks form within which the BLR fragments are created and accelerated. The mass-fluxes required to explain the BLR are equal to the mass-energy equivalent of the luminosity. Increasing the mass flux to provide full thermal pressure cloud confinement conflicts with optical depth constraints and requires excessive mass fluxes. Since shocks are just as effective in creating cool clouds in high mass flows as in more reasonable low mass flows too many BLR clouds are created in thermal pressure confining flows; furthermore, the ionisation parameter of such dynamically confined clouds does not agree with observation.

CONCLUSIONS A picture of the formation of the BLR is emerging in which the BLR is seen to be a consequence of the creation of an energetic ISM by stellar mass loss in the centre of a QSO-galaxy. In the process of injecting the gas destined to feed the central engine, the stars provide both the observed heavy elements and the energy necessary to create the temporary regions of high pressure within which short-lived BLR clouds form. Detailed solutions of the gas dynamical problems raised here can be expected to increase our understanding of the QSO-galaxy interaction. Undoubtedly revisions in details of the picture will become necessary as detailed solutions of mass-loaded flows become available. The shocks which mediate the mass injection require careful attention, as does the evolution of the cool fragments which form in their shadow.

REFERENCES
Begelman,M.C., 1985, Ap. J., **297**, 492.
Beltrametti,M., 1980, Astron. Astrophys., **86**, 181.
Beltrametti,M., 1981, Ap. J., **250**, 18.
Beltrametti,M., Perry,J.J., 1980, Astron. Astrophys., **82**, 99.
Collin, S., 1985, Proc. Meudon Wkshp Model Nebulae (July 85) ed. Pequignot
Dyson,J.E., & Perry,J.J., 1982, Third European I.U.E. Conf., Madrid, 592.
Dyson,J.E., & Perry,J.J., 1985, Proc. Meudon Wkshp Model Nebulae
Hartig,G.F., & Baldwin,J.A., 1985, Ap. J., in the press.
Hutchings,J.B., Crampton,D., Campbell,B., 1984, Ap. J., **280**, 41.
Kippenhahn,R., Mestel,L., Perry,J.J., 1975, Astron. Astrophys., **44**, 23.
Kippenhahn,R., Perry,J.J., Roser,H.-J., 1974, Astron.Astrophys., **34**, 211.
Krolik,J.H., McKee,C.F., Tarter,C.B., 1981, Ap. J., **249**, 422. (KMT)
Kwan,J., & Carroll,T.J., 1982, Ap. J., **261**, 25.
Lada, C. J., 1985, Annual Rev. Astron. & Astrophys., **23**, 267.
MacKenty,J.W., & Stockton,A., 1984, Ap. J., **283**, 64.
Malkin,M.A., Margon,B., Chanan,G.A., 1984, Ap. J., **280**, 66.
Mathews, W.G., 1974, Ap. J., **189**, 23.
Mathews, W.G., 1982, Ap. J., **252**, 39.
Mathews, W.G. and Baker, J. C., 1971, **170**, 241.
Mestel, L., Moore, D.W., Perry, J.J., 1976, Astron. Astrophys., **52**, 203.
Mushotzky,R.F., Solomon,P.M., Strittmatter,P.A., 1972, Ap.J., **174**, 7.
Perry,J.J. & Dyson,J.E., 1985a, M.N.R.A.S., **213**, 665. (PD)
Perry,J.J. & Dyson,J.E., 1985b, Proc. Meudon Wkshp Model Nebulae
Rees, M. J., 1985, Proc., IAU Symposium No. 119.
Turnsek,D.A., 1984, Ap. J., **280**, 51.

DISCUSSION

Abramowicz : Mass estimations of the central black hole are based on the observed velocities of the clouds. It is assumed that these veclocities are Keplerian (free motion). In your picture the clouds are not moving freely. How does this influences the estimation of the central mass ?

Perry : Such estimates may well over-estimate the mass of the central black hole by an order-of-magnitude. The mass distribution of the central star "cluster" must be included in analyses of line profiles and self consistent stellar and gas dynamic models are needed before accurate determinations of M_h can be made.

Kundt : What are your objections to replacing the confining IGM by a relativistic pair plasma (which later-on is focussed into the jets by a magnetic pinch) ?

Perry : The size of radio jets which have been resolved places them within the BLR - therefore we would expect focusing to have occurred interior to the BLR in most cases. The most serious objection to a relativistic plasma confining the BLR (to $r \gtrsim 1$ pc) is that plasma would itself have to be confined to the nuclear region by a thermal plasma, or be continuously resupplied.

Wampler : We know that the broad line clouds give about the same line width for objects with nuclear luminosities that range over factors of $10^5 (-30 < M_B < -18)$. We also know that in the upper end of this range $(-30 < M_B < -24)$ the total luminosity in the CIV $\lambda 1548$ line is roughly constant, independent of luminosity (Baldwin effect). Can your model produce both the constant flow velocity and constant $\lambda 1548$ luminosity in the high luminosity quasars ?

Perry : In principle, yes, and I would like to refer you to our Monthly Notices (213, 665) paper for details of the model. In practice, detailed line profiles cannot be computed until the flow behind the radiating cooled shock is analysed numerically with an accuracy not yet available. Such modeling should be available in the next few years, and we shall then be able to give you a reliable answer.

Schilizzi : In a radio loud quasar there will be a radio jet passing through the Broad line Region. What effect might that have on the dynamics of the Inter Cloud Medium ?

Perry : The effect of a jet on a global flow will only be through surface (local) interactions - we would not expect any global effects. There may well be some entrainment and mixing in the surface layer.

BROAD ABSORPTION LINE QSOs

David A. Turnshek
Space Telescope Science Institute
Baltimore, MD, 21218
USA

Abstract. Observational constraints on models for broad absorption line (BAL) QSOs are discussed. The picture which emerges is one in which the broad absorption line region (BALR) contributes to the broad emission line region (ELR) as a high ionization component. Depending on QSO luminosity, the limits on the distance of the BALR from the central source are 1 - 1000 pc. A highly metal enriched and disk-like geometry for the region is indicated. The similarity between the absorption in some radio quiet BAL QSOs and the associated complexes of absorption in some radio loud QSOs may indicate that the BALR outflow can affect the region surrounding a QSO out to distances in excess of several hundred kpc.

1. Introduction

Several review-type papers dealing exclusively with BAL QSOs have appeared recently (Weymann and Foltz 1983; Turnshek 1984a; Weymann, Turnshek and Christiansen 1985). As a result, this review paper makes only a brief summary of past observations of BAL QSOs, but it does try to put them into perspective. Newer observations are reviewed in more detail. In particular, I discuss recent observations which pertain to the fraction of BAL QSOs as a function of redshift, observations concerning the ELs of BAL QSOs, the derivation of column densities of BAL clouds which are relevant to abundance determinations and observations which constrain possible changes in BAL profiles with time. I discuss in more detail how the observations are capable of placing constraints on models of QSOs and BALRs. Imposed constraints have implications for geometry, size scales and abundances. I also review some interesting observations of radio loud QSOs with associated complex absorption and discuss how this absorption may be related to the BAL phenomenon. Finally, I discuss problems and the needs for future work.

2. Review of Observations

2.1. General Characteristics

A BAL QSO shows a substantial amount of resonance line absorption blueshifted with respect to the emission redshift. I suspect that at the present time over 100 BAL QSOs may be known to various observers, but only half that number are

G. Swarup and V. K. Kapahi (eds.), Quasars, 317–330 .

suggested from data in the literature. A small fraction of these could be considered well-studied. Ninety-five percent of the time, the absorption has only a high level of ionization and transitions of CIVλ1549, NVλ1240 and OVIλ1034 are dominant; SiIVλ1397 is moderate and Ly-α is moderate to weak. Five percent of the time, the absorption also has low ionization components and transitions of MgIIλ2798 and sometimes CIIλ1335 are present. Other reported weak to intermediate strength transitions include AlIIIλ1857 and CIIIλ977. There is less information on the frequency of appearance of these other transitions because they fall in wavelength regions not always observed, but they do appear more often than MgII and CII. Although hydrogen Lyman limit absorption has never been observed, BALs have been detected in the Lyman continuum (e.g. see Alec Boksenberg's contribution at this symposium). BALs generally only appear in radio quiet QSOs (Stocke et al. 1984), but see section 4 for some qualifications on this effect. The BAL profiles can have a smooth appearance or a very broken-up appearance, with multiple troughs and/or individual absorption line doublet components present. Very often the absorption is detached with respect to the peak emission redshift. Absorption-absorption line-locking is sometimes present (e.g. see Foltz et al. 1986a). The absorption outflow velocity typically ranges up to 20,000 - 30,000 km s^{-1} and in one case it may range up to 60,000 km s^{-1}. However, cases where the absorption has an extent of only a few thousand km s^{-1} also exist. Multiple troughs are said to be present when several discrete complexes or troughs of absorption exist, separated by several thousand km s^{-1}. The existence of multiple troughs is significant because of the potential constraints they may impose on model geometry (see section 3.1). In cases where the absorption has an extent of a little less than a few thousand km s^{-1}, whether the absorption is fundamentally similar to the BALs or whether it is due to a related or completely different phenomenon is unclear (see section 4).

In their review article Weymann and Foltz (1983) present a nice compendium of BAL QSO spectra. For convenience I illustrate two typical types of spectra in Figures 1 and 2. The wavelength scales in the Figures have been shifted into

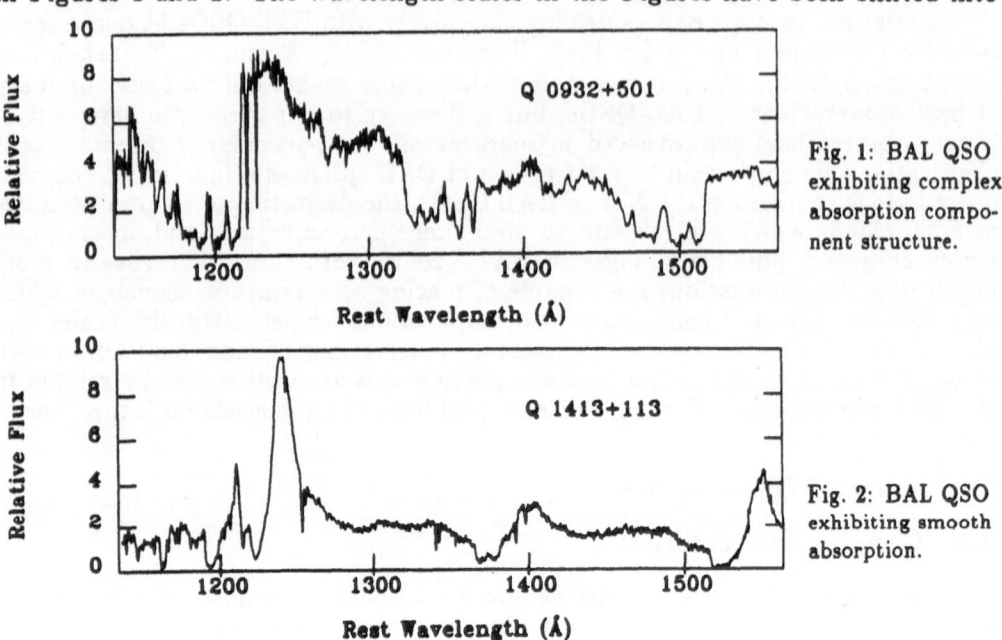

Fig. 1: BAL QSO exhibiting complex absorption component structure.

Fig. 2: BAL QSO exhibiting smooth absorption.

the QSO restframe. As opposed to some of the most complex BAL QSO spectra (e.g. Q1414+087 in Foltz et al. 1983a and UM141 in Turnshek et al. 1980), these spectra are fairly easy to assess because outflow velocities are not high enough to cause much confusion due to overlapping transitions from different ionic species. The object in Figure 1 has detached absorption showing considerable structure in the trough, whereas the object in Figure 2 shows considerably smoother absorption which may also be somewhat detached. See Turnshek (1984a) for a discussion of how the combination of optical depth effects, cloud sizes and kinematics can influence the observed characteristics of BAL profiles.

BAL QSOs are also the only class of QSOs which are radio quiet and can have a polarization in excess of a few percent. Finally, a small amount of data suggests that BAL QSOs are more optically variable than their radio quiet non-BAL QSO counterparts. See Turnshek (1984a) for more details about these two effects.

2.2. Fraction of QSOs with BALs as a Function of Redshift

Hazard et al. (1984) has estimated that BAL QSOs comprise 0.03 - 0.10 of the QSO population near redshifts of 2.3; there is no evidence that this fraction varies significantly with redshift. I emphasize this latter point to an even greater extent than I did in my talk because shortly after this symposium, a search of the IUE archives for other purposes resulted in the discovery of two new low redshift BAL QSOs (Turnshek and Grillmair 1986). Both of these objects are identified in the Palomar-Green Bright Quasar Survey (BQS). In addition, Matt Malkan reported at this conference (see the discussion following this paper) that during the course of his IUE observations, he recently discovered a very low redshift BAL QSO identified in the BQS. Coupled with previous identifications of BAL QSOs in the BQS (Wilkes 1985; Turnshek et al. 1985), this means that there are at least 5 low redshift BAL QSOs in the BQS sample. I estimate that only about 50 objects identified in the BQS have been observed in sufficient detail to have discovered BALs. The mean redshift of the five known BQS BAL QSOs is about 0.5. Hence, the fraction of QSOs with BALs at low redshift is compatible with the 0.03 - 0.10 fraction given by Hazard et al. (1984). However, Cyril Hazard (private communication) has noted that in his survey work he finds a greater tendency for BALs to be present in the spectra of the highest redshift QSOs. Since selection effects are so controversial and hard to evaluate at the highest redshifts, more work will have to be done to explore this important possibility.

2.3 Emission Lines of BAL QSOs

One interesting statistical observation concerning BAL QSOs is that the distribution of their emission line (EL) properties differs from those of non-BAL QSOs (Turnshek 1984a; Hartig and Baldwin 1986). For instance, after the effects of absorption are taken into account, CIV emission tends to be weaker in BAL QSOs and, although Hartig and Baldwin do not concur, I have said that NV emission is stronger in BAL QSOs. BAL QSOs also appear to have an excess of AlIII and iron emission which complicates the CIII] spectral region. I emphasize that these results on differences between the ELs of BAL QSOs versus non-BAL QSOs refer to differences in the distribution of EL properties between the two classes. For example, for a particular BAL QSO with weak CIV emission, one is likely to have little trouble finding a non-BAL QSO with even weaker CIV emission, but this does not negate the fact that weak CIV emission is a more common property of

BAL QSOs than of non-BAL QSOs.[1] Apparently, Turnshek (1984a) and Hartig and Baldwin (1986) reached different conclusions concerning NV emission in BAL QSOs versus non-BAL QSOs simply because the Hartig and Baldwin (1986) non-BAL QSO comparison sample had no objects with very weak NV emission. Since the detailed resolution of the NV question depends so much on understanding selection effects, I believe that the best clarification which can be made at this time is to note that there are non-BAL QSOs which show practically no evidence for NV broad emission (e.g. see Foltz et al. 1983b), while the same is not true for BAL QSOs.

In addition, Hartig and Baldwin (1986) have presented observations of the BAL QSO Q0335-336, a very unusual object which exhibits unusually narrow emission ($1000 - 2000$ km s^{-1}) and components of FeIII emission near or in place of the CIII]λ1909 EL. This object also has strong FeII emission and an emission feature near 2080 Å that is probably FeII. Non-BAL QSO counterparts to this object have not been discovered to date.

There also appear to be correlations between a QSO's BAL profile type and the corresponding characteristics of the adjacent EL. For example, when comparisons of BAL QSOs are made, one finds that as BAL profiles become less smooth and more detached from the emission, showing multiple troughs and sometimes narrow component structure, the peak emission line height to continuum ratio is generally smaller and the emission line width is usually broader. There may also be a tendency for the characteristic width of the BALs to match the width of the ELs. These general effects can be seen by comparing Figure 1 and Figure 2.

Clearly the systematics of QSO EL and BAL properties are very intricate and large sample sizes are needed to explore the types of systematics I have noted above. This is particularly obvious when one considers that there are many different types of BAL profiles to be explored and yet BAL QSOs comprise only 0.03 - 0.10 of the QSO population.

2.4. Detailed Studies of BAL Profiles

One of the more interesting aspects of studing BAL QSOs is the determination of column densities and comparison of these column densities to predictions of models for the level of ionization in the BALR. In certain circumstances column densities can be determined fairly accurately (Turnshek 1984a). One object (Q0932+501) that I have studied with Frank Briggs, Craig Foltz and Ray Weymann (Turnshek et al. 1986) is worth mentioning because it appears to have a region where the C^+ column density is only a factor of 2 or so less than the H^0 column density. Explaining this observation is difficult unless abundances in the BALR are significantly enhanced relative to solar values (section 3.3).

Foltz et al. (1986a) recently reported on the comparison of two sets of observations of Q1303+308 separated by two years. As Q1303+308 has considerable absorption structure, they were able to put a 3σ upper limit on any change in velocity of the absorption components of 30 km s^{-1}. This can be used to put a lower limit on the BALR size scale (section 3.2). Comparison of these observations did show that, at a low level of significance, the intensity of some high velocity smoother BAL features did appear to get weaker during the two year time period.

---

[1] I should point out as an aside that BAL QSOs appear to show no evidence for the Baldwin effect and I believe that this is also true for radio quiet QSOs in general.

3. Constraints on Models

3.1 BALR Covering Factor and Geometry

Before I begin reviewing the arguments which constrain BALR covering factor and geometry, I will note two constraints which are dealt with more specifically in section 3.2: (1) that the BALR lies outside the broad Ly-α emitting region and (2) that the BALR can contribute a high ionization component to a QSO's broad EL spectrum.

Comparison of observations with resonance line scattering (RLS) models are capable of placing constraints on the global value of the BALR covering factor. If the global covering factor is too large, RLS produces more emission than is observed in certain parts of the spectrum. The global value of the BALR covering factor is simply the fraction of the QSO covered by BALRs as determined by integrating over all possible lines of sight. This should be distinguished from the local BALR covering factor which is determined by integrating only over lines of sight which differ from the line of sight to the observer by small angles. Generally, the value of the global BALR covering factor is constrained to be less than 0.1 - 0.2 (Turnshek et al. 1980; Turnshek 1981; Junkkarinen 1983; Turnshek 1984b). The CIV and NV regions of a spectrum provide the most powerful constraints on the maximum value of the covering factor.

Working in the CIV region has the advantage of being able to make a fairly accurate assessment of the unabsorbed continuum level. For the CIV case, if the BAL clouds are distributed in a spherically symmetric manner or in an axially symmetric disk-like manner, the value of the global BALR covering factor is usually constrained by the amount of emission present in the region 5,000 - 20,000 km s^{-1} to the red of the CIV emission peak and also by the residual intensity in the trough region on the blue side of the CIV emission peak. However, if the BAL clouds are distributed in a jet or fan pointed toward the observer, the residual intensity in the trough is the only constraining factor. There are some cases where the total equivalent width of the CIV emission is so much smaller than typical BAL equivalent widths that the global covering factor must be small ($< 0.1 - 0.2$), independent of EL profile or residual intensity considerations.

Working in the NV region has the advantage that the radiation scattered by the NV BALs includes not only the inner continuum radiation, but also the inner broad Ly-α EL (see section 3.2). However, since the inner Ly-α EL is attenuated by the NV BAL, one has to assume a value for the Ly-α EL equivalent width in order to calculate the resulting EL equivalent width of NV RLS radiation as a function of global BALR covering factor. If one assumes reasonable values for the unattenuated Ly-α EL equivalent width, one finds that the observed NV EL equivalent also sets upper limits on the global BALR covering factor of 0.1 - 0.2. This is a very powerful constraint in a statistical sense, but its value for any one object is unclear.

The most obvious consequence of the constraints on global BALR covering factor is that many QSOs must have BALRs which are not observable via their absorption signature and, since the constraint is an upper limit, all QSOs might have BALRs. Of course, the idea that the global BALR covering factor is constant in QSOs and that a QSO either does or does not have a BALR may be unrealistic. In fact, if QSOs form any kind of a continuous sequence of objects, the most practical way to think of the global BALR covering factor parameter is that it varies from 0.0 to 0.2 and it applies to all QSOs.

Mechanisms for the acceleration of BALR material were discussed in detail in the Weymann, Turnshek and Christiansen (1985) review article and so they will not be discussed here. However, one point that must be emphasized is that constructing a model for the acceleration where multiple absorption troughs originate from a single ejection episode or region in space is difficult. For example, in the context of the discussion of Weymann, Turnshek and Christiansen (1985), a single trough may result from one's line of sight passing through a main cloud which has been accelerated by ram pressure, followed by smaller ablated 'cloudlets' which have been torn off the main cloud and accelerated to a higher speed. Thus, when observing BAL QSOs with multiple troughs, our line of sight is most probably passing through multiple BALRs, each of which, by itself, is of sufficient extent to have the QSO classified a BAL QSO. Since multiple troughs are observed so frequently (approximately 30 - 40 percent of the time), interpreting each trough as arising in a separate BALR suggests that the optical depth to BALRs is about 0.7 - 0.9. This requires a fairly large local BALR covering factor of 0.5 - 0.6. To see this more explicitly, define the global BALR covering factor to be q_c and the solid angle containing BALR clouds to be $\frac{\Omega}{4\pi}$. If for simplicity $\frac{\Omega}{4\pi}$ is assumed to be a single valued quantity, i.e., it does not vary from one QSO to the next, then the value of q_c can be computed as the product of two probabilities:

$$q_c = \frac{\Omega}{4\pi}(1 - e^{-\eta})$$

where η is the optical depth of isolated BALRs in space and $(1 - e^{-\eta})$ is the value of the local BALR covering factor which is the probability that a QSO will be classified as a BAL QSO if our line of sight lies within the solid angle containing BAL clouds. One can see that if $\frac{\Omega}{4\pi}$ is unity, then the BALR must be spherically symmetric by definition, but if it is less than unity, then a non-spherically symmetric distribution holds. For the case of a spherically symmetric BALR and the realistic range $0.03 < q_c < 0.20$, $0.03 < \eta < 0.22$ follows. From Poisson's distribution, the expected distribution of trough multiplicities for $\eta = 0.03$ is computed to be: 98.5 percent single troughs, 1.5 percent double troughs and 0.015 percent triple troughs; for $\eta = 0.22$: 89.3 percent single troughs, 10.0 percent double troughs and 0.7 percent triple troughs. As noted, the percentage of double and triple troughs observed is much higher than this, suggesting that the solid angle containing BALRs might typically be only 5 - 35 percent of 4π steradians.

Taken together the evidence for a small global BALR covering factor along with a large local BALR covering factor suggests a non-spherically symmetric BALR geometry. The two simplest possible geometries for the BALR would be disk-like and jet- or fan-like. Since jet or fan geometries could not produce symmetric emission, a disk geometry is preferred.

3.2 Size Scale of the BALR and Emission from the BALR

There are basically three size scale regimes, corresponding to three different models that have been proposed, that are worth noting: (1) the wind model of Drew and Gidding (1982) and Drew and Boksenberg (1984) in which BALR distance scales are much less than 1 pc (typically 0.01 - 0.003 pc), (2) the ELR component model of Turnshek (1984b) and Turnshek et al. (1985) in which BALR distance scales are greater than 1 pc, but less than 1 kpc and (3) the interstellar cloud - thermal wind model of Weymann et al. (1982) and Scott et al. (1984) in which BALR distance scales are in excess of 1 kpc. I will not go into the detailed attributes of each model, but rather discuss constraints on size scales which tend to place the BALR 1 - 1000 pc from the central source, ruling out models (1) and (3). The constraints are generally luminosity dependent.

Models with size scales less than about 1 pc have difficulty explaining the deduction that the BALR occults the broad Ly-α emitting region (Turnshek, Foltz and Weymann 1986). The effects of this occultation can be seen in Figure 2; note that the bump in the otherwise smooth NV trough is the result of NV absorption attenuating inner continuum plus Ly-α emission and not just attenuating the continuum as is the case near CIV. Since the region giving rise to the broad Ly-α EL is generally thought to have $n_e \approx 10^{11}$ cm^{-3} (Canfield and Peutter 1981), the Ly-α broad ELR must generally be no more than ~ 1 pc from the central source and the BALR lies beyond this. These limits are constrained even further by the recent observations of Foltz et al. (1986a). Assuming the BALR acceleration is constant, the lack of a change in velocity of any of the absorption components requires that the BALR be at least 7 pc from the central source. This lower limit is derived from luminosity independent considerations, but one should remember that it applies to a QSO with a specific luminosity. The assumption that the BALR acceleration is due to radiation pressure leads to the requirement that the BALR be at least 4 pc from the central source.

Upper limits on the size scale of the BALR are inferred from deductions about the emissivity of BAL clouds. As detailed in Turnshek (1984b), collisional excitation in the BALR is clearly capable of providing a significant amount of observable high ionization broad emission. The two possible caveats to this conclusion concern adopting the appropriate value for the global BALR covering factor and the validity of estimating a lower limit for $n_e r^2$ in the absence of a working model for the BALR level of ionization (see section 3.3). Since the fraction of QSOs with BALRs is so large, appealing to the first caveat is not a way out of having the BALR contribute to a QSO's broad EL spectrum in some objects, but for QSOs having a very small global BALR covering factor the contribution would be very small. The second caveat is more of a caution than anything else as I believe the techniques used to put a lower limit on $n_e r^2$ have been conservative. For $n_e < 10^6$ - 10^7 cm^{-3} the strongest EL that would be produced in the BALR would be [OIII]λ5007 by a large factor. A lower limit on n_e in the BALR then results from the lack of broad [OIII]λ5007 emission in QSOs, which in turn can be used to infer an upper limit for the size scale of the BALR. This argument was applied to a specific low redshift BAL QSO (PG1700+518) in Turnshek et al. (1985) and now that other low redshift BAL QSOs are known to exist, the argument is even stronger. In PG1700+518, since the Lyman limit luminosity is relatively small and the effective spectral index between the Lyman limit and the local continuum level[2] near [OIII]λ5007 is so steep, the limits placed on n_e versus the global BALR covering factor are not extremely tight. The upper limit derived for the size scale of the BALR in PG1700+518 is 30 pc, but this would scale up to 100 - 1000 pc in the most luminous BAL QSOs.

Concerning collision excitation giving rise to other ELs, as previously noted, there is the potential for the BALR to produce a significant contribution to the higher ionization lines. However, the fact that the flux detected in the NV BAL troughs are often very small generally demands that no broad Ly-α emission be produced in the BALR. This could either be a temperature effect (Turnshek 1984b) or an abundance effect (section 3.3). Flux that is seen in the NV trough is consistent with being flux from the inner broad Ly-α EL, but attenuated by the NV BALs, or possibly some contribution from the narrow ELR which is unattenuated. Note

---

[2] The local continuum level near [OIII] is high due to the presence of substantial FeII emission (Wampler 1986).

that the observations seldom demand that the CIV BALR occults the adjacent CIV broad ELR.

One compelling piece of evidence which suggests that different ELs often originate in the same region occurs when EL velocity profiles match. As shown by Wilkes and Carswell (1982) and Gaskell (1982) there are differences between low and high ionization lines. One conclusion which might be drawn from the observations of EL profiles is that when the profile of a high ionization EL matches exactly that of the broad Ly-α EL coming from the inner and denser region, the high ionization EL is also likely to arise in this inner part of the broad ELR which is occulted by the BALR. However, in cases where profiles do not match very well, some high ionization broad emission may be arising in the BALR.

The situation for RLS is different. For global BALR covering factors in the 0.1 - 0.2 range, all of the observed NV broad emission could be accounted for by the BALR. A much smaller, but still noteworthy, fraction of CIV broad emission would be caused by this effect. See Turnshek (1984b) for a discussion of how aspect angle might effect the situation. Aspect angle and radiative transfer effects could offer an explanation for why NV is stronger in BAL QSOs. A general conclusion one might reach from RLS considerations is that a QSO with little or no NV broad emission in its spectrum may have a BALR with very small global covering factor, whereas a QSO with significant NV broad emission in its spectrum may have a much larger global BALR covering factor. The velocity extent of the BALR is also an important parameter in estimating the amount of RLS NV emission.

3.3 Abundances in the BALR

Simple models for the level of ionization in the BALR are not very successful at predicting derived column densities. The basic problem which applies to all BAL QSOs is that central source photoionization models adopting solar abundances and assuming either a power law relation for the distribution of ionizing photons or a power law plus blackbody for the distribution of ionizing photons generally fail to produce observed Si^{+3} column densities without producing much more H^0 than is observed (Turnshek 1981; Weymann, Turnshek and Christiansen 1985). When the parameters describing the power law and blackbody contribution to the ionizing photons are adjusted optimally so that the Si^{+3} to H^0 column density ratio is in fair agreement with the observations, the parameterization of the ionizing spectrum is not what one really expects it to be and, additionally, the C^{+3}, N^{+4} and O^{+5} column densities are factors of 10 - 1000 lower than their observed values. Simple collisional ionization models characterized by solar abundances and a small range of temperatures also offer no hope of explaining BALR column densities.

At the present time, one must try to decide whether abundance or model assumptions are lacking. On the one hand, none of the photoionization models that have been calculated take into account the possibility of absorption in the Lyman continuum. This was because the observations demonstrate that BAL clouds are not optically thick at the hydrogen Lyman limit. However, observations in the Lyman continuum of BAL QSOs are now telling us that, although BAL clouds are not optically thick at the Lyman limit, BALs in the Lyman continuum can be present, and this will alter the effective ionizing spectrum that the BALR sees. Any new work utilizing photoionization models for the BALR should try to account for absorption in the Lyman continuum by BALs. Also, there are conceivably many other sources of energy input for such an energetic outflow. Furthermore, adjusting the abundances is not too attractive because the Si, C, N and O all required different factors.

On the other hand, based on the observation discussed in section 2.4, I believe the situation should now be viewed from a different perspective when attempting to assess whether abundance assumptions or model assumptions are lacking. In particular, any model for the level of ionization of BALR clouds will have difficulty explaining an H^0 column density only a factor of 2 or so larger than the C^+ column density without having carbon abundances enhanced 10 - 100 times solar values. Observations of FeII emission in the BAL QSO PG1700+518 (Wampler 1986) may be telling us that iron is similarly enhanced.

4. Radio Properties of QSOs with Absorption

BAL QSOs are known to be radio quiet. However, a number of radio loud QSOs having extensive associated complexes of absorption come close to being included in the BAL class. The point is, while these associated complexes of absorption are not as extensive as the BAL absorption, the absorption is extensive enough to raise questions about whether the associated complexes of absorption can be explained by so-called intervening material, even if the material is taken to reside in clusters (Morris et al. 1986). Briggs, Turnshek and Wolfe (1984) have discussed the situation and possible implications. In particular, these authors point out that past work on the so-called marginal BAL QSO PKS1157+014 was somewhat misleading in that broad Ly-α absorption was taken to have a BAL nature. In fact, the broad Ly-α absorption is due to a damped Ly-α line in the wing of Ly-α emission. Systems characterized by damped Ly-α are most easily explained by intervening galaxy disks (Wolfe et al. 1986). However, even if the damped Ly-α absorption is excluded from consideration, the absorption in PKS1157+014 is still quite extensive. Similarly, Morris et al. (1986) have analyzed the spectrum of the radio source GC1556+335 and have shown that, although Weymann et al. (1979) initially considered this object to be a possible member of the BAL class, the absorption is not quite extensive enough and the physical conditions in the GC1556+335 absorbing clouds are not the same as in the BAL clouds. Finally, a survey by Foltz et al. (1986b) has shown that z_{abs} near z_{em} systems, which are similar to the associated complexes of absorption, are generally found in radio loud QSOs, but not in radio quiet QSOs.

The main result of the above referenced work is that the properties of the clouds producing the BALs in radio quiet QSOs are quite different from the properties of the clouds producing the associated complexes of absorption in radio loud QSOs. There are kinematic, level of ionization and abundance differences. In addition, as Morris et al. (1986) discuss, from considerations of CII and CII f.s. one finds that the clouds responsible for the associated complexes of absorption in GC1556+335 are at least several hundred kpc from the central QSO. Yet, the similarity between the spectra of some BAL QSOs with absorption component structure and the spectra of some radio loud QSOs with associated complexes of absorption gives one the impression that there may be continuity between the absorption line morphology of the two classes of objects. Thus, remnant ejecta from the BALR may be the cause of associated complex absorption systems. One speculation is that the mass swept up in the outflow may become more diluted with hydrogen, explaining why the BALR appears metal enriched, while the clouds causing the associated complexes of absorption do not. If this is the case, either a radio structure becomes visible at about the same time BALs dissipate, or when viewed from other aspect angles that are out of the plane of the BALR disk, both the radio structure and the complex absorption are visible. This latter possibility is more consistent with a picture in which QSOs have steady-state BALRs. One can not rule out that free-free absorption in the BALR causes the radio structure to be absorbed. If such a speculation holds, the implication is that the BALR is capable of affecting regions surrounding a QSO in excess of a few 100 kpc.

5. Summary of Implications

In section 4, I discussed constraints on the global and local values of the BALR covering factor. The observations of BAL QSO EL profiles and residual intensities in BAL troughs necessitates a small global covering factor; however, in order to have a viable model for the acceleration of BALR clouds, a large local covering factor is required. I discussed why a high ionization broad EL component may arise in the BALR; the lack of broad [OIII]λ5007 emission in QSO spectra then requires that electron densities in the BALR be less than 10^6 - 10^8 cm^{-3}. These densities generally imply upper limits on the distance to the BALR of a 100 - 1000 pc. The constraint that the BALR lies beyond the broad Ly-α emitting region and the recent observations which show no change in the velocity of individual BAL components put a lower limit on the distance to the BALR of 1 - 7 pc. All of these constraints, coupled with the need to have only symmetric emission components arising from the BALR, strongly indicate a disk-like distribution for the BALR.

Column density determinations for BAL clouds show levels of ionization that are difficult to understand using solar abundances and simple models for the ionization. In particular, however, the observation that the H^0 column density can be only a factor of 2 or so greater than the C$^+$ column density in a specific absorption component suggests BALR abundances which are very enhanced (10 - 100 times) relative to solar values. Observations of iron emission in BAL QSOs may also be telling us that abundances are enhanced.

The further observation that many radio loud QSOs have associated complexes of absorption that are not quite extensive enough to have the QSO placed in the radio quiet BAL class suggests that the BALR may affect the region surrounding a QSO out to at least several hundred kpc. If this were the case, some connection between the observed presence of radio emission and either the viewing angle needed to observe BALs or the evolutionary state of the BALR must exist.

Collectively, the observations and constraints which have been discussed offer a snapshot of what the BALR may look like. The notion that it is a high ionization component of a QSO's broad ELR is probably correct, but the source of the BALR clouds need not be the same as the other clouds giving rise to a QSO's ELs. Indeed, the properties of the clouds producing the broad Ly-α emission, BALs and the narrow ELs are very different in terms of the constraints which I have discussed. In particular, the clouds producing the broad Ly-α emission are known not to be able to occult the continuum source either individually or collectively, which is in contrast to the BALR clouds. The symmetric profiles that are characteristic of the Ly-α emission may be an indication that rotation is important rather than outflow in the inner, denser, lower ionization broad ELR. Also, as the observations show, the BALR must be incapable of producing Ly-α emission. As far as other broad ELs are concerned, the extent to which the BALR contributes to the emission is unclear. Certainly, there is a very efficient RLS mechanism for producing NV in the BALR and collisional excitation may be important to varing degrees. As discussed by Bev Wills and Alexei Fillippenko at this symposium, all of these different ELRs may blend into one another to some degree.

6. Problems and Future Work

One of the main problems in this area is the lack of a suitable model for the BALR level of ionization. Until a reliable model is found, many of these results must be evaluated carefully. Of course, the situation is complicated because deciding on the ionization mechanism is only part of the problem; the other part is adopting the correct abundances. If models rely on energy input into the BALR clouds via central source photoionization, they should try to properly take into account attenuation of the Lyman continuum by any BAL resonance line transitions that may be located in the Lyman continuum. The most straightforward way to do this will be to wait for UV observations in the Lyman continuum of BAL QSOs. This will not only tell us how the Lyman continuum is chopped-up by the BALs, but it will also give us new and important information on the level of ionization in the BAL clouds. One starting point for considering models for the level of ionization would be to try to produce a region where the neutral hydrogen density is not too much greater that the singly ionized carbon density. However, one should keep in mind that low ionization components are not typical of BALR clouds. Other important theoretical work needs to be done in the areas of acceleration of the clouds and the actual source of the clouds. Two possible sources for the clouds are the region which gives rise to the inner and denser Ly-α emitting region and supernovae of massive stars.

Important observational work is needed in the areas of detailed studies of individual BAL QSOs and on the statistics of the EL and BAL properties of QSOs. In particular, placing limits on possible changes in BAL profiles with time, placing limits on the amount of broad [OIII] emission present in a QSO's spectrum, the derivation of BALR column densities and detailed analysis of the Ly-α/NV region in BAL QSO spectra for the purpose of studing RLS, covering factor, aspect angle and radiative transfer effects are all very necessary. As studies of polarization and optical variability properties are sparse, more work in these areas is needed. The statistical questions fit into two catagories. First, there is the question of whether there are changes in the incidence of BAL QSOs as a function of redshift or luminosity. Second, there is the question of how the EL properties of BAL QSOs compare with non-BAL QSOs within a specific redshift range and luminosity interval. Such a study should consider the absorption and EL properties of steep and flat spectrum radio sources as well. To answer both of these questions, one would need samples in which the selection criteria as a function of absorption and emission line properties are well understood. Results dealing with the second question would tell us something about aspect angle effects and the value of the BALR covering factor as a function of QSO EL properties. Of course, in the simplest case where there is a spherically symmetric geometry and the BALR covering factor is a constant in all QSOs, one expects the distribution of EL properties in BAL QSOs to be identical to the distribution of EL properties in non-BAL QSOs. As was the point of most of the discussion presented here, this possibility is almost certainly ruled out. The observations now challenge us to put together a self-consistent snapshot of the distribution of QSO EL and BAL properties as a function of geometry, aspect angle and radiative transfer effects.

Acknowledgements

My work on BAL QSOs has benefited from discussions with Frank Briggs, Craig Foltz, Cyril Hazard and Art Wolfe; I would especially like to acknowledge continuing discussions with Ray Weymann on this subject.

References

Briggs, F.H., Turnshek, D.A., and Wolfe, A.M. 1984,*Ap.J.(Letters)*, **287**, 549.

Canfield, R.C., and Puetter, R.C. 1981, *Ap.J.*, **226**, 1.

Drew, J., and Gidding, J. 1982, *M.N.R.A.S.*, **201**, 27.

Drew, J., and Boksenberg, A. 1984, *M.N.R.A.S.*, **211**, 813.

Foltz, C.B., Wilkes, B.J., Weymann, R.J., and Turnshek, D.A. 1983a, *P.A.S.P.*, **95**, 341.

Foltz, C.B., Weymann, R.J., Hazard, C., and Turnshek, D.A. 1983b, *P.A.S.P.*, **95**, 117.

Foltz, C.B., Weymann, R.J., Morris, S.L., and Turnshek, D.A. 1986a, in preparation.

Foltz, C.B., Weymann, R.J., Peterson, B.M., Sun, L., Malkan, M.A., and Chaffee, Jr., F.H. 1986b, *Ap.J.*, submitted.

Gaskell, C.M. 1982,*Ap.J.*, **263**, 79.

Hartig, G.F., and Baldwin, J.A. 1986, *Ap.J.*, in press.

Hazard, C., Morton, D.C., Terlevich, R., and McMahon, R.G. 1984, *Ap.J.*, **282**, 33.

Junkkarinen, V.T. 1983, *Ap.J.*, **265**, 73.

Morris, S.L., Weymann, R.J., Foltz, C.B., Turnshek, D.A., Schectman, S., Price, C., and Boroson, T.A. 1986, *Ap.J.*, submitted.

Scott, J.S., Christiansen, W.A., and Weymann, R.J. 1984, unpublished preprint.

Stocke, J.T., Foltz, C.B., Weymann, R.J., and Christiansen, W.A. 1984, *Ap.J.*, **280**, 476.

Turnshek, D.A., Weymann, R.J., Liebert, J.W., Williams, R.E., and Strittmatter, P.A. 1980, *Ap.J.*, **238**, 488.

Turnshek, D.A. 1981, Ph.D. Thesis, University of Arizona, Univ. Microfilm Inc., Ann Arbor, Michigan.

Turnshek, D.A. 1984a, *Ap.J.*, **280**, 51.

Turnshek, D.A. 1984b, *Ap.J.(Letters)*, **278**, L87.

Turnshek, D.A., Foltz, C.B., Weymann, R.J., Lupie, O.L., McMahon, R.G., and Peterson, B.M. 1985, *Ap.J.(Letters)*, L1.

Turnshek, D.A., Foltz, C.B., and Weymann, R.J. 1986, preprint.

Turnshek, D.A., and Grillmair, C. 1986, in preparation.

Turnshek, D.A., Briggs, F.H., Foltz, C.B., and Weymann, R.J. 1986, in preparation.

Wampler, J.S. 1986, preprint.

Weymann, R.J., Williams, R.E., Peterson, B.M., and Turnshek, D.A. 1979, *Ap.J.*, **234**, 33.

Weymann, R.J., Scott, J.S., Schiano, A.V.R., and Christiansen, W.A. 1982, *Ap.J.*, **262**, 497.

Weymann, R.J., and Foltz, C.B. 1983, *Proc. of the 24th Liege Symposium: Quasars and Gravitational Lenses*.

Weymann, R.J., Turnshek, D.A., and Christiansen, W.A. 1985, *Astrophysics of Active Galaxies and Quasi-Stellar Objects*, ed. Miller, J.S.

Wilkes, B.J., and Carswell, R.F. 1982, *M.N.R.A.S.*, **201**, 645.

Wilkes, B.J. 1985, *Ap.J.(Letters)*, **288**, L1.

Wolfe, A.M., Turnshek, D.A., Smith, H.E., and Cohen, R. 1986, *Ap.J.(Supplement)*, in press.

DISCUSSION

Blandford : If, as you suggest, reasonance line scattering and anisotropic distribution of the absorbing material are both important features of BAL Quasars, then you expect to see some substantial polarization changes through the line profiles and this might give a unique diagnostic for the cloud distribution and optical depths. What prospects are there for doing high resolution spectropolarimetry on these objects ?

Turnshek : Under certain circumstances, you would indeed expect to see a polarization signature of some sort. However, if you had multiple scatterings, this would cause depolarization and the prospects for detecting anything would be poor. I have done resonance line scattering polarization calculations for the case of a disk geometry. The calculations show that for the case where you have photons typically scattered once, you might expect the resonance line scattered part of an emission line to be polarized by as much as 20-25%. From the observation you could determine the plane of the disk on the sky. For the case where BALs are seen (i.e. your viewing angle passing through the plane of the disk) you would have maximum polarization. As your line of sight started to miss the disk (i.e. viewing angles out of the plane) you would not expect to see BALs and you would get an interesting "W" shaped polarization profile as a function of wavelength. At polar angles there would be no polarization. The observation would not be trivial because the continuum of BAL QSOs in sometimes polarized. In anycase, you would have to subtract this continuum contribution off. There might also be an unpolarised part to the emission line. The prospects are not good for detecting this with instruments that exist today. It may be that Space Telescope offers the best hope of detecting this because integration times would be long and night sky polarization from the ground could cause a problem. However, it would be good to get an expert opinion in this area.

Burbidge : I have two questions: First, you suggested that from the absorption lines you could derive heavy element abundances ~100 times solar. If the absorption and emission arises in the same (contiguous) region, how do you reconcile this with the fairly normal (solar) abundances deduced from the emission lines ? Second, if the broad absorption line QSOs are part of the normal QSO population should you not see the same incidence of sharp absorption features in them that you see in the non BAL QSOs in the same emission-line redshift range ?

Turnshek : For the first question, my answer is that in the case of BALs I believe you can more accurately determine column densities as long as you do not have unocculted emission at the wavelength of the absorption features. In the case I have discussed, you have a fairly isolated absorption component at 14,100 km s^{-1}. The problem with abundance determinations from emission lines is that there is a collection of clouds which, as we have heard at this meeting, have a wide range of properties in terms of densities and levels of ionization. There is kinematic evidence, and evidence from BAL QSOs as well, that different emission lines can come from different regions or clouds. Therefore, I would argue that you need a very good model to properly constrain abundances from emission line observations.

For the second question, my answer is that BAL QSOs are necessarily excluded from spectroscopic surveys which try to determine the distribution and statistics of narrow, isolated absorption lines. This is because you can not adequately assess the path length covered by the survey and because there could be contamination from narrow components which make up the BALs. On the other hand, I do know of some BAL QSOs which have narrow absorption lines and these lines are consistent with the properties of the so-called intervening population of absorption lines. Of course, you would expect this.

Malkan : To agree with your statement that the BAL phenomenon is independent of z : of the < 100 PG quasars with z < 1.5, we now know that at least <u>three</u> are BAL QSO's. The newest one, found by IUE observations, has a redshift of 0.089, with strong, relativity narrow NV and CIV troughs.

Turnshek : That is very interesting. Of course, there is still the question of the fraction of QSOs with BALs at the highest redshifts, z > 3.

Wampler : Harting and Baldwin (1986) have pointed out that the spectra of BAL quasars are different from non-BAL quasars. The region to the red of about $\lambda_0 \sim 2000$ A is generally free of broad absorption lines and the corresponding regions can be compared. They find that BAL quasars have much stronger FeII emission than non-BAL quasars. This strongly suggests that many quasars are not BAL quasars.

Turnshek : Yes, this is true and very important. However, it is not completely clear what the statistics are yet. If there are some BAL QSOs which have weak FeII emission, these objects may just have systematically smaller BALR covering factors. Of course, it is also possible that emission line properties are aspect angle dependent. If the BALR is confined to a disk, emission line properties may be diffrerent when viewed in the plane of the disk.

Wilkes : Geoff Burbidge expressed concern with the discrepant abundances between BALR clouds and BELR clouds especially if they are the same clouds. I understood you to say that BALR clouds contribute to the broad emission but not that they <u>are</u> the BELR clouds. Is this correct ?

Turnshek : Yes. The BAL clouds certainly have some very different properties from the "standard" BEL clouds. For one thing the BAL clouds are capable, at least collectively, of completely covering a region the size of the inner $L\alpha$ emitting region, which is probably $>10^{18}$ cm. However, the clouds giving rise to this inner $L\alpha$ emitting region are not even collectively capable of covering a region which is probably 10^{14} cm. It may be necessary to have these two types of clouds originating from two completely different sources. For example, the inner $L\alpha$ emitting clouds may come from the accretion disk, but the BAL clouds may come from supermassive stars.

EMISSION-LINE KINEMATICS AS A PROBE OF THE CENTRAL ENGINE IN QSOs

Amri Wandel
Astromomy Program
University of Maryland
College Park, MD 20742
USA

ABSTRACT. We investigate the constraints on dynamic models for the line-emitting regions in quasars and AGN. The parameters characterising the central energy source (Mass, efficiency, accretion rate) are calculated in terms of the physical conditions in the line emitting gas. In a large sample the central mass (calculated assuming the emission-line clouds are bound) is proportional to the continuum luminosity. We find typical values of $L/L_E \sim 10^{-2 \pm 0.5}$, $e \sim 0.1$-1%, and $\dot{M}/\dot{M}_E \sim 1$-10.

1. INTRODUCTION

The two most typical features of quasars and AGN are their prominent emission lines and their powerful nonthermal continuum radiation source, the "central engine". While the former is fairly well explained in terms of a photoionization model (e.g. Davidson and Netzer 1979), there is little consensus on the mechanism producing the nonthermal continuum. Most of the recent models, however, invoke a massive compact object, in the potential well of which energy is extracted from accreted matter (Rees 1983). The currently favoured picture of the broad line-emitting region (BLR) features a size of 0.1-1pc, an ensemble of fast ($\lesssim 10^4$km/s), dense ($n \sim 10^9$-10^{11} cm^{-3}) cloudlets or filaments, partially photoionized (ionization parameter in the range $0.3 \lesssim \Xi \lesssim 3$) by the UV continuum.

Evidently the physical conditions in the BLR are governed by the central source: the ionization state and the temperature are fixed by the continuum radiation, while the velocity dispersion of the clouds, which is reflected in the line width, is induced by the central object, via radiative acceleration (Blumenthal and Mathews 1979), gravity (Kwan and Carroll 1982; Wandel, Milgrom and Yahil 1985), a wind or other mechanisms, such as shocks, relativistic particles or jets. If the primary energy source is indeed release of rest mass energy of matter accreted by the central object, the fuelling matter presumably comes through (or from) the BLR, which provides an additional link between the line emission and the central source. It is therefore attractive to use the extensive emission-line data available, in the context of a kinematic model, in order to constrain the parameters characterizing the central engine (such as M, e and dM/dt), which are essential (though, to date, quite undetermined) for any model of the central source.

331

G. Swarup and V. K. Kapahi (eds.), Quasars, 331–336 .

2. RADIATION PRESSURE VERSUS GRAVITY

The forces which the radiation and gravity of the central source exert on a cloud have an opposite direction and the same functional dependence on the distance from the central object, r, hence their ratio for a given cloud is independent of r. The radiation force on an optically thick cloud is $F_R = L_{ion} S/4\pi r^2 c$, where S is the cloud's cross section and L_{ion} is the ionizing radiation (between 1 and 100 Rydberg). The gravitational attraction by the central mass is $F_c = GMM_c/r^2$, where $M_c \sim SNm_H$ is the cloud's mass and N - its column density (N_{23} in units of $10^{23} cm^{-2}$). The ratio of the two is

$$F_G/\ F_R = 4\pi m_H cGMN/L_{ion} = \sigma_T N \left(-\frac{L_{ion}}{L_E}\right)^{-1} \approx 0.06\ N_{23} \left(-\frac{L_{ion}}{L_E}\right)^{-1}\ , \qquad (1)$$

where $L_E = 4\pi GMm_H c/\sigma_T$ is the Eddington limit, and σ_T - the Thomson cross section. As this ratio depends only on the column density, for a given value of L/L_E there is a critical column density, $N_{cr} = 1.6 \times 10^{24} (L_{ion}/L_E)$ cm^{-2}; clouds with $N > N_{cr}$ are bound, while smaller clouds are radiatively accelerated outwards. Note that the relevant parameter is the column density of the whole cloud, not only that of the ionized part. Photoionization models require an ionized column density of $N_{23} \approx 1$ (e.g. Kwan and Krolik 1982). The above result branches into two different self consistent scenarios: radiatively accelerated clouds with $L/L_E \approx 1$, and gravitationally bound clouds with $L/L_E \ll 1$. Models in which the line width is induced by gravity imply high central masses, with $L/L_E \ll 1$ (Wandel and Yahil 1985; Joly et al. 1985). This in turn gives $F_G \gg F_R$, consistent with the basic assumption of those models, that gravity dominantes the clouds' dynamics. Radiative acceleration, on the other hand, is possible if $L \approx L_E$ and if the clouds do not have large neutral parts beyond the photoionized front, which are both postulated by the radiative acceleration model (Mathews 1982).

3. CONSTRAINTS ON THE BLR

3.1 Size. The distance of the emission-line gas from the central source can be expressed in terms of the the ionization parameter, defined as $\Xi = L_{ion}/4\pi r^2 cknT = 2.3\ P_{rad}/P_{gas}$,

$$r_{pc} = 0.44\ (L_{ion,45}/\Xi n_9)^{1/2}\ , \qquad (2)$$

where subscripts indicate units and $T = 10^4 K$ has been assumed. A different method is to estimate the size of the emitting region, using the luminosity in a specific emission line I, $L(I) = 4\pi r^3 f_v j_I n^2$. Here j_I is the emissivity and f_v is the volume filling factor. Expressing the later in terms of the covering factor f_a ($f_v \approx f_a N/rn$) (here N is the column density of the photoionized part of the cloud), yields (for the H_β line)

$$r_{pc} \approx 0.082\ [L(H_\beta)_{43}/\ f_a n_9 N_{23}]^{1/2}\ , \qquad (3)$$

where we have assumed the emissivity is enhanced by a factor of 3 over the case B value, due to collisional exitation (Mathews, Blumenthal and Grandi 1980).

3.2 <u>Intercloud medium.</u> In order to confine the clouds by thermal- or ram pressure, a hot intercloud medium (HIM) must be invoked. A "standart" quasar spectrum would heat the HIM to typically 10^8K (Krolik, McKee and Tarter, 1981), but this value depends on the assumed UV continuum. Equating the ram pressure to the thermal pressure in the clouds gives $n_h \sim 2.3$ $knT_c/m_H v^2 \approx 2\times10^3 n_9 T_{c4} v_4^{-2}$ cm^{-3} , where subindices c and h refer to the cool (cloud) and hot (HIM) phases, respectively. In this case, however, the cloud's trailig end is not supported, and eventually the clouds will diffuse. On the other hand, confinement by the thermal pressure of the HIM gives $n_h = 10^5 n_9 T_{c4}/T_{h8}$ cm^{-3} . In the case of radiative accelration, radiation pressure must balance the ram pressure,

$$L_{ion}/4\pi r^2 c \gtrsim \rho_h v^2 . \qquad (4)$$

For ram-pressure confined clouds this yields $\Xi > 2.3$, which excludes this combination for most objects. For thermal-pressure confinement of radiatively accelerated clouds, on the other hand, eq. (4) gives $v < 10^3 (\Xi T_{h8})^{1/2}$ km s^{-1} , or, $T_h > 10^{10} v_4^2/\Xi$ K, requiering an unplausibly high temperature. Observations of continuum variability and sharp lines constrain the BLR to be optically thin to electron scattering, giving a lower limit for T_h; $\tau_{es} = n_h r \sigma_T \approx 0.1 (L_{ion,45} n_9/\Xi)^{1/2} T_{c4}/T_{h8} < 1$.

3.3 <u>Mass flow and efficiency.</u> Unless the HIM is a part of a wind or in global outflow, in which case the mass supply to the central energy source is undetermined, the hot gas can be assumed to have a net inward velocity component (for example, quasispherical accretion, turbulent disk-like flow, or shocked infall). As we have seen, the clouds must be nearly comoving with the HIM, so it is not unreasonable that the later has velocities of the order of those inferred from the line width. In the absence of significant sources (for example, stellar mass loss) or sinks (reejection, for example by jets, wind or outflowing clouds) of matter between the BLR and the central source, the HIM-inflow can provide an estimate of the fuelling rate of the central engine,

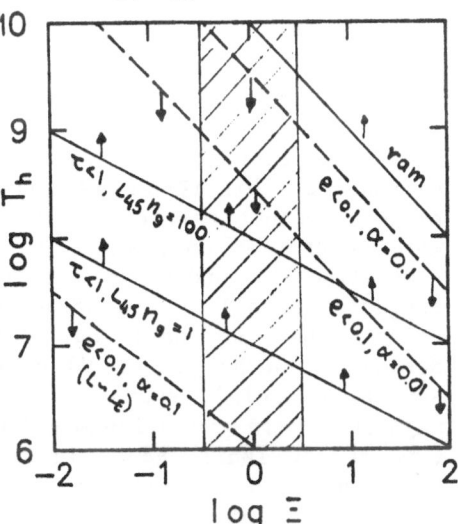

Fig. 1. Constraints on the HIM imposed by the ram pressure, efficiency ($e < 0.1$) and optical depth ($\tau_{es} < 1$) conditions. The observed range of Ξ is hatched.

$$\dot{M} \approx 4\pi \alpha r^2 \rho_h v \sim 68 \; \alpha v_4 L_{ion,45}/\Xi T_{h8} \quad M_\odot yr^{-1} , \qquad (5)$$

where α is the fraction of the flow that actually reaches the central source. The efficiency of mass conversion into <u>ionizing</u> radiation is

$$e_{ion} = L_{ion}/\dot{M}c^2 \sim 2.8\times10^{-4} \; \alpha^{-1} \Xi T_{h8} v_4^{-1} . \qquad (6)$$

For example, if $0.03 < \alpha < 0.3$, $L_{ion} = 0.3 \; L_{bol}$, and $\Xi = T_{h8} = v_4 = 1$, the total efficiency e is in the range 0.1-1%. Since e$\lesssim$10%, equation (6) gives a constraint on T_h, namely $T_{h8}<(10-100)v_4/\Xi$. For radiatively accelerated clouds one can derive an even more stringent constraint (Wandel 1986, in preparation), $T_{h8}< 0.5\ \alpha\Xi^{-3/4}(L_{ion,45}n_9)^{1/4}$, which is certainly mutually exlusive with the drag constraint. Fig 1. shows the constraints derived above in the Ξ-T_h-plane.

4. BOUND CLOUDS AND THE CENTRAL MASS

The line width is induced by the gravitational field of the central object in several models of the cloud dynamics: orbital motion, radial inflow, or outflow close to the escape velocity. In either one of these cases, the central mass can be related to the line width by

$$M_8 \approx G^{-1}v_{eff}^2 r \sim 58\ v_4^2 r_{pc}\ , \tag{7}$$

where $M_8=M/10^8 M_\odot$ and $v_{eff}= (\sqrt{3}/2)\ v(FWZI)$.

4.1 <u>Broad lines.</u> Combining this with the estimates for r for the BLR (section 3.1) we have

$$M_8 \sim 72\ (L_{ion,45}/\Xi n_9)^{1/2}v_4^2\ , \tag{8.a}$$

or

$$M_8 \sim 14\ (L_{H\beta,43}/\ f_a n_9 N_{23})^{1/2}v_4^2\ . \tag{8.b}$$

As demonstrated for NGC 4151 (Ulrich <u>et al.</u> 1984), the BLR may be stratified, with different lines originating at different radii and having different widths. For this reason we consider the emission-volume method, eq. (8.b), as a more consistent one, since it uses the same line in order to determine the distance as well as the velocity. Using this method we have calculated the masses for a sample of 90 quasars and Seyfert 1 nuclei, spanning 4 orders of magnitude in continuum luminosity (Wandel and Yahil 1985). In Fig. 2 the the continuum luminosity (in the B band) is plotted versus the mass. A linear regression gives $\log M_8 \approx (\log L_{45}-1.5)\pm0.5$, with a correlation coefficient of 0.9, which gives $L_B/L_E\sim0.001-0.01$. The bolometric L/L_E is of course higher by a factor of $L_{bol}/L_B\approx3-10$.

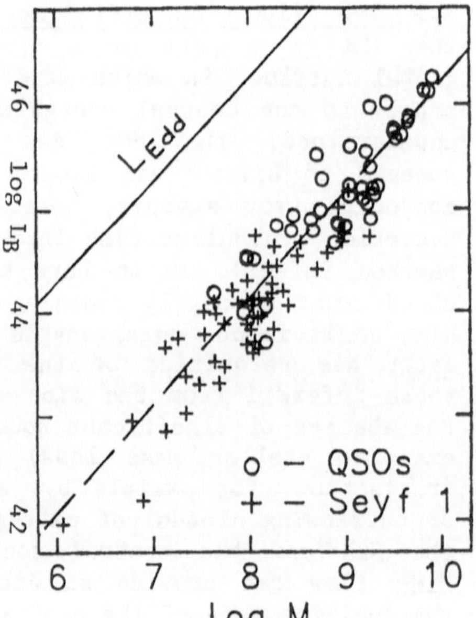

Fig. 2. Luminosity vs. mass calculated from the Hβ line

4.2 <u>Narrow lines.</u> The same method may be applied to the narrow line region (NLR). The later has the advantage that it is more likely to be dominated by gravity (i.a. the width of the forbidden narrow lines seems to be correlated with the stellar velocity dispersion, cf. Wilson and Heckman 1984), but on the other hand, it may be affected by the mass of stars of the galactic nucleus, as its size is much larger than that of the BLR. An expression for the mass inside the NLR, analogous to eq.(8.b), can be derived by appying the emission-volume method to the [OIII] line. For typical parameters one has

$$M_8 \approx 32 \ (L_{OIII,43}/ \ f_a n_5 N_{20})^{1/2}(v/300 \text{ km s}^{-1})^2 \ , \qquad (9)$$

where v is the half-maximum velocity width of the [OIII] line and v_{eff} has been taken as $\sqrt{3}v$. Calculating the masses with the OIII method for a sample of 50 objects (Wandel 1985) yields a correlation similar to the one found for the BLR (with masses larger by a factor of 1-5; the normalization depends on the choise of parameters, which are less certain for the NLR than for the BLR), and $L_B/L_E \sim 10^{-3\pm0.5}$. Combining this result with the estimate of the efficiency (sec. 3.3), it is possible to estimate the dimensionless accretion rate. For $\alpha \sim 0.1$ and assuming $L_B/L_{ion} \sim 0.3$ we get $\dot{M}/\dot{M}_E \equiv L/eL_E \approx 3L_B/L_E e_{ion} \approx 1-10$.

4.3 <u>Comparison with variability method.</u> The mass of the central object may be estimated from X-ray variability (Barr and Mushotzky 1985). The masses found by this method closely match those calculated by the volume emission-line method (Wandel and Mushotsky, in preparation). The agreement of these completely independent methods provides strong evidence that the velocity dispersion of the line-emitting material is indeed induced by the gravity of the central object.

5. CONCLUSIONS

The physical conditions in the emission-line regions are governed by the radiation and gravity of the central source. This relation imposes constraints on models of the line emitting regions, as well as on the basic parameters defining the central engine. The dimensionless parameters of the energy source - L/L_E, e and $\dot{M}/\dot{M}_E$ - seem to be restricted to rather narrow ranges over many orders of magnitude in the continuum luminosity. More statistical research on large samples, spanning a wide range in luminosity is needed in order to overcome the spread introduced by differences between individual objects, confirm the conclusions of this work and bring out further characteristics of the quasar phenomenon.

REFERENCES
Barr, P. and Mushotzky, R.F. 1985, preprint.
Blumenthal, G.R. and Mathews, W.G. 1979, Ap. J., 233, 479.
Davidson, K., and Netzer, H. 1979, Rev. Mod. Phys., 51, 715.
Joly, M., et.al. 1985, Astron. and Astrophys., 152, 282.
Kwan, J., and Krolik, J.H. 1981, Ap. J., 250, 478.
Kwan, J., and Carroll, T.J. 1982, Ap. J., 261, 25.
Mathews, W.G., 1982, Ap. J., 232, 59.
Mathews, W.G., Blumenthal, G.R., and Grandi, S.A. 1980, Ap. J. 235, 971.
Rees, M.J. 1984, Ann. Rev. Astr. Astrophys., 22, 471.
Ulrich, M., et. al. 1984, M.N.R.A.S., 206, 221.
Wandel, A. 1985, Proc. Trieste Meeting on AGN, ed. Giuricin et.al.
Wandel, A., Milgrom, M., and Yahil, A. 1985, Ap. J., 292, 206.
Wandel, A., and Yahil, A. 1985, Ap. J. (Letters), 295, Ll.
Wilson, A.S., and Heckman, T.M. 1984, Proc. of Santa Cruz workshop.

DISCUSSION

Filippenko : 1) You have concluded that $L \simeq 0.01 L_{Edd}$, but surely this must refer only to the <u>optical</u> luminosity. If one considers the bolometric luminosity, one has a much higher fraction of the Eddington luminosity. In this case, shouldn't radiation pressure play an important role in galactic nuclei ?

2) Your correlation between L and M seems almost <u>too</u> tight to be true, and leads me to suspect that circular reasoning <u>is</u> involved. I must admit, however, that I have not discovered where the flaw is (if it indeed exists).

Wandel : 1) That is correct. Since the expression for the ratio of radiation pressure to gravity is derived for optically thick clouds, the ionizing luminosity (~1-100 Ryd), not the bolometric, is the relevant one. This luminosity could still be significantly larger than the visual luminosity we have used, yielding a larger L/L_E ratio. The L/L_E ratio would also be larger if a larger density is used, since $M \propto n_c^{-1/2}$

2) It should be kept in mind that the mass is actually a combination of two observables, namely, $M \propto V_{FWZ1} L^{1/2} (H\beta)$. The tight correlation between M and L is therefore a reflection of the known correlation between L and L ($H\beta$). The slope of the M-L correlation ($M \propto L$), however, is a direct result of the independent correlation between the line width and the luminosity. If they were uncorrelated, L/M would not be constant, but rather vary as $L^{1/2}$. Finally, the correlation between log M and Log L(r=0.92) is significantly better than expected from the linear combination of the $L\beta$ -L correlation (r=0.93) and the L-V correlation (r=0.93) and the L-V correlation (r=0.4).

PHOTOIONIZATION MODELS AND LOW IONIZATION LINES IN AGN

M. Joly
DAF
Observatoire de Meudon
92195 Meudon Cedex
France

ABSTRACT. A two-component photoionization model where the high and the low ionization lines are allowed to arise from different kinds of clouds is explored. A spectral distribution of the incident radiation which tends to overestimate the X-ray flux around 3 keV is adopted in order to enhance the extent and heating of the excited HI region. It is concluded that the model is still unable to explain the observed ratio FeII/Hβ.

1. INTRODUCTION

Up to now one-component models of the broad line region of quasars fail to explain the strength of low ionization lines. In particular, FeII lines are too weak compared to Hα, Hβ and Lα in the photoionization models computed by Kwan and Krolik (1981), Kwan (1984), Wills et al. (1985). However, Collin-Souffrin et al. (1985) have ultimately tried to find if, under very favourable conditions it is possible to account for the low ionization lines in the framework of photoionization models.
   It is allowed that the low and high ionization lines are emitted in two different kinds of photoionized clouds :
- the clouds emitting the high ionization lines are those usually described by standard models.
- the clouds responsible for the low ionization lines should have a larger excited HI zone (HI^{*}) produced by an intense X radiation (compared to the UV one), a high density and possibly an overabundance of heavy elements.
   The code, mainly designed to study optically thick cases (up to $N_H \sim 10^{25}$ cm^{-2}) is described in Collin-Souffrin and Dumont (1985).

2. RESULTS

Table I summarizes the inputs and the main outputs of each run : n_0 and T_0 are the density and the temperature of the illuminated boundary, U is the ionization parameter, Z/Z_0 is the abundance of heavy elements,

G. Swarup and V. K. Kapahi (eds.), Quasars, 337–339.
© *1986 by the IAU.*

TABLE I : Characteristics of the models

Model	0	1	2	3	4	5
	Input data					
n_o (cm^{-3})	$3.6\ 10^9$	$3.6\ 10^9$	$3.6\ 10^9$	$3.6\ 10^{10}$	$3.6\ 10^{10}$	$3.6\ 10^{10}$
U	.03	.003	.03	.003	.03	.03
Z/Z_0	1	1	1	1	1	3
H (cm)	$6.1\ 10^{12}$	$1.57\ 10^{14}$	$1.53\ 10^{14}$	$5.3\ 10^{12}$	$5.4\ 10^{12}$	$5.4\ 10^{12}$
	Out put data					
T_o (°K)	22400	17150	22500	17500	24400	24000
P (bars)	$2.2\ 10^{-2}$	$1.7\ 10^{-2}$	$2.2\ 10^{-2}$	$1.7\ 10^{-1}$	$2.4\ 10^{-1}$	$2.4\ 10^{-1}$
$<T>$ HI*	7700	5650	6500	6500	7400	6900
$<n_H>$ HI*	2.10^{10}	2.10^{10}	$2.4\ 10^{10}$	$1.9\ 10^{11}$	$2.3\ 10^{11}$	$2.5\ 10^{11}$
$<n_e>$ HI*	4.10^9	$3.5\ 10^8$	$1.5\ 10^9$	6.10^9	$3.5\ 10^{10}$	$2.5\ 10^{10}$
τ_L	$4.6\ 10^5$	$2.5\ 10^7$	$2.6\ 10^7$	$7.2\ 10^6$	$6.0\ 10^6$	$8.3\ 10^6$
τ_{Bac}	1.2	1.7	5.5	1.6	5.9	4.6
τ_{Pas}	.05	.01	.26	.08	.83	.64
τ_{2343}	$5.3\ 10^4$	$2.3\ 10^6$	$2.5\ 10^6$	$1.9\ 10^6$	$6.6\ 10^5$	$2.5\ 10^6$

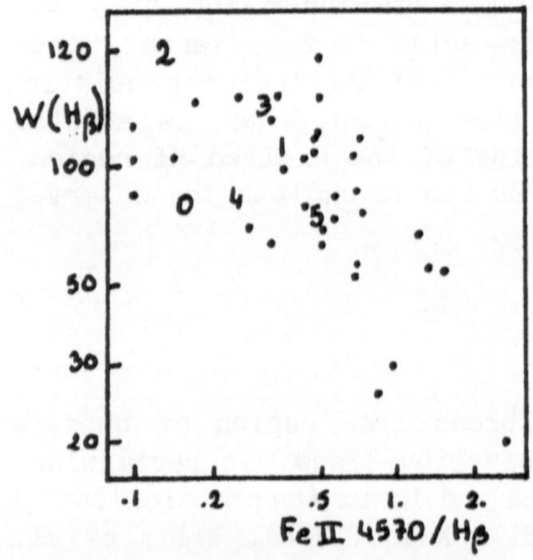

Figure 1

Figure 1 : W(Hβ) in Å versus
FeII 4570/Hβ observed in 34 AGN
plus the relevant theoretical
values obtained with models
0 to 5 assuming a covering
factor of the BLR $\Omega/4\pi = 0.1$.

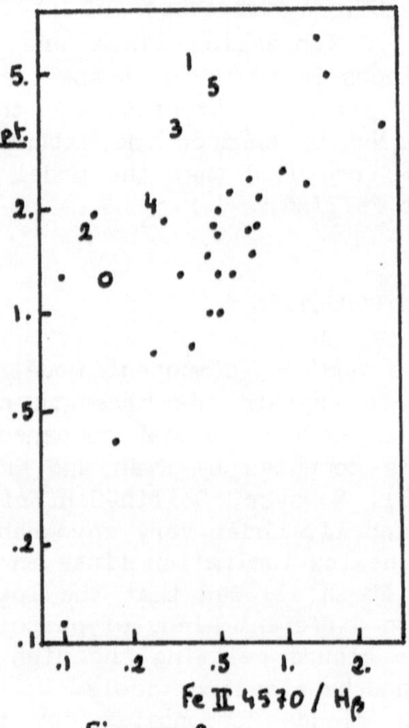

Figure 2

Figure 2 : FeII opt ($\lambda \sim 3000$–6000)/Hβ versus FeII 4570/Hβ observed in
27 AGN. Numbers are as in Figure 1.

H is the thickness of the clouds in cm, P is the pressure, $<T>$, $<n_H>$ and
$<n_e>$ are the average values of the temperature, atomic hydrogen density
and electron density in the HI*, τ_L, τ_{Bac}, τ_{Pas} the optical thickness at
the Lyman, Balmer and Paschen limits and τ_{2343} is the optical thickness
at the centre of FeII UV3.

Line ratios published for about 30 well observed AGN are gathered on

Figure 1 and 2 together with the corresponding ratios obtained from our models.

Figure 1 shows an anticorrelation between W(Hβ) and FeII 4570/Hβ. Following Gaskell (1985) a constant coverage factor $\Omega/4\pi = 0.1$ is assumed. Models neither encompass the whole range of W(Hβ) and FeII 4570/Hβ nor explain the anticorrelation, in particular the coexistence of large FeII 4570/Hβ and small W(Hβ). The high density – high column density overabundant model yields FeII 4570/Hβ = 0.5, far below the ratio observed in I Zw1 (∿1.5), Mk 231 (∿1.4), 3C 232 (∿2.7) or PHL 1092 (∿6). Another interpretation of the correlation would be that in objects having a low coverage factor, there is a non photoionized region which could emit a lot of FeII and almost no Hβ (see for example Clavel et al. 1983).

Figure 2 shows the correlation between FeII opt/Hβ and FeII 4570/Hβ (FeII opt is the sum of all optical multiplets) a straightforward implication of the similar variations of all the optical FeII lines with physical parameters once the optical thickness is large (Joly 1981). The theoretical prediction for FeII opt($\lambda > 3000$ Å)/ FeII 4570 is 4 times larger than the observed ratio, showing that both the near UV lines (∿ 3000–4000 Å) and, generally weak lines escape detection.

3. CONCLUSION

Two component photoionization models cannot account for the observed intensity of FeII. The average temperature in the HI[*] zone is always lower than 8000 K and the ratio FeII 4570/Hβ never exceeds 0.5. One needs to consider another kind of clouds to account for high FeII intensities. Furthermore, the ratio FeIIopt/FeII 4570 is always underestimated by observations. FeIIopt/Hβ which is observed to be ∿ 5–10, should be in reality 20–50.

REFERENCES

Clavel, J., Joly, M., Collin-Souffrin, S., Bergeron, J., Penston, M.V., 1983, Mon. Not. R. Astr. Soc. 202, 85.
Collin-Souffrin, S., Dumont, S., 1985, submitted to Astron. Astrophys.
Collin-Souffrin, S., Joly, M., Dumont, S., Péquignot, D., 1985, submitted to Astron. Astrophys.
Gaskell, C.M., 1985, Astrophys. J. 291, 112.
Joly, M., 1981, Astron. Astrophys. 102, 321.
Kwan, J. Krolik, J.H., 1981, Astrophys. J. 250, 478.
Kwan, J., 1984, Astrophys. J. 283, 70.
Wills, B.J., Netzer, H., Wills, D., 1985, Astrophys. J. 288, 94.

HIGH DENSITIES IN QSO AND SEYFERT BROAD-LINE CLOUDS *

R. C. Puetter
Center for Astrophysics and Space Sciences
The University of California, San Diego
La Jolla, California 92093
The United States of America

ABSTRACT. While it is widely held that the densities of QSO and Seyfert broad-line regions lie below 10^{11} cm^{-3}, there is a growing body of evidence, indicating that at least some of the broad-line gas has densities equal to or exceeding 10^{11} cm^{-3}. Such evidence includes (1) the presence of a 3μm "bump" in roughly half of all QSOs, (2) the depression of Brγ in several Seyfert galaxies, (3) the broad-line variablity in some objects, and (4) the ability to produce the hydrogen spectrum theoretically.

1. INTRODUCTION

Recently, great strides have been made toward understanding the broad-line emission from QSOs and active galaxies. A fundamental aspect of all recent models is a high density, optically thick emission line region. The most often quoted argument against "extremely" high densities ($>10^{10}$ cm^{-3}) is the presence of broad semi-forbidden emission (e.g. C III] λ1909, N IV] λ1486, O IV λ1402, and O III] λ1663--c.f. Kwan 1984). However, as pointed out by Hubbard and Puetter (1985), the "normal" critical density arguments against the efficient production of these lines relative to permitted lines are inappropriate since the permitted lines also suffer significant collisional effects due to large optical depths. Furthermore, theoretical calculations of the strengths of the semi-forbidden lines have not been done for high density, low ionization parameter conditions, conditions under which acceptable hydrogen line emission may be produced. Thus ruling out high densities may be premature.

2. EVIDENCE OF HIGH DENSITIES

A number of theoretical works (Canfield, Puetter, and Ricchiazzi 1981, Collin-Souffrin, Dumont, and Tully 1982, and Hubbard and Puetter 1985) demonstrate that acceptable hydrogen line emission can result from high density gas, and Hubbard and Puetter (1985) have pointed out that

---

* Discussion on p.356

G. Swarup and V. K. Kapahi (eds.), Quasars, 341–342 .
© *1986 by the IAU.*

while the multi-element calculations (e.g. Kwan 1984) indicate that the observed semi-forbidden line strengths are inconsistent with high densities and "normal" ionization parameters, trends apparent in these same works indicate that acceptable emission might be obtained at high densities and lower ionization parameter. However, aside from these purely theoretical suggestions that the densities may be high, there are a number of observational results that suggest this as well.

Probabily the most direct indication of high densities comes from time variability studies of the broad-line emission. In Ark 120, for example, Peterson et. al (1985) demonstrate that the line variability constrains the broad-line region to be less than 30 light-days in size. This coupled with the large continuum brightness in this object, requires the densities to lie above 5×10^{16} cm^{-3} in order to have an ionization parameter which will not evaporate the broad-line clouds.

A second piece of observational evidence is the presence of the 3μm bump seen in QSO spectra. This bump could be naturally explained by free-free emission from the broad-line clouds if the densities lie at or above roughly 10^{11} cm^{-3} (Puetter and Hubbard 1985).

A final observational result is the observation by Cutri, Rieke, and Lebofsky (1984) that the Brγ line is depressed relative to Pα in a number of Seyfert galaxies. This observation could also be explained as due to free-free opacity effects. As described by Puetter and Hubbard (1985), preferential destruction of Brγ lines relative to Pα would occur upon conversion of the Brγ line to either 7→6 or 7→5 transitions followed by free-free continuum absorption.

3. CONCLUSIONS

There is mounting evidence of high densities in the broad-line regions of QSOs and active galaxies. The remaining pertinent question is whether all or only part of the emission line gas is at high density. If it becomes clear that most of the gas is at high density, then we may wish to question a number of fundamental aspects of our "cartoon" picture of the broad-line clouds. Indeed, in this case, the outer atmospheres of stars might make a quite viable candidate for the broad-line clouds.

REFERENCES

Canfield, R.C., Puetter, R.C., and Ricchiazzi, P.J. 1981, Ap.J., 249, 383.
Collin-Souffrin, S., Dumont, S., and Tully, J. 1982, Astr.Ap., 106, 362.
Hubbard, E.N., and Puetter, R.C. 1985, Ap.J., 290, 394.
Kwan, J. 1984, Ap.J., 283, 70.
Peterson, B.M., Meyers, K.A., Cappriotti, E.R., Foltz, C.B., Wilkes, B.J., and Miller, H.R. 1985, Ap.J., 292, 164.
Puetter, R.C., and Hubbard, E.N. 1985, Ap.J., 295, 394.

SPECTROPHOTOMETRY OF 3C 232 AND 3C 249.1

D. Dultzin-Hacyan
Instituto de Astronomía
Universidad Nacional Autónoma de México
Apartado 70-264
México, D. F. 04510, MEXICO

ABSTRACT. We find indication of variability of the ratio Mg II 2934/2798 for 3C 232 and discuss the excitation mechanism of Mg II 2934. For 3C 249.1 we identified the following new lines: [Ne V] 3346, [Ne V] 3426 and He I 3884. Comments are made on continuum shapes; in both cases optical continuum variability is confirmed.

1. OBSERVATIONS

The observations were carried out in March 1982 with the 2.1 m telescope of the OAN, San Pedro Mártir, B. C. (México), using a low dispersion spectrograph coupled with the Optical Multichannel Analyzer described by Firmani and Ruíz (1981).

2. EMISSION LINE SPECTRA

2.1 TON 469 (3C 232)

TABLE 1

Line	λ_{rest} (Å)	I_{obs} (x 10^{-14}) (erg cm^{-2}s^{-1})	W_{obs} (Å)
MG II	2798	23 ± 5.0	148 ± 26
MG II	2932	13 ± 3.0	119 ± 15
[O II]	3727	5 ± 0.1	70 ± 1

Grandi and Phillips (1978) have suggested two possible fluorescence mechanisms for the excitation of Mg II λλ2929, 2936: Lyβ an N V Fluorescence. Several observational predictions have not been confirmed for the case of Lyβ fluorescence: the presence of Mg II λ1752 and λ1737

343

G. Swarup and V. K. Kapahi (eds.), Quasars, 343–344 .

with a strength comparable to λ2798 and also the presence of OI λ1302.
(See Dultzin-Hacyan, Salas and Daltabuit, 1982).

Our observed ratio Mg II 2934/2798 (see Table 1) is 0.56±.20, a factor
of two higher than that observed by Grandi and Phillips (1978). If this
ratio is varying with time, simultaneous UV and optical observations
are needed to establish whether N V fluorescence is responsible for the
excitation of Mg II λ2934 (Dultzin-Hacyan, 1985).

2.2 3C 249.1

TABLE 2

Line	λ_{rest} (Å)	I_{obs} (x 10^{-14}) erg cm^{-2} s^{-1}	W_{obs} (Å)
[Ne V]	3346	0.9 ± 0.20	5 ± 1.0
[Ne V]	3426	1.1 ± 0.20	6 ± 1.0
[O II]	3727	2.7 ± 0.90	16 ± 5.0
[Ne III]	3869	2.2 ± 0.30	13 ± 2.0
He I + H$_8$	3889	1.8 ± 0.40	12 ± 2.0
[Ne III] + H$_7$	3967	1.2 ± 0.08	8 ± 0.5

3. CONTINUUM

With these two cases we have an example of locally different continuum
energy distribution shapes, in the sense that 3C 232 shows the typical
optical flat spectrum and the UV is much steeper, whereas for 3C 249.1
the UV (corrected for galactic reddening) is an extrapolation of the
optical continuum (Dultzin-Hacyan, 1985). For both quasars, previous
data on variability are confirmed.

REFERENCES
Dultzin-Hacyan, D., Salas, L. and Daltabuit, E., 1982, Astr. and Ap.,
    111, 43.
Dultzin-Hacyan, D., 1985, Rev. Mex. Astron. y Astrof., 11, in press.
Firmani, C. and Ruíz, E., 1981, Recent Advances in Observational Astro-
    nomy, eds.: H. L. Johnson and C. Allen (México: UNAM), p. 25.
Grandi, S. A. and Phillips, M. N., 1978, Ap. J., 220, 426.

ON THE [OIII] EMISSION OF PG 1501+106

G. M. Van Heerde
Sterrewacht Leiden
Postbus 9513
2300 RA LEIDEN

ABSTRACT. Narrow-band imaging of PG 1501+106 revealed no extra-nuclear [OIII] emission above a level of ~ 20 μJy. The disagreement with the result of Boroson et. al., 1982 (who detected 40 μJy) can be explained by assuming a slight difference in angular size between the broad and narrow line emitting regions.

1. INTRODUCTION AND OBSERVATIONS

To investigate the nature of emission line regions in the nebulosity surrounding low-redshift QSOs we started a narrow-band imaging survey to determine the morphologies (if present) of emission line regions surrounding a sample of low-redshift QSOs. We included in our survey a source (PG 1501+106) of which a published spectrum indicated the presence of [OIII] emission at a distance of a few arcsec from the nucleus (Boroson et. al., 1982).

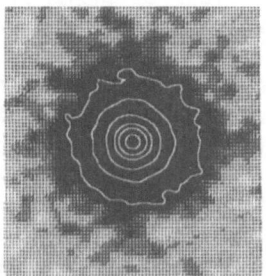

Fig.1 [OIII] image

10"

The observations of PG 1501+106 (= Mkn 841; z = 0.036) were obtained 17 April 1985 at ESO, La Silla using the ESA/PCD at the 2.2m telescope. The PCD (Photon Counting Detector) has been described in detail by di Serego Alighieri et. al., 1985. We obtained two 1800 sec narrow-band (~ 40 A FWHM) exposures : one centred on [OIII]λ5007 and one on the nearby emission-line free continuum (5080 A in the rest frame of the QSO).

2. DISCUSSION OF THE RESULTS

After careful aligning both reduced images (see Fig.1 and 2) the continuum image is subtracted from the line image. The resulting image (Fig.3) does not show (as we had hoped) the type of very extended

Fig.2 Continuum image

G. Swarup and V. K. Kapahi (eds.), Quasars, 345–346 .
© *1986 by the IAU.*

[OIII] emission that has been found surrounding a
number of QSOs (Stockton and MacKenty, 1984; di
Serego Alighieri et. al., 1984). On the contrary the
deviation from a point source is very small. To
derive an upper limit to the extra-nuclear [OIII]
emission we scaled and subtracted a point source
from the pure [OIII] image (Fig.4). The mismatch
between [OIII] and star image is probably largely
due to a residual of the geometrical distortion
(unavoidable in image tubes) causing a slight
deviation from the axial symmetry. Note, however,
that the wings have disappeared leaving just a noise
background outside the central area.

Fig.3 Pure [OIII]
image
(continuum
subtracted)

Boroson et. al.,1982 reported the detection of
extra-nuclear [OIII] emission of PG 1501+106. The
flux detected within the 8.1 arcsec^2 slit (indicated
in Fig.4) was 2.4 x 10^{-15}ergs cm^{-2}s^{-1} which would
correspond to 40 μJy (= 40 x 10^{-29}ergs cm^{-2}s^{-1}Hz^{-1})
flux within our filter bandwidth. In the area
covered by the slit we measure a flux of -11 ±
15 μJy. In spite of the uncertainty in subtracting a
point source we feel confident to give an upper
limit of ~ 20 μJy . A possible explanation of the
disagreement in the two results is the following.

The nebulosity spectrum was obtained by
subtraction of a carefully defined fraction of the
nuclear spectrum from the raw nebulosity spectrum
which was severely contaminated by scattered nuclear
light. This fraction was determined to completely
remove the broad component of Hβ which is known to

Fig.4 Residual
[OIII] image
(point source
subtracted)

be of nuclear origin (the Broad Line Region or BLR is typically less
than a parsec in size). But also is known that the Narrow Line Region
(NLR) is typically a few hundred parsecs across which at the distance of
PG 1501+106 corresponds to a few tenth to one arcsec.

The fraction of the emission from such a NLR that is scattered into
the off-nuclear slit will be larger than that from the BLR. So we
propose that the flux detected by Boroson et. al. is scattered light
from the NLR which has an angular size of a few tenths of an arcsec.
From Fig.4 it is obvious that this interpretation is consistent with our
result but that we cannot make a statement about the angular extent of
the extra-nuclear emission.

3.   REFERENCES

Boroson, T.A.; Oke, J.B. and Green, R.F., 1982, Ap.J., 263, 32
di Serego Alighieri, S.; Perryman, M.A.C. and Macchetto, F., 1984,
    Ap.J., 285, 567
di Serego Alighieri, S.; Perryman, M.A.C. and Macchetto, F., 1985,
    Astron. Astroph., 149, 179
Stockton, A. and MacKenty, J.W., 1983, Nature, 305, 678

FUV AND OPTICAL SPECTROPHOTOMETRY OF X-RAY SELECTED SEYFERTS

J. T. Clarke, S. Bowyer, and M. Grewing
Code TA02 Space Telescope Project
NASA Marshall Space Flight Center
Huntsville, AL   35812 USA

ABSTRACT. Nearly simultaneous FUV and optical spectrophotometry of
X-ray selected Seyfert galaxies has revealed an average Ly α/H β ratio
of 22, a positive correlation between the ratio Ly α/H β and the width
of the lines, and additional Ly α emission in the wings of one source
which is not matched by emission in the Balmer line wings. However, we
find no distinguishing features in the continuum emission from these
X-ray selected objects compared with other samples. If the correlation
between Ly α/H β and the width of the lines is found to apply to larger
samples of Seyferts, it may be that our objects appear Ly α bright
because they are also broad-lined compared with other samples.

1. OBSERVATIONS

Observations of five X-ray selected Seyferts from the sample of
Reichert et al. (1982) were performed in May 1983 with the Lick 120"
telescope and a CCD spectrograph and roughly two weeks later with the
IUE. The objects observed were E0849+08, E1426+01 (also known as
Mrk 1383), E1530-08 (a Seyfert II), E1556+27, and E1613+65 (a.k.a Mrk
876). We have compared the continuum and line emission properties of
these objects with the larger sample presented by Wu, Boggess, and
Gull (1983) and found significant discrepancies only in the areas to be
discussed.

2. RESULTS

Indications of variability in these sources, even over the two week
interval between our optical and FUV observations, stresses the
importance of nearly simultaneous observations in the study of
correlations between emissions in different wavelength bands. The X-ray
to optical continuum spectra and most emission line properties of the
Seyfert I's appear very similar to those of previous samples of
Seyfert I's, but the observed ratio Ly α/H β in our sample averages 22,
which is significantly higher than that observed in previous samples
(although individual sources with high ratios have been observed). In
our limited sample there also appears a positive correlation between

347

G. Swarup and V. K. Kapahi (eds.), Quasars, 347–348.

Ly α/H β and the width of the lines, coupled with the knowledge that existing larger samples of sources have, on the average, lower Ly α/H β values by a factor of 4 and narrower H lines by roughly a factor of 2. Comparisons between the Ly α and H α line profiles in these sources further indicate that while part of this effect may be due to an increasing Ly α/H β ratio in the wings of the lines, there must also be a generally higher ratio Ly α/H β throughout the line profiles to explain the difference between our average value of Ly α/H β = 22 and averages of roughly 5 from previous studies. Finally, from our data and the larger sample of Wu, Boggess, and Gull (1983) it is known that while the individual line strengths correlate well with continuum luminosity, the ratio Ly α/H β does not appear to correlate with either continuum luminosity or slope.

3. DISCUSSION

Since the explanation for the previously observed anomalously low ratio Ly α/H β = 5 from Seyfert I's has been the "trapping" of Ly α photons by a high optical depth of H, possibly coupled with some additional extinction by dust or variation in electron density, it may be that the broad-lined objects that we have observed are least affected by this phenomenon and therefore have ratios Ly α/H β much closer to those predicted by the early photoionization calculations. Further observations to check the strength and slope in this correlation will have to use nearly simultaneous observations, however, since variations in the strengths and shapes of the lines (as observed from NGC 4151) would tend to wash out this correlation.

4. REFERENCES

Reichert, G., Mason, K., Thorstensen, J. and Bowyer, S. 1982, Ap. J., 260, 437.

Wu, C., Boggess, A. and Gull, T. 1983, Ap. J., 266, 28.

VARIABILITY OF EMISSION LINES IN QUASAR SPECTRA[*]

P M Gondhalekar[1], P O'Brien[2] and R Wilson[2]

[1]Rutherford Appleton Laboratory, Chilton, Didcot,
 Oxon, OX11 0QX, United Kingdom
[2]Department of Physics & Astronomy, University College London,
 Gower Street, London, WC1 6BT, United Kingdom

ABSTRACT. Analysis of ultraviolet spectra of quasars obtained with IUE
indicate that the permitted emission lines in quasar spectra are
variable. The upper limit for the delay between a change in the
continuum and the consequential change in the emission lines is about 9
months. This rapid response of the permitted emission lines to changes
in the continuum leads to a number of problems in understanding the
nature of quasars.

1. INTRODUCTION

Analysis of the integrated intensities of permitted emission lines in
quasar spectra indicate that in the broad line region (BLR) of quasars
the electron density is of the order of 10^9 cm^{-3}, the ionisation
parameter (ratio of photon to gas density) is of the order of 0.02 and
the covering factor is of the order of 0.1. These parameters along with
the luminosity of ionising radiation determine the distance of the BLR
from the source of the continuum radiation. In low luminosity (10^{41} -
10^{42} ergs s^{-1}) sources like Seyfert 1 galaxies, the BLR should be a few
tenths of a parsec away from the source of ionising radiation and the
variation of permitted lines over a few weeks have been detected (Ulrich
et al 1984; Peterson et al 1985). In high luminosity (10^{46} - 10^{47}
ergs s^{-1}) sources like quasars, the BLR should be light decades in size
and the emission lines should not be rapidly variable. Evidence for
rapid line variability in a small sample of quasars is presented here.

2. DISCUSSION

The details of observations and data reduction have been given by
Gondhalekar et al (1985). The fractional change in the emission line
intensities and the period over which these changes were observed have
been given in Table 1. The intensities of strong lines like Ly α and
C IV have been measured with an accuracy of better than 25% and the

---

[*] Discussion on p.356

G. Swarup and V. K. Kapahi (eds.), Quasars, 349–350 .
© *1986 by the IAU.*

intensities of weaker lines like Ly β + 0 VI and Si IV + 0 IV] have been
measured with comparatively lower accuracy.

TABLE 1   Changes in Emission Line Intensities in Quasar Spectra

ID	Interval Months	Changes %			
		Ly β + 0 IV	Ly α	Si IV + 0 IV]	C IV
Q0414 - 06	10	+ 59	- 38		- 28
Q1317 + 28	18		0	- 80	
Q1512 + 37	8		- 42		
Q2201 + 31	9	- 45	0	- 70	
Q2344 + 09	11		+ 63		+ 41

The line intensities can change significantly over periods as short as 9
months. Gondhalekar et al (1985) have shown that changes in line
intensities are accompanied by changes in line profiles. The present
data are limited by the frequency of observations and it has not been
possible to accurately establish the delay between a change in the
continuum and the corresponding change in the line flux. At present it
is only possible to say that the upper limit for this delay is 9 months.
This would suggest that the BLR was less than 0.25 pc from the source of
ionising radiation unless a contrived geometry was adopted.

The rapid variability of emission lines or the small extent of BLR
poses severe problems in understanding the physics of BLR. For example,
in a BLR 0.25 pc from the source of ionising radiation, the ionisation
parameter would be ~ 1 for an electron density of 2×10^9 cm^{-3} and the
luminosity of ionising radiation given by Gondhalekar et al (1985).
This ionisation parameter is two orders of magnitude higher than the
ionisation parameter generally assumed for photoionisation models of BLR
(Kwan and Korlik, 1981). For such a high ionisation parameter high
ionisation lines should be observed in the ultraviolet and these have
not been detected. It is possible that the permitted lines are formed
in the region of BLR where the electron density is high (and the
ionisation parameter is low) and lines like C III] λ1909Å are formed in
a low density region further from the source of ionising radiation. It
has not been possible to study in detail this stratification of BLR.

REFERENCES

Gondhalekar P M, O'Brien P & Wilson R, 1985. Mon Not R astr Soc
   (submitted).

Kwan J & Korlik J H, 1981. Astrophys J 250, 478.

Peterson B M, Mayers K A, Capriotti E R, Foltz C B, Wilkes B &
   Miller H R, 1985. Astrophys J 292, 164.

Ulrich M H, Boksenberg A, Bromage G E, Clavel J, Elvius A, Penston M V,
   Perola G C, Pettini M, Snijders M A J, Tanzi E G & Tarenghi M, 1984.
   Mon Not R astr Soc 206, 221.

A DETAILED STUDY OF THE CIV 1550 LINE PROFILE AND ADJACENT SPECTRAL FEATURES IN NGC 4151 FROM 1978 UP TO 1983

J.Clavel[1], A.Altamore[2], A.Boksenberg[3], G.E.Bromage[4], A.Elvius[5], D.Pelat[6], C.Perola[2], M.V.Penston[3], M.A.J.Snijders[3], M.H.Ulrich[7]

[1] ESA IUE Observatory, Apartado 54065, Madrid, Spain.
[2] Instituto Astronomico dell 'Universita, Roma, Italy.
[3] Royal Greenwich Observatory, East Sussex, UK.
[4] Rutherford & Appleton Laboratory, Oxfordshire, UK.
[5] Stockolm Observatory, 1330 Saltsjobaden, Sweden.
[6] Observatoire de Meudon, 92195 Meudon, France.
[7] ESO, Garching bei Munchen, FRG.

ABSTRACT. Fitting technics are applied to study the variations of the 1550 line profile and adjacent features in the IUE spectrum of the Seyfert galaxy NGC 4151, from 1978 up to 1983. New important results are found concerning the kinematics and physics of the emission line gas at a few light-days from the continuum source.

1. THE ANALYSIS

Fitting technics are used to study the variations of the CIV 1550 line profile and adjacent features in the spectrum of the Seyfert galaxy NGC 4151 from 1978 up to 1983. The data-base consists of low resolution (900 km/s) high S/N IUE spectra taken at 69 different epochs. The profile of the various features is well represented by a gaussian and for most of them the full-width-at-half-maximum (FWHM) can be held fixed; NIV 1485 (FWHM = 1600 km/s), L1 1516 (1600 km/s), CIV 1550 in emission (FWHM variable from 2400 to 5500 km/s) and in absorption (1300 km/s), L'2 1576 (1600 km/s), L2 1597 (1600 to 3000 km/s) HeII 1640 (1600 km/s) and OIII] 1663 (2500 km/s). No good fit can be achieved without allowing for the presence of two additional <u>ultra-broad</u> (UB) emission components : CIV 1550 (FWHM = 14600 km/s) & <u>HeII 1640</u> (8600 km/s). The continuum is approximated by a straight line and its intensity at 1450 and 1720 A are free parameters of the model.

2. EVIDENCE FOR A DECELERATED WIND

The most significant result is that emission from the highest velocity material shows extremely rapid variations which mimic closely those of the continuum. Indeed, the variations of intensity of the UB CIV 1550 component lag by only ~5 days behind those of the continuum. This result supports the hypothesis that the high velocity gas is photoionized and

351

G. Swarup and V. K. Kapahi (eds.), Quasars, 351–352.

lies at 5 light-days (1d) from the source of continuum. The absence of
such UB emission in the profile of the other UV & optical lines - in
particular MgII 2800 - implies that the high velocity clouds are ionized
throughout. The gas density must also exceed $7 \ 10^{10} \ cm^{-3}$, as can be
inferred from the absence of NIV] 1485 and OIII] 1663 UB emission. An
important correlation is also found between the wavelength and the intensity of the UB CIV 1550 component in the sense that this feature has a
more pronounced blue-shift when it is weak than when it is strong. This
strongly suggests the existence of outflowing motions at 5 1d's from the
nucleus. However, the above correlation also requires that the innermost
receeding gas is partially hidden to the observer. We note that a tilted
accretion disk naturally provides both the necessary obscuration and the
tank from where the wind can be continuously replenished. The velocity
of the wind decreases outward from ~ 4000 km/s at 5 light-days from the
nucleus to ~ 800 km/s at ~ 120 light-days where the absorption line gas
is located (Bromage et al 1985).

3. A JET-WIND INTERACTION

The intensity of the unidentified L1 1516, L'2 1576 & L2 1597 lines varies on a time scale of a few days. However, their variations are not
 correlated with those of the continuum. This constrasts strongly with
the behaviour of the other lines and suggests an emission mechanism
distinct from photoionization. Ulrich et al (1985) have interpreted
these features as Doppler shifted CIV 1550 lines resulting from the
interaction of a two-sided jet with the broad line gas. The present analysis shows that the variations of L2 correlate well with those of L1 on
short time scales. Moreover, any lag, if present, must be shorter than
4 days. This constrains the jet to be almost in the plane of the sky,
and therefore strengthens the conclusion reached from independent arguments by Ulrich et al that the jet is inclined at 75° with respect to
the line-of-sight. This also requires that the L1 & L2 lines originate
within 5 1d's from the origin of the jet. It is therefore likely that
they arise from an interaction of the jet with the wind.

REFERENCES

Bromage, G.E. et al, 1985, Mon. Not. R. Astr. Soc 215, 1.
Ulrich, M.H. et al, 1985, Nature 313, 745.

RECURRENT OUTBURSTS IN THE BROAD LINE REGION OF NGC 1566

D.M. Alloin
Observatoire de Paris-Meudon
Place J. Janssen
92195 Meudon-Principal Cedex
France

ABSTRACT. From a set of spectrophotometric data covering 15 years, we observe a recurrent activity in the BLR of NGC 1566. Each period of activity lasts for 1300 days. It consists of successive bursts characterized by a rise-time less than 20 days and an exponential decay on a time scale of around 400 days. The total energy involved in one 1300 day period of activity is 10^{51} ergs. These variations can be understood in the framework of accretion disc models for AGN, as resulting from a limit cycle which operates in the inner regions of the disc.

The aim of the present study is to investigate further the frequency of the rapid and intense outbursts occuring in the BLR of the galaxy NGC 1566 (Alloin et al., 1985a).

We have collected spectrophotometric data on this active galactic nucleus covering the period 1970-1985 (Alloin et al., 1985b). We discuss with care the uncertainties introduced on the emission line measurements by the use of different instrumentations and slit sizes in particular. The intensities of both the $H\alpha$ and $H\beta$ lines arising from the broad line region are found to vary substantially with time. Interestingly we do not detect any conspicuous Balmer profile changes. This can be understood readily in the frame of ionization bounded clouds distributed in a stable geometric configuration throughout this region. Owing to the small size of the broad line region, less than 20 light-days, the temporal behaviour of the Balmer lines reflects without significant smearing genuine variations of the ionizing source itself. Over this 15 year time-base we observe four successive periods of activity having approximately the same duration of 1300 days (Figure 1). The time spent by the nucleus in a low state is between a few months and a year. In contrast, the highest states do not last for more than a few weeks. The most recent period of activity (1981 January to 1985 August) shows a series of rapid bursts. A burst has a typical rise time of 20 days but a longer exponential decay of around 400 days. The continuum both increases and steepens together with the Balmer line enhancement (Figure 2). Assuming the ionizing flux increase

G. Swarup and V. K. Kapahi (eds.), Quasars, 353–355 .

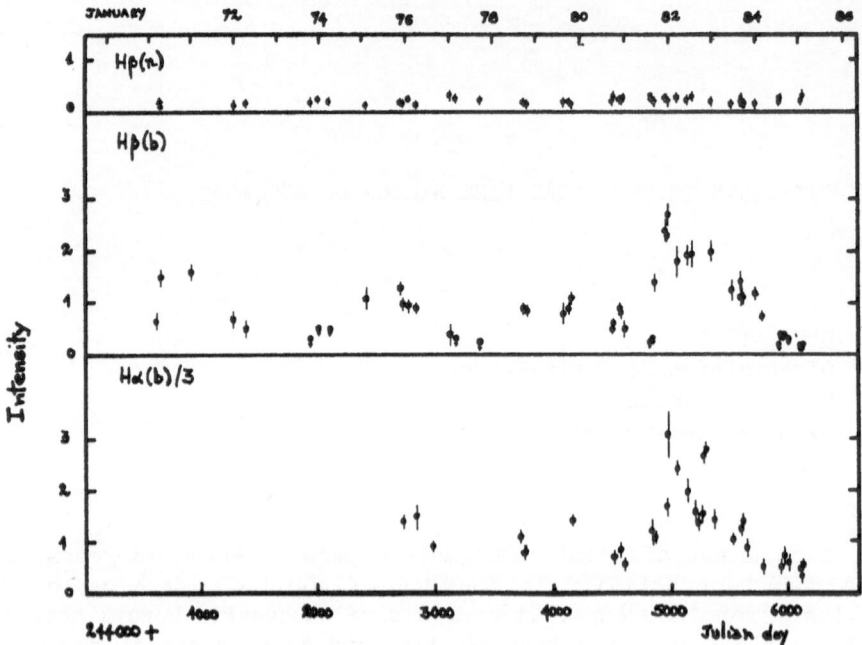

Figure 1. The temporal behaviour of the line emission from the BLR.
The intensities of Hβ(b) and Hα(b)/3 are displayed in the medium and
low parts of the diagram, respectively. For comparison, we present the
intensity of the narrow line, Hβ(n) in the upper part. The intensities
are in unit 10^{-13} ergs cm^{-2} s^{-1}.

is linked to the non-stellar continuum, the amount of energy released
during one period of activity is of 10^{51} ergs. The energies related
to the three successive bursts occuring in the most recent active
period appear to follow an exponential decrease.

Theoretical arguments are presented (Abramowicz et al., 1985) in favour
of the hypothesis that this variability can be explained by a limit
cycle which works due to thermal and viscous instabilities present in
the innermost regions of an accretion disc orbiting a black hole in
the active nucleus of NGC 1566.

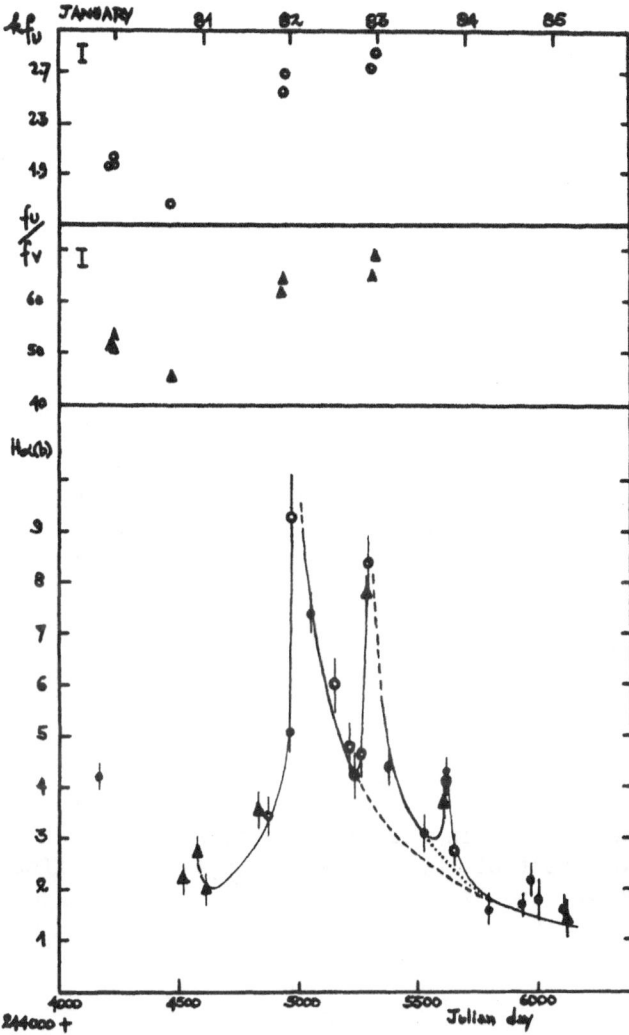

Figure 2 . Enlargement of the most recent period of activity in the Hα(b) line emission. Black dots correspond to low resolution data : Empty circles to high resolution data : Empty triangles to a weighted mean of low and high resolution data for the same epoch. The unit is 10^{-13} ergs cm^{-2} s^{-1}. A sketch of the burst activity has been drawn on top of the data points. In the upper part, we have plotted the ultraviolet flux as well as the ultraviolet to visible light ratio from photometric data by de Ruiter and Lub (1985) in an 11.6" entrance diaphragm. The two main bursts in Hα are correlated with both an enhancement and a steepening of the continuum in the ultraviolet by around 40%.

References

Alloin, D., Pelat, D., Phillips, M., Whittle, M., 1985a Astrophys. J. **288**, 205.

Alloin, D., Pelat, D., Phillips, M., Fosbury, R., Freeman, K., 1985b Astrophys. J. submitted.

Abramowicz, M., Alloin, D., Lasota, J.P., Pelat, D., 1985, Nature submitted.

De Ruiter, H., Lub, J., 1985, Astron. Astrophys, in press.

DISCUSSION ON THE PAPER BY **PUETTER** (p.341)

Malkan : Edelson and I looked at the data on the "near-IR bump", which you attribute to free-free emission. We found that it always peaks around 5 microns, with a rather narrow spread (± 20%). This would seem to require a surprisingly narrow range of frequencies at which BLR clouds go optically thick to free-free absorption. What do you think could explain this seemingly narrow range ?

Puetter : I have several ideas. However at this time they are not fully thought out. On the other hand, this problem is not unlike that of requiring a relatively tight range of ionization parameter or emission line cloud density. Hence, this problem is not unique to my models and if one solves this problem for ionization parameters, one solves the problem of the constant wavelength of the free-free peak simultaneously.

DISCUSSION ON THE PAPER BY **GONDHALEKAR** et al. (p.349)

Wilkes : I am concerned about NV contamination of the Lyα profile in IUE spectra. Can you explain how you allow for this.

Gondhalekar : NV is not very strong in the majority of spectra we have analysed. When the broad component of Lyα weakens the NV line affects the intensity of Lyα line at less than 10% level.

G. Swarup and V. K. Kapahi (eds.), Quasars, 356.
© *1986 by the IAU.*

IV

PRIME MOVERS, MODELS AND MECHANISMS

*"But a clean cut refutation, leading to abandonment
of some theory, happens only rarely in astrophysics; many
models have persisted unrefuted for a long time.A cynic
might argue that they have survived only because they
don't go beyond generalities, or else because their proponents have been adept, like the motor mechanics who
must abound here in India, at replacing or modifying
faulty parts to keep shaky old models 'roadworthy'."*

- Martin Rees (p.7)

Martin Rees cautioning Roger Blandford about black holes ?

"The generic black hole model is not infinitely flexible, and not invulnerable."

— Martin Rees (p.6)

BLACK HOLE MODELS OF QUASARS

R. D. Blandford
Theoretical Astrophysics
130-33 Caltech
Pasadena
CA 91125, U.S.A.

ABSTRACT. Observations of active galactic nuclei are interpreted in terms of a theoretical model involving accretion onto a massive black hole. Optical quasars and Seyfert galaxies are associated with holes accreting near the Eddington rate and radio galaxies with sub-critical accretion. It is argued that magnetic fields are largely responsible for extracting energy and angular momentum from black holes and disks. Recent studies of electron-positron pair plasmas and their possible role in establishing the emergent X-ray spectrum are reviewed. The main evolutionary properties of active galactic nuclei can be interpreted in terms of a simple model in which black holes accrete gas at a rate dictated by the rate of gas supply which decreases with cosmic time. It may be worth searching for eclipsing binary black holes in lower power Seyferts.

1. INTRODUCTION

As this topic has been extensively discussed in several recent conference proceedings and review articles, (e.g. Phinney 1983, Blandford 1984, Begelman, Blandford, and Rees 1984, Begelman in Miller 1985, and especially Rees 1984), I shall emphasize recent developments. More complete references can be found in the cited articles.

The original idea that quasars contain black holes (Zel'dovich and Novikov 1964, Salpeter 1964, Lynden-Bell 1969) has developed into a fairly strong hypothesis, namely that essentially all galaxies contain massive black holes which power most of the non-stellar activity that we observe in their nuclei.

This hypothesis is motivated by observational and theoretical considerations which include:

(i) Rapid variability in several Seyferts and BL Lac objects on timescales that have been reported to be as short as 30s.(Feigelson et al. 1985, preprint).

(ii) Observations of linear radio features in all types of active nucleus. This suggests the presence of a stable gyroscope like a spinning black hole.

G. Swarup and V. K. Kapahi (eds.), Quasars, 359–369.

(*iii*) Dynamical estimates of central masses. In the case of M87, this is $3 \times 10^9 M_\odot$. Masses of $\sim 10^6 M_\odot$ have been inferred in the nuclei of M31, M32 (Tonry 1984, Dressler 1984) and a similar mass may be present in our galactic centre. (Crawford *et al.* 1985).

(*iv*) Low level radio and emission line activity in most galactic nuclei (e.g. Keel in Miller 1985). Broad wings to these emission lines are commonplace and have been used to infer the presence of central black holes (Filippenko and Sargent 1985).

(*v*) Superluminal expansion, which suggests the presence of a relativistic gravitational potential well.

(*vi*) Inferred efficiencies of conversion of rest mass energy into radiant energy up to $\sim 10\%$ which suggests that gravitational energy rather than nuclear energy is involved.

(*vii*) Evolutionary arguments which suggest that alternative prime movers like dense star clusters or spinars will soon form black holes.

It hardly needs emphasizing that although this is all good circumstantial evidence, we have no *proof* that black holes exist in active galactic nuclei or indeed anywhere else.

2. PROPERTIES OF BLACK HOLES

For the sake of completeness, I will list some key features of massive black holes. For a hole of mass $10^8 M_8 M_\odot$. the gravitational radius is $r_g \sim 1.5 \times 10^{13} M_8$ cm and the associated time is $500 M_8$ s. A fiducial power is the Eddington luminosity $L_E \sim 10^{46} M_8$ erg s^{-1} at which the repulsion of the radiation balances the gravitational attraction. If the efficiency of conversion of rest mass into energy is $0.1 \varepsilon_{-1}$ then the associated mass accretion rate is $\dot{M}_E \sim 2 \varepsilon_{-1}(L/L_E)M_8 M_\odot$ yr^{-1}. If the escaping radiation can be thermalized (i.e. $\tau_{es} \gtrsim 1000$), then the equivalent black body temperature for an effective emitting area A is

$$T_{BB} = 3 \times 10^5 M_8^{-1/4} \left[\frac{L}{L_E} \right]^{1/4} \left[\frac{A}{100 r_g^2} \right]^{-1/2} K \qquad (1)$$

In a standard thin disk model, $A \gtrsim 100 r_g^2$ and the flux peaks at a frequency $\sim 3 k T_{BB}/h$ which lies in the UV.

In a spinning black holes, described by the Kerr metric, up to 10% (i.e. $\sim 2 \times 10^{61} M_8$erg) of the mass can be extracted by electromagnetic fields supported by exterior currents in the form of a relativistic electromagnetic wind. (Thorne, 1985).

3. INTERPRETATION

These basic considerations suggest a qualitative interpretation of the different classes of AGN. Let us suppose that there are three separate channels through which an accreting black hole can radiate; a thermal channel in the UV, a non-thermal channel that produces power law spectra from the far IR to gamma ray energies and a relativistic hydromagnetic channel that is responsible for the jets and the distant double radio sources. We then hypothesise that the branching ratios for these three channels are mainly determined by the ratio of the mass supply rate to the critical accretion rate. When this ratio is large, we expect an accretion disk to form that is both geometrically and optically thick. A large fraction of the power will be released through the thermal channel at about the Eddington limit. However the gas flow near the black hole is unlikely to be steady and a substantial power will probably also be released through the non-thermal channel via shocks, flares electromagnetic cascades etc. We identify this case with the radio-quiet quasars (when $M_8 \gtrsim 0.1$) and the Seyfert galaxies (when $M_8 \lesssim 0.1$). By contrast, when the accretion rate is low relative to the mass of the hole, the accreting gas is optically thin and the power is dominated by the hydromagnetic extraction of energy from the spinning hole. This gives rise to a radio galaxy. Quasars associated with extended radio sources occur when the accretion rate changes from being low to critical.

Our classification of an AGN is also influenced by our orientation. Many of the properties of compact radio sources can be accounted for if they are relativistically beamed in our direction (c.f. Saikia, this volume). Similarly, the observed differences between Seyfert 1 and Seyfert 2 galaxies, principally the absence of broadlines and powerful x-ray emission from the latter, can interpreted as the result of absorption by a thick, equatorial disk (e.g. Osterbrock in Miller 1985).

Lower power LINERS, emission line galaxies etc can be associated with low mass black holes, or in some instances intermediate mass holes that are accreting sub-critically but have lost their spin energy in an earlier incarnation as a radio source. The marked preference of extended radio sources for elliptical galaxies and of Seyfert nuclei for spirals can be understood as a reflection upon the likely history of the gas supply.

4. STABILITY OF RADIATION TORI

Equilibrium models of thick accretion disks or radiation tori have been invoked to explain radio-quiet quasars and type 1 Seyferts. In most instances, these models adopt a polytropic equation of state and specify arbitrary angular momentum distributions (commonly that the angular velocity Ω varies with cylindrical radius r, or its relativistic counterpart, according to $\Omega \propto r^{-q}$). They do not treat the internal circulation and heat transport self-consistently and, perhaps partly for this reason, they are radiatively inefficient when the mass accretion rate is high, $\dot{M} \gg \dot{M}_E$.

In an important development, Papaloizou and Pringle (1984, 1985) have shown that orbiting tori are susceptible to global, non-axisymmetric, dynamical instabilities. The physical character of these instabilities has recently been elucidated in the case of slender tori (Goldreich, Goodman, and Narayan 1985, preprint, Blaes and Glatzel, 1985, preprint). It now appears that the unstable modes can be interpreted as edge waves that are coupled around an unstable region around corotation. For, modes in which the azimuthal wavelength is long compared with the minor radius of the torus, rapid growth in a few orbital periods is found as long as $\sqrt{3} < q < 2$. When $1.5 < q < \sqrt{3}$, there are modes with azimuthal wavelength shorter than the minor radius which grow more slowly and may provide an effective viscosity for the inflowing gas. 3D numerical simulations will be necessary to understand the non-linear evolution of these instabilities. (The existence of radio jets assures us that dynamically unstable flows need not be catastrophically disrupted.)

5. MAGNETIC EXTRACTION OF ENERGY FROM DISKS AND BLACK HOLES

Several arguments suggest that magnetic fields mediate nuclear activity.

(*i*)   Radio jets frequently appear to have minimum internal equipartition pressures that exceed the maximum permissible external thermal pressure allowed by X-ray observations. This suggests that the pinching action of toroidal magnetic fields amplifies the gas pressure and confines the jets. The current that supports this toroidal field presumably flows along the jet and is believed to be generated by a rotation-induced e.m.f. near where the jet is collimated.

(*ii*) Recent observations of the galactic center (e.g., Morris and Yusef-Zadeh 1986, Sofue and Handa, 1984) exhibit straight "filaments" and "threads", (presumably magnetic flux tubes) extending perpendicular to the galactic plane and appearing to twist. (The sense of twist of the field lines should reflect that of the differential rotation in the underlying disk and should oppose the general galactic rotation.)

(*iii*) Magnetic fields allow the angular momentum that must be liberated by the inwardly spiralling gas to be carried off (e.g., Pudritz and Norman 1983). If the distribution of mass within a galactic nucleus is generally axisymmetric, then there is no efficient way to extract the angular momentum using gravitational forces. This is in contrast to an accretion disk in a binary system. (A non-axisymmetric bar may however be effective.)

(*iv*) Variability data suggests that the non-thermal continuum originates from a highly compact region of high radiation energy density. Relativistic electrons lose energy by Compton scattering on timescales short compared with the crossing time. They must therefore be re-accelerated locally in a highly efficient manner. Electrostatic acceleration associated with rapidly changing magnetic fields, such as might be found in the corona of a differentially rotating disk, is probably the most efficient way of doing this (e.g., Burbidge and Burbidge 1967). The equipartition field strength near the inner edge of an accretion disk is envisaged to be $\sim 10^4$G. The associated electric fields induced by the differential rotation are $\sim 1 \mathrm{MVcm}^{-1}$ and strong enough to accelerate ~ 100GeV electrons.

Arguments like this, developed at slightly greater length in Coroniti(1985) and Blandford in Mihalas and Winkler (1986), suggest that magnetic fields are an important feature of accretion disks and that most astrophysical jets are launched and collimated hydromagnetically. Serious axisymmetric numerical compilations are only just starting to be made (e.g., Uchida and Shibata 1984), but we can anticipate steady progress in this area over the next few years.

6. ELECTRON-POSITRON PLASMAS

There have been several recent studies of non-thermal pair plasmas in galactic nuclei (cf. Novikov this volume and an excellent review by Svennson in Mihalas and Winkler 1986). These studies are partly motivated by observations of radio sources. However, the most compelling argument for the existence of pair plasmas within AGN's is that the X-ray spectra in type I Seyferts and radio loud quasars (Elvis, these proceedings) have spectral indices in the range $0.5 \lesssim \alpha \lesssim 1$. If we extrapolate these spectra into the gamma-ray region, ($\varepsilon_\gamma \gtrsim 0.5$ MeV) and use the observed X-ray variability time scale t_{var} as an indication of the source size $R \sim ct_{var}$, then the inferred γ-ray compactness parameter

$$l_\gamma = \frac{L_\gamma(\varepsilon_\gamma \geq m_e c^2)\sigma_T}{R m_e c^3} \tag{2}$$

$$\simeq 4\pi \left[\frac{L_\gamma}{L_E}\right]\left[\frac{m_p}{m_e}\right]\left[\frac{r_g}{R}\right] \tag{3}$$

typically exceeds unity (Barr and Mushotzky 1986 preprint). This implies that the optical depth to pair production in the source region $(\gamma + \gamma \to e^+ + e^-)$ exceeds unity and a Thomson-thick pair plasma must be produced. (This extrapolation need not apply to the emergent γ-rays because which may be absorbed.) The optical - X-ray continua can also be re-processed by Compton scattering by the pair plasma.

Fabian, Guilbert, Phinney, Cuellar, and I have been calculated some simple models of pair plasma in source regions and find that various features of the observed spectra can be reproduced. Although our calculations are of more general applicability, we envisage that gamma rays are created when relativistic electrons inverse Compton scatter thermal UV photons in a corona above the innermost parts of a disk accreting at a near critical rate. We have developed a time-dependent code that evolves the electron and photon distribution functions in such a way as to conserve particle number and energy explicitly. This has enabled us to determine the dependence of the emergent radiation spectrum on the nature of the photon and electron spectra and to explore the time-dependent response to changes in the particle acceleration rate. We find that our results are most sensitive to the treatment of the radiative transfer and that this sets the necessary level of sophistication required to treat the various microphysical processes. The simplest treatment of the radiative transfer is to give every photon an escape probability $c\,\Delta t / R(1 + \tau_c)$ in a time step Δt, where $\tau_c(\varepsilon)$ is the relevant Compton scattering optical depth (using a simple approximation to the the Klein-Nishina cross section. The microphysical processes that have to be included are inverse Compton scattering by the relativistic electrons, photon-photon pair production, electron-positron annihilation and Comptonisation by the thermal electrons (Fig. 1). Under the conditions of immediate interest, the electrons and positrons cool to non-relativistic energies through inverse Compton scattering before there is any significant annihilation. In a steady state, the Thomson depth is then a direct measure of the total production rate of relativistic pairs.

Two limiting cases can be handled analytically. When secondary pair production is ignorable and electrons are injected mono-energetically with energy $\gamma = \gamma_{max}$, inverse Compton cooling establishes a steady state electron distribution function $N_\gamma \propto \gamma^{-2}$ for $\gamma < \gamma_{max}$. The X-ray spectrum of the inverse Compton scattered ultra-violet photons therefore has a spectral index $\alpha = 0.5$. In the opposite limit, when secondary pair production dominates the steady electron spectrum, $\alpha = 1$.

We find that when we include the reprocessing by the thermal electrons, the observed X-rays spectral index has an intermediate value $0.7 \lesssim \alpha \lesssim 0.9$ for a substantial range of soft photons to relativistic electron power ratios (0.1 - 1.0) and compactness parameters $1 \lesssim l_\gamma \lesssim 100$. More compact sources have steeper X-ray spectra and this may possibly be related to the observations of radio-quiet quasars (Elvis, this volume). The emergent flux above $\sim 100\,\text{keV}$ is generally much smaller than given by an extrapolation of the X-ray spectrum an account of Compton recoil on the cool thermal electrons and the high pair production

opacity. An annihilation line is also generally present, although it must be borne
in mind that dynamical expansion of the source region will probably smear it
out.

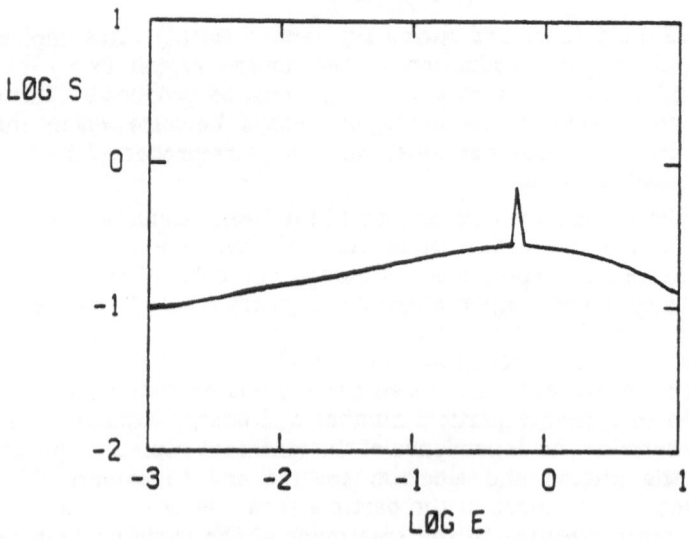

Figure 1. Stationary X-ray and γ-ray spectrum from an electron-positron
plasma. Soft photons are injected with a compactness parameter (see text) of 1
and 500 MeV electrons are also injected with a similar compactness parameter.
S measures the power per logarithmic frequency interval of the emergent radiation at energy E MeV. Note the slope ~0.2 corresponding to a spectral index
α~0.8 in the 2-10 keV electron range and the 0.5 MeV annihilation line.

7. A SIMPLE EVOLUTIONARY MODEL

Much effort has been devoted to inferring the evolutionary properties of
AGN purely from the observations. It is generally agreed that powerful radio
sources and quasars were far denser (per unit comoving volume) in the past and
that weaker sources evolved far less. There has been a parallel effort to try to
unify the different types of AGN and to distinguish them by hole mass, accretion
rate, orientation, spin, galactic environment, etc. Most of these unifying
schemes contain, at least implicitly, an assumption that an individual nucleus
can transform from one type to another and so they should predict the evolutionary behavior. This more physically based approach to evolution has been
pioneered by Cavaliere, Giallongo, and Vagnetti (1985, preprint). Here, I develop
a simple model that is appropriate to the particular unification scheme that I
outlined above and which, in contrast to the models of Cavaliere et al. has active
galaxies becoming systematically brighter with time.

We make the following simple assumptions.

(*i*) Essentially all galaxies of moderate or greater mass grow black holes in their nuclei starting from initial masses $M_i \sim 10^3 M_\odot$. These grow at a rate dictated by the local supply of gas which is supplied in fairly large parcels independent of the mass of the hole. The average rate of supply of gas in this form declines with cosmic time t at a prescribed rate $\langle \dot{M} \rangle (t)$.

(*ii*) This gas is accreted by the hole at a critical rate $(\dot{M} = M / t_E \sim M_8 M_\odot yr^{-1}$, corresponding to a radiative efficiency of 0.1 and a total power of half the Eddington value. The radiation pressure of the escaping photons prevents accretion at a faster rate, although physically, this need not be the case.

(*iii*) Holes that are currently accreting (at the critical rate) are identified with quasars and bright Seyferts whose luminosity depends upon the hole's mass.

(*iv*) Dormant holes, in which the accretion is subcritical are identified with low luminosity Seyferts, LINERS, starburst galaxies and, in the case of large holes in ellipticals, radio galaxies.

We must specify two functions, the galaxy (and hence by assumption) the black hole birth rate as a function of cosmic time $(S(t)$, and the mean fuelling rate $(\langle \dot{M} \rangle)$. Given these assumptions, we find that the fraction of holes of mass M that are active at any time is given by

$$\delta(t) = \min[1, \langle \dot{M} \rangle t_E / M] \qquad (4)$$

In other words, holes accrete continuously until the mean rate of gas supply falls below the Eddington rate, whereafter a decreasing fraction of them are observed as bright AGN's. (In fact equation(4) is somewhat inaccurate because it ignores the reservoir of gas whose build up depends upon the history of the accretion. However, as the critical rate of gas supply rises exponentially with time and $\langle \dot{M} \rangle$ declines rapidly, this turns out to be a good approximation.) We now change the independent variable to L, the bolometric luminosity of the hole when it is accreting. If the fraction of all galaxies with associated luminosity L lying in unit interval of $d\ln L$ is denoted by $Y(L)$, then $Y(L)$ satisfies the partial differential equation

$$\frac{1}{\delta} \frac{\partial Y}{\partial t} + \frac{1}{t_E} \frac{\partial Y}{\partial \ln L} = S\delta(L - L_i) \qquad (5)$$

where L_i is the critical luminosity corresponding to M_i. Equation(5) is straightforward to solve given $S(t)$ and $\langle \dot{M} \rangle$ and it turns out to be possible to choose these functions so that the evolution of the luminosity function over the redshift range $0 < z \lesssim 4$ is qualitatively consistent with what we observe. (See Fig. 2.)

Of course this prescription is overly simple ignoring as it does fluctuations and additional factors, listed above. Furthermore there is no particularly good reason for choosing the birth and fuelling rates that we have used. Nevertheless, the solution does demonstrate that it is possible to account for the principal features of apparent AGN evolution in a fairly natural manner. Some aspects of this solution with observational ramifications should be noted.

(*i*) Most bright galaxies should contain holes with $M \gtrsim 10^6 M_\odot$. In addition most spirals should have been Seyferts for a few percent of their life.

(*ii*) The co-moving density of low luminosity AGN's does not evolve rapidly with redshift; indeed it declines beyond $z \sim 2$. We know that this is probably the case; otherwise the X-ray emission from Seyfert 1 galaxies would exceed the observed X-ray background.

(*iii*) The co-moving density of intermediate power AGN's also declines beyond
$z \sim 3$. This may be related to the apparent lack of quasars in high redshift
surveys discussed by Osmer in these proceedings.

(*iv*) The differential evolution of radio sources can also be interpreted in terms
of this model if we further assume that the total power of a radio source is
roughly proportional to the reducible mass of the central black hole which
will spin down on a timescale $\sim 10^9$yr unless reactivated by a fresh period of
accretion. Most of the massive holes that are responsible for the most
luminous radio sources stopped accreting by $z \sim 2$ and were braked by magnetic torques soon afterwards.

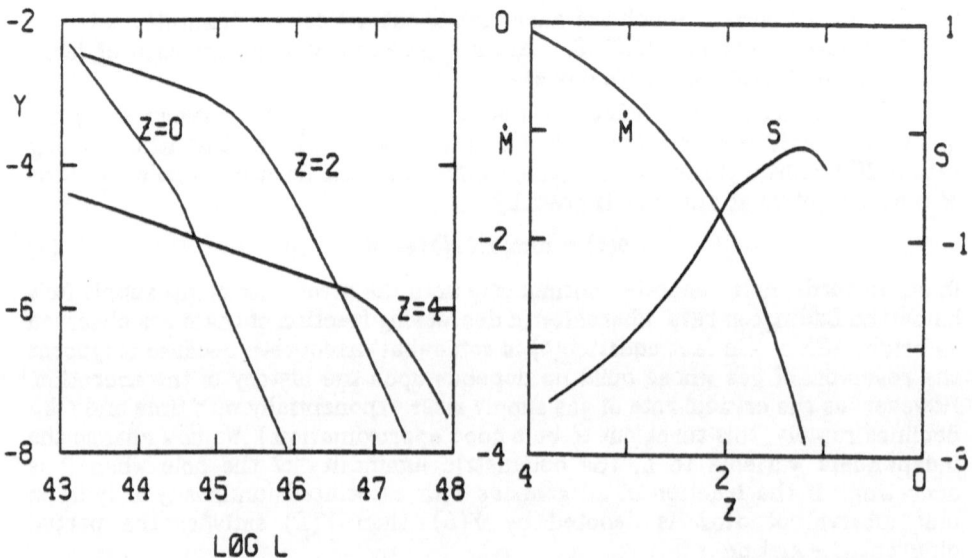

Figure 2. Differential luminosity function (Y) for power L erg s^{-1} (plotted
logarithmically) at redshifts $z = 0,2,4$ and associated galaxy birth rate S
and mean fuelling rate $(\dot{M}$ M$_\odot yr^{-1})$ also plotted logarithmically.

8. OBSERVATIONS OF BLACK HOLES IN AGN'S

Independent of these higher order considerations, the basic black hole
hypothesis is still without satisfactory observational verification. The most
immediate hope for changing this lies with space telescope which should be able
to perform spectroscopy on the innermost ~ 10pc of nearby galaxies and
thereby determine their central masses dynamically. Further indirect evidence
may be forthcoming from GRO if radio sources turn out to be detectable gamma
ray sources.

However, the most persuasive evidence would probably be the detection of a
black hole binary. As pointed out by Begelman, Blandford and Rees (1980), when
two galaxies (each with black holes in their nuclei) merge, the hole associated
with the smaller galaxy will sink into the nucleus of the larger galaxy in a few

orbital periods. Thereafter, the dynamical evolution time will increase to a value $\gtrsim 10^9$ yr until gravitational radiation by the black hole binary takes over. Now, the lifetime in years of a binary comprising holes of masses $10^8 M_{18} M_\odot$ and $10^8 M_{28} M_\odot$ with orbital period P yr is $T \sim 4 \times 10^7 P^{8/3} M_{18}^{-2/3} M_{28}^{-1}$, where for simplicity, we have assumed that $M_2 \lesssim M_1/3$. Now, contrary to the assertion of Gaskell(1983), it is not very likely that we observe black holes as spectroscopic binaries because the radius of the broad emission line region is typically larger than the radius of the orbit. However, if the black holes are endowed with accretion disks, then there is good chance that they will appear as an eclipsing binaries. We have argued that most galaxies have $\sim 10^8 M_\odot$ black holes in their nuclei and that they are currently active for $\sim 3 \times 10^8$ yr. Suppose that there are N mergers per galaxy and that an eclipsing binary can be recognised if its period is shorter than a year. Then the probability that an individual black hole nucleus harbor an observable binary is the joint probability that $P \lesssim 1$ yr i.e. $\sim 10^{-10} TN$ and the probability that it be currently active ~ 0.03. If these two probabilities are independent, then a sample of $\sim 10^4/N$ active galaxies must be searched to find one example. However if, as is more likely, binaries are always quite active then they may be up to 30 times more common. Of course this estimate is quite speculative, but it does suggest that a monitoring program of the optical continua of nearby Seyfert and LINER nuclei could be fruitful.

ACKNOWLEDGEMENTS. I thank Drs. Kapahi and Swarup for their kind invitation to attend this symposium and Drs. Phinney, Rees and Schmidt for helpful discussions of AGN evolution. I also thank the US national committee of the IAU for the award of a travel grant and the National Science Foundation for support under grant AST84-15355.

REFERENCES

Begelman, M. C., Blandford, R. D., and Rees, M. J., 1980, *Nature*, 287, 307.

Begelman, M. C., Blandford, R. D., and Rees, M. J., 1984, *Rev. Mod. Phys.*, 56, 255.

Blandford, R. D., 1984, *Ann. N. Y. Acad. Sci.*, 422, 303.

Burbidge, G. R. and Burbidge, M., 1967, *Quasi-stellar Objects*, (Freeman, San Francisco).

Coroniti, F. V., 1985, *Proc. IAU Symp. No.112*, (ed. Kundu and Holman), (Reidel, Holland).

Crawford, M. K., *et al.*, 1985, *Nature*, 315, 467.

Dressler, A., 1984, *Astrophys. J.* 286, 97.

Filippenko, A. V., and Sargent, W. L. W., 1985, *Astrophys. J. Suppl.*, 57, 3.

Gaskell, C. M., 1983, *Proc. 24th Liege Int Astrophys. Symp.*,(ed. Swings).

Lynden-Bell, D., 1969, *Nature*, 223, 690.

Mihalas, D. and Winkler, K.-H., (ed.), 1986, *Proc IAU Coll. No. 89*,

Miller, J., (ed.), 1985, *Astrophysics of Active Galaxies and Quasi-stellar Objects*, (University Science Books, Mill Valley, California).

Morris, M., and Yusef-Zadeh, F., 1986, *Astronom. J.*, (in press).

Papaloizou, J. C. B. and Pringle, J. E., 1984, *Mon. Not. R. astr. Soc.*, 208, 721.

Papaloizou, J. C. B. and Pringle, J. E., 1985, *Mon. Not. R. astr. Soc.*, 213, 799.

Phinney, E. S., 1983, unpublished thesis, University of Cambridge.

Pudritz, R. E. and Norman, C. A., 1983, *Astrophys. J.*, 274, 677.

Rees, M. J., 1984, *Ann. Rev. Astron. Astrophys.*, 22, 471.

Salpeter, E. E., 1964, *Astrophys. J.* 140, 796.

Sofue, Y., and Handa, T., 1984, *Nature*, 310, 568.

Thorne, K. S., 1986, in *Highlights of Modern Astrophysics*, (ed. Shapiro and Teu-
kolsky), (Wiley, New York).

Tonry, J., 1984, *Astrophys. J. Lett.*, 283, L27.

Uchida, Y., and Shibata, K., 1984, *PASJ*, 36,105.

Zel'dovich, Ya. B., and Novikov, I. D., 1964, *Dokl. Akad. Nauk. SSSR*, 158, 811.

DISCUSSION

Burbidge : There is an old problem concerned with the origin of the
rather strong magnetic fields in the extended lobes of radio galaxies.
In your theory does the field energy and flux come from the nucleus or
is it already present in the intergalactic medium ?

Blandford : I believe that the flux is supplied by the nucleus through
the jet. Note that the ordered field component along the jet is irrelevant because if flux is conserved, $B \propto A^{-1}$, (A is the area), pressure
$\propto A^{-2}$. Fixing the nuclear pressure at a plausible value gives a negligible parallel field in the outer jets and radio lobes. However toroidal
field will vary as $B \propto A^{-1/2}$ and is consistent with observed equipartition values. Small velocity gradients across the jet will in addition
produce reversing parallel field which may dominate.

Elvis : There is a tight connection between IR and X-ray continua in
AGN (Malkan 1984, Fabbiano et.al., this meeting), probably tighter than
the UV-X-ray correlation. The infrared does not appear in your sketch
of a 'Cauldron' spectron. How can you fit this observation into the
Cauldron model ?

Blandford : I have always been worried by this connection because the
IR and X-ray spectral indices are different. However if we accept it
at face value then the energy at ~ 2 keV may be optimal for seeing an
underlying power law, being too high to be contaminated with "thermal"
photons associated with the "big bump" and too low to be strongly influenced by the X-rays inverse Compton scattered by the secondary, relativistic pairs. The underlying IR-X-ray power law is most naturally
interpreted in this type of model as syncrotron radiation that has been
Comptonised by the non-relativistic pair plasma.

Vinod Krishan : (1) Could you specify the electromagnetic fields which
change into electric field to accelerate electrons ? (2) There is an

equivalent process to Compton scattering wherein the soft photon is converted to hard photon by the collective behaviour (plasma wave) of the hard electrons, called Raman scattering which is much faster and more efficient. Why are we not talking about it ? (3) If we can accelerate electrons continuously or in situ and get around the problem of life time of electrons we can also solve the problem of life time of plasma waves in the same manner.

Blandford : (1) Stationary magnetic fields frozen into an orbiting condensation (e.g. an accretion disk) induce a large EMF and the electric field will occur in regions of higher effective resistance. Changing magnetic fields from shearing flux loops in the disk also produce large inductive electric fields.

   (2) I agree that this process may be important under the conditions that you specify.

Sturrock : Surely the question of the confinement of a jet by a self-produced magnetic field is not as simple as you suggest. We know from the virial theorem that, from a global view point, the self-produced magnetic field of a plasmoid will always act in the sense of causing expansion rather than contraction.

Blandford : Yes, that's correct but not relevant if the jet is surrounded by thermal gas at a lower pressure (p_g) than that inside the jet (p_j). To give a simple example, suppose that a current I flows axisymmetrically along the walls of the jet, then the toroidal field outside the jet will be given by $B_\phi = 2I/r$. Static equilibrium requires that the jet radius r_j equal $(2\pi p_j)^{-1/2}$ I. There must be a return current however. Suppose that this all flows at the radius r_g where the gas pressure balances the outwardly directed magnetic stress. In this case, $(p_j/p_g) = (r_g/r_j)^2$. This simple example shows a hydromagnetic equilibrium can be established without violating the virial theorem. As we generally only need to amplify the gas pressure by factors $\lesssim 100$, values of $r_g \lesssim 10 \, r_j$ should be adequate.

"And one of the few opinions about quasars where the consensus has been steady since 1963 is that the energy is gravitational in origin. (An un-weighted majority view is in itself, however, worth little in this subject ! "

- Martin Rees (p.4)

VARIABILITY AND INSTABILITIES OF ACCRETION DISKS

M.A. Abramowicz[1], J.P. Lasota[1,2,3] and C. Xu[1]

1. International School for Advanced Studies,
   Strada Costiera 11, 34014 Trieste, Italy
2. Departement d'Astrophysique Fondamentale
   Observatoire de Paris Meudon, 92195 Meudon, France
3. Institut d'Astrophysique
   98 bis Bd. Arago, 75014 Paris, France

ABSTRACT. In the Seyfert galaxy NGC 1566 and quasars 3C 345, 3C 446 long term ($\sim 10^3$ days) variations have been observed. We propose to explain them as a limit cycle behaviour, similar to that well studied in the case of dwarf novae outbursts. The key element of our analysis is the relationship between the accretion rate $\dot{M}$ and the surface density Σ.

1. OBSERVATIONAL BASIS AND FUNDAMENTAL THEORETICAL IDEAS

This lecture gives a review of only some of the properties of accretion disks which are relevant to quasars, Seyfert galaxies and other active galactic nuclei. Most of the facts presented here belong to the standard theory of accretion (see e.g. an excellent review of Pringle 1981). The new theoretical discovery is the possibility of a limit cycle behaviour of the accretion flow.

In several active galactic nuclei quasi periodic bursts have been observed. They have a steep rise time of the order of 20 days and a period of about 1500 days. Alloin, Pelat, Phillips, Fosbury and Freeman (1985) have published data on the Seyfert galaxy NGC 1566. The data is also presented at this Symposium (Alloin, poster session, published in this book).

A schematic light curve of NGC 1655 is shown in Fig. 1. The observed variability of the broad emission lines H_α and H_β is interpreted as a variability of the accretion rate in the accretion disk orbiting the black hole in the centre of NGC 1655. The shaded areas are equal. This defines both the characteristic accretion rate $\dot{M}_0 = 3 \times 10^{-1} \dot{M}_E$ and the two timescales of 900 and 400 days. Two regimes occur along the rising branch: first

371

a mild increase on a timescale of 310 days followed by a steep rise on a
timescale less than 20 days. The decreasing part of the bursts corresponds to an exponential decay between 390 and 475 days. The three successive sub-bursts observed in the last cycle are separated by 310 days.

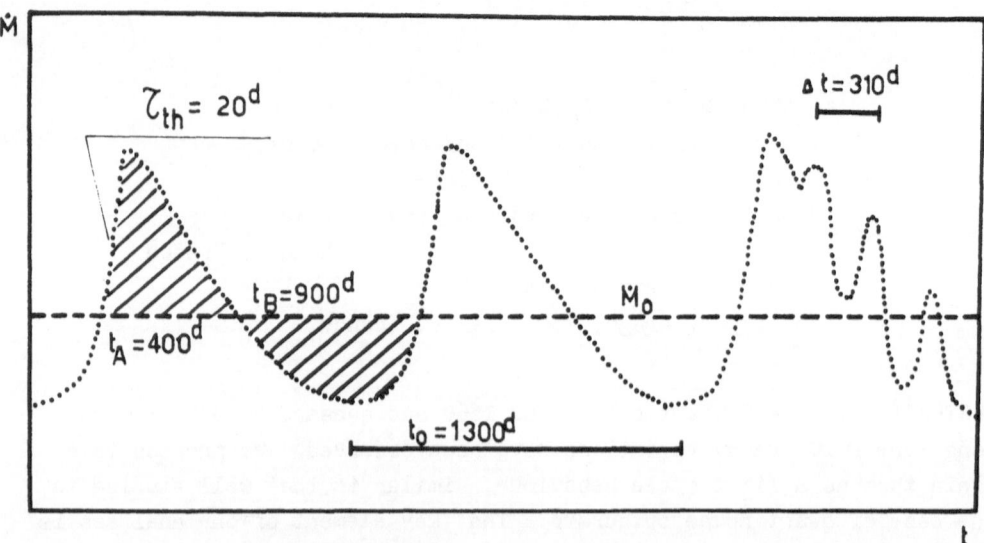

FIGURE 1. The light curve of NGC 1655 (schematic).

Abramowicz, Alloin, Lasota and Pelat (1985) have suggested an explanation for this variability: a limit cycle which works due to thermal
and viscous instabilities present in the innermost region of an accretion
disk orbiting a black hole. This explanation uses the following properties of accretion disks models:
    a) There is a relation between the shape of the $\dot{M}(\Sigma)$ curve, des-
cribing the dependence of the accretion rate M on surface density Σ,
and the stability of the disk: a negative slope corresponds to unstable
disks. An S-shaped $\dot{M}(\Sigma)$ curve indicates the possibility of a cyclic
outburst behaviour in accretion disks (Bath and Pringle 1981).
    b) The equilibrium states of the innermost regions of an accretion
disk around a black hole can be described by an S-shaped $\dot{M}(\Sigma)$ curve:
for small accretion rates the disks
are stable (positive slope), while for higher accretion rates well-known
thermal and viscous instabilities develop (negative slope), and for still
higher accretion rates they are stabilized by efficient cooling by advection (positive slope).
    The rest of the paper discusses these two points. Because defi-
nitions and equations relevant to the thin accretion disk theory are
basic for this discussion, we collect them in Tables 1 and 2.

TABLE 1.: The structure of thin accretion disks

M=black hole mass	Q^+=viscous heat production rate
$\dot{M}$=accretion rate	Q^-=radiation loss rate
R= distance from the hole	q=horizontal heat flux
α=viscosity coefficient	γ=heat balance coefficient
P=pressure	R_{in}=inner edge of the disk
ρ=density	H=half thickness
T=temperature	f=radial function
V_S=sound velocity	v=accretion velocity
Σ=surface density	$v_K=(GM/R)^{1/2}$=Keplerian velocity
ν=kinematic viscosity	$\ell_K=(GMR)^{1/2}$=Keplerian momentum
κ=opacity coefficient	ℓ=specific angular momentum

(Gravitational radius) $= R_G = 3 \times 10^{11} (M/10^6 M_\odot)$ [cm]

(Eddington accretion rate) $= \dot{M}_E = 1 \times 10^{23} (M/10^6 M_\odot)$ [g/sec]

R_G and $\dot{M}_E$ give convenient scaling factors for the distance and accretion rate. The heat fluxes Q^+, Q^- and q are defined per unit surface area.

TABLE 2.: The structure equations of thin accretion disks

(1) $\Sigma = 2H\rho$ ....definition of surface density
(2) $H/R = V_S/V_K$ ....vertical hydrostatic equilibrium
(3) $V_S^2 = P/\rho$ ....sound velocity
(4) $\dot{M} = 2\pi R \Sigma V$ ....mass conservation
(5) $Q^+ = f(3GM\dot{M})/(4\pi R^3)$ ....viscous heat generation rate
(6) $Q^- = (4acT^4)/(3\rho H \kappa)$ ....radiation heat loss
(7) $Q^+ = Q^- + q$ ....heat balance
(8) $q = (1-\gamma)Q^+$ ....horizontal heat flux
(9) $3\pi\nu\Sigma = \dot{M}f$ ....angular momentum balance
(10) $\nu = \alpha V_S H$ ....kinematic viscosity
(11) $P = (\mathcal{R}/\bar{u})\rho T + (a/3)T^4$ ....equation of state
(12) $f = [\ell(R) - \ell_K(R_{in})]/\ell(R)$ ...radial function
(13) $\ell_K^2 = GMR$ ....specific Keplerian angular momentum
(14) $V_K^2 = GM/R$ ....Keplerian velocity
(15) $R_{in} = 3R_G$ ....inner edge
(16) $\ell = \ell_K$ ....angular momentum distribution
(17) $\kappa = \kappa(\rho, T)$ ....opacity coefficient
(18) $\alpha = \alpha(\rho, T)$ ....viscosity coefficient
(19) $\gamma = \gamma(\rho, T)$ ....heat balance coefficient

Additional definition: $\beta \equiv P_{gas}/P = (\mathcal{R}/\bar{u})\rho T/P$
The physical constants $a, c, G, \bar{u}, \mathcal{R}$ have the standard meanings (e.g. Tassoul 1978, Appendix A).

2. THE $\dot{M}(\Sigma)$ CURVE

Consider a sequence of accretion disks for which the mass of the central
black hole M, the equation of state $P=P(\rho,T)$, the opacity law $\kappa=\kappa(\rho,T)$,
and the viscosity law $\nu=\nu(\rho,T)$ are fixed. The distance from the centre,
R is also fixed. We put $R=5R_G$ as most of the radiation produced by the
disk originates there.

In Fig. 2 we present (after Bath and Pringle, 1981) an intuitive
proof that

$$\frac{d\ln\dot{M}}{d\ln\Sigma} > 0 \Rightarrow \text{stability}, \qquad \frac{d\ln\dot{M}}{d\ln\Sigma} < 0 \Rightarrow \text{instability} \qquad (2.1)$$

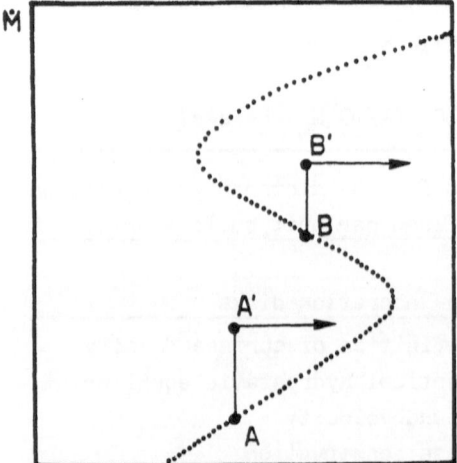

FIGURE 2. Stability of the accre-   FIGURE 3. The limit cycle behaviour
tion disks sequence (Bath and   connected with an S-shaped sequence.
Pringle criterion).

The point A' lies off the equilibrium sequence and can be considered
as the perturbed state of the model A. In the perturbed state A' the
accretion rate is higher than in the equilibrium state A. Thus, the
model A' is over-supplied with matter and its surface density will increase. This brings the model A' back to the equilibrium curve - the
model A is stable. A similar argument shows that model B is unstable.

Linear stability of accretion disks has been formally studied only
in the case of thin disks ($\dot{M} \ll \dot{M}_E$). Two types of instabilities are present: thermal instabilities (Pringle et al. 1973, Shakura and Sunyaev
1976) and viscous instabilities (Lightman and Eardley 1974, Shakura and
Sunyaev 1976).

Using the phenomenological coefficients defined by Piran (1978):

$$\mathcal{K} \equiv \left(\frac{\partial\ln Q^-}{\partial\ln H}\right)_\Sigma \; , \; \mathcal{L} \equiv \left(\frac{\partial\ln Q^-}{\partial\ln\Sigma}\right)_H \; , \; \mathcal{M} \equiv \left(\frac{\partial\ln Q^+}{\partial\ln H}\right)_\Sigma \; , \; \mathcal{N} \equiv \left(\frac{\partial\ln Q^+}{\partial\ln\Sigma}\right)_{\dot{H}} \qquad (2.2)$$

and slightly modifying his original results, one can write the criteria
for thermal and viscous instability as

$$\text{thermal:} \quad \left(\frac{\partial \ln Q^-/Q^+}{\partial \ln H}\right)_H = \mathcal{X} - \mathcal{M} > 0 \;,$$ (2.3a)

$$\text{viscous:} \quad \left(\frac{\partial \ln \Sigma \nu}{\partial \ln \Sigma}\right)_{Q^+=Q^-} = \frac{\mathcal{X}\mathcal{N} - \mathcal{M}\mathcal{L}}{\mathcal{X} - \mathcal{M}} > 0 \;.$$ (2.3b)

Criteria (2.3a) and (2.3b) are together a sufficient and necessary condition for stability. We have computed that

$$\left(\frac{d\ln\dot{M}}{d\ln\Sigma}\right) = \frac{\mathcal{X}\mathcal{N} - \mathcal{M}\mathcal{L}}{\mathcal{X} - \mathcal{M}} \;.$$ (2.4)

This proves that (2.1) is sufficient and necessary for the viscouse stability, so $(d\ln\dot{M}/d\ln\Sigma) < 0$ means that the disk is unstable with respect
at least to viscous perturbations.

In the case of the standard accretion disk model:

$$\mathcal{X} = 4\,\frac{1+\beta}{4-3\beta} \;, \quad \mathcal{L} = -\,\frac{\beta}{4-3\beta} \;, \quad \mathcal{M} = 2, \quad \mathcal{N} = 1$$ (2.5)

and the criterion (2.3) gives $\beta > 2/5$ for stability: radiation pressure
supported think disks are unstable (Shakura and Sunyaev, 1976).

Bath and Pringle (1981) also noticed that an S-shaped $\dot{M}(\Sigma)$ curve indicates a possibility of cyclic variation in the accretion flow. Fig. 3
provides an elementary explanation of this. Let the basic accretion rate
in the system be $\dot{M} = \dot{M}_0$. This corresponds to an unstable equilibrium
model E. It cannot be realised by the flow. Consider point A. The equilibrium accretion rate $\dot{M}(A)$ is smaller than the actual accretion rate $\dot{M}_0$.
The model A is oversupplied and will evolve, along the path AB, in the
direction of increasing surface density, as indicated by the arrow. From
the point B further evolution is only possible after a jump to C. Here,
however, the equilibrium accretion rate is higher than the actual one.
The surface density now decreases down to the point D, where another jump,
to A, must take place. This closes the cycle.

We proceed further by noticing that the turning points of the $\dot{M}(\Sigma)$
curve are given by $\mathcal{X} - \mathcal{M} = 0$.

If a particular $\dot{M}(\Sigma)$ has N turning points, then the equation
$\mathcal{X} - \mathcal{M} = 0$ has N different solutions, which means that $\mathcal{X} - \mathcal{M}$ is an N-th
degree polynomial in $\dot{M}$. Because both $\mathcal{X}$ and $\mathcal{M}$ are defined through logarithmic derivatives of physics quantities, this means that there exists
a physical quantity X with a very strong dependence on the accretion rate,
$X \sim \exp(k\dot{M}^N)$, where k=const.

Suppose, that a particular $\dot{M}(\Sigma)$ curve has one stable and one unstable branch, as shown in Fig. 4. This means that one of the physical
quantities varies according to $X \sim \exp(k\dot{M})$. To change this into
$X \sim \exp(k\dot{M}^2)$, which is needed for the existence of an S-shaped $\dot{M}(\Sigma)$ curve,
one can assume that a free parameter, e.g. the viscosity coefficient,
varies like an exponential function of $\dot{M}$. (Fig. 4).

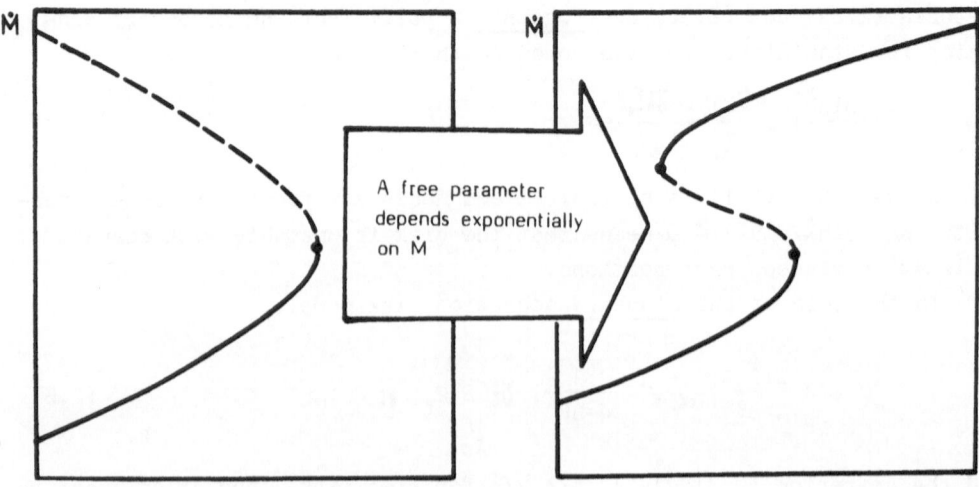

FIGURE 4. How to get an S-shaped $\dot{M}(\Sigma)$ curve.

3.   THE INNERMOST REGION OF THE DISK

Abramowicz and Lasota (in preparation) noticed that for accretion rates
$\dot{M} \leq 0.02 \, \dot{M}_E$ the accretion is stable. This is reflected by the positive
slope of the $\dot{M}(\Sigma)$ curve, which is well approximated by Shakura-Sunyaev
(1973), formula valid for $\beta \gg 2/5$:

$$\frac{\dot{M}}{\dot{M}_E} = 2.5 \times 10^{-4} \, (\frac{M}{10^6 M_\odot})^{\frac{5}{3}} \, \alpha^{\frac{4}{3}} \, (\frac{R}{R_G}) \, f^{-1} \Sigma^{\frac{5}{3}} . \qquad (3.1)$$

At $\dot{M} = 0.02 \dot{M}_E$ the $\dot{M}(\Sigma)$ curve bends and its slope becomes negative. This
corresponds to the thermal and viscous instabilities found by Pringle,
Rees and Pacholczyk (1974), Pringle (1976), Lightman and Eardley (1974),
and Lightman (1974). The instabilities result because of the insufficient
cooling. The analytic approximation of Shakura and Sunyaev, valid for
$\beta \ll 2/5$, reads:

$$\frac{\dot{M}}{\dot{M}_E} = 1.3 \times 10^1 \alpha^{-1} (\frac{R}{R_G})^{\frac{3}{2}} f^{-1} \Sigma^{-1} . \tag{3.2}$$

At still higher accretion rates the $\dot{M}(\Sigma)$ curve bends again and accretion disks become stable. The stability is restored because of a very efficient cooling by advection (Abramowicz, 1981). This results from the fact that the accretion flow onto a black hole is transonic: its sonic point lies close to the inner edge (cusp) of the disk. The heat is thus removed by advection at a speed close to the speed of sound. This cooling mechanism works for both small ($\dot{M}/\dot{M}_E \ll 1$) and large ($\dot{M}/\dot{M}_E > 1$) accretion rates, but in the former case it is relevant only in the immediate vicinity of the inner radius. Cooling by advection can be efficient only if its characteristic time scale is not longer than the thermal timescale,

$$t_{th} = 1.6 \times 10^{-3}/\alpha (R/R_G)^{3/2} (M/10^6 M_\odot) \quad [days]$$

In a thin, gas pressure dominated disk the characteristic time for radial motion is the viscous time, $t_{vis} = t_{th}/(H/R)^2$, which is much longer than the thermal one (as $H/R \ll 1$). At higher accretion rates, close to $\dot{M}_E$, this is no longer true. The angular momentum distribution changes from being almost Keplerian to being almost constant (see e.g. Muchotrzeb and Paczyński, 1982). To move a matter element from place to place in a fluid with constant angular momentum no viscous torque is necessary. In addition black holes accrete matter with non-zero angular momentum. This is a relativistic effect, since in a Newtonian potential there is an infinite potential barrier for non-zero angular momentum. As a result when accretion rate is high, $\dot{M} > \dot{M}_E$ advection is not related to viscous processes and may proceed on a much faster timescale than t_{vis}. For such high accretion rates, however, the disk cannot be considered to be thin and the standard approximations are no longer valid.

Below we give a purely phenomenological description of the advective cooling.

The total amount of heat dQ carried in the horizontal direction by the advection form a volume $d\Upsilon = HdS = H2\pi RdR$ in a unit time is $dQ = (dV_S^2/dR)\rho V \, d\Upsilon$. Therefore the heat flux per unit surface, $q = dQ/dS$, equals:

$$q_{ad} = \frac{dV_S^2}{dR} H V \approx (\frac{H}{R}) V_S^2 V\rho . \tag{3.3}$$

The sign of the gradient dV_S^2/dR is fixed in (3.3) in such a way that q_{ad} describes cooling. Using the equations in Table 2 this may be expressed as:

$$q_{ad} = \frac{2}{3} f^{-1} (\frac{H}{R})^2 Q^+ \tag{3.4}$$

The horizontal heat flux carried by radiation (c.f. Maraschi et al. 1976; Muchotrzeb and Paczyński, 1982) can be estimate to be:

$$q_{th} = (\frac{H}{R}) \; Q^-. \tag{3.5}$$

Note, that for the equilibrium structure the horizontal radiation flux is more important than the advection flux in the case of thin disks, $H/R \ll 1$. However, for stability the advection flux is more important. This is because of the stability condition $\mathcal{X} - \mathcal{M} > 0$. (See Table 3).

Note that because of the factor f^{-1} in equation (3.4) advection flux always dominates close to $R=R_{in}$ as $f(R_{in})=0$. This factor becomes dominant also when $\ell \approx$ const, because $f=[\ell(R)-\ell(R_{in})]/\ell(R)$. In precisely this situation the advection flux becomes efficient, because it acts on a timescale much shorter than the viscous one.

TABLE 3.:   Stability properties of radiation pressure supported disks.
In all three cases $\mathcal{M} = 2$.

Q^-	$\mathcal{X}$	Stability condition $\mathcal{X} - \mathcal{M} > 0$
vertical radiation	1	violated
horizontal radiation	2	marginally fulfilled
advection	4	fulfilled

The key element is therefore the angular momentum distribution which determines the factor f. Computing $\ell=\ell(R)$ is not easy for $\dot{M} > \dot{M}_E$ (c.f. Muchotrzeb and Paczyński, 1982). The main difficulty is due to the transonic nature of the flow, which in turn implies the eigenvalue character of the mathematical model of it: the eigenvalue is the location of the inner edge, R_{in}. For small accretion rates $R_{in}=3R_G$ is a very good approximation for the eigenvalue, and this problem is not acute. The coefficient for the heat balance γ can, with the help of equations (8) and (3.4) be expressed as:

$$\gamma = \frac{2}{3} \; f^{-1} (\frac{H}{R})^2 + 1. \tag{3.6}$$

Because there is no way today to compute f, the coefficient γ is unknown. However, as was explained before, for the S-shaped $\dot{M}(\Sigma)$ curve, it is sufficient

$$\gamma = \exp(-k \frac{\dot{M}}{\dot{M}_E}) \quad , \tag{3.7}$$

with k=const. We have explained that this behaviour is very likely, because $\gamma \sim f^{-1}$ and the function f depends very strongly on $\dot{M}/M_E$, going to zero when $\dot{M}/M_E \gg 1$. Thus, we have assumed (3.7) and computed the resulting $\dot{M}(\Sigma)$ relations. They are shown in Fig. 5 and Fig. 6.

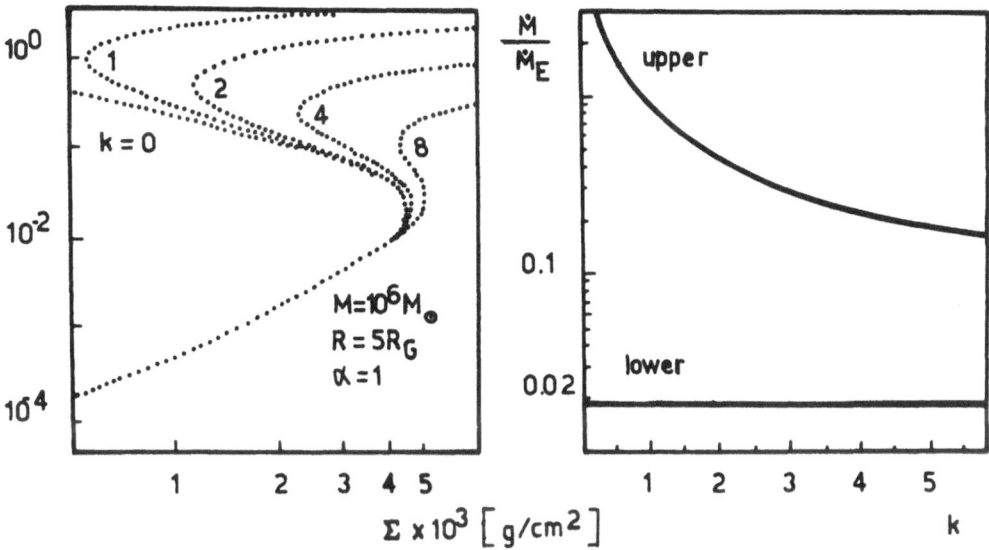

FIGURE 5. The sequences of equi-   FIGURE 6. The location of the lower
librium accretion disks models   turning point does not depend on the
with advection heat flux (the   advection strength, but the location
strength of advection is measured   of the upper one does.
by k).

Results would be similar or identical if instead of (3.7) an exponential dependence on $\dot{M}$ of the α (viscosity) or κ (opacity) were assumed.

4.   APPLICATION TO NGC 1566 AND CONCLUSIONS

Since the average accretion rate is $\dot{M}_0 = 0.3\dot{M}_E$, hence larger than at the lower turning point, $\dot{M} = 0.02 M_E$, the inner regions of the accretion disk in NGC 1566 cannot settle to a steady state. The steep rising branch $\tau \stackrel{\sim}{\scriptstyle\sim} 20$ days is the rising time for thermal instability (for this α must be greater than 10^{-2}). The three sub-bursts observed during the decay time may be connected with the fact that viscous instability breaks the disk into concentric rings. According to the interpretation of the varia-bility observed in NGC 1566 as due to a limit cycle connected with the S-shaped $\dot{M}(\Sigma)$ curve one can estimate the accretion rate on the upper stable branch as

$$\dot{M}_{high} \approx \frac{t_A + t_B}{t_A} \; \dot{M}_O \approx \dot{M}_E \; , \tag{4.1}$$

by assuming that on the lower branch

$$\dot{M}_{low} << \dot{M}_E \; . \tag{4.2}$$

This gives a good agreement with the theoretical S-curves in Fig. 5.

The hypothesis that a limit cycle instability operates at the centre of NGC 1566 requires more theoretical investigations. Of special importance are time dependent disk models for $\dot{M} > \dot{M}_E$. Observationally, statistical studies of long term variability would be relevant to the question whether this variability indeed occurs only in the range $10^{-2} < \dot{M}/\dot{M}_E \lesssim 1$ as our analysis predicts. More observational tests are described in the papers quoted in the introductory paragraph.

REFERENCES

Abramowicz, M.A. 1981, Nature 294, 235.
Abramowicz, M.A., Alloin, D., Lasota, J.P., and Pelat, D. 1985, preprint
    submitted to Nature.
Alloin, D., Pelat, D., Phillips, M.M., Fosbury, R.A.E., and Freeman, K.
    1985, preprint submitted to Astrophys. J. see also Alloin, D., this
    book .
Bath, G.T. and Pringle, J.E. 1981 Mon. Not. R. astr. Soc. 194, 967.
Lightman, A.P. 1974, Astrophys. J. 194, 429.
Lightman, A.P., and Eardley, D.M. 1974, Astrophys. J. Lett. 187, L1.
Maraschi, L., Reina, C., and Treves, A. 1976, Astrophys. J. 206, 295.
Muchotrzeb, B. and Paczyński, B. 1981, Acta Astron. 32, 1.
Piran, T. 1976, Astrophys. J., 221, 652.
Pringle, J.E. 1976, Mon. Not. R. astr. Soc. 177
Pringle, J.E. 1981, Ann. Rev. Astron. Astrophys. 19, 137.
Pringle, J.E., Rees, M.J., and Pacholczyk, A.G. 1973, Astron. Astrophys.
    29, 179.
Shakura, N.I. and Sunyaev, R.A. 1973, Astron. Astrophys. 27, 337.
Shakura, N.I. and Sunyaev, R.A. 1976, Mon. Not. R. astr. Soc.175, 613.
Tassoul, J.-L. 1978, Theory of Rotating Stars, Princeton University Press:
    Princeton.

DISCUSSION

Pranab Ghosh : 1) Comment : The fact that the turnover value of $\overset{\bullet}{M}$ for a Shakura-Sunyaev disc is practically independent of α and M is entirely as expected, since the physical variables in such a disc depend <u>very</u> weakly on α and fairly weakly on M.

(2) Question : All of us have been trying to understand the nature of the viscosity coefficient α for more than a decade now. Part of these efforts have come from the work of Bath and Pringle related to that which you just described, namely, their modelling of the time evolution of accretion discs and comparison with dwarf novae outbursts. Where, do you think, we stand today in our understanding ?

Abramowicz : I think, theoretical estimations of α are as bad today as they were ten years ago. However, there are very convincing <u>observational</u> estimations of α based on light curves of the dwarf novae. (e.g. Smak 1984 review article in PASP). In particular, it is now obvious that the parameter in the "low" and "high" state has different values. Because I mentioned again dwarf novae let me stress that although mathematical theory of limit cycle in dwarf novae case <u>is</u> similar to what I have described in my lecture, the physical nature of instability is different. However, dwarf nova mechanism may be relevant for outer, cool parts of accretion discs in AGN. Katy Clark in Oxford is working on this. She is getting periodicities of the order of a few years.

Burke : If the discs are stable, then they will tend to axisymmetry on the average. If the magnetic fields are generated by dynamo action, does not Cowling's theorem present difficulties ?

Abramowicz : Roger Blandford is perhaps the right person to answer that question.

Blandford : These objects are not exactly axially symmetric.

Sturrock : I wish to address the question of whether compact radio sources and jets are due to activity in the vicinity of the black hole or in the magnetosphere of the accretion disc. If the ejecta are produced near the black hole, it is hard to see why ejection is so frequently onesided. On the other hand, one might consider the possibility that activity is due to flare activity resulting from dynamo action in the accretion disc. If there is a probability P that this process will occur, at any time, on either surface of the disk, there is a probability P^2 of seeing two-sided ejection, $2P(1-P)$ of seeing one-sided ejection, and $(1-P)^2$ of seeing no ejection (and no compact central radio source). For quasars, $P \ll 1$ since most quasars are radio quiet, and this helps one understand why ejections from quasars are usually one-sided.

Leonid Matveyenko listens to Ken Kellermann

THE ELECTRON-POSITRON CAULDRON AND THE FORMATION OF THE HARD RADIATION OF QUASARS AND AGN

N.S. Kardashev, I.D. Novikov
Space Research Institute
Academy of Sciences of the USSR
Profsoyuznaya 84/32, MOSCOW 117810, USSR

and

B.E. Stern
Institute for Nuclear Research
Academy of Sciences of the USSR
60th October Anniversary Prospect 7a
MOSCOW 117312, USSR.

Active Galactic Nuclei (AGN) and quasars have unique physical parameters among all the objects in the Universe. Undoubtedly it is the uniqueness of the physical conditions in these systems that gives rise to the peculiar physical processes in them.

It is the compactness of these objects - the large ratio of luminosity to size - that is, probably, their most striking and well known feature. The lower limit on the compactness L_{hard}/R in AGN ranges from 10^{26}-10^{29} erg $s^{-1}cm^{-1}$. (Here R is determined through the characteristic time t of the luminosity variability $R = ct$). On the other hand, if one assumes that the luminosity is of the order of the Eddington luminosity, and the dimensions of the radiating region are some few gravitational radii then $L_{hard}/R = 10^{32}$ erg $s^{-1}cm^{-1}$.

Such an extreme compactness leads invariably to $e^{\pm}$ pair production due to photon-photon interactions. The processes which involve electron-positron pairs seem to be among the main ones that crucially influence the formation of X- and γ -radiation spectra in quasars and AGN.

Indirect evidence about the presence of $e^{\pm}$ pairs in these objects comes from the discovery of the annihilation line from the centre of our Galaxy, where a less powerful version of quasar's and AGN's energy generator may be working. From these observations we can estimate the effectiveness of the $e^{\pm}$ pair generating mechanism. It turns out that the ratio of the rest mass energy of the generated $e^{\pm}$ pairs to the total energy released should be no less than 10^{-3}, or greater. The greater the ratio for a model, the easier it is for the model to describe the observational data.

G. Swarup and V. K. Kapahi (eds.), Quasars, 383–393 .
© *1986 by the IAU.*

Furthermore, some of the details of the X- and γ -spectra of AGN and quasars seem to tell us that these hard spectra are probably not of thermal origin. This seems natural in those models in which the energy source in AGN and quasars is a massive black hole rotating in an external magnetic field. Such a black hole operates as an electric battery which accelerates charged particles. In the models being discussed in this paper, it is assumed that the emission from AGN and quasars in the hard part of the spectrum originates due to reprocessing of the energy of particles accelerated to relativistic velocities.

The crucial point in these models is the $e^{\pm}$ pair creation. The most effective mechanism of $e^{\pm}$ pair creation is the $\gamma + \gamma \rightarrow e^{+} + e^{-}$ reaction. Besides this reaction $e^{\pm}$ pairs can be created by the interaction of γ-photons with matter. Novikov and Stern (1985) describe the method of numerical simulation of the physical processes for these models in which $\gamma + \gamma$ reaction is the main source of the $e^{\pm}$ pairs, and discuss the results of these simulations. In the present paper we shall discuss our new results on these models. The alternative model in which γ + matter reaction is used to produce $e^{\pm}$ pairs and to form the hard spectrum was discussed by Kardashev et al., (1983). We do not give here a complete list of references. They can be found in Novikov and Stern (1985) and in the excellent review by Svensson (1985) (see also Blandford, 1986; Rees, 1986; Zdziarsky and Lightman, 1985).

Let us consider models of the first type. By now many models of this kind have been constructed starting with the well known work of Bonometto and Rees (1971). A typical representative of such a model is the following (Svensson, 1985): hard particles with energy ϵ = 2.7 10^4 (in units of $m_e c^2$) are injected into a region filled with soft photons with ϵ = 2.7 10^{-5} (e.g., due to thermal radiation from an accretion disk). Then a variety of cascade processes involved are investigated, and the resultant spectrum of photons which will be observed is calculated.

Our main idea was to develop the simplest self-consistent model with the least possible number of assumptions, which would explain the observational data. Furthermore, the observational predictions in the model should not be a result of a fine tuning of some parameters but must be a natural consequence of the model's general fundamental properties.

Let us summarize briefly and schematically the main set of observational data on the hard radiation from AGN and quasars, which any model should explain.

1) in the X-ray region (up to 100 KeV) the spectral index α of AGN seems to be "universal" and $\alpha_x \approx 0.7 \pm 0.15$(Rothschild,et al., 1983; Petre et al., 1984). Quasars exhibit wider variations of the index α around the value $\alpha \approx 1$.

2) In some cases, e.g., NGC 4151, there is a cut off in the spectrum at energies of the order of MeV. There is an indication that this cut off is a general feature of AGN spectra (see Bignami et al., 1979).

A cut off of the same type is seen in the spectrum of the centre of our Galaxy. In this case, when the annihilation line has sufficient intensity, an increase in radiation at energies of a few MeV and then a cut off at higher energies is observed.

Here we deal with the following model. (For details see Novikov and Stern, 1985.)

1) We assume that in the centres of quasars, the charged particles are being accelerated. The details of the mechanism of such acceleration are not important.

These particles are injected with some pitch angle into the magnetic field region. Alternatively, the energy could be injected in the form of hard γ quanta which could be created by accelerated particles due to curvature radiation or inverse Compton scattering.

After that we provide a numerical simulation, by the Monte-Carlo method, of all processes in this region without any additional assumptions. We call this region - the $e^- e^+$ cauldron. In this region the processes of multi-particle creation and interaction occur, and we take all of them into account. It should be emphasized that our model does not involve any other particles or radiation sources besides those primarily injected. The problem is solved in a completely self-consistent manner.

We calculate the evolution with time of the photon spectra and the energy distribution of $e^{\pm}$ pairs.

Our final purpose is to demonstrate that in the case of large optical depths (τ) of the whole $e^{\pm}$ cauldron the photon spectrum evolves to some specific shape with a cut off near 2 MeV and a power-law slope at energies < 100 KeV. This spectrum is to be confronted with the observational data.

We have studied the dependence of the spectrum on the parameters of the electron-positron cauldron and have shown that for a wide range of these parameters the photon spectrum approaches a "universal" form which is the one observed from AGN and Quasars.

In our model, the injected γ-rays interact with soft photons producing high energy pairs. The soft photons can arise as synchrotron radiation from pairs produced before. Multiple interactions of $e^{\pm}$ with the magnetic field and photons and of photons with photons lead to a cascade increase in the number of particles. The energy of hard particles diminishes until electrons and positrons become semi-relativistic and all photons with energy $> m_e c^2$ are converted into pairs or scattered into the soft region. The particle spectra should relax to a rather soft quasi-equilibrium shape, and further evolution of the system should be much slower than the energy degradation. Our method of numerical modelling of the $e^{\pm}$ cauldron is described in Stern (1985). We recall here the main idea.

Numerical simulation of a realistic steady state cauldron would be rather difficult. The energy injected at a constant rate into the centre of the cauldron spreads to its outer parts, and as a consequence one must solve a spatially non-homogeneous problem. But it suffices to solve a simpler problem.

Let us suppose that field lines stretch from the injection region to the outer parts of the cauldron and that injected particles in one element and their descendants move along these lines and do not interact with those in another element. In this case we can introduce a reference frame co-moving with the injected particles of an element and calculate the evolution of this system of particles with time from the initial stage of injection upto the time of reaching the boundary of the cauldron where the radiation escapes. In our first version of the calculation (Novikov and Stern, 1985), to simplify the problem, we had further assumed that the momentum distribution of particles in the comoving frame is isotropic at all times. But that it is not true was emphasized by Petrosian (in the discussion of Novikov and Stern, 1985). In our present calculations we do take into account that charged particles rapidly lose energy when the pitch angles are large. The momentum distribution of particles then becomes anisotropic. (The process of isotropization may be possible.) It turns out that even if we take the anisotropization into account, the results change only slightly.

Under the above simplifying assumptions the problem can be solved numerically with the Monte-Carlo method. It is natural to define the age t of the $e^{\pm}$ soup as the number of collisions of a particle with others until a given moment. The definition depends on the energy of the particle, but we can take a typical particle with energy $m_e c^2$ and with the classical Compton cross-section $\sigma_0 = 2 \ 10^{-25} \ cm^2$.

We shall characterize the optical depth τ of the whole cauldron by the age t at the moment of escape. This parameter in fact is the dimensionless measure of the cauldron compactness. Indeed

$$\tau = \frac{L}{R} \frac{\sigma}{4 \ m_e c^3} \approx (L/R)/(10^{30} \text{erg s}^{-1} \text{cm}^{-1})$$

Thus, the results of the calculations for different moments of time t may be treated as the results for $e^{\pm}$ cauldrons with different values of the compactness L/R.

The parameters of $e^{\pm}$ cauldron are:

1) The magnetic field H, and (2) the primary energy of injected particles E_o. It is the upper bound $\varepsilon_m = (H/H_o) E_o^2$ (where $H_o = 4 \ 10^{13}$ Gs) of the synchrotron spectrum from primary pairs rather than the primary energy E_o that is important in the problem and we use it as a parameter. (3) the ratio Ω of the energy density of injected particles to the energy density of the magnetic field H. This parameter describes the relative value of Compton processes with respect to synchrotron emission, and (4) the optical depth τ of the whole cauldron.

The outcome of these processes are different for different cases, $\epsilon_m \gg 1$, $\epsilon_m \approx 1$ and $\epsilon_m < 1$ for the following reasons. If $\epsilon_m \gg 1$ then a cascade degradation of energy occurs because the descendant synchrotron photons are energetic enough to produce pairs. If $\epsilon_m \approx 1$ then the energy does not degrade in the cascade way. All pairs are produced by the radiation from the primary particles. In the case $\epsilon_m < 1$, synchrotron photons are too soft and cannot produce pairs, but the pair production goes through comptonized photons if Ω is not small.

Figures 1-5 demonstrate the results of our new calculations.

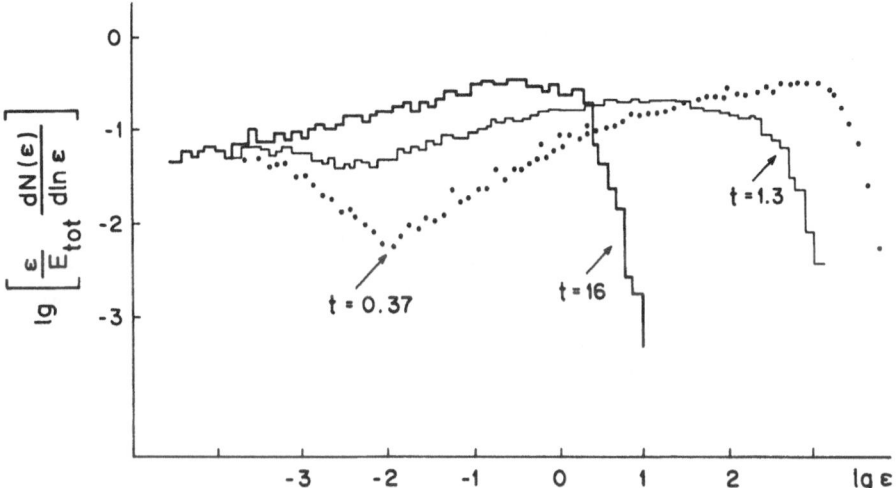

Fig. 1: *The evolution of the photon energy distribution with age.*
$H = 10^3$ Gs, $E_0 = 10^7$, $\epsilon_m = 3\,10^3$, $\Omega = 5$.

Figure 1 demonstrates the evolution of the photon spectrum with age. Note that the photon spectra are presented in figures with the scale $\dfrac{\epsilon}{E_{tot}} \dfrac{dN(\epsilon)}{d\ln\epsilon}$ which is useful for theoretical consideration, but inconvenient for presenting the observational data. In the latter case, one uses the scale with the spectral luminosity $L_\epsilon = dL/d\epsilon$. In case of a power-law spectrum, $L_\epsilon \sim \epsilon^{-\alpha}$, and our spectral index is then $\beta = 1 - \alpha$. Figure 1 corresponds to the case when Ω is of the order of unity, which means that there is an equipartition in energy between particles and the magnetic field. The parameter $\epsilon_m = 3\,10^3$ which is $\gg 1$.

At small t (t $\ll 0.37$) there is a typical synchrotron spectrum of a e^- or e^+ decelerating due to its radiation and $\alpha \approx 0.5$ or $\beta = 0.5$. At t = 0.37, the intensity rises for $\epsilon < 10^{-2}$ due to the synchrotron radiation of the next generation of pairs. For t > 1 the minimum is smeared out due to componization of synchrotron photons and for t $\gg 1$, the spectrum becomes flatter.

The hard part of the spectrum has a cut off due to the pair production. This cut off moves to the region of energy ~ 0.5 MeV with age.

The spectral index of the resultant spectrum lies near the value $\alpha \overset{\sim}{\sim} 0.75$.

Analogous results are obtained in the case $\epsilon_m \gg 1$ for a wide range of variation of parameters. This can be easily understood. It was already noted by Bonometto and Rees (1971) that if the number n of pair production cascades grew and became much larger than 1, then the spectral index of particle radiation tended to $\alpha = 1$. These authors dealt with particle energy loss due to Compton scattering, but the same arguments are valid in case of synchrotron energy loss.

It can be shown easily (see e.g., Svensson, 1985) that the spectral index of radiation emitted by the cooling particles of each subsequent generation relates to the spectral index of the previous generation in the following way

$$\alpha_{n+1} = \frac{\alpha_n + 1}{2}$$

The spectral index of the radiation of injected particles, $\alpha_0 = 0.5$. The first generation pairs will have $\alpha_1 = 0.75$. If it is the first generation that is responsible for the radiation from the cauldron (after the Compton smearing-out of the minimum) then $\alpha_{observed} = \alpha_1 = 0.75$. It is this case that is realised for a wide range of values of $\epsilon_m \gg 1$. That is why the "universal spectrum has the slope $\alpha \overset{\sim}{\sim} 0.7$. If later generations are important then $\alpha \to 1$.

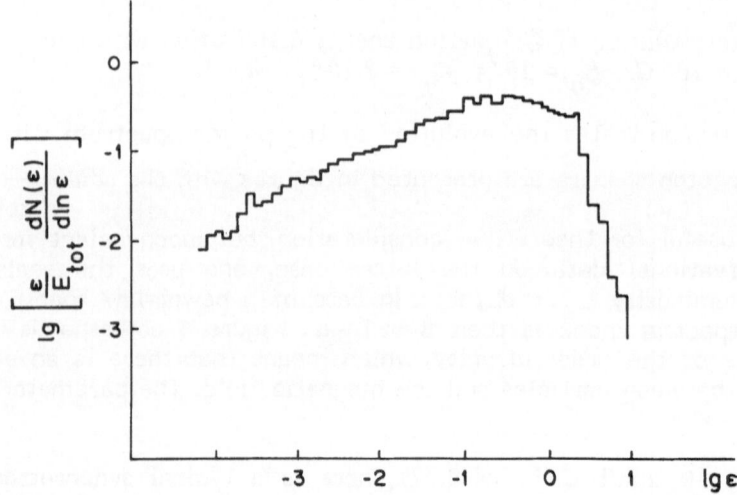

Fig. 2: The final photon energy distribution (t = 16) when the magnetic field is turned off at t = 0.12.

$H = 10^4$ Gs at $t < 0.12$,
$H = 0$ at $t < 0.12$
$E_0 = 10^6$, $\Omega = 5$.

Figure 2 demonstrates another result of our simulation. In this case we try to imitate the inhomogeneity of the magnetic field in the cauldron. For that we turn off the magnetic field at t = 0.12. As a result the spectrum becomes steeper because in this case synchrotron radiation is absent after t = 0.12 and only some part of the energy is transferred to the soft part of the spectrum.

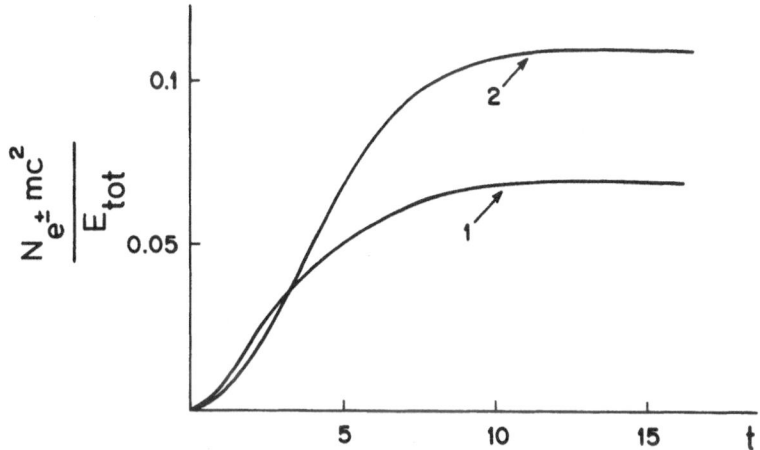

Fig. 3: The yield of pairs versus age.
 1. $H = 10^3 Gs$, $E_o = 10$, $\Omega = 5$.
 2. $H = 10^4 Gs$ at $t < 0.12$; $H = 0$ at $t > 0.12$,
 $E_o = 10^6$, $\Omega = 5$.

Figure 3 demonstrates the evolution of the number of pairs with age. The efficiency of $e^{\pm}$ pair rest-mass-energy generation could be equal to 0.1.

The natural question which arises is what the shape of the final spectrum would be in case there is no magnetic field in the region. (This question was raised by M. Rees at the discussion of our previous paper, Novikov and Stern, 1985.) We have done a numerical simulation for this case. The results of our simulation are presented in Figures 4 and 5. Obviously, in this case, we need some soft photon targets for injected particles. But it should be noted that in our simulation the energy in the form of soft photons is only a small part of the total energy, $E_{soft} = 5 \cdot 10^{-3} E_{tot}$. In case $E_{soft} \approx E_{tot}$ the final photon spectrum should be flatter. However, we feel that the $e^{\pm}$ cauldron model with the magnetic field seems more realistic. This model can explain the main features of the hard radiation from AGN and quasars. Thus an electron-positron cauldron may be a widespread phenomenon in the universe.

One of the authors (I.N.) acknowledges the hospitality of the Raman Research Institute and many discussions with its staff.

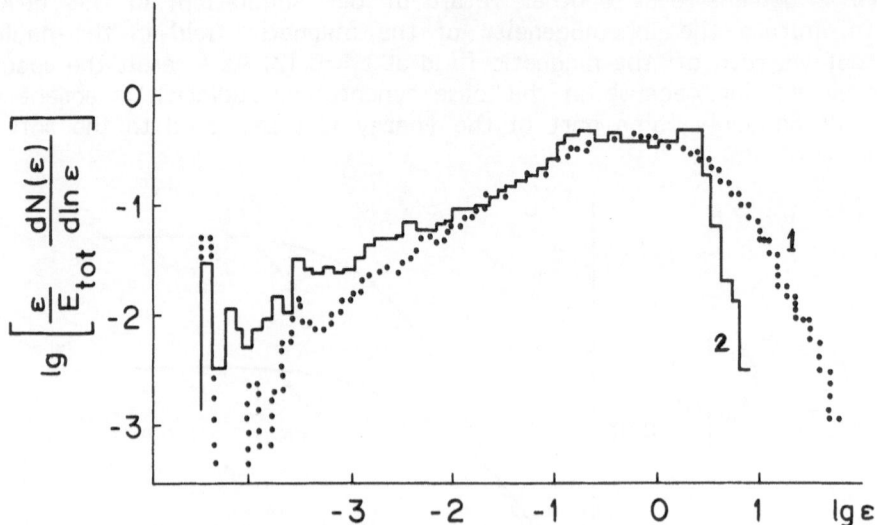

Fig. 4: *The final photon energy distribution (t = 18) for the cauldron without the magnetic field,* $E_{soft} = 5.10^{-3} E_{tot}$

$1 - E_0 = 2.7 \ 10^4, \quad \varepsilon_{soft} = 2.7 \ 10^{-5}$

$2 - E_0 = 470, \quad \varepsilon_{soft} = 2.7 \ 10^{-5}$

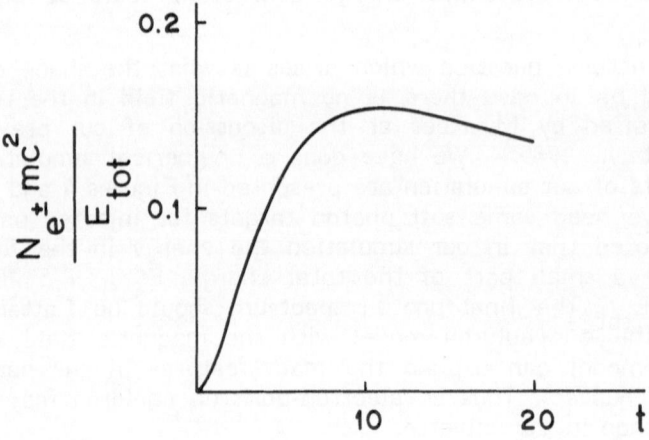

Fig. 5: *The yield of pairs versus age* $H = 0$, $E_0 = 470$, $E_{soft} = 2.7 \ 10^{-5}$, $E_{soft} = 5.10^{-3} E_{tot}$.

REFERENCES

Bignami, G.R., Fichtel, C.E., Hartman, R.C. and Thomson, D.J. 1979, Astrophys. J. 232, 649.
Blandford, R.D. 1986, this volume.
Bonometto, S. and Rees, M.J. 1971, Mon. Not. Roy. Astr. Soc. 152, 21.
Kardashev, N.S., Novikov, I.D., Polnarev, A.G. and Stern, B.E. 1983, Astron. Z. 60, 209.
Novikov, I.D. and Stern, B.E. 1986, in Triest Symp: Structure and Evolution of AGN, eds. G. Giuricin et. al. (Reidel).
Petre, R., Mushotzky, R.F., Krolik, J.H. and Holt, S.S. 1984, Astrophys. J. 280, 499.
Rees, M.J. 1986, this volume.
Rothschild, R.E. et. al. 1983, Astrophys. J. 269, 423.
Svensson, R. 1985, in IAU Collq. 89: Radiation Hydrodynamics in Stars and Compact Objects, eds. D. Mihalas & K-H. Winkler (Springer, Lecture Notes in Physics Series, to be published).
Stern, B.E. 1985, Astron. Z. 62, 529.
Zdziarsky, A. and Lightman, A. 1985, Astrophys. J. 294, L79.

DISCUSSION

Svensson : Besides the Monte Carlo calculations by Dr.Novikov et. al. it has also been possible to solve certain cases of nonthermal pair cascades analytically. Many of the properties of such cascades are then explained easily. (See review in IAU Coll.Nr.89, Radiation Hydrodynamics, and references therein.)

Novikov : I want to say in the review of Dr.Svensson many problems are solved analytically, and it is very important. But only numerical modelling of the processes can demonstrate their evolution with time and clear up the important details.

Wandel : In your model a single generation gives a spectral index of 0.75, appropriate for Seyferts, while for more than one generation the spectral index quickly converges to unity, which is more appropriate for optical quasars. Can the model account for the apparent difference between Seyferts and optical quasars ?

Novikov : Yes, it can. The slope of the spectrum depends on the parameters of the $e^{\pm}$ Cauldron, first of all on the initial ε_m. In the case $\varepsilon_m < 1$ the slope is $\alpha \sim 0.5$. If $\varepsilon_m \gg 1$ the slope is $\alpha \sim 0.75 \div 1$. So it is a test for the energy of the injected particles. In the case of quasars some additional mechanism of the formation of hard spectrum may be important also. For example, it could be the mechanism of the γ-beam model (see Kardashev et.al., 1983).

Rees : The slope of the X-ray continuum spectrum is influenced not only by the secondary relativistic pairs produced in the cascade, but also by Comptonisation due to the optically thick subrelativistic (quasi-thermal) pairs.

Blandford : I would comment that our (Fabian et.al. 1986) preliminary results indicate that steeper ($\alpha \sim 1$) X-ray spectra are produced when the compactness parameter is increased. This may be relevant to interpreting the optical quasars analysed by Dr.Elvis. However, before we rush to this sort of interpretation, we must convince ourselves that the three approaches (analytic, Monte Carlo and Kinetic) give consistent and robust results.

Segal : You and others are modelling the dynamics of a large-scale many-particle system in terms of 2- or 3-particle processes. Might it not be more natural to use a fluctuation in a statistical state of a QED field ? In particular, could a random fluctuation in the CBR be significantly involved in starting the process you describe ?

Novikov : May be, I do not know. But it is another problem.

Burbidge : Your unified scheme for the whole nonthermal spectrum is quite attractive. Can you relate it to the observations since presumably different parts of the spectrum arise in different sized volumes ?

Novikov : We have considered mainly the hard part of the continuum, but it is quite possible that the soft part of the continuum has the same origin. The Cauldron can operate if τ is great enough. For quasars it means that R is less than $\sim 10^{15}$ cm. The line spectrum arises on the larger scales.

Rees : One general statement, however might be that the pair plasma is restricted to length scales below $\sim 10^{17}$ cm (for quasar luminosities). It's important that the optically thick cloud of thermal pairs can have effects on the optical continuum as well as on X-rays - for instance, high synchrotron-type polarization would be destroyed. One interesting possible test, pointed out by Fabian and collaborators, is that the X-ray variability should display rises in hard X-ray flux that are stronger than the falls.

Canizares : Many compact galactic X-ray sources have "compactness" parameters, $\tau \gg 10^{30}$. Should not these processes then be more evident and detectable in these galactic sources than in AGN ?

Novikov : Probably only pulsars with appropriate parameters and accreting black holes can produce a large acceleration power in a small volume to give rise to an electron-positron Cauldron. May be some of the galactic X-ray sources can be an $e^{+} e^{-}$-Cauldron.

Rees : White <u>et. al.</u>, and other authors, have attributed the transition between two well-defined states in (for instance) Cygnus X1 - a high luminosity state with soft spectrum, and a lower luminosity state with harder spectrum - to the presence and absence, respectively, of an optically thick pair plasma.

Burke : Imaging studies on the scale suggested by your "Cauldron" calculations would be useful for verification. The necessary resolution scale could be reached by aperture-synthesis imaging at optical wavelengths. No such instruments are in prospect at the moment, but could be built if the programs can be started.

Rees : Perhaps gravitational lensing can do it, as suggested by Canizares and others ? That is the poor man's method, anyway.

Burke : One method depends on luck, the other on planning.

Dultzin-Hacyan : The cloud with which the central beam interacts (in the model of Kardashev et.al. (1983)) has very specific physical conditions (T and P). To what extent does the predicted observed radiation depend on these physical conditions of the cloud ?

Novikov : The physical parameters of the cloud in our γ-beam model could be considered as physically reasonable in the case when a gas-cloud target is a result of a disruption of a star by the tidal forces of a black hole or as a result of a collision with another normal star. We predict the strong variations of the continuum in the range 300-500 keV due to variations of thermal photon density in the cloud. Variability of the continuum and the annihilation line are the result of the motion of the cloud and the change of irradiation condition of the cloud by the directed beam.

THE REVIVED PENROSE PROCESS CAN POWER THE CENTRAL ENGINE IN ACTIVE GALACTIC NUCLEI

Sanjay M. Wagh and N. Dadhich
Department of Mathematics
University of Poona
Pune - 411007
India

ABSTRACT. Using the fact that the efficiency of the revived (Wagh et al 1985) Penrose process of energy extraction from black holes immersed in electromagnetic fields can be very high (Parthasarathy et al, 1986) we show that this process can comfortably power the 'central engine' in Active Galactic Nuclei. The microphysical Penrose process energized particles will be ultrarelativistic in the asymptotic frame. Hence the kinematical analysis of escaping photons by Piran and Shaham (1977) will be a good approximation to the kinematics of these particles. From this analysis one expects the energized particles to emerge within an angle~ 40° above and below the equatorial plane. These energetic particles, which are collimated in the funnel of an accretion disk and further on by the magnetic field, then, form supersonic, relativistic, bilateral jets. The relativistic γ factor for such jets can be expected to be ~ 2 since these ultrarelativistic particles will effectively mimick radiation in 'dragging' the matter already injected inside the funnel. Various implications of high energy extraction efficiency are illustrated.

Recently we have shown that the Penrose process of extracting energy from rotating black holes in electromagnetic fields, i.e., the magnetic Penrose process (MPP) is not only astrophysically potent (Wagh et al, 1985) but a very highly efficient way of doing it also (Parthasarathy et al, 1986). The efficiency per (relativistic or non-relativistic) event of energy extraction is given by

$$\eta \sim e_3 \Phi/m_1 \sim (e_3/m_1)M_8 B.10^{-8} . \tag{1}$$

In (1) $\Phi = - \underset{\sim}{A}.\underset{\sim}{\partial/\partial t}$, where $\underset{\sim}{A}$ is the four-potential of the axisymmetric, perturbative, and stationary electromagnetic field, and $\underset{\sim}{\partial/\partial t}$ is the static Killing vector of the Kerr geometry, is the electrostatic

G. Swarup and V. K. Kapahi (eds.), Quasars, 395–398.

potential as measured by the statictic observer. Also e_3 is the

charge of an escaping particle while m_1 is the mass of an incident particle. The last step of (1) follows from the dipole field approximation. Here M_8 is the black hole mass in units of $10^8 \, M_\odot$ and B is the asymptotic magnetic field in guass. As is evident $\eta \gtrsim 1$ for

realistic situations, $B \sim 10^{-6}$ gauss, $e_3/m_1 \sim 10^{17-14}$ esu/g- the

specific charge of an electron/proton. This efficiency estimate is valid near the statictic limit.

The luminosity that can arise due to the MPP is

$$L_{MPP} \sim \beta <n> \; BM_8 \dot{M}.10^{55} \text{ erg/s} \quad (r \sim 2M) \; . \tag{2}$$

Here β is the photon production efficiency, $\dot{M}$ -accretion rate in units of $M_\odot$ per year, and $<n>$ denotes 'average MPP events' that generate the luminosity. If the photons are produced by the synchrotron mechanism, then, $\beta \sim 0.1$. By way of comparison, the Blandford-Znajek (1977) luminosity is

$$L_{BZ} \sim 2.8 M_8^2 B^2 \; 10^{37} \text{ erg/s.}$$

By demanding that $L_{MPP} \sim L_{BZ}$ we obtain

$$< n > \; \sim (M_8 B/\dot{M}) \; . \; 10^{-17}$$

implying that even if the MPP is rare it still <u>can</u> power the 'central engine' in active galactic nuclei.

If the 'collisional MPP' (CMPP) of the type envisaged in non-magnetic scenarios by Piran and Shaham (1977) is to be important, then

$$\text{mean free path} = \lambda_{CMPP} \sim 6.64 \; (ve^\nu/\sigma) \; (M_8^2/\dot{M}) \; . \; 10^{-12}$$

$$\lesssim \; \text{proper thickness of target zone}$$
$$\text{(around } r \sim 2M) \; = \epsilon \, M \tag{3}$$

where $v \; (= v/c)$ is the mean velocity of the accreting particles in the local frame, $\exp(\nu)$ is the metric coefficient of the Kerr spacetime $(ds^2 = \exp(2\nu)dt^2 + ---)$, and $\sigma = \sigma_T \, f(\alpha)$ is the cross-section,

α being a suitable dimensionless parameter. This places a constraint on the accretion rate as

$$vM_8 \, e^{\nu}/\epsilon \, f(\alpha) \leq \dot{M} \quad , \quad \sigma_T = 0.67 \times 10^{-24} \text{ cm}^2 \tag{4}$$

Because $v \sim 10^{-2} -- 10^{-3}$ in astrophysical plasmas (and other factors of order unity) it is evident that unduly high accretion rates are not required. The CMPP, thus, can be expected to solve the problem of scarcity of fuel (Frank, 1978) without placing severe constraints on the galactic structure and evolution.

Demanding further that

$$t_{CMPP} = \lambda_{CMPP} / c \sim t_{sy} \lesssim t_{in-C} \tag{5}$$

where t_{sy} is the synchrotron lifetime and t_{in-C} is the inverse - Compton lifetime of electrons, we obtain

$$8\pi U_{ph} / B^2 \lesssim 1 \tag{6}$$

where U_{ph} is the photon energy density, as a consistency relation.

Thus, the inverse-Compton losses do not render the CMPP ineffective as an acceleration mechanism. That this Hoyle-Burbidge-Sargent (1966) condition is compatible with the requirements of the CMPP is not surprising. In view of (3) it can always be arranged to have a sufficiently small mean free path of collisions for sufficiently small velocity v. With a hindsight we can say that from (6) a very high energy extraction efficiency should have been suspected.
Since the CMPP energized particles will be ultrarelativistic in the asymptotic frame, the kinematical analysis of escaping photons by Piran and Shaham (1977) will be a good approximation to the Kinematics of these particles. From their analysis one expects the energized particles to emerge within an angle $\sim 40^{\circ}$ above and below the equatorial plane. These energetic particles, which are finely collimated inside the funnel of an accretion disk and further on by the magnetic field, then, form supersonic, relativistic bilateral jets. The relativistic γ factor for such jets can be expected to be ~ 2 since these ultrarelativistic particles will effectively mimick radiation in 'dragging' the matter 'already' injected inside the funnel of an accretion disk. This expectation is not at variance with the value of γ as observed in jets of active galactic nuclei.
In conclusion, we have shown that the CMPP can be expected to solve two of the major problems of modelling active galactic nuclei, namely, the fuel problem and the inverse-Compton-catastrophe problem. Although a detailed model is yet to be worked out, these above indications strongly suggest that the CMPP is indeed 'the candidate' for the 'prime mover' in active galactic nuclei.

ACKNOWLEDGEMENTS

   A research grant from the Indian Space Research Organization
under its RESPOND program is gratefully acknowledged.

REFERENCES

Wagh, S.M., Dhurandhar, S.V., and Dadhich, N. 1985, Ap. J., **290**, 12
Parthasarathy, S., Wagh, S.M., Dhurandhar, S.V., and Dadhich, N. 1986,
      Ap. J. , in the press.
Dhurandhar, S.V., and Dadhich, N. 1984, Phys. Rev. D., **30** , 1625.
Blandford, R.D., and Znajek, S.L. 1977, MNRAS, **179**, 433.
Piran, T., and Shaham, J. 1977, Phys. Rev. D, **16**, 1615.
Frank, J. 1978, MNRAS, **184** , 87.
Hoyle, F., Burbidge, G.R., and Sargent, W.L.W. 1966, Nature, **209**, 751.

NOTE

   A detailed version is submitted to   Nature for publication
(Wagh, Dhurandhar, and Dadhich 1986, preprint).

WIND-TYPE FLOWS IN QUASARS

Attilio FERRARI and Edoardo TRUSSONI
Istituto di Fisica Generale, Università di Torino
Istituto di Cosmo-geofisica del C.N.R., Torino, Italy

Jets are found in many astrophysical phenomena, from young stellar objects and collapsed stellar cores to quasar and radiogalaxies. Therefore they must represent a common dynamical phenomenon, while different morphological characteristics arise from the interaction with specific environments (Ferrari and Tsinganos 1985). In order to investigate the basic aspects of jet dynamics and morphologies, we have proposed a simple analytical treatment based on the fluid theory of stellar winds (Parker, 1963), which allows a direct test of different physical effects (Ferrari et al. 1985, 1986). We discuss here implications for the physical theories of quasars.

Quasar jets do not differ substantially from jets associated with radio galaxies (Bridle and Perley 1984). However a peculiar distinction in connection with the central core from which they emerge exists: while radiogalaxies have often weak cores, quasars are characterized by larger luminosities, both radio and optical, and this suggests a strong contribution of radiation pressure for jet acceleration.

An aspect of AGN, also relevant to the physics of collimated outflows, are the spatial distribution and width of optical emission lines. In some cases the spatial distribution is elongated and aligned with large scale radio jets, suggesting a direct connection between outflow hydrodynamics and cloud formation and/or excitation (Keel 1985). In addition line widths bear direct indication that AGN do in fact contain high-velocity clouds ($T \sim 10^4$ K) immersed in an optically-thin, hotter background, from sub-pc to kpc scales.

STEADY WIND EQUATIONS

Current AGN models assume that a galactic core consists of a compact massive body, $\gtrsim 10^8 M_\odot$, collapsed below its gravitational radius and surrounded by an accretion disk (cfr. reviews by Rees and Blandford in these Proceedings). The disk is heated via collisional dissipation of the inflow angular momentum and ionizes the surrounding optically emitting (coronal) plasma. Outflows are accelerated from the funnels along the spin axis; they also interact with the ionized plasma before escaping from the core.

The hydrodynamic equations for steady outflows in different physical and geometrical configurations have been presented in previous papers; for a general discussion we refer to the review by Ferrari and Tsinganos at the Toronto Conference on *Jets from Stars and Galaxies* (1985). We have adopted the wind theory to describe (i) the jet acceleration inside the accretion funnels and (ii) the jet propagation outside AGN with formation of various morphologies. The geometrical parameters of the funnels were assumed from standard models: the funnel length ranges from a few times up to a hundred times the

G. Swarup and V. K. Kapahi (eds.), Quasars, 399–403.

gravitational radius of the central mass, and above this scale the cross-section of the channel undergoes a sudden expansion. As an example, the flow acceleration is taken from electromagnetic radiation fields generated by disk's walls and luminosities up to L_{Edd} are considered. Other mechanisms (acceleration by plasma waves, electrodynamic effects, etc.) do not produce qualitatively different results in the outflow dynamics. Similarly, although we refer to an optically thin proton/electron outflow, a positron/electron and/or optically thick plasma can be treated, as we shall discuss in the following.

The basic physical elements of the model are: (i) the gravitational force exerted on the outflow by the concentrated mass of the galactic core and by the extended potential well of the associated galaxy; (ii) the jet plasma thermal pressure force; (iii) the pressure force determined by the geometry of the confining channel; (iv) the non-thermal momentum deposition by external fields (i.e. radiation pressure from the accretion disk's walls). We have discussed separately all these effects in terms of which the various phases of the propagation of jets from AGN can be reasonably reproduced. We refer to Figs. 1 for discussing the results.

The relevant solutions are the so-called wind solutions, connecting a subsonic outflow close to the center (in the figure $Z = 1$ is the base of the flow) with a supersonic outflow at infinity, even when the thermal velocity at the base of the flow is well below the escape speed. The classical topology for isothermal flows is illustrated in Fig. 1a, where the sonic transition occurs through an X-type critical point. Figs. 1b and 1c illustrate the case of momentum addition upstream the Parker critical point. O-type critical points appear in the regions where momentum is deposited. Wind solutions do not cross these points, but their shape is modified. Multiple wind solutions are possible for the same physical parameters and boundary conditions: one solution is continuous, while one or more discontinuous wind solutions with shocks appear (vertical dashed lines).

Multiple wind solutions with shocks are again present where the channel's cross-section changes sharply. Fig. 1d shows the specific result of the divergence of the funnel above the disk ($L = 0$). Fig. 1e shows, for $L \sim 0.8L_{Edd}$, the occurrence of a focal point. Fig. 1f shows how a very strong radiation field ($L = L_{Edd}$) can produce a supersonic flow close to the base.

When the mass distribution around AGN is non-point-like, i.e. $M_{grav} = M_{grav,0} Z^n$, continuous outflow solutions are possible only for $n < n_1$, with $n_1 = (5 - 3\Gamma)/(\Gamma + 1)$, where Γ is the plasma polytropic index; in the opposite limit the flow cannot escape to infinity, because it does not have sufficient internal energy, nor receive sufficient energy from external sources to escape with supersonic speed at infinity: the corresponding solutions are called trapped solutions (Fig. 1g).

For the phase of propagation of the jet through the galaxy and its extended halo, the interaction of pressure-confined outflows with environment may excite fluid instabilities of Kelvin-Helmholtz type: perturbations modulate the cross-section of the propagation channel. In the case of a sinusoidal perturbation of wavelength $\lambda \gtrsim 2\pi R$, where R is the channel radius, which corresponds to the most unstable mode of the instability theory, the sequence of equally spaced compressions and rarefactions gives rise to a sequence of alternating O-type and X-type critical points (see Fig. 1h); the continuous wind solution Mach number oscillates due to the (equivalent) momentum additions and subtractions. For sufficiently large perturbation amplitude (but still in the linear limit), shock transitions may occur between the first upstream transonic solution and the downstream transonic solutions.

A time-dependent treatment is required to define which solution is actually adopted by the outflow. The complexity of the topologies discussed arises as soon as one modifies the standard Parker wind model; this agrees with the result of numerical simulations which display sequences of shocks and channel's walls perturbations. A time-dependent code is being used to discuss the evolution of solutions and shall be presented in a forthcoming paper. On the other hand we stress that the present discussion on static wind solutions allows a clear understanding of the physical elements which cause the various morphologies.

APPLICATIONS TO QUASARS

(a) Jets are accelerated and collimated well inside the quasar core. In fact outflows become supersonic very close to the disk. If only geometrical effects are considered, this transition occurs at the exit of the accretion disk; when momentum addition by radiation fields is included, the flow becomes supersonic inside the funnels. This agrees with VLBI radio data suggesting that jets are formed inside the parent AGN.

(b) Stationary shocks at the exit of the accretion funnel can be responsible for AGN short-period variability. The occurrence and position of shocks in discontinuous wind solutions depend critically upon the shape of the funnel and the radiation field distribution. Slight variations of the disk parameters can actually enhance or inhibit the dissipation of the outflow energy inside the core.

(c) Critical funnel luminosities ($L \sim L_{Edd}$) yield asymptotic outflow velocities $\beta_\infty \lesssim 0.28$; velocities close to the velocity of light can be reached for supercritical luminosities. The large luminosities required to produce high velocities are allowed in the case of quasars, if not for all radiogalaxies. Larger velocities ($\gamma \sim 5$) can be obtained for electron/positron jets. A detailed numerical study of the interaction of photons and particles inside thick disk radiation funnels yields $\beta_\infty \lesssim 0.7$ for the case of optically thick plasma (Nobili et al., 1985). In both cases a radiation-driven-wind theory appears inadequate to explain superluminal jets, if these actually imply $\gamma \sim 10$.

(d) The formation of jets is forbidden for certain disk and radiation field parameters. In certain configurations with focal points, no continuous wind solutions are allowed; and if the shock conditions cannot be satisfied, no supersonic wind is produced outside the galactic core. This situation, as well as those in which the shock at the exit of the funnel is very strong, suggests that the physical parameters of the core itself can actually forbid the formation of a collimated jet. The directional energy transported by the flow is randomized and gives rise to bright galactic nuclei with broad emission lines, but without extended jets.

(e) The distribution of matter in galaxies and extended galactic halos defines the length of large-scale jets. Diffused matter distributions around the central cores do not allow jets to escape from the gravitational well, because the flow topologies are trapped when $n < n_1$: this may correspond to AGN which show some indication of collimated outflow in their nuclei (spatial distribution of optically emitting regions), but are not associated with extended radio jets. In addition this process must occur at some distance from the core also for non-trapped solutions when heavy halos are present; a termination shock is formed where the jet dynamical pressure is balanced by the external pressure and represents a way to understand the formation of extended lobes in radiogalaxies. The Space Telescope will provide relevant informations about matter distribution in AGN and halos, allowing a test of this suggestion.

(f) Knots of enhanced brightness along jets at all scales correspond to compressions and transverse shocks in the flow generated by the flow's channel modulation by fluid instabilities. The oscillations of the Mach number correspond to plasma compressions where the flow expands under the effect of the instability; shocks are the nonlinear evolution of compressions. These dynamical phenomena can be associated with localized particle acceleration and non-thermal radiation emission (Ferrari et al. 1986).

(g) Thermal instabilities are responsible for cloud formation in galactic cores. In our framework it is possible to suggest that hot jets and their boundaries are unstable to thermal instabilities and break up into clouds which cool down to low temperatures in agreement with the suggestions derived from optical lines observations (Bodo *et al.* 1985).

Support by Italian MPI and CNR is acknowledged.

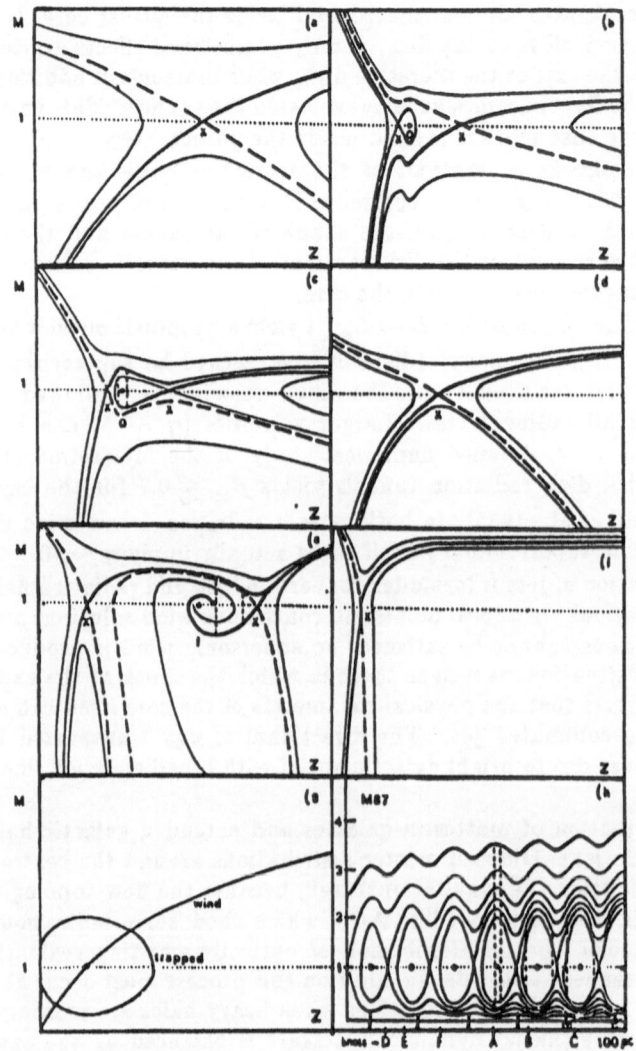

Figure 1. Typical topologies of wind-type flows (see the text for discussion)

Bodo, G., *et al.*, 1985, *Astron. Astrophys.* **149**, 246.
Bridle A.H., and Perley, R.A., 1984, *Ann. Rev. Astr. Ap.* **22**, 319.
Ferrari, A., and Tsinganos, K., 1985, in *Jets from Stars and Galaxies*, in press.
Ferrari, A., *et al.*, 1985, *Ap. J.* **294**, 397.
Ferrari, A., *et al.*, 1986, *Ap. J.* in press, January 15.
Keel W., 1985, *Nature*, **318**, 43.
Nobili, L., *et al.*, 1985, *M.N.R.A.S.*, **214**, 161.
Parker, E.N., 1963, *Interplanetary Dynamical Processes*, Interscience.

DISCUSSION

Perry : Multiple and irregular sonic points in flows with radiation pressure and mass injection were found by Beltrametti (1980 A and A.<u>86</u> 181), and she also discussed shock transitions from sub to supersonic flows in that paper.

Ferrari : Beltrametti's paper does not contain a full description of flow topologies and therefore cannot really discuss the existence of shock transitions although she argues that they may occur but are actually unstable. For understanding shocks, one needs a time-dependent approach that we have now adopted. Preliminary results do infact confirm the existence and stability of stationary shocks connecting different transonic solutions.

Abramowicz : I would like to stress that the existence of more than one critical point is a very important general property of relativistic flows with angular momentum. We found this also in accretion flows (Abramowicz Zurek, Lu and Livio). However both in the case described by Dr. Ferrari and in our works the model of the flow is rather simple. It seems that complicated and more realistic numerical studies (e.g. Smarr and Co.). give far more complicated shock structure.

Ferrari : The significance of quasi-analytic approaches as the one I presented, is not in representation of all details of the outflow pattern, but in interpreting complicated features of numerical simulations - The topologies I have discussed allow to trace back morphological effects to specific physical processes and this is essential to interpret simulations.

Blandford : These interesting calculations demonstrate that a relatively modest (compared with what must actually be required) increase in the input physics of an outflow leads to a formidable increase in the complexity of the critical point structure. Now critical points are, in a sense, fossil relics of the initial conditions - the only places from where transients cannot propagate away. Isn't this then telling us that the correct way to understand these flows is not to seek stationary solutions but to carry out time - dependent numerical calculations ?

Ferrari : Thank you for stressing this aspect of the solutions - We have performed time-dependent calculations starting from a Parker-type outflow and adding momentum at a given rate or considering specific geometries of the outflow channel. In fact the rate of momentum addition and the cloud's expansion scale length define which solution is undertaken by the flow - one can follow the evolution of transient structures, and define in which cases they relax to stationary shocks. In this sense stationary solutions are useful to understand the asymptotic flow pattern.

D.Dultzin-Hacyan, V.Petrosian, J.Bergeron and G.Swarup

A SHOCK MODEL FOR QUASARS AND ACTIVE GALACTIC NUCLEI [*]

I. Pérez-Fournon[1] and P. Biermann[2]

[1]Instituto de Astrofísica de Canarias, 38071 La Laguna,
Tenerife, Spain
[2]Max-Planck-Institut für Radioastronomie, Auf dem Hügel 69,
5300 Bonn 1, F. R. of Germany

ABSTRACT. First order Fermi acceleration is an efficient mechanism to
produce relativistic particles in strong astrophysical shocks. We have
developed a model using this theory to explain the non-thermal emission
from non-relativistic jets in Quasars and AGNs. Fluid dynamical simulations of cosmic jets predict shock formation in them at quasi-periodic
intervals. This is also suggested by observations of jets. We identify
the shocks as the sites of reacceleration of the energetic particles
required to explain the non-thermal emission of the jets. In order to
facilitate the comparison with observations, we have done numerical
simulations of the expected observational properties at high and low
spatial resolution.

1. THE MODEL

The basic assumption of our model is to consider jet knots, and compact
radio sources, planar shock waves of circular cross section with magnetic field parallel to the flow and constant diffusion coefficients upstream and downstream of the shock. Radiation losses are important in
extragalactic jets and in compact nuclei for high energy electrons.
Typical lifetimes for the electrons emitting at optical and higher frequencies are much lower than the light travel time from the nucleus.
Therefore, cut-offs in the electron distribution function and in the
corresponding synchrotron spectrum are expected (Webb, Drury and Biermann, 1984). Strong spectral steepening is actually observed in some
sources (Rieke et al., 1982). The first solution of the cosmic ray
transport equation in shocks is not strictly valid in this type of astrophysical shocks since it does not consider energetic losses. The solution in the presence of strong synchrotron and inverse Compton losses
has been obtained by Webb, Drury and Biermann (1984), hereafter WDB,
Schlickeiser (1984) and Bregman (1985).

The electron energy distribution as a function of distance to the
shock was obtained numerically (following the solution of WDB) and used
to calculate the synchrotron emission and absorption coefficients in all

[*] Discussion on p.424

G. Swarup and V. K. Kapahi (eds.), Quasars, 405–406 .

Stokes parameters for different frequencies. The synchrotron radiation transfer equations were solved to obtain the expected observational properties.

2.   RESULTS

The main results of our simulations for a flow direction perpendicular to the line of sight are: 1) the local synchrotron spectrum is a power-law with a cut-off at high frequencies (the spectral index at low frequencies is the same as the one obtained in the absence of losses), 2) the cut-off frequency decreases with distance to the shock, 3) the upstream emission is negligible whereas downstream the size of the emitting region decreases with increasing frequency, 4) there is a sharp edge at the position of the shock front, 5) the spectrum steepens, and the degree of linear polarization increases, with distance to the shock, and 6) a flip in the linear polarization position angle is expected for sources which are optically thick near the shock front and become optically thin with increasing distance to the shock. When finite observing resolution as well as projection effects are considered the results of our simulations resemble the general core-jet morphology of compact radio sources.

For shocks with compression ratio higher than 4 the results are slightly different due to the pile-up in the electron distribution function below the cut-off momentum. The main difference is that the position of the surface brightness maximum at low frequencies does not coincide with the shock front but is displaced downstream of it. This can contribute to the wavelength dependence of the position of the maximum observed in the Quasars 1038+528 A,B (Marcaide et al., 1985)

The overall continuum properties of unresolved sources such as BL Lacs and red Quasars were obtained from the electron energy distribution integrated over the whole shock wave region. The integrated spectral index is,below the shock front cut-off frequency, steeper by 0.5 than the local one.

ACKNOWLEDGEMENTS

I. Pérez-Fournon would like to thank the IAU Symp. 119 Organizing Committee for financial support.

REFERENCES

Bregman, J.N.: 1985, Ap. J. 288, 32
Marcaide, J.M. et al.: 1985, Astron. Astrophys. 142, 71
Rieke, G.H., Lebofsky, M.J., Wiśniewski, W.Z.: 1982, Ap. J. 263, 73
Schlickeiser, R.: 1984, Astron. Astrophys. 136, 227
Webb, G.M., Drury, L.O´C., Biermann, P.: 1984, Astron. Astrophys. 137,
    185

TRANSPORT OF RADIATION & ENERGETIC PARTICLES IN ACCRETION FLOWS[*]

R. Cowsik
Tata Institute of Fundamental Research
Bombay 400 005, India

M.A. Lee
University of New Hampshire
Durham, NH 03824, USA

ABSTRACT. The equations describing the transport of suprathermal charged particles and electromagnetic radiation across accretion flows onto compact objects are solved analytically, including the effect of shocks in the flows. These solutions indicate (a) accretion flows with shocks accelerate particles very efficiently upto ultra-relativistic energies. (b) the emergent spectra of electromagnetic radiation from such flows reproduce the observed spectra of quasars from the infrared to the hard X-ray region.

1. MATHEMATICAL FORMALISM

We consider an idealised spherically symmetric supersonic accretion flow with a shock transition at $r=r_s$. Parametrising the velocity at $r > r_s$ as $V=-V_0(r/r_s)^{-\alpha}$ and the diffusion constant $\kappa_{rr}=\kappa_0(r/r_s)^\beta x \ (p/p_0)^\gamma$, the transport equation for the distribution function of energetic particles (and quanta reads)

$$-\kappa_0\left(\frac{p}{p_0}\right)^\gamma \frac{1}{r^2}\frac{\partial}{\partial r}\left[r^2\left(\frac{r}{r_s}\right)^\beta \frac{\partial f}{\partial r}\right]-V_0\left(\frac{r}{r_s}\right)^{-\alpha}\frac{\partial f}{\partial r}+\frac{1}{3}V_0\frac{1}{r^2}\frac{\partial}{\partial r}\left[r^2\left(\frac{r}{r_s}\right)^{-\alpha}\right]p\frac{\partial f}{\partial p} = 0.$$

Defining σ through $V(r=r_s-\epsilon) = V_0(1-3\sigma^{-1})$, the boundary conditions at $r=r_s$ may be written as

$$-\kappa_0\left(\frac{p}{p_0}\right)^\gamma \frac{\partial f}{\partial r}+\kappa_-(r_s,p)\frac{\partial f_-}{\partial r}+\frac{V_0}{\sigma}p\frac{\partial f}{\partial p} = N(4\pi p_0 r_s)^{-2}\delta(p-p_0), \ \ f(r_s,p) = f_-(r_s,p),$$

where N is the total number of particles injected per second, and $\kappa_-(r,p)$ and $f_-(r,p)$ are the downstream diffusion coefficient and distribution function, respectively. Analytic solutions for $\alpha+\beta=1$ and for

---

[*] Discussion on p.424

G. Swarup and V. K. Kapahi (eds.), Quasars, 407–408.

$\alpha+\beta=2$ (applicable to photons) are obtained; these exibit power law behaviour $f \approx p\Gamma$. With $\eta = r_s V_o/\kappa_o$

$$\Gamma_{particle} = \alpha\{1+(1+\beta)\eta^{-1}\},(\alpha+\beta)=1 \; ; \; \Gamma_{photons} = \sigma\{1-2\beta\,\eta^{-1}(\tfrac{1}{3}\sigma-1)\},(\alpha+\beta)=2$$

The observed radiated power will have an index $\Gamma_i = \Gamma_{ph}+3$. These are shown in figs.1,2.

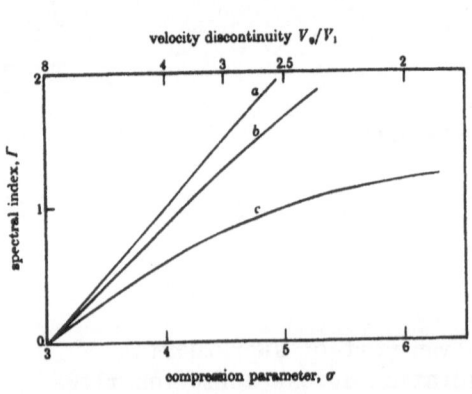

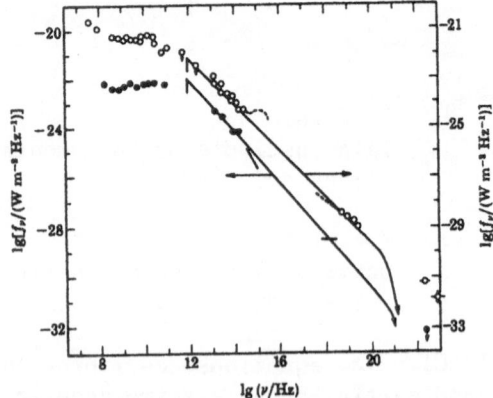

Fig.1. Γ_{ph} from IR to hard x-rays Typical $\eta=20\text{-}100$, $\sigma=4\text{-}5$ (a)$\eta=\infty$ (b) $\eta=30$ (c) $\eta=10$.

Fig.2. Fits to 2 quasars (1)3C120 (●) (2)3C272 (0). $\Gamma_1=0.97$, $\eta_1=30$, $\sigma_1=4.05$; $\Gamma_2=1.07$, $\eta_2=30$, $\sigma_2=4.22$.

Comparison between planar and accretion shocks under astrophysical conditions

Planer Shocks	Accretion Shocks
1) Shock dissipates its energy.	Shock is continuously supplied with energy by the accretion flow.
2) Particle acceleration occurs only at the shock	Particle acceleration every where since $\nabla.V$. is negative.
3) Acceleration rather inefficient since only particles crossing the shock front are accelerated.	Particles are convected into the shock and are accelerated efficiently.
4) Source of injection of super-thermal particles & photons at the shock not clear.	Strong post shock turbulence can be a good injector.

Also, it is instructive to compare the spectral index, Γ obtained by us (R.Cowsik and M.A.Lee, Proc. Roy. Soc. Lond. A383, 409, 1982) with that obtained by Payne and Blandford (MNRAS, 194, 1033, 1981) without any shock in the flow: $\Gamma = -3\beta^{-1}(1+\beta)$. The hardest spectrum they get is for $\beta = 2$ ($\Gamma = 9/2$) and for free-fall $\beta = 3/2$, $\Gamma = 5$. Both these spectra are too steep to reproduce the spectra of quasars. One may conclude shocks are important in giving the requisite hardness to the emergent spectra.

MODELS OF STEADY THICK ACCRETION DISCS AROUND A SCHWARZSCHILD HOLE *

M. R. Anderson
Institute of Astronomy
Cambridge
CB3 OHA
England

ABSTRACT. An outline is given of a project to model thick accretion discs around a black hole. A simple shear viscosity is assumed and the effects of radiation transport and thermal conduction neglected. Preliminary results are given for the equatorial structure of the model discs.

1. INTRODUCTION

Accretion onto a black hole differs in at least two important respects from accretion onto white dwarfs and neutron stars. Firstly, the black hole has no hard surface, so there is no guarantee that the potential energy of the infalling matter will ultimately be dissipated by viscous forces and radiated away. This makes plain the need for a dissipation mechanism which is efficient enough to account for the observed luminosities of active nuclei without postulating an unreasonably high rate of infall.

The second difference is that the poloidal velocity of the fluid, as measured by a co-rotating observer, is necessarily supersonic as it crosses the event horizon. Models of black hole accretion discs which, following the standard thin disc recipe, assume a subsonic poloidal velocity therefore ignore possible dissipative mechanisms - particularly multiple shocks - in the supersonic region.

Also, as Bondi (1952) pointed out in one of the pioneering works on stellar accretion, the condition that a flow pass through a sonic point removes one degree of freedom from the set of possible flow parameters. If the ambient conditions at infinity are specified, it is not in general possible to impose an arbitrary mass flux and obtain a steady solution. The value of the flux, in other words, is an eigenvalue of the problem. This is true for spherical accretion in the Schwarzschild metric (Michel, 1972), and for the equatorial plane of an inviscid, polytropic disc with rotation (Abramowincz and Zurek, 1981; Lu, 1984a,b).

* Discussion on p.424

G. Swarup and V. K. Kapahi (eds.), Quasars, 409–410.

2. THE MODEL DISCS

The author is presently engaged on a project which aims to generalise
the work referred to above to include thick discs with viscosity, in
both the Schwarzschild and Kerr metrics. For simplicity, the only
contribution to the viscous stress tensor is assumed to be rotational
shear. Heat conduction and radiative transfer are assumed to have
negligible effect on energy transport in the disc, and the fluid is
taken to be non-self-gravitating and to have a polytropic equation
of state.

When complete, it is hoped the analysis will yield the shape
of the disc surface, the locus of sonic points, the run of specific
angular momentum through the disc, and the maximum possible efficiency
for a given viscosity law. (The maximum efficiency is taken to be
the maximum value of the specific energy in the disc.)

So far, models of equatorial flows in the Schwarzschild metric
have been constructed which match onto the axisymmetric Newtonian
thick disc solutions due to Gilham (1981) at infinity. The specific
angular momentum diverges as the square root of the radial coordinate at infinity, and drops to less than 4GM/c at the event horizon
(where M is the mass of the hole). In all the models investigated to
date, the sonic point is located at between 1.2 and 1.5 Schwarzschild
radii. However, work by Abramowicz and Zurek (1981) suggests that
there may be a family of discs with sonic points located tens of
Schwarzschild radii from the event horizon. The best maximum
efficiency achieved so far is 1.5%, but this is anticipated to
increase to about 6% as the full range of the parameters is explored.

Further work will extend the analysis away from the equatorial plane,
and to the full Kerr metric.

3. REFERENCES

Abramowicz, M. and Zurek, W.   *Ap.J.* 246, 314 (1981)
Bondi, H.   *MNRAS* 112, 95 (1952)
Gilham, S.   *MNRAS* 195, 755 (1981)
Lu, J.   *ISAS preprint* 70/84/A (1984a)
Lu, J.   *ISAS preprint* 71/84/A (1984b)
Michel, F.   *Astrophys. Space Science* 15, 153 (1972)

ACCRETION DISK FLARES

P. A. Sturrock and W. Yang
Center for Space Science and Astrophysics
Stanford University, Stanford, California, USA

Flare activity on the sun may be attributed to the distortion of coronal magnetic field caused primarily by vortical (Shearing) photospheric motion. If an accretion disk is permeated by a magnetic field, the strong differential rotation of the disk will lead to similar stressing of the magnetosphere, and one may expect similar flare activity to occur. It is likely that the density of plasma in the maganetosphere is sufficiently large that the "frozen flux" condition is satisfied, and sufficiently small that the magnetic field will be substantially in the force-free state.

We have calculated the evolution of a model force-free magnetic-field configuration. For simplicity, the configuration is assumed to be of cylindrical symmetry. Differential rotation will tend to develop cylindrical symmetry, but any convection will tend to destroy it. It is assumed that the foot points of the magnetic field experience the differential rotation to be expected in an accretion disk. We consider the simple case that, on one side of the disk, the source of the magnetic field is provided by two narrow rings of opposite magnetic polarity. We assume that the field is initially in a current-free state and study the effect of progressive differential rotation.

We find that the principal effect of differential rotation is an "inflation" of the magnetic field that can be attributed to the pressure of the toroidal component produced by the differential rotation. We find that differential rotation can lead to magnetic free energy (excess energy compared with the current-free state) that is comparable with or even larger than the energy of the original current-free configuration.

The limiting state of the configuration is the open field configuration. Since the open field configuration contains an extended current sheet, the tendency to produce an open field is also a tendency to produce intense current densities. Such a situation is susceptible to magnetic reconnection by the tearing-mode instability The resulting particle acceleration will depend critically on the environmental conditions: magnetic field strength, plasma density, and photon density. The effect of reconnection on the magnetic-field topology is to restore the original current-free configuration but, in addition, to create a closed toroidal magnetic-field configuration (a "smoke ring").

G. Swarup and V. K. Kapahi (eds.), Quasars, 411–412 .

In a real situation, the reconnecting flux will be only a small part of the total flux of the magnetosphere: In this case, the toroidal plasmoid will be subject to the stress of the ambient magnetospheric field in such a way as to cause ejection of the plasmoid by the "melon-seed mechanism" (Schluter 1957).

In the real situation arising in an accretion disk, it is likely that the magnetic-field configuration is not cylindrically symmetrical. In this case, it is likely that plasmoids will be ejected at an angle to the rotation axis, although the average direction of ejection must lie along the axis. It is proposed that any single ejection may be responsible for a compact radio source which, in certain circumstances, may be detected as a "superluminal" source. The cumulative effect of flare-produced ejections may be a sequence of plasmoids moving more or less along the rotation axis: this may be the origin of the jets that are produced by some quasars and active galactic nuclei.

The generation and destruction of toroidal magnetic field also constitutes a mechanism for exchanging angular momentum between different regions of an accretion disk, contributing to the process normally attributed to viscosity (Pringle 1981). It therefore represents a procedure for converting gravitational energy into nonthermal electromagnetic energy. The efficiency of this process depends primarily on the comparative importance of angular momentum coupling by magnetic field with respect to angular momentum coupling by viscosity (including turbulent viscosity).

Although the large-scale structure of quasars and radio galaxies may be symmetric, the small-scale structure (compact radio sources and small-scale jets) is typically one-sided. Moreover, it is known that most quasars are radio-quiet. These considerations suggest that the magnetic field of an accretion disk develops episodically, by sporadic dynamo action, and that the two sides of the disk operate independently. We may suppose that, at any moment, there is probability p that dynamo action is occurring on either side of the disk. Then the probability that dynamo action is occurring on neither side is $(1-p)^2$, the probability that dynamo action is occurring on one side or the other is $2p(1-p)$, and the probability that no dynamo action is occurring is p^2. Since only a few percent of quasars are "radio-loud," we would infer that p is of order a few percent. Then p^2 is sufficiently small to explain why no quasar produces ejections in both directions. If dynamo action comes and goes on a time scale short compared with the time scale for development of the large-scale radio structure, we can understand why the large-scale structure may be more or less symmetric.

This work is supported in part by NASA grants NGL 05-020-272 and NGR 05-020-668.

REFERENCES

Pringle, J.E.: 1981, Ann. Rev. Ast. Astrophys. 19, 137.
Schluter, A.: 1957, IAU Symp. No. 4, 356.

DECOLLIMATION OF A JET DUE TO SPACETIME CURVATURE

R.C. Kapoor
Indian Institute of Astrophysics, Bangalore 560034

Central cores of compact radio sources are believed to contain super-massive black holes accreting material from their vicinity which produce fast moving plasma in the form of directed beams (jets), with apparent opening angles $\gtrsim 5°$ (Rees et al. 1981). In case, the jets are produced and their collimation is established on scales few times the Schwarzschild radius ($2m$; $m=GM/c^2$) of the central engine, deflection in particle trajectories in the curved spacetime (φ_0) would be large enough to widen the beam and thereby reduce the particle density and effective luminosity in the beam (Fig.1).

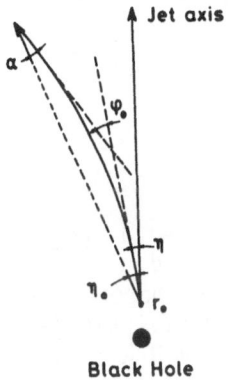

Fig. 1. Schematic Illustration of particle trajectory starting from $r=r_0$ near the black hole.

We have made a numerical study of particle motion in a Schwarzschild background to estimate the general relativistic contribution to the beaming process for various values of initial bulk velocities ($v= 0.8-0.999c$), for different opening angles ($\eta=1°-5°$) and for mean radii (r_0) varying from 2 to 8 time the Schwarzschild radius where jet material might be accelerated. We describe the degree of collimation

413

G. Swarup and V. K. Kapahi (eds.), Quasars, 413–414.

and attenuation of density by a quantity $\epsilon = (\sin \eta / \sin \eta_0)\,(d\eta/d\eta_0)$ where η_0 is the new value of the opening angle: $\eta_0 = \nu_0 + \eta - \iota (\alpha \to 0$ as $r \to \infty)$ and ρ_0 is determined from a knowledge of particle trajectories in the Schwarzschild spacetime; ϵ is found to be larger for smaller values of the initial bulk velocities and the mean radius considered (see Fig.2).

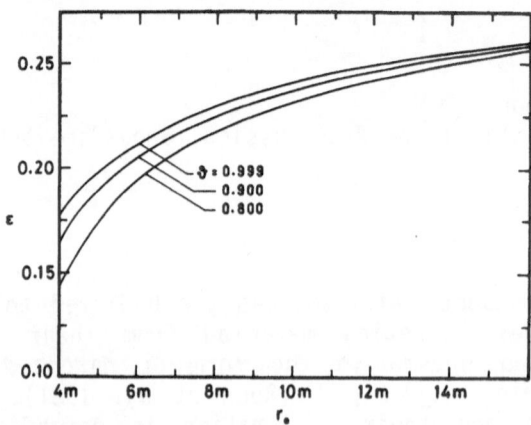

Fig.2.   Attenuation factor for various values of r_0 and ν.

The general conclusion that can be drawn from the calculations is that in the local rest frame the beam starts out with an opening angle less than half and a particle density as well as an effective luminosity in the beam ϵ^{-1} (~ 10) times larger than the values implied by beam models provided $r_0 \lesssim 20m$; for $r_0 > 20m$, the effect becomes comparatively less in magnitude (c.f. Abramowicz and Piran 1980; Sikora and Wilson 1981).   Details of this work will be published elsewhere (Kapoor 1986).

References

Abramowicz, M.A., Piran,T. 1980 Ap.J. <u>241</u>, L7
Kapoor, R.C., 1986 to be published.
Rees, M.J., Begelman, M.C. Blandford, R.D. 1981 Ann.N.Y.Acad. Sci.
        10th Texas Symp. Relativistic Astrophysics p.254.
Sikora. M, Wilson, D.B., 1981, M N R A S, <u>197</u>, 529

MODELS FOR EXTRAGALACTIC RADIO SOURCES

W.Kundt
Institut für Astrophysik der Universität
Auf dem Hugel 71
D-5300 BONN
WEST GERMANY

ABSTRACT. Doubt is cast on the reality of the following four assumptions commonly made in the treatment of extragalactic radio sources : (1) The central engine is a black hole; (2) Electrons can be accelerated in situ in the knots and heads of the jets, to large Lorentz factors $\gamma \gtrsim 10^2$, with an efficiency exceeding 30%; (3) The (non-thermal) radiation emitted by the beam fluid is isotropic in some (comoving) Lorentz frame; and (4) The flow velocity in the jets of SS 433 is v = 0.26 c.

1. THE CENTRAL ENGINE

A black hole represents the ultimate state of any high mass concentration. On the other hand, hydrogen-burning can re-expand a massive rotator, of compactness $u:=R_s/R \lesssim 5.10^{-3}$ (R=radius, R_s=Schwarzschild radius), in the form of a strong stellar wind (Kundt, 1979). Such winds are observed throughout the BLR (Stoner & Ptak, 1985), NLR and perhaps even in the form of large-radius emission shells (Kundt & Krause, 1985). If such winds had not blown, the accumulated consumption integrated over cosmic times of a putative black hole would be in marginal conflict with upper bounds on the unresolved core masses of nearby galaxies (Soltan, 1982, who assumes a high conversion efficiency $\varepsilon \gg 1\%$ and very conservative core-mass upper bounds).

The black-hole model has also difficulties in explaining rapid UV-variabilities (Stoner & Ptak, 1985) and polarization, which ask for strong deviations from axial symmetry of the driving engine.

Another indication against black holes in AGN is given by their similarity to the binary neutron-star sources Sco X-1 (cf. Kundt & Gopal-Krishna 1984) and SS 433 (Margon,1984) as well as to young stellar objects (YSO; cf.Konigl,1982) and a few white-dwarf systems.In all these cases, supersonic twin-jets are apparently produced by a non-collapsed magnetised rotator at the center of a feeding disk (Kundt, 1984a).

2. IN SITU ACCELERATION

It is often argued that the short lifetimes of the synchrotron-emitting electrons in the optical knots of the jets ask for in situ acceleration.

G. Swarup and V. K. Kapahi (eds.), Quasars, 415–416.

On the other hand, a TV screen shows that in situ deceleration is an
alternative possibility to explain bright spots in a jet : in the absence of significant obstacles (such as filamentary obstructions, or photons), extremely relativistic charges can move loss-free ($\vec{E}\times\vec{B}$-drift).
The high efficiency claimed by workers on shock acceleration models is
by no means established (Kundt, 1984b).

3. RELATIVISTIC BEAMING

Almost all publications on relativistic beaming have assumed that the
jet-fluid radiates isotropically in its comoving frame (e.g. Kellermann
& Pauliny-Toth, 1981; Begelman et.al.,1984). This assumption implies
different values of the bulk Lorentz factor γ (between 2 and 20) for
different questions asked (Scheuer & Readhead,1979). The assumption
lacks plausibility and is inconsistent with a relativistic power-law
distribution of the supersonically streaming charges (Kundt &
Gopal-Krishna, 1980, 1981). As demonstrated clearly at this Symposium,
it leads to all kinds of difficulties. These difficulties go away when
(intensity-dependent) wide-angle beaming patterns of the collision-free
jet-material are permitted. No tight correlaiton then exists between γ
and the inclination angle of a jet. All values derived for γ are strict
inequalities and consistent with $\gamma \gtrsim 10^2$ (Kundt, 1982a).

  An extremely relativistic bulk velocity of the streaming pair plasma is a natural consequence of the fact that on its way from the BLR to
the outer head, this plasma reduces its pressure from $\gtrsim 1$ dyn cm^{-2} to
less than 10^{-8} dyn cm^{-2} without excessive dissipative losses (Kundt,
1982b). Not even SS 433 need be a counter example to this 'unified
scheme' (Kundt, 1985).

4. References

Begelman,M.C., Blandford,R.D., Rees,M.J., 1984 : Rev.Mod.Phys. 56, 255
Kellermann,K.I., Pauliny-Toth, I.I.K.,1981 : Ann.Rev.Astron.Astrophys.
  19, 373
Konigl,A., 1982 : Astrophys.J. 261, 115
Kundt,W., 1979 : Astrophys. Sp. Sci. 62, 335
Kundt,W., 1982a : IAU 97, eds. Heeschen & Wade, Reidel, Dordrecht, p.265
Kundt,W., 1982b : Mitt. Astron. Ges. 57, 65
Kundt,W., 1984a : Astrophys. Sp. Sci. 98, 275
Kundt,W., 1984b : J. Astrophys. Astr. 5, 277
Kundt,W., 1985 : Astron. Astrophys. 150, 276
Kundt,W., Gopal-Krishna, 1980 : Nature 288, 149
Kundt,W., Gopal-Krishna, 1981 : Astrophys. Sp. Sci. 75, 257
Kundt,W., Gopal-Krishna, 1984 : Astron. Astrophys. 136, 167
Kundt,W., Krause,M., 1985 : Astron. Astrophys. 142, 150
Margon,B., 1984 : Ann. Rev. Astron. Astrophys. 22, 507
Scheuer,P.A.G., Readhead,A.C.S., 1979 : Nature 277, 182
Soltan,A. 1982 : Mon. Not. Royal Astron. Soc. 200, 115
Stoner,R., Ptak,R. 1985 : Astrophys. J. 297, 611

SOME CONSTRAINTS ON QUASAR PROGENITORS AND QUASAR EVOLUTION

C.SIVARAM
Indian Institute of Astrophysics
Bangalore 560034
India

There is strong evidence that quasars are powered by the gravitational energy released by in fall of matter into a compact super-massive object which is most likely a black hole with mass $M > 10^6$ to $10^9 M_0$. The question then arises as to how such massive black holes came to be formed. What sort of objects were the progenitors? While looking for an answer we have to bear in mind another bit of intriguing evidence that even high Z quasars show lines of heavier elements (Oxygen, magnesium, etc) with solar abundances, strongly suggesting that matter had already undergone nuclear processing at even earlier epochs. Again even the oldest stars in our galaxy show evidence of metal content. All this suggests that atleast part of the metal content could have been generated by pregalactic supermassive stars, sometimes referred to as Pop.III objects. Such stars with masses ranging from $10^2 - 10^6 M_\Theta$ could have formed in clusters consisting of about a few hundred or thousand of these objects at $Z \simeq 10$. Now objects $> 10^6 M_\Theta$ would have central temperatures not high enough for nuclear reactions the reason being a well known instability of general relativistic origin which is reached when $GM/R \approx 9c^2/16 \ (M_0/M)^2$; $M_0 \approx M_\Theta$ giving rise to a maximum central temperature T_c reached at the instability, i.e. just before collapse. This is given by $T_c = 2 \times 10^{13}/(M/M_0)$ giving $T_{max} = 10^7 K$ (necessary for CNO reactions) for $M \simeq 10^6 M_0$ and for $M > 10^6 \ M_\Theta$, T_c is too low. It is believed that 10^{-5} of the primordial material might have been processed in Pop III objects which would have undergone nuclear reactions and evolved over some fraction of a Salpeter time, $c\sigma_T/4\pi Gm_p \simeq 10^8$ yrs; (σ_T = Thompson cross section, m_p is proton mass)., producing heavier elements in the process. It turns out that objects of mass $M < 300 \ M_\Theta$ can explode and scatter the heavy elements. For instance a 116 M_Θ oxgen core of a $200 M_0$ object can completely disrupt and the evolution of such objects can explain anomalies like the O/Fe enrichment in metal poor stars and the G dwarf problem. Again there can be substantial mass loss from these superstars (a typical empirical relation like $M \simeq (t_D/t_k H)^{1/2}$. L/GM/R (where t_D is the dynamical time, and $t_k H$ is the Kelvin-Helmholtz scale, L is the luminosity) predicting $\dot{M} \simeq 10^{-3} M_0/yr$) again leading to enrichment of the medium. Stars $> 300 \ M_0$ do not explode but would collapse. The relaxation time scale for a cluster of these objects is again 5×10^8 to 10^9 yrs,

G. Swarup and V. K. Kapahi (eds.), Quasars, 417–418.
© *1986 by the IAU.*

so that they would all collapse to form a central black hole of $M \sim 10^8 - 10^9 \, M_0$, the surrounding earlier ejected matter enriched in heavier elements being now accreted onto it.   There would have been atleast a few times 10^8 of these black holes formed. Assuming near Eddington luminosity the rate of accretion given by $\dot{M}(t) \simeq 8\pi G M(t)/kc \simeq 16\pi G^2 M^2(t) X$ $\rho(t)V/c^4$, $\rho(t)$ being a function of Z) would put a constraint on when Pop III stars formed and accreted enough matter to trigger quasar activity as quasars do not seem to be present before $z = 4$.   This gives the result that they could not have formed before $z \simeq 8-10$. Another possibility of forming a central black hole is by the collapse of a dense star cluster (consisting mainly of neutron stars or white dwarfs), the central relaxation time is $t_R \sim V_c^3 / m_c^2 \ln(0.4 \, N_c) \sim 10^9$yr.   One obtains the result that for a core of compact stars to evolve to a relativistic state in a Hubble time it should have a minimal velocity dispersion given by V_c (min)$\simeq 10^3 (m_c)^{(1-A/B)} (7-3A/B)$km/s,   A, B are constants. This corresponds to clusters roughly having $10^7 - 10^9$ compact stars and the gravitational collapse of such a cluster would then lead to black holes in the required mass range.   An observed example is the central 3.5 pc core of NGC 4151 which has a mean stellar density of $2 \times 10^8 \, M \, \text{pc}^{-3}$ corresponding to $V_c \simeq 10^3$ km/s.   As the central black hole grows the relative tidal force decreases and quasar activity can stop when the hole can no longer break up stars. (Tidal break up is also producing and scattering heavy elements.) For a central black hole accreting white dwarfs $M < 10^6 M_0$ and for it to break up neutron stars $M < 2 \cdot 10^3 M_0$. The break up of a neutron star by such a hole would give a burst of energy $\sim 10^{52}$ergs/sec which has so far not been recorded in any quasar or AGN, the upper limit to the luminosity being $\sim 10^{49}$ ergs/s.   This may be indirect evidence that the mass of the central black hole is well above $10^4 \, M_0$ in all cases.   It is a pleasure to thank Professor Martin Rees for valuable discussions.

STIMULATED RAMAN SCATTERING IN ACTIVE GALACTIC NUCLEI

V. Krishan
Indian Institute of Astrophysics
Bangalore 560034, India

P. J. Wiita
Department of Astronomy & Astrophysics
University of Pennsylvania
Philadelphia, PA 19104-6394, USA

ABSTRACT. Stimulated Raman scattering (SRS) processes offer an attractive and efficient method for producing both essentially the entire non-thermal continuum as well as fast electrons in active galactic nuclei (AGN). In this picture, electrons are accelerated by Langmuir waves which are generated by Raman forward scattering (RFS); these electrons then rapidly radiate their energy by means of Raman back scattering (RBS) off of spatially periodic magnetic fields. Such periodic fields can be produced by magnetic modulational instabilities of the Langmuir field. The emission is envisaged to arise from an expanding region, with the highest frequency radiation originating from the smallest volumes at the core of the AGN. Time variability is dominated by density fluctuations in these magnetohydrodynamic flows.

1. INTRODUCTION

We argue that the bulk of the power emitted by active galactic nuclei, from the radio through gamma-rays, could be coherent SRS radiation produced when a pump field is scattered off the collective mode of a relativistic electron beam. Such a pump could be either electromagnetic waves or static periodic electric or magnetic fields. In more standard models of AGN synchrotron photons are the pump and produce higher energy photons via inverse Compton scattering (CS; Wiita 1985). In SRS the scattering is off of Langmuir waves, which are collective excitations of high energy electrons. SRS is faster and more efficient than Compton scattering, since the Compton scattering rate is proportional to the square of the amplitude of the pump while the SRS rate is linear in that quantity (Hasegawa 1978). We propose RFS as the acceleration process and RBS as the emission mechanism in AGN, where the rate of loss of energy of the electrons via RBS is balanced by the rate of gain of energy in the Langmuir field generated by RFS. Details are in Krishan (1983, 1984, 1985) and Krishan and Wiita (in preparation).

G. Swarup and V. K. Kapahi (eds.), Quasars, 419–420.

2.   ACCELERATION AND LOSS MECHANISMS

Langmuir fields can be generated when two electromagnetic waves beat at
frequency $\omega_p = \omega_0 - \omega_1$ producing a plasma wave of phase
velocity $v_p \approx c[1-(\omega_p/\omega_0)^2]^{1/2}$. This wave saturates by
trapping electrons, and since $v_p \approx c$ the electrons tend to remain in
phase with the wave and are highly accelerated, via an induced electric
field given by $E_p \approx m\omega_p cn'/(en)$; here the maximum value of the
density fluctuation, $n' \approx n$, the ambient plasma's density. The time
needed to gain energy γmc^2 is $t_a \approx \gamma/\omega_p$ (Krishan 1985).
      Plasma oscillations with inhomogeneous phase distributions produce
vortical currents which can increase the spontaneously generated
magnetic field through magnetic modulational instabilities (Bel'kov and
Tsytovich 1979). These fields have spatial periodicities between
$\omega_p/c < \kappa_0 < \omega_p/v_{th}$ with amplitude $B = eE_p^2/(4mc\omega_p)$.
      RBS occurs when the difference of the frequencies and wave vectors
of the pump wave and the scattered wave equal those of the plasma
frequency of the electron beam. Equating the resulting rate of loss of
energy to the gain from RFS one finds $n_b/n \approx 8/\gamma$, with n_b the
electron beam density. While the ambient plasma takes part in the
acceleration, only beam particles are responsible for the radiation.

3.   APPLICATION TO AGN

The maximum amplitude of the scattered field is limited by the shift
from RBS to CS, and yields the luminosity $L \sim 10^7 nA/\gamma^2$ at the
frequency $\omega = 2\gamma^2\omega_p$, where A is the cross-sectional area of
the beam. A plot of L/ω vs ω gives a spectral index of -1 as
long as (nA) and γ are constant for all frequencies. This would
define a density variation law $n \propto R^{-2}$, corresponding to an uniformly
expanding region. If, in general, $n \propto R^{-\beta+2}$ and $\gamma^2 \propto R^{-\alpha}$ we find
that the spectral index is given by $\Gamma = (2\alpha+1-\beta/2)/(\alpha+1+\beta/2)$.
      Observations of quasars and other AGN (Wiita 1985) can be fit into
this framework if we take $L_x = 5\times10^{45}$ erg/s at $\omega_x = 5\times10^{17}$Hz and
assume $\alpha = 1$. Then, $n_x \approx 7\times10^{12}$cm^{-3} and $R_x \approx 10^{16}$cm. Extensions to
the optical give $R_0 \approx 22R_x$, $n_0 \approx 10^{10}$cm^{-3}, and $L/\omega \approx 6\times10^{31}$erg/Hz/s
at 10^{15}Hz; while in the radio, $R_r \approx 50R_0$, $n_r \approx 6\times10^2$cm^{-3} and $L_r/\omega \approx$
10^{35}erg/Hz/s at 5×10^9Hz. An AGN's typically flat spectrum in the radio
implies dominance of SRS by synchrotron emission there; the usual cutoff in gamma-rays can be attributed to absorption by pair production.

REFERENCES

Bel'kov, S. A. and Tsytovich, V. N. 1979, Sov. Phys. JETP, 49, 656.
Hasegawa, A. 1978, Bell System Tech. J., 57, 3069.
Krishan, V. 1983, Astrophys. Lett., 23, 133.
Krishan, V. 1984, Astrophys. Lett., 24, 21.
Krishan, V. 1985, Astrophys. Space Sci., 115, 119.
Wiita, P. J. 1985, Phys. Reports, 123, 117.

A NOVEL INTERPRETATION OF THE COSMIC CONSPIRACY IN AGNs[*]

Colin S. Coleman
Department of Astrophysics,
South Parks Road,
Oxford, OX13RQ
England

ABSTRACT. The flat radio spectra of core dominated AGNs are often attributed to a superposition of optically thick synchrotron components. This is known as the cosmic conspirac A radio index $p_r \sim 0$ is not a natural consequence of this model, and an alternative model proposed here interprets the source components as regions in which relativistic electrons stream along a magnetic field. Stream collimation is determined by plasma instabilities, and components are assumed to be optically thin, with cyclotron turnover at low frequencies. An appealing feature of this model is that radio photon index depends only upon electron energy distribution, which may be obtained from the photon index above the turnover frequency. An energy power-law with index $s \gtrsim 3$ is typical for these sources, and implies a flat radio spectrum with $p_r = (4/3) - p \sim 0$.

Relativistic particles streaming through a non-relativistic plasma are most effectively scattered by Alfven (A) mode oscillations propagating parallel to the magnetic field (Achterberg, 1980). This remains true for a strongly magnetized plasma, in which magnetic energy density exceeds rest mass energy density (Coleman, 1986a). The A-mode dispersion relation for parallel propagating oscillations is given by:

$$n^2 = 1 + \omega_P^2 / \omega_B (\omega_B - \omega) \tag{1}$$

with $n = kc/\omega$ the refractive index, ω_P the plasma frequency and ω_B the proton cyclotron frequency. This dispersion relation breaks down near the proton cyclotron frequency due to thermal proton motion, where A-mode waves are strongly damped by proton cyclotron resonance absorption.

---

[*] Discussion on p.424

G. Swarup and V. K. Kapahi (eds.), Quasars, 421–423.
© 1986 by the IAU.

Proton cyclotron damping certainly precludes A-mode propagation in the frequency band (Akheizer et al., 1975):

$$(\omega_B^2 - \omega^2) \; / \; 2k^2\beta_p^2c^2 \; < 1 \tag{2}$$

where $\beta_p c$ is the average thermal proton velocity. In this band, real and imaginary parts of the refractive index are of a similar magnitude, and hence damping is strong. Damping is not effective over a much broader frequency band because the co-efficient falls off exponentially with decreasing frequency. Substitution of dispersion relation (1) provides an estimate of the maximum weakly damped A-mode frequency:

$$\omega_d \sim \omega_B \; (1-\sqrt{2}\beta_p) \tag{3}$$

which corresponds to a propagation velocity specified by the Lorentz factor:

$$\Gamma_d^2 \sim 1 + \sqrt{2}\beta_p \; (\omega_B^2/\omega_P^2) \tag{4}$$

Nuclear jets in AGNs may be sufficiently hot and strongly magnetized for all undamped waves to propagate at a high Lorentz factor. The physical parameters required do not contravene virial limits on the field strength near a 10^8 $M_\odot$ compact object, nor the density estimates obtained from Faraday depolarization measurements and other arguments. In such a plasma relativistic electrons with pitch angles $\alpha \gtrsim 1/\Gamma_d$ are not scattered by plasma waves, and the force due to a field gradient causes them to develop into streams propagating with approximate Lorentz factor Γ_d. During this process the energy index s of an initial power-law distribution is increased by unity to s+1 (Coleman, 1986b).

The optically thin non-thermal spectrum of relativistic electrons streaming at small pitch angles is a power-law with low frequency cyclotron turnover. The turnover frequency ν_t is given by (eg. Pacholczyk, 1977):

$$\nu_t \sim \nu_B \; / \; \sin(\theta) \tag{5}$$

The angle θ between magnetic field and line of sight is identified with the average electron pitch angle $\bar{\alpha} \simeq 1/\Gamma_d$. The strong fields and small pitch angles envisaged here are responsible for promoting the cyclotron turnover to high radio frequencies, where the cosmic conspiracy operates.

Equations (4) and (5), and the fact plasma density scales with field strength B in the case of perfect MHD flow, indicate that turnover frequency $\nu_t \sim B^{3/2}$. For a superposition of several electron stream sources in regions of different field strengths, flux densities at the turnover frequency scale as B^2 and the composite low frequency spectrum has the general form:

$$F_\nu \sim B^2 \nu_t^{-p} \sim \nu^{\{(4/3)-p\}} \tag{6}$$

A flat radio spectrum clearly requires a high frequency spectral index $p \simeq 4/3$, which is consistent with the mean infrared spectral index of blazars (Gear, et al., 1985). If electrons are initially injected with energy index $s \gtrsim 2$ ($p \gtrsim 1/2$), the effect of a field gradient is to produce a steepening to $s \gtrsim 3$ ($p \gtrsim 1$), and the typical continuum spectrum of strong core AGNs is obtained.

In any synchrotron self-absorbed source, circular polarization undergoes a marked sign change near the turnover frequency. It is a firm prediction of this model that no such sign change be found in core dominated AGNs. Furthermore, individual source components (which have been identified with superluminal knots in several objects) must not exhibit strong spectral evolution, since synchrotron and adiabatic cooling processes have little effect upon the energy distribution of relativistic electrons streaming with small pitch angles. Both these predictions could be settled by making careful observations using current or planned instrumentation.

References:

Achterberg, A., 1980 Astron. Astrophys., 98, 161.

Akhiezer, A.I., Akhiezer, I.A., Polovin, R.V., Sitenko, A.G. and Stepanov, K.N., 1975 "Plasma Electrodynamics", Vol. I, Pergamon Press.

Coleman, C.S., 1986a "Structure and Evolution of AGNs", ed. Giuricin, G. et al., Reidel.

Coleman, C.S., 1986b Mon. Not. R. astr. Soc. (in press)

Gear, W.K., Robson, E.I., Ade, P.A.R., Griffin, M.J., Brown, L.M.J., Smith, M.G., Nolt, I.G., Radostitz, J.V., Veeder, G. and Lebofsky, L., 1985 Astrophys. J., 291, 511.

Pacholczyk, A.G., 1977 "Radio Galaxies", Pergamon Press.

DISCUSSION ON THE PAPER BY **FEREZ-FOURNON** AND **BIERMANN** (p.405)

Kundt : Every TV-set is a counter example to in-situ acceleration; which is an assumption, not a must.

Cowsik : The observations on jets and knots for eg. in M87 and in 3C273 do not compellingly support models with in-situ acceleration. A relativistic plasmoid moving with bulk speed of $\gamma \simeq 10$ can also explain the observations. (cf. Cowsik et.al. Proc. International Cosmic Ray Conf. 1983, Bangalore, Eds. Ramanamurthy et.al.)

Blandford : Did you use an energy independent diffusion coefficient ? If so this might be rather unrealistic.

Perez-Fournon : I used a momentum and position - independent diffusion coefficient. This is possibly unrealistic but it is the best you can do as a first approach to the problem.

DISCUSSION ON THE PAPER BY **COWSIK** AND **LEE** (p.407)

Rees : Is this paper similar to the earlier work by Blandford and Payne ?

Blandford : Yes, similar in spirit, though Cowsik and Lee considered photon acceleration at a spherical shock front whereas Payne and I computed the emergent flux for a smooth spherical inflow.

DISCUSSION ON THE PAPER BY **ANDERSON** (p.409)

Rees : Does your estimate of efficiency depend on any assumptions about the radiation mechanism ?

Anderson : The "efficiency" of the models is the maximum possible efficiency - that is, the energy dissipated as a fraction of the rest mass energy. The models do not take radiation transport into account. Whether all the dissipated energy can escape the disc will depend on the actual radiation mechanism and the disc geometry.

DISCUSSION ON THE PAPER BY **COLEMAN** (p.421)

Rees : Self-absorption must occur in your model at sufficiently low

G. Swarup and V. K. Kapahi (eds.), Quasars, 424–425.
© *1986 by the IAU.*

frequencies (unless you invoke stimulated emission effects arising from anisotropic pitch angle distributions or cyclotron maser action in the comoving frame). Is this always at sufficiently low frequencies that you can ignore it ?

Coleman : It is likely that synchrotron self-absorption is responsible for low frequency turnover in regions of particle acceleration, where the pitch angles are presumably isotropic. Under the conditions which I consider here, however, the strong fields and small average pitch angles promote the cyclotron turnover to well above the self-absorption frequency. It is easy to check that the kinetic temperature of these streaming electrons is far higher than the observed brightness temperature, and hence the source is optically thin.

Peacock : Do you think your particle streaming model provides a good explanation of the occurence and statistics of superluminal motion ?

Coleman : Yes. In a field dominated plasma the Alfven velocity is relativistic, and particles (e^+, e^-) may stream at this velocity. It is then natural to consider the superluminal knots as regions of enhanced relativistic particle number density, which propagate along the field lines at a high Lorentz factor. If the field lines occupy a large solid angle, superluminal motion may be observed far more frequently than is implied by relativistic ejection along a single axis. The model avoids the necessity for bulk relativistic flows of a proton dominated plasma.

Joe Wampler and Maarten Schmidt arguing about quasar evolution?

V

COSMOLOGICAL STUDIES, CLUSTERING, ISOTROPY ETC.

"The fact that a conference on quasars cannot find more than two percent of total available time for a discussion of the question as to how far these objects are located, indicates that most astronomers have already made up their minds about the answer."

- Jayant Narlikar (p.463)

Giancarlo Setti appears unconvinced by Richard Green's arguement

EVOLUTION OF THE LUMINOSITY FUNCTION

Richard F. Green
Kitt Peak National Observatory
National Optical Astronomy Observatories
P. O. Box 26732
Tucson, Arizona 85726-6732

ABSTRACT. In this review, the currently published, complete,
spectroscopically identified samples of quasars are assembled to
produce a composite luminosity function, independent of evolutionary
assumptions. Two interpretations of the change with cosmic time
provide reasonable fits to the data. Luminosity evolution implies a
fixed population of host objects, with nuclear luminosity that fades
with advancing cosmic time; some dependence of the timescale on
intrinsic luminosity is required. Density evolution traces objects of
comparable luminosity to find the change in space density, without a
requirement of long lifetime. The change in co-moving volume density
depends on luminosity; newer data suggest that somewhat stronger
evolution is required at the low luminosity end than the models of
Schmidt and Green allowed. Caution is advised in drawing direct
physical conclusions about the evolution of individual quasars from
mathematical representations of ensemble properties.

The aim of this review is to examine the methodology of deriving
the quasar luminosity function and its change with cosmic time, as
well as the current data and the success of various model fits. A
caveat is offered about taking the step from mathematical model to
physical interpretation. Illustrative examples are presented from the
extensive work in this area; therefore, the review will not be
entirely comprehensive.

A fundamental question underlying the validity of a luminosity
function exercise is whether the redshifts of quasars can be used as
cosmological distance indicators. There are two general arguments in
favor of the cosmological interpretation. The first is the steep
count slope, by which the integral surface density of quasars changes
by a factor of eight per magnitude in the brighter magnitude range
(Schmidt and Green 1983; hereafter SG83). If quasars were relatively
local, this result implies a steep radial gradient of objects, with a
local deficit centered on the Galaxy, in violation of Copernican
notions of uniformity. The other general argument is based on the
association of quasars with normal galaxies for which the redshift is
assumed to be a valid distance indicator. In their imaging survey,

G. Swarup and V. K. Kapahi (eds.), Quasars, 429–438.
© *1986 by the IAU.*

Yee and Green (1984) found that fields centered on quasars showed an
excess of faint galaxies over the background, and that those objects
are radially concentrated around the quasars. This excess was
detected only for quasars with redshifts that would place normal
luminous galaxies within the survey limiting magnitude; beyond that
redshift no excess was found. Narlikar (these proceedings) gives an
extensive review of the arguments against acceptance of a cosmological
interpretation of quasar redshifts. The most compelling involve
objects of significantly different redshifts in close angular
proximity on the sky, with very low probability of chance
occurrence. For the purpose of this discussion, the conclusion is
drawn that most quasars are at the cosmological distances implied by
their redshifts, which will be used for calculations of luminosities
and volumes.

     A second major question is whether gravitational lensing so
distorts the distribution of apparently high luminosity objects in a
magnitude-limited sample that correct conclusions cannot be drawn
about that end of the luminosity function. The Palomar Bright Quasar
Survey would be particularly susceptible to this effect, and provides
an interesting example. Only 1 of 15 of the most luminous objects in
the BQS was found to be lensed, the triple quasar PG 1115+08, so the
fraction of objects affected is small. The slope of the luminous end
is proportional to the luminosity to the -3.5 power, somewhat too
shallow to be produced entirely by a reasonable distribution of
lensing material (see, e.g., Turner et al., 1984).

     At this point, I make the claim that we have insufficient
physical understanding of the quasars themselves to test a
cosmological model and the change in the collective properties of
quasars simultaneously. For example, Segal and Nicoll (1986) compare
the size of cosmological volume elements in the chronometric and
Friedman cosmologies on the assumption of no number density evolution
for the quasar probes. The present discussion will assume a Friedman
cosmological model, in order to investigate the changes in the
luminosity function with cosmic time. It must be recognized that this
world model is an assumption, and involves at least three parameters.

     It is important to distinguish the luminosity function and its
change with cosmic time from the source and evolution functions. The
luminosity function is the instantaneous distribution of quasar space
densities as a function of luminosity, while the source and evolution
functions describe the birthrate and change of luminosity with time of
individual objects (Petrosian 1986, Cavaliere et al. 1985). The
derivation of the source and evolution functions are the ultimate goal
of luminosity function research, but that step is a difficult one to
make. Ideally the entire magnitude-redshift plane would be filled
with counts from complete samples. Slices in redshift would then
yield the luminosity function directly as a function of look-back
time. The purpose of mathematical modeling of the luminosity function
itself has been primarily to fill in deficiencies in the m-z plane
coverage and to make testable predictions for surveys with new limits
in magnitude or redshift. Discrepancies with new observational
results then teach us about the ensemble properties of the quasar
population.

A critical aspect to this investigation is the completeness of
the samples used. A summary of the discussion by Green et al. (1986)
about the completeness of the BQS highlights the critical issues in
color-selected samples. Since the count slope for all objects shows
no inflection down to the instrumental magnitude limit, completeness
to that limit is not in doubt. The photometric transformations show
no scale or zero-point shifts when compared to a photoelectrically
measured sample of white dwarfs, so systematic problems are also
small. There remain three sources of error: measuring inaccuracies
near the magnitude limit and near the color limit, and accidental
errors. The measuring errors are large for the BQS, with one standard
deviation being 0.29 mag in the B magnitude and 0.38 mag in the U-B
color. With a steep count slope, more objects will be included in the
survey in error for being measured to be brighter than the magnitude
limit than are lost for appearing too faint. This effect leads to an
overcounting of 18% for the BQS. An assessment of the color error is
more complicated because the intrinsic colors of quasars are a strong
function of redshift, in particular, in response to the passage of the
3000 A bump through the B filter. From assuming a Gaussian error
distribution, we can compute the fractional losses as a function of
the actual U-B colors of the objects. From a large independent sample
of (mostly radio selected) quasars the distribution of U-B vs.
redshift is constructed. The losses as a function of redshift then
follow from the BQS detections, producing an average value of 15% for
the whole sample, but ranging from 8% at high and low redshifts to 30%
in the redshift range 0.6 - 0.8. Accidental losses of 8% were derived
from a sample of 120 previously known white dwarfs, by computing
analytically the color and magnitude losses based on the
photoelectically measured colors and magnitudes. The net result is an
average undercounting by 5%, with a redshift dependence. The Medium
Bright Quasar Survey (Mitchell et al. 1984) shows a count slope
similar to that of the BQS, but a 20% upward displacement in surface
density. The cause could be a lower measuring error in magnitude for
the BQS, which is unlikely, substantially higher color and accidental
errors, or a scale mismatch between the two surveys.

A different set of considerations operates for slitless
spectroscopic surveys. In this case, the limiting magnitude is a
function of limiting equivalent width. It is necessary to know the
intrinsic distribution of equivalent widths, found from a sample
selected by an independent criterion, in order to estimate the
incompleteness of the slitless survey. Wampler (1985) points out a
significant systematic difference between color-selected and slitless
spectroscopic samples in the way that magnitudes are derived. The
color-selected samples use total magnitudes, while the slitless
samples use continuum magnitudes in well-defined intervals. As an
example, Wampler gives the distribution of C IV equivalent widths for
flat spectrum radio sources, which shows a median value of ~60 A in
the rest frame. The addition of this line to a broad-band magnitude
can brighten it by 0.1 to 0.2 mag. In the steep count slope regime,
this systematic effect can lead to an overestimate of the numbers of
color-selected objects with respect to spectroscopically selected
quasars by as much as 50%.

    To examine the luminosity function itself, the redshift-magnitude
diagram was populated with objects drawn from those complete samples
available to me at the time of this meeting. The value of the work
of Shanks and his collaborators (this volume and Boyle et al., 1986)
soon becomes apparent. In deriving the luminosity function, no
assumptions were made about evolution; H_o = 50 km/s/Mpc, q_o = 0.5, and
a spectral index of -0.5 were assumed. Spherical shells were defined
by redshift intervals, and the contribution of each object from the
various surveys to the volume density within its shell was computed.
When objects from more than one sample appeared in the same redshift-
luminosity bin, the contributions from each survey were combined in a
weighted average, with the weighting inversely proportional to the
Poisson error of the counting statistics. The result is independent
of the binning in luminosity, but does depend in detail on the choice
of redshift boundaries. The quasar samples are listed in Table 1;
they are all statistically complete, fully spectroscopically
identified, and contain objects with absolute B magnitudes brighter
than -23.0. An additional sample is that of Seyfert galaxies from the
CfA redshift survey by Cheng et al. (1985).

Table 1
Complete Quasar Samples

Name	Number	B lim	Area(sq.deg.)	z	Ref.
BQS	92	16.16	10,714	0 - 2.15	SG83
AB	16	18.0	36	0 - 2.15	1
US - SA 29	12	18.5	7.67	0 - 2.15	2
Curtis Schmidt	15	18.0	340	1.8 - 2.5	SG83
4-M Grism	19	19.5	7.8	1.8 - 2.5	SG83
(Hoag-Smith, Sramek-Weedman)					
BF	31	19.8	1.72	0 - 2.15	1
Koo-Kron SA 57	8	20.53	0.27	0 - 2.5	3

References for Table 1
1) Marshall, et al. 1984
2) Usher, et al. 1983
3) Koo, et al. 1986.

    The results are shown in the Figures. All the redshift intervals
except the lowest represent equal volume shells in the q_o = 0.5
cosmological model. It can be seen that there is a marked change in
the luminosity function between redshifts 0.2 and 1.3, after which the
functions remain essentially identical. In this world model, the
fractional look-back time is 0.75 for z=1.5, while it is 0.85 for
z=2.5, so evolutionary effects over that small time interval are
negligible within our means to discriminate them. Note also that
there is rather good agreement between the Seyfert galaxy luminosity
function and that of the low redshift quasars.

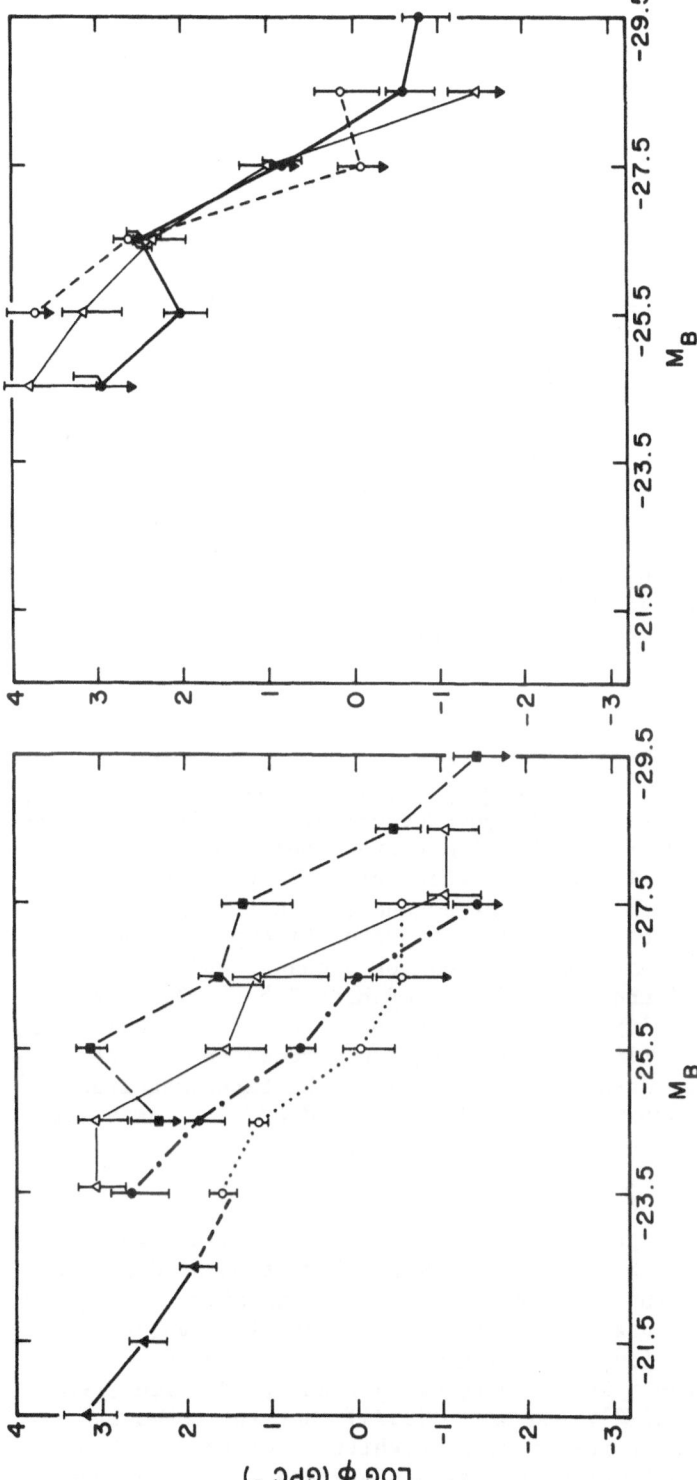

The Luminosity Function of Quasars. The figures show the comoving volume density vs. absolute blue magnitude for different epochs. In the left-hand panel, filled triangles are the local Seyfert galaxy function of Cheng et al. (1985). Open circles are quasars from redshifts of 0.1 to 0.3; filled circles show the LF from z of 0.3 to 0.8; open triangles are 0.8 to 1.2; and filled squares show the range from 1.2 to 1.5. In the right-hand panel, open triangles represent redshifts from 1.5 to 1.8, filled circles from 1.8 to 2.2, and open circles show the redshift range from 2.2 to 2.5.

Two different mathematical representations of the change in the
luminosity function have been investigated. A luminosity evolution
treatment arises from a natural description for the galaxy luminosity
function, in which there is a population of objects with constant
comoving space density that fades in brightness with advancing cosmic
time. Pure luminosity evolution requires a luminosity function of
constant shape and normalization that slides (in the Figures,
horizontally) to higher luminosity with increasing look-back time. In
this case, the decay of brightness with cosmic time is independent of
the intrinsic luminosity of the source. The application of this
technique to the quasar luminosity function has been investigated by
Mathez (1978), Braccesi et al. (1980), Cheney and Rowan-Robinson
(1981), and more recently by Weedman (1986), Koo (1986) and Marshall
(1985). Marshall approximates the high luminosity portion with a
power-law, then cuts off at the inflection with an epoch-dependent
luminosity cut-off. He finds that for an exponential representation
of source dimming with cosmic time, the e-folding time is about 1/6 of
the Hubble time. The major test of the validity of this simple model
is whether the characteristic luminosity or inflection point
translates at constant density with look-back time. Extensive deep
surveys are required to address this point.

An argument in favor of pure luminosity evolution is that it
gives an adequate fit to a portion of the data with a single
evolutionary parameter. If the mathematical description were
interpreted physically, it would imply long lifetimes for quasar
activity in a constant population of hosts. Some evidence in favor of
this picture is offered by Crampton et al. (1986), who find from a
CFHT blue grens selected sample of spectroscopically confirmed quasars
that there is no evidence for a changing nearest neighbor distance
with increasing look-back time. Some stronger trend would have been
expected if there were much higher quasar space densities at earlier
epochs.   Objections to the pure luminosity evolution picture arise
from the large power consumption required for a quasar which fades
from an absolute B magnitude of -29.5 at redshift 2.5 to -24.0 at the
present epoch. A luminosity-independent propagation of the luminosity
function also creates difficulties with the zero-redshift Seyfert
luminosity function over-shooting the counts at the faint magnitude
end, as pointed out by Koo (1986). A luminosity-dependent luminosity
evolution, as suggested, for example, by Cavaliere et al. (1985) is
required for a fit consistent with all the data.

An alternative view of the problem is to apply the concept of
density evolution. This approach is motivated by galactic structure
investigations, in which the space density as a function of distance
is derived for a population of tracer objects of constant
luminosity. In the presentation of the Figures, density evolution
means a vertical propagation of the luminosity function with
increasing look-back time. It was originally found by Schmidt (1968,
1974), Lynds and Wills (1972), and Wills (1974) that the co-moving
space density of quasars increases with redshift.   On the basis of
much more complete redshift-magnitude plane coverage, SG83 and then

Marshall (1985) noted that pure density evolution is not an adequate representation of the data, because the space density of lower luminosity objects increases much less rapidly than that of high luminosity objects. In the Figures, we see a factor of 30 increase in the space density at $MB = -23.5$, while there is a factor of over 1000 increase at $MB = -25.5$ between $z = 0.2$ and $z > 1.2$. Before the availability of the Koo and Kron or BF samples, SG83 saw even less increase at the low-luminosity end in constructing their models of luminosity-dependent density evolution. The composite luminosity functions derived here suggest that a small revision in the SG83 parameters would be required to fit this data set, but that the sense of the original models is still an adequate description of the data.

The physical implications of this mathematical model are much less stringent than those of luminosity evolution, because short lifetimes are allowed, and there need not be the identical set of host objects from epoch to epoch. Some evidence in favor of this interpretation is presented by Yee and Green (these proceedings) from their imaging survey. Radio quasars with $z > 0.55$ are often found in a rich cluster environment that is never observed at lower redshifts. The availability of new sites for quasar activity at earlier cosmic times implies that there is definitely some number density evolution for one sub-population of quasars. Objections to the luminosity-dependent density evolution model are based primarily on the choice of the functional form and fit parameters of the SG83 models; these predictions can be improved with increasing availability of high-quality survey data.

The promised caveat in interpretation is based on a well-known luminosity function. It has a steep high-luminosity portion, flattening off at the low-luminosity end. There is a sharp, well-defined feature at a rather bright absolute magnitude. The faintest end of the function is poorly determined observationally, while the luminous end is sparsely populated. As we look back in cosmic time, the characteristic luminosity propagates toward higher values, while the bump-like feature stays roughly constant in amplitude and position, and the faint end normalization remains constant. The change in the shape of the more luminous end of the luminosity function can be interpreted successfully by either luminosity or density evolution, with a non-evolving component populating the bump. However, I have just described the luminosity function of a globular star cluster; from the physics of stellar evolution, we know that stars move through that function with cosmic time in a non-monotonic way, including a stay in the bump feature, which is the horizontal branch. Until we reach a deeper understanding of the physical evolution of quasars, we should be very cautious in drawing physical conclusions from successful mathematical representations of the quasar population ensemble properties.

References

Boyle, B.J., Fong, R., Shanks, T. and Peterson, B.A. 1986, Trieste
    Symposium on "Structure and Evolution of Active Galactic Nuclei",
    April, 1985.
Braccesi, A., Zitelli, V., Bonoli, F. and Formiggini, L. 1980, Astr.
    Ap., 85, 80.
Cavaliere, A., Giallongo, E. and Vagnetti, F. 1985, Ap.J., 296, 402.
Cheney, J.E. and Rowan-Robinson, M. 1981, M.N.R.A.S., 195, 497.
Cheng, F.Z., Danese, J., De Zotti, G. and Franceschini, A. 1985,
    M.N.R.A.S., 212, 857.
Crampton, D., Cowley, A.P., and Hartwick, F.D.A. 1986, preprint.
Green, R.F., Schmidt, M. and Liebert, J. 1986, Ap. J. Suppl., in
    press.
Koo, D.C. 1986 Trieste Symp. on "Structure and Evolution of Active
    Galactic Nuclei", April, 1985.
Koo, D.C., Kron, R.G. and Cudworth, K.M. 1986, P.A.S.P., in press.
Lynds, R. and Wills, D. 1972, Ap. J., 172, 531.
Marshall, H.L. 1985, Ap.J., 299, 109.
Marshall, H.L., Avni, Y., Braccesi, A., Huchra, J.P., Tananbaum, H.,
    Zamorani, G. and Zitelli, V. 1984, Ap. J., 283, 50.
Mathez, G. 1978, Astr. Ap., 68, 17.
Mitchell, K.J., Warnock, A. III and Usher, P.D. 1984, Ap. J. Lett.,
    287, L3.
Petrosian, V. 1986, IAU Proc. of the 19th Genl. Assembly, Comm. 47.
Schmidt, M. 1968, Ap. J., 151, 393.
Schmidt, M. 1974, ESO/SRC/CERN Conf. on Research Programs for the New
    Large Telescopes, p. 253.
Schmidt, M. and Green, R.F. 1983, Ap. J., 269, 352.
Segal, I.E. and Nicoll, J.F. 1986, Ap. J., in press.
Turner, E.L., Ostriker, J.P. and Gott, J.R. III 1984, Ap. J., 284, 1.
Usher, P.D., Green, R.F., Huang, K.L., and Warnock, A. III 1983, in
    "Quasars and Gravitational Lenses", Proc. of the 24th Liege
    International Astr. Colloq., Universite de Liege.
Wampler, E.J. and Ponz, D. 1985, Ap. J., 298, 448.
Weedman, D. 1986 Trieste Symp. on "Structure and Evolution of Active
    Galactic Nuclei", April, 1985.
Wills, D. 1974, ESO/SRC/CERN Conf. on Research Programs for the New
    Large Telescopes, p. 275.
Yee, H.K.C. and Green, R.F. 1984, Ap. J., 280, 79.

DISCUSSION

Shanks : The fact that you observe more galaxies around high redshift QSOs than around low redshift QSOs; could this alternatively be interpreted as evidence for non-cosmological redshifts ?

Green : As was the case with our first sample, when we observe quasars with redshifts for which the magnitudes of brightest cluster members would be beyond our limiting magnitude, we see no excess. The magnitudes of the "excess" galaxies are considerably fainter for the higher redshift objects.

Segal : A statistical study of the Seyfert I sample of Cheng et.al. indicates that the luminosity evolution model has a problem fitting observations with $z < 0.1$ as well as at high redshifts. Could you comment on the situation at low redshifts ?

Green : It seems that both the optical counts and the X-ray background would be exceeded by a direct brightening of the faint end of the Seyfert I luminosity function. Cavaliere and his collaborators have fit the data by using a luminosity-dependent luminosity evolution with more success.

Narlikar : If redshifts are not cosmological but intrinsic to the objects, then a luminosity that depends on z is not unexpected. This is because both L and z are intrinsic to the object and would be related. Do you agree with this alternative interpretation of an evolving luminosity function ?

Green : The steep source counts certainly imply some kind of evolution in the luminosity function. I would have to understand how to compute the intrinsic luminosity in this interpretation, before evaluating whether the data support it.

Machalski : Do you have any comment on evolution of the radio luminosity function ?

Green : We would like to use the radio material presented by Ken Kellermann to answer that question. One difficulty with our individual object method calculation is that the very few most luminous radio sources dominate the counts at faint flux levels, so that we have poor leverage on the evolution of the faint end of the radio/optical flux ratio distribution.

Shanks : Does the Cheng Seyfert luminosity function not turn over at fainter luminosities ?

Green : It does; those points were not plotted on the figure. The peak amplitude for that function is somewhat higher than the densities shown for lower luminosity quasars.

Margon : You have commented that you find that L_x/L_{opt} must decrease with redshift to avoid overpredicting X-ray counts. Does this imply that you claim to unambiguously distinguish a redshift dependence from a luminosity dependence ? As you know, in a flux limited sample this is quite difficult to do.

Green : Our technique is to tag each member of the Bright Quasar X-ray sample with its observed f_x/f_{opt} ratio, then to evolve the sources according to the prescription of the optical evolution model. Any dependence of L_x/L_{opt} on L_{opt} is implicitly taken into account. To reconcile our predictions with the counts and redshift distribution in the EINSTEIN Medium Sensitivity Survey, we then parameterize L_x/L_{opt} as a decreasing function of increasing z.

Sapre : Can you comment on the search for standard candles and distance indicators in quasars with a view to determining q_o ?

Green : Work by Baldwin, Wampler and collaborators has shown a correlation between CIV equivalent width and luminosity for certain classes of quasars. It remains difficult to distinguish source (spectral) evolution from cosmological effects.

DO QUASARS EVOLVE OVER COSMOLOGICAL TIME SCALES?

E.J. Wampler, D. Ponz
European Southern Observatory
Karl-Schwarzschild-Str. 2
D-8046 Garching bei München
Federal Republic of Germany

ABSTRACT. Systematic biases that are redshift dependent can influence the optical discovery of quasars and the evolution laws derived from counts of quasars. New data and their interpretation for quasars brighter than $M_B = -24$ in the Palomar Bright Quasar Survey (BQS) (Schmidt and Green, 1983) are consistent with no evolution. A comparison of BQS quasars with the brightest quasars from the CTIO Schmidt Telescope Survey (Osmer and Smith, 1980) shows that if q_o is near zero, the co-moving density of bright quasars in a Friedmann cosmology is about 15 times higher for the CTIO survey quasars (mean $z \approx 2.8$) than for the BQS quasars (mean $z \approx 1.8$). In this case spectral evolution is also required since the CTIO quasars have stronger CIV $\lambda 1548$ lines than the BQS quasars of similar luminosity. Alternatively, if q_o is taken to be near 1, the CTIO survey quasars would then have lower luminosity than the BQS quasars and these data would be consistent with no evolution. Strong CIV $\lambda 1548$ lines for the CTIO quasars would then fit the general correlation between absolute quasar luminosity and emission line strength (Wampler, Gaskell, Burke and Baldwin, 1984).

1. INTRODUCTION

Much of the material given in this talk will soon be published in the Astrophysical Journal (Wampler and Ponz, 1985). Here we summarize the main points of that article and give a comparison of the scanner IDS magnitudes with broad band blue magnitudes obtained by Peter Allan and William Keel using the Mark II computer controlled photometer at the 1.3-m telescope on Kitt Peak. These data provide an important check on the IDS photometric system and were kindly provided by W. Keel prior to publication.

2. BIASES THAT COULD AFFECT QUASAR STATISTICS

1.1. Line Emission: In the ultraviolet Ly-α $\lambda 1215$ and CIV $\lambda 1548$ can become very strong. For quasars with absolute continuum magnitudes of

G. Swarup and V. K. Kapahi (eds.), Quasars, 439–445.

about -24 at λ1550 these two lines can have a combined equivalent
width of nearly 600 Å in the quasar rest frame. For brighter quasars
the equivalent width is much less. This rough anti-correlation between
equivalent width and quasar luminosity is called the Baldwin effect
(Baldwin, 1977; Wampler et al., 1984). At a redshift of 2, quasars
with λ1550 continuum magnitudes of -24 have apparent magnitudes (which
depend on the exact cosmological model) near mag 21. At this redshift
both strong emission lines are in the UV-blue spectral region and can
increase the apparent magnitude of the blue continuum of faint quasars
on blue survey plates. To the red, strong FeII emission can also
substantially increase the apparent brightness of the continuum
magnitude (Wills, Netzer and Wills, 1984; Wampler, 1986). These
effects would not be important if samples of quasars were actually
observed at a given rest wavelength and a careful correction were made
for the effects of line emission. Unfortunately this has not usually
been the case. The steep increase in the space density of quasars with
decreasing luminosity can result in substantial redshift dependent
errors in determining the space density of the objects.

1.2. Photometric Errors: Accurate photometry exists for only a few
"complete samples" of quasars. And for those observed with broad band
filters the published magnitudes are affected in an unknown way by the
line emission described above. Two types of photometric errors exist:
accidental errors and systematic errors. The accidental errors usually
increase with decreasing apparent magnitude. Also several surveys seem
to be affected by systematic errors that result in faint quasars being
assigned a magnitude that is too bright (Wampler et al., 1984; Wampler
and Ponz, 1985). Both of these effects result in a steepening of the
log(N)-log(mag) curve. Since, roughly, the fainter quasars have a
higher z this effect results in the high z quasars being assigned a
magnitude that is actually too bright.

1.3. Survey Incompleteness: All surveys are incomplete in some sense.
If the characteristics of the survey are well understood this
incompleteness can be taken into account and corrections made for the
errors that develop when working near the survey limit. But often the
characteristics of a survey are rather poorly understood and redshift
dependent effects are usually not correctly addressed. Most biases
seem to favor the discovery of high redshift quasars over those of low
redshift. Great care must be taken to ensure that all the lower
redshift quasars that correspond to the identified high redshift ones
have in fact been discovered.

3. THE PALOMAR BRIGHT QUASAR SURVEY

Because the Palomar Bright Quasar Survey (BQS) was one of the most
carefully determined complete surveys of quasars and because the
objects in it are bright and easy to observe we decided to re-examine
the survey to assess the extent to which the possible errors listed
above might have affected the conclusions of Schmidt and Green (1983).

Image Dissector (IDS) scans were obtained for 67 of the 69 BQS quasars which Schmidt and Green give absolute magnitudes brighter than -24. The two unobserved quasars are both low redshift, low luminosity objects and their neglect does not affect the conclusions reached here. From these scans U,B,V Blue magnitudes were constructed using the B filter response and a computer program due to Jack Baldwin. These Blue magnitudes were compared to the observed magnitudes listed by Schmidt and Green (1983). A systematic effect seems to be present in the comparison of these two sets of data. For the fainter quasars the magnitudes listed by Schmidt and Green (1983) are about 0.5 mag brighter than the B magnitudes found from the IDS scans. IDS magnitudes of faint white dwarf stars with broad band filter photometry by Eggen (1968) showed a slight non-linearity in the IDS system but this was corrected for in the photometry published by Wampler and Ponz (1985) (see also Wampler, 1985). Fortunately, after this work was completed, Peter Allan and William Keel (private communication) sent us filter photometry magnitudes obtained from observations at KPNO for many of the BQS quasars we observed. In general their results agree rather well with ours although there is a systematic seasonal effect of a few tenths of a magnitude (see Fig. 1).

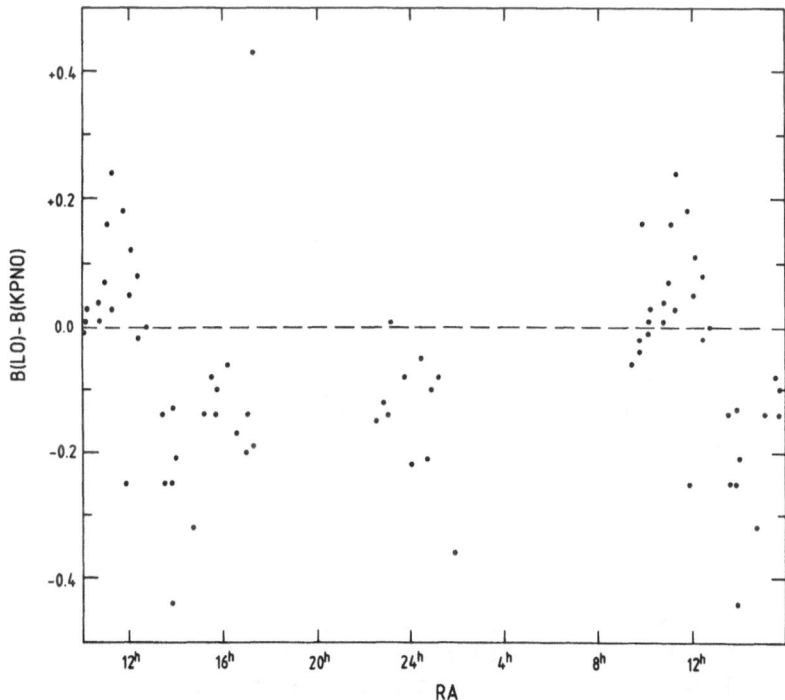

Fig. 1. - The difference between the blue magnitudes obtained at Lick Observatory (LO) with the IDS scanner and those obtained at Kitt Peak National Observatory (KPNO) using a broad band filter photometer. The comparison is shown as a function of right ascension.

After the Lick Observatory data was corrected for this seasonal
effect the Lick and the KPNO photometry are in satisfactory agreement
(see Fig. 2). In particular there is no systematic effect with
magnitude similar to that seen in the comparison with BQS data.

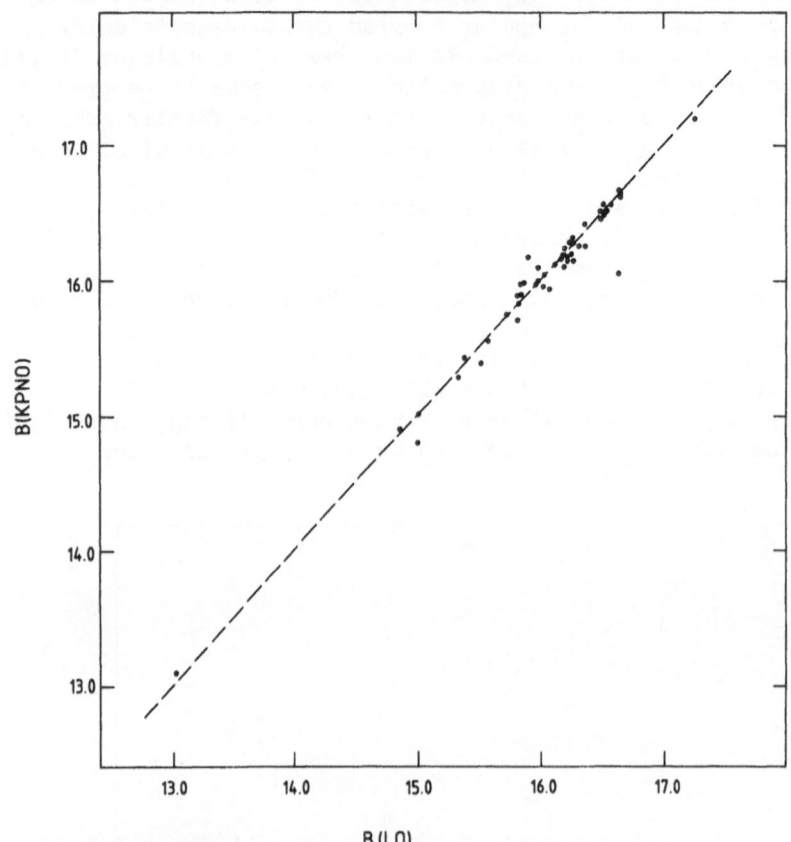

Fig. 2. - A comparison of the LO and KPNO Blue magnitudes after
          correction for the systematic seasonal effects shown in
          Fig. 1.

Rather than convert the measured Blue magnitude to a magnitude at
some fixed rest wavelength it was decided to actually measure the
continuum magnitude near $\lambda 2800$, a wavelength that is accessible, or
nearly so, to the IDS system for all of the quasars in the survey. In
the process of obtaining these scans it was found that two of the
quasars in the BQS that has listed redshifts near $z = 0.5$ in fact had
redshifts near 1.0. This change accentuates the decrease of
sensitivity of the BQS that was noted by Schmidt and Green (1983) for
quasars with redshifts near 0.5. The drop in the survey sensitivity
near $z = 0.5$ is caused by the appearance of the MgII $\lambda 2798$ line and
strong FeII bands in the blue filter. This causes the quasar
candidates to appear redder than normal and apparently resulted in
their elimination from the survey.

The result of all these changes is that for the brightest quasars in the survey there is no strong evidence for evolution. This conclusion, which differs very markedly from that of Schmidt and Green (1983), mostly results from differences in the interpretation of the data although differences in the measured magnitudes and the correction of the redshift for two objects contribute to this different conclusion.

4. COMPARISON WITH OTHER SURVEYS

The only other survey of radio quiet quasars with accurate monochromatic magnitudes is a CTIO objective prism survey (Osmer and Smith, 1980). Most of these objects have redshifts greater than the $z = 2.05$ cut-off in the BQS that is occasioned by the presence of Ly-α in the Blue filter passband for quasars with redshifts slightly greater than 2.05. Therefore a comparison of these two surveys requires a cosmological model. It was found that in Friedmann universes with $q_0 = 0$ the space density of the bright CTIO survey quasars was about 15 times the space density of BQS quasars with similar luminosity. However if this model is chosen, then for the same luminosity the bright CTIO survey quasars have about twice the CIV equivalent width as the BQS quasars. Also the CTIO survey quasars now no longer fit the relationship Baldwin (1977) found between absolute luminosity and CIV line strength.

However if a Friedmann model with $q_0 = 1$ is chosen the CTIO survey quasars fit the Baldwin relationship and are sufficiently fainter than the BQS quasars that their increased space density is consistent with the steep quasars' luminosity functions that have been determined from other observations. Thus the CTIO and BQS data are consistent with no evolution of quasar density from $z \approx 3.2$ to the present if q_0 is taken to be near 1. This value of q_0 is in agreement with other indications that $q_0 \approx 1$ (Wampler et al., 1984).

5. CONCLUSIONS

Even the most carefully studied complete samples of quasars seem to have sufficient discrepancies as to throw into doubt the assertion that they show strong quasar evolution over cosmic time scales. In fact it is possible to assert that the quasar data show no evidence for evolution and favor a high value for q_0. This high value (close to the Euclidean result) and the apparent lack of evolution is not understood in the context of popular cosmological theories.

We would like to thank Drs. Allen and Keel for supplying their BQS blue magnitudes to us prior to publication.

REFERENCES

Baldwin, J.A., 1977, Ap.J., 214, 679.
Eggen, O.J., 1968, Ap.J. Suppl., 16, 97.
Osmer, P., and Smith, M.G., 1980, Ap.J. Suppl., 42, 333.
Schmidt, M., and Green, R.F., 1983, Ap.J., 269, 352.
Wampler, E.J., 1985, ESO Messenger, No. 41, 11.
Wampler, E.J., 1986, Astron. Ap. (in press).
Wampler, E.J., Gaskell, C.M., Burke, W.L., and Baldwin, J.A., 1984,
     Ap.J., 276, 403.
Wampler, E.J., and Ponz, D., 1985, Ap.J., 298, 448.
Wills, B.J., Netzer, H., and Wills, D., 1985, Ap.J., 288, 94.

DISCUSSION

Shanks : In terms of answering the question of whether QSOs evolve or
not, I think if both the bright and faint surveys are taken together
there is stronger evidence for evolution than taking any survey individually; for example in the steep slope of the QSO number counts found
brighter than B = 19.5 mag. So although the effects you have discussed
must be considered, in my opinion the evidence for evolution may be too
strong to be explained away.

Wampler : You may well be right ! I have only compared the CTIO survey
with the BQS. Because the CTIO survey may be incomplete, even for bright
quasars, other surveys may be better. But it is important to prove that
the two surveys are on the same photometric system and that the redshift
or luminosity difference in the populations are properly accounted for.

Veron : You said that the photometric errors have an important effect
on the log N/B curve because the slope of this curve is so steep. But
if you accept this high slope, does not that mean that you accept evolution ?

Wampler : There are two slopes that are important: the luminosity function and the intrinsic log N/B curve. If strong emission lines from many
faint, high redshift quasars overpopulate the fiant end of the log N/B
curve then its observed slope would be steeper than the unevolving slope
even in a non-evolving Universe. Photometric errors that increase with
magnitude increase this slope even more. I think it is in general dangerous to use B magnitudes to derive cosmological parameters. The "K"
corrections are large and, in my view, uncertain.

Kapahi : Since the radio luminosity function is well known to evolve
strongly with epoch, wouldn't the absence of evolution in the optical
luminosity function imply a strong epoch dependence of the radio emission from optically selected quasars, contrary to observations ?

Wampler : First, I'm not sure how tightly the observations constrain the ratio $R = P_{opt}/P_{radio}$. But I am willing to accept the working hypothesis that R is independent of z. I have been told that the observed evolution of the radio luminosity function is somewhat uncertain. In any case it seems that now people believe that the evolution is not as strong as was first reported.

Green : I disagree with your statement that the higher luminosity objects show less evolution than those of lower luminosity. Taking the data in my figure at face value, the lower luminosity objects show an increase in number density of a few hundred, while the higher luminosities increase by more than 2000. Marshall's formal statistical test showed a 5σ difference in the density evolution rate for the lower and higher luminosity quasars.

Wampler : I had intended only to say that it seems to me that your current figure is not in very good agreement with Fig. 4 of Schmidt and Green (1983). In that figure there was little evolution of quasars fainter than about -25 but there was rapid evolution of bright ($M_B \sim -30$) quasars. Also for the bright quasars the evolution for $0 < z \leq 0.5$ is comparable to that for $0.5 < z < 2.5$. In the figure you just gave the low luminosity quasars have substantial evolution and there is little evolution for the high luminosity quasars until $z \sim 1$. I think that these different results only reflect the fact that these evolution "laws" are not well constrained statistically, or perhaps changing sample biases are affecting the results. Are your new curves statistically compatible with the old ones or do they only appear different ?

" .... despite some tantalizing possible inhomogeneities in some quasar surveys, there is as yet no evidence
based on carefully calibrated, quantitative searches for
significant large-scale anisotropies in the distribution of
quasars."

- Patrick Osmer (p.451)

CLUSTERING, ISOTROPY, AND REDSHIFT CUTOFF

Patrick S. Osmer [*]
National Optical Astronomy Observatories
Cerro Tololo Inter-American Observatory [1]
Casilla 603, La Serena, Chile

The topics of clustering, isotropy, and redshift cutoff are in reality just different aspects of the problem of the three-dimensional distribution of quasars, assuming, of course that the redshifts are cosmological and therefore an indication of radial distance. The distribution in redshift has additional interest because of the substantial lookback times involved, up to four fifths of the age of the universe. The radial variation of quasar density between redshift zero and two, and the attendant questions of density and luminosity evolution, are discussed elsewhere in this symposium by Green and shall not be treated here. Rather we shall concern ourselves with the behavior at redshifts larger than two and the specific question of a steep decline of quasar density at redshifts near three. For simplicity we may characterize the problem as one of studying either the formation of quasars as we normally see them in a cosmologically short time or of the properties of the universe and its optical depth, should intergalactic absorption contribute significantly to blocking our view at redshift three. Of course other hypotheses are also possible.

The topic of quasar clustering is analogous to that of clusters of galaxies and superclusters at low redshift. One wishes to see if quasars at large redshifts exhibit clustering properties, and, if so, what bearing they have on the development of clustering with time in the universe. Quasars provide our only chance of seeing the state of clustering at redshifts or cosmic epochs intermediate between those of the microwave background, which is very smooth, and the present, which is very clumpy on scales up to at least those of superclusters. The association of low redshift quasars with galaxies is an important topic that is to be discussed by Yee at this symposium. Here we shall limit the discussion to that of quasar clustering itself.

The topic of isotropy concerns the distribution of quasars on much larger scales than those considered for clustering and therefore bears on the structure of the universe at very large scale. If it can be established that the density of quasars in regions of the sky separated by tens of degrees varies significantly, then the implications for cosmology would be profound indeed.

Let us examine recent results on these three topics one by one.

---

[*] Presented by M.Schmidt

G. Swarup and V. K. Kapahi (eds.), Quasars, 447–453.
© *1986 by the IAU.*

Previous symposia volumes may be consulted for earlier work (e.g. Abell and Chincarini 1983, Mardirossian, Giuricin, and Mezzetti 1984).

REDSHIFT CUTOFF

The concept of a significant decrease in the space density of quasars at redshifts greater than two was discussed fifteen years ago by Schmidt (1970, 1972) and Sandage (1972), who noted that it could represent the epoch of quasar formation. With the subsequent discovery of quasars with redshift three and greater, the question was reopened. After the development of the objective prism technique (see review by Smith 1978) enabled a new approach to the problem, Osmer (1982) claimed that the absence of quasars in a deep survey over the redshift interval 3.7 to 4.7 was definite evidence for a decrease in their space density. Then PKS2000-330, with redshift 3.78 was discovered by Peterson, Savage, Jauncey, and Wright (1982), and questions were raised about possible selection effects in Osmer's survey (Savage and Peterson 1983). Subsequently more quasars with redshifts greater than 3.5 were discovered by a variety of techniques (see Table 1), so that it is timely to re-examine the observational state of the so-called redshift cutoff.

TABLE 1

Quasars with z > 3.5

Object	z	Mag.	Technique	Reference
1159+123	3.51	18	obj. prism	Hazard et al.1984
OQ172 (1442+1011)	3.53	17.8	radio source id.	Wampler et al. 1973
DHM0054-284	3.61	19.6	multi-color photom.	Shanks et al. 1983
0055-2659	3.7	17	obj. prism	Hazard & McMahon 1985
1351-018	3.71	19	radio source id.	Dunlop et al. 1985
PKS2000-330	3.78	19	radio source id.	Peterson et al. 1982

Obviously quasars exist with redshifts up to 3.78. The questions are what is their space density and how is it varying with redshift. Hazard and McMahon (1985) consider from an analysis of UK Schmidt fields at 00 h 53 m -28° and 15 h 0 m +11° that the decline in space density with increasing redshift sets in already near redshift two. They give arguments in addition that the failure to find faint quasars at high redshifts can be explained by a combination of luminosity and density evolution as proposed by Koo (1983).

Dunlop et al. (1985) discuss the properties of faint radio sources in the Parkes survey. Even though their discovery of 1351-018 as a faint, high-redshift quasar with weak emission lines raises the possibility that other such objects are missed in optical surveys, they conclude that the luminosity function for radio quasars declines for redshifts larger than two.

Recently Schmidt, Schneider, and Gunn (1985) have completed two major surveys with the Palomar 5m telescope that add significantly to our knowledge of quasar densities at faint magnitudes. They used a grism technique and CCD to search 0.9 sq deg to a considerably fainter limit than can be reached photographically. Spectra of more than 9000 objects were examined, and 45 emission-line candidates selected with observed equivalent widths greater than 50A and high signal-to-noise ratio. Followup spectroscopy confirmed 10 as quasars, with redshifts between 0.9 and 2.7 (but none higher), with the remainder of confirmed objects being emission-line galaxies of lower redshift. The incidence of the observed quasars is in good agreement with the luminosity function arguments of Schmidt and Green (1983), but the complete absence of quasars with z > 2.7 is in strong contradiction to the expected numbers of 40 to 121. Thus these new results also indicate a cutoff in the space density of quasars with absolute magnitudes of -25 at a redshift less than three.

In a related survey with the 5m telescope used in a transit mode, the same group has surveyed 7.5 sq deg and covered 43000 objects. Subsequent slit spectroscopy of the 52 emission-line candidates confirms 24 to be low redshift galaxies and 8 to be quasars with 1.0 < z < 2.8. Again the number of quasars found is consistent with expectations, while the lack of quasars at z > 2.8 contrasts with the predicted discovery of 30 - 62 such quasars. This survey provides additional convincing evidence for their absence, now at a higher luminosity.

Thus, independent studies by three different groups now point toward a decline in the space density of quasars at redshifts below three. This is in contrast to the first results for bright quasars in the Curtis Schmidt survey, where enough quasars with redshifts up to 3.3 were found to be consistent with a constant space density in the interval 2.0 to 3.3 (Osmer and Smith 1980, MacAlpine and Feldman 1982). Whether this result is only a statistical fluctuation in the small sample of objects or whether it represents a luminosity dependence of the space density is a matter that needs further study. In any case, the recent findings for fainter quasars definitely point to a decline in their space density setting in below redshift three.

What are the possible explanations for these observed effects? Obviously we do not yet have adequate data, let alone certain explanations, but three possibilities are being actively explored. One is that quasars first formed and became visible at redshifts near three. This would be remarkable, as the cosmological epoch is rather late. In this case, the data could be interpreted to mean that the formation time depends on luminosity. A second hypothesis is that the universe is becoming opaque near redshift three because of increasing intergalactic absorption. Ostriker and Heisler (1984) have shown that dust in the universe could well cause an apparent drop in the number of quasars observed above redshift three. This hypothesis immediately suggests that observable effects on the reddening of quasars should occur at high redshifts, effects that should be looked for. A third hypothesis is that the nature of quasars changes near redshift three. For example, if their emission lines weakened, they would not be detectable with the grism technique. Their colors could be difficult to distinguish from faint stars (see, for example, Dunlop et al. 1985).

Now that more than fifty quasars with z > 3 are known, observational programs can be designed to investigate these hypotheses in significant detail, so that we may expect further progress in the next few years.

CLUSTERING

The topic of quasar clustering is important for the reasons discussed in the introduction, but so far it has proved a difficult observational problem. The main reasons are that extreme care must be taken to have properly calibrated and uniform data and to eliminate the various selection effects that inevitably creep in. Unfortunately, nearly all selection effects tend to make clsutering falsely appear, not disappear, in quantitative analyses, examples being non-uniformity in sensitivity from field to field, non-uniform sensitivity to all redshifts in the interval under study, and patchy interstellar absorption in the galaxy. The steep increase in the apparent luminosity function of quasars at the magnitudes so far studied is also a problem, since a change of only 0.3 mag in the limiting sensitivity can change the apparent surface density by a factor of two. For these reasons, studies to data of clustering of the quasar population as a whole have not turned up convincing evidence (Osmer 1981, Webster 1982, Clowes 1985). Nonetheless, a variety of interesting groups and pairs have been found that merit further study regardless of the behavior of the overall population (op. cit.). It is certainly possible that subsets of the quasar population cluster.

Indeed the strongest claims in favor of clustering made to date are those by Shaver (1984) for quasar pairs. As he is discussing the topic elsewhere in this symposium, we will not pursue the matter here, except to say that the association of quasar groups or pairs with superclusters and clusters of galaxies will become a very important topic once Space Telescope or large groundbased telescopes can reliably image galaxies at redshift 2.

In a completely different approach to the problem, Burbidge et al. (1985) argue that the observed number of quasar pairs with different redshifts is significantly in excess of expectations and therefore a challenge to the cosmological hypothesis for redshifts. Shaver (1985), however, comes to the opposite conclusion, that the sky distribution of quasars with different redshifts is random.

Now that automated quasar detection techniques have been developed (Clowes et al. 1984, Hewett et al. 1985), we may expect definite progress in the analysis of existing and new surveys for clustering (e.g. Osmer 1983). Given the difficulties inherent in tying different photographic plates together, the application of automatic techniques to single UK Schmidt plates offers a very attractive approach. In any case, the acquisition of followup spectroscopic data is key to all work on the subject, so that reliable redshifts can be established.

ISOTROPY

A related topic to clustering is isotropy, which we may define as the quasar distribution on scales larger than those investigated for clustering. Let us adopt scales of tens of degrees or more on the sky as the province of the isotropy question. Fliche, Souriau, and Triay (1982) have called attention to possible anisotropies in the quasar distribution over the whole sky. Obviously the confirmation of effects such as they describe would have extremely important cosmological consequences. However, it is vital to establish the observational basis first. Considering the inhomogeneous and generally uncalibrated manner in which different quasar surveys have been carried out, I think it is premature to ascribe significance to inhomogeneities found in the all-sky quasar compilations produced to date.

Another discussion of possible inhomogeneities in the quasar distribution has been published by Arp (1984a,b), who calls attention to non-uniformities in the area of the Sculptor group of galaxies. He discusses the deficit of quasars in the 2 -4 hour region of the -40° zone of the Curtis Schmidt survey (Osmer and Smith 1980). He notes that the weak-lined quasars in the -40° zone have a uniform distribution with right ascension, but the strong-line ones concentrate toward the Sculptor group. Because the weak-lined quasars are more difficult to detect, and therefore more sensitive to selection effects, he argues for the reality of the non-uniformity. However, he also noted a deficit of fainter quasars in the 2 -4 hour region, which could be a result of the lowered sensitivity originally proposed by Osmer (1981) as the reason for the deficit.

In contrast, the one survey for anisotropy that has been carried out in a well-calibrated manner is that of Hawkins (1985). He analyzed two UK Schmidt fields separated by 13° on the sky and found no significant differences in the number of quasars in the two fields with a sensitivity at the 6% level.

Thus, we may conclude that despite some tantalizing possible inhomogeneities in some quasar surveys, there is as yet no evidence based on carefully calibrated, quantitative searches for significant large-scale anisotropies in the distribution of quasars.

[1] Cerro Tololo Inter-American Observatory, National Optical Astronomy Observatories are operated by the Association of Universities for Research in Astronomy, Inc., under contract with the National Science Foundation.

REFERENCES
Abell, G.O., and Chicarini, G. (eds.) 1983, in IAU Symposium No. 104, Early Evolution of the Universe and Its Present Structure, (Dordrecht: Reidel).
Arp, H. 1984a, Ap. J. 285, 547.
_____. 1984b, ibid. 285, 555.
Burbidge, G.R., Narlikar, J.V., and Hewitt, A. 1985, Nature, 317, 413.
Clowes, R.G., Cooke, J.A., and Beard, S.M. 1984, M.N.R.A.S. 207, 99.
Clowes, R.G. 1985, M.N.R.A.S. in press.
Dunlop, J.S., Downes, A.J.B., Peacock, J.A., Savage, A., Lilly, S.J., Watson, F.G., and Longair, M.S. 1985, Nature, in press.
Fliche, H.H., Souriau, J.M., and Triay, R. 1982, Astron. Astrophys. 108, 256.
Hawkins, M.R.S. 1985, M.N.R.A.S. 216, 589.

Hazard, C., Terlevich, R., McMahon, R., Turnshek, D., Foltz, C., Stocke,
      J., and Weymann, R. 1984, M.N.R.A.S. 211, 45P.
Hazard, C., and McMahon, R. 1985, Nature, 314, 238.
Hewett, P.C., Irwin, M.J., Bunclark, P., Bridgeland, M.T., Kibblewhite,
      E.J., He, X.T., and Smith, M.G. 1985, M.M.R.A.S. 213, 971.
Koo, D.C. 1983, in Quasars and Gravitational Lenses, 24th. Liege
      Colloquium, (Liege: Institute d'Astrophysique), p. 240.
MacAlpine, G.M., and Feldman, F.R. 1982, Ap. J. 261, 412.
Mardirossian, F., Giuricin, G., and Mezzetti, M. (eds.) 1984, in Clusters
      and Groups of Galaxies, (Dordrecht: Reidel).
Osmer, P.S. 1981, Ap. J. 247, 762.
Osmer, P.S. 1982, Ap. J. 253, 28.
Osmer, P.S., and Smith, M.G. 1980, Ap. J. Suppl. 42, 333.
Osmer, P.S. 1984, in Clusters and Groups of Galaxies, eds. F.
      Mardirossian, G. Giuricin, and M. Mezzetti (Dordrecht: Reidel),
      p. 579.
Ostriker, J.P., and Heisler, J. 1984, Ap. J. 278, 1.
Peterson, B.A., Savage, A., Jauncey, D.L., and Wright, A.E. 1982,
      Ap. J. Lett. 260, L27.
Sandage, A. 1982, Ap. J. 178, 25.
Savage, A., and Peterson, B.A. 1983, in IAU Symposium No. 104, Early
      Evolution of the Universe and Its Present Structure, eds. G.O. Abell
      and Chincarini (Dordrecht: Reidel), p. 57.
Schmidt, M. 1970, Ap. J. 162, 371.
_____. 1972, ibid. 176, 273.
Schmidt, M., and Green, R.F. 1983, Ap. J. 269, 352.
Schmidt, M., Schneider, D., and Gunn, J. 1985, Ap. J. in press.
Shanks, T., Fong, R., and Boyle, B.J. 1983, Nature, 303, 156.
Shaver, P.A. 1984, Astron. Astrophys. 136, L9.
_____. 1985, Astron. Astrophys. 143, 451.
Smith, M.G. 1978, Vistas in Astronomy, 22, 321.
Wampler, E.J., Robinson, L.B., Baldwin, J.A., and Burbidge, E.M. 1973,
      Nature, 243, 336.
Webster, A. 1982, M.N.R.A.S. 199, 683.

DISCUSSION

Malkan : Ostriker & Heisler's explanation of the high-z cutoff by intergalactic reddening is unlikely. When you look at optical/UV colors of 2 dozen (mostly radio-selected) high-z (2.5 to 3.5) quasars, such as the flux ratio from rest wavelengths of 4200 and 1770 A, you find they are the <u>same</u> as the colors of lower-z QSOs. This limits their diffuse intergalactic reddenings to $E_{B-V} \leq 0.^m1$. As Ned Wright has pointed out (1986), you cannot get off the hook by saying that reddened QSO's are simply invisible, because any realistic absorbing clouds would not have hard edges. So we ought to find moderately-reddened QSO's at high-z, but we do not

Kembhavi : Would the scarcity of quasars beyond $z \approx 3$ still remain significant if there were no evolution in the luminosity function as claimed by Wampler ?

Schmidt : Yes, I believe the observations require a decrease in co-moving density at constant luminosity, for increasing redshift.

Rees : Regarding the possible quasar anisotropies of scales $>10°$, probably the X-ray background isotropy constraints are stronger than any we are likely to get from direct surveys of optical quasars, if we accept that quasars are important contributors to that background.

Schmidt : The constraint would be tempered if quasars contribute only around 10% of the 2 keV background, as suggested by Schmidt and Green.

Wandel : In light of the possibility suggested in the talk, that quasars may be forming as late as $z = 3$, it would be physically plausible to consider a "reversed" density evolution (in the sense of space-density increasing with cosmic time) for say, $z \geq 3$, imposed on the ordinary luminosity evolution. Could such a combination fit the data, and in particular, could it account for the knee in the luminosity function ?

Burke : With respect to clustering on large scales, the MIT-Green Bank (MG) radio source survey was examined by J. Mahoney. About half the sources are quasars, so it is a sample of high luminosity quasars with a small statistical corruption by radio galaxies. No significant pair correlation was found, and the sample, covering two steradians between $0°$ and $20°$ declination appears to be isotropic on a $10°$ scale.

"Even the most carefully studied complete samples of quasars seem to have sufficient discrepancies as to throw into doubt the assertion that they show strong quasar evolution over cosmic time scales."

- E.J.Wampler and D.Ponz (p.443)

THE STATISTICS OF RADIO GALAXIES & QUASARS AT HIGH REDSHIFT

J.A. Peacock
Royal Observatory, Edinburgh.

J.S. Dunlop
Department of Astronomy, University of Edinburgh.

ABSTRACT.   Evidence is now accumulating that the most powerful
extragalactic radio sources display a cutoff in comoving density at
relatively low redshift. For both radio galaxies and quasars, the
luminosity function falls by a factor $\geqslant 3$ between $z = 2$ and 4. We
describe the data which lead to this conclusion, and discuss the
prospects for studying the behaviour of low-luminosity active galaxies
at high redshift. The strong evolution of highly luminous radio-quiet
quasars at $z > 2$ may be an artifact of gravitational lensing.

1.   EVOLUTION OF POWERFUL RADIO SOURCES

This paper is concerned with recent studies of the population evolution
of extragalactic radio sources. We are now approaching a point where
the empirical change of the Radio Luminosity Function with epoch is
known: this will provide a firm constraint on physical mechanisms for
the evolution. A good review of this field is given by Wall (1983); at
that time the most recent study of RLF evolution was that by Peacock &
Gull (1981). Since then papers by Condon (1984) and Peacock (1985) have
appeared. All these workers use model fitting to derive their
conclusions: there is no other way of combining partial data such as
source counts with complete redshift data from bright samples. The
problem with this approach is the non-uniqueness of the resulting model
RLF. Condon (1984) gives only one model: from this, it is impossible to
know whether any given feature is required by the data or whether it is
an artifact of the model formulation. In contrast, Peacock (1985)
considers an ensemble of 5 different consistent RLFs to estimate the
uncertainty allowed by the data. Where these differ, we may be sure the
data are inadequate. Where they agree, this may be fortuitous, but is
more likely to be a hint that some feature is required by the data: we
can then construct a specific test to see if this is the case.
  Both Condon (1984) and Peacock (1985) confirm the main conclusions
of Peacock & Gull (1981): the change with epoch of the RLF is
differential in radio power, being greatest at high power. Also, both
steep- and flat-spectrum sources behave similarly. Peacock & Gull,

G. Swarup and V. K. Kapahi (eds.), Quasars, 455–462.

however, were unable to draw any conclusions about behaviour at high redshifts, z ≳ 2: new data have improved this situation in Peacock's (1985) study. Figure 1 shows cuts through the RLF for two different values of q_0. The pairs of lines shown are the limits on ρ (in units of density per unit $\log_{10}(P)$, $H_0 = 50$ km s^{-1} Mpc^{-1}).

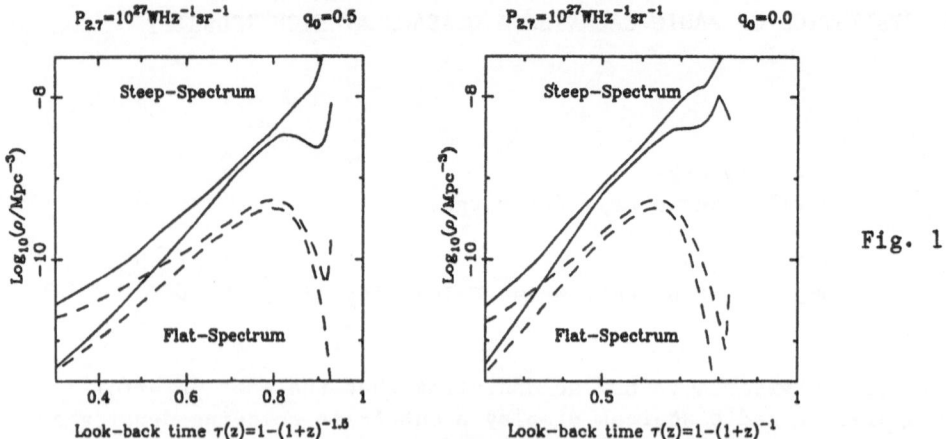

Fig. 1

Beyond z ≃ 2, all models agree that the flat-spectrum RLF falls, independent of q_0; conversely, the steep-spectrum RLF is still uncertain at these redshifts. This difference may be understood when we consider the redshift data used by Peacock (1985). Figure 2 shows a plot of radio power against flux density at 2.7 GHz ($\Omega_0 = 1$ assumed). The vertical lines distinguish different flux-density limits; the steep-spectrum data go no lower than 1.5 Jy, whereas complete flat-spectrum samples exist down to 0.5 Jy. In this context, the term 'complete' means ≳ 95 percent of the sources are identified and ≳ 70 percent have spectroscopy. The missing redshifts can generally be estimated reliably from apparent magnitude.

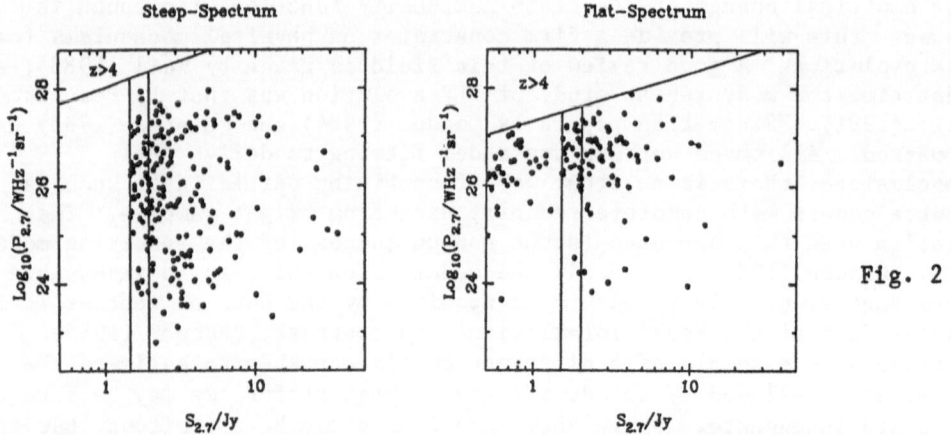

Fig. 2

Figure 2 makes clear the extra strength of the flat-spectrum data: objects of a given luminosity at z ≃ 1 can be compared with their counterparts at z ≳ 4. As none are seen, we deduce a limit on the density at high redshift. This can be quantified through the V_e/V_a test

(Avni & Bahcall 1980). Considering $z > 1.9$ only, the combined flat-spectrum data yield $\langle V_e/V_a \rangle$ = 0.315 (Ω_0 = 1) or 0.352 (Ω_0 = 0). With 24 sources in this range, these results lie respectively 3.1σ and 2.5σ below the expectation of 0.5 for a constant comoving density.

2. THE PARKES SELECTED REGIONS

To test the above conclusions and investigate whether powerful steep-spectrum sources also display a redshift cut-off we need redshift data on a deeper sample. This has been tackled by a collaboration at Edinburgh and Cambridge. The Parkes Selected Region surveys provide 178 sources (allowing for confusion) complete to 100 mJy at 2.7 GHz. VLA maps and sky-survey identifications are reported by Downes et al. (1985): even using UKST material reaching $J > 22$, only 56 percent of the objects were identified, with only 22 percent having redshifts in the literature. This is much as expected from previous RLF models: samples such as this were expected to be biased to high redshifts.

During 1985, we have greatly improved the data base. We have CCD data on all fields in two colours, reaching $B \simeq 25$ and $R \simeq 24$; at this level, $\gtrsim 96$ percent of the sample is identified (3 sources are affected by bright stars). We have also added 31 new redshifts. The remaining sources are almost all galaxies, for which photometric redshift estimation should be possible. The most spectacular result of the spectroscopy programme was the spectrum of 1351-018 (Dunlop et al. 1985). This flat-spectrum quasar has a redshift of 3.71 - the highest in any of the complete samples considered here.

With this exception, however, the highest redshift quasar in the Selected Regions is at $z = 2.8$ - which fits in with our previous idea of a fall in density for $z > 2$. We can test this by performing the V_e/V_a test again. (Now considering quasars only). There are a total of 163 quasar candidates in the complete samples used by Peacock (1985) plus the Selected Regions (60 steep- and 103 flat-spectrum). Figure 3 shows the analogous plot to Figure 2 for these objects.

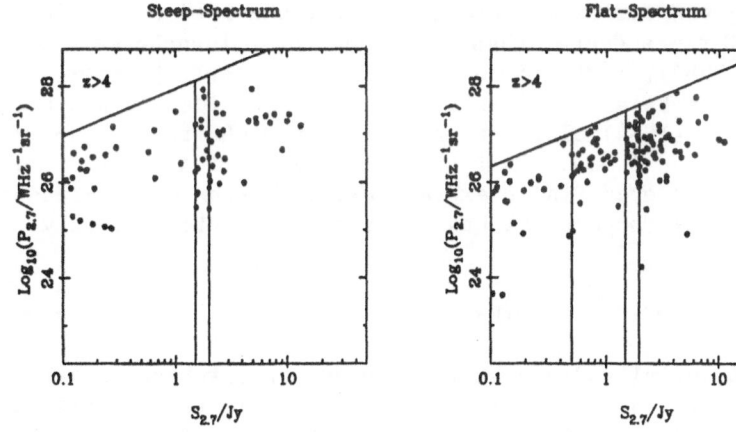

Fig. 3

The results of the V_e/V_a test are given in Table 1.

Table 1: $\langle V_e/V_a \rangle$ values

	Flat-Spectrum				Steep-Spectrum		
	n	$\Omega_o=1$	$\Omega_o=0$		n	$\Omega_o=1$	$\Omega_o=0$
z > 1.8	30	0.391	0.431		12	0.501	0.543
z > 1.9	27	0.362	0.395		12	0.442	0.459
z > 2.0	22	0.351	0.399		10	0.448	0.465

The statistics for the steep-spectrum case are poorer, but the overall trend is similar. However, we do not yet have conclusive evidence for a cutoff for steep-spectrum quasars alone (particularly considering that ~ 15 percent of our candidates still have unknown redshifts, which may affect the result). Also, the V_e/V_a test is imperfectly efficient, as it rejects all low-redshift data; a fuller analysis is needed.

We are now left with important question of whether similar behaviour applies for the radio galaxies. We have seen that essentially all objects are detected; according to the study of 3CR galaxies by Djorgovski & Spinrad (1985) this should not be a surprise. They found a convergence of apparent magnitude at $z \simeq 1$, largely attributable to stellar evolution, so that no galaxy was fainter than B = 24 or R = 23. Better constraints will come from infrared observations (cf Lilly et al. 1985); we have so far obtained K photometry for > 50 percent of the sample, and found no objects fainter than K = 19. As discussed by Lilly et al. (1985), this probably corresponds to $z < 3$. In fact, Peacock (1985) showed that the steep-spectrum RLFs in Figure 1 without a turndown predicted ~ 20 percent of selected-region sources to have z > 3. If they do not, we will have evidence for a cutoff for powerful radio galaxies.

Following the refinement of the above conclusions, the next step must be to compare with the behaviour of low-luminosity sources. This will be possible given complete optical data for the 5C12 survey (Benn et al. 1984), ~ 50 times deeper than here, and the Leiden-Berkely Deep Survey (Windhorst 1984), ~ 300 times deeper. At present, however, > 50 percent of these objects are unidentified so firm conclusions are not possible. Windhorst (1984) claims to show that the RLF for $P_{2.7} \sim 10^{24}$ WHz^{-1} sr^{-1} falls beyond $z \simeq 1$, which would be important if true. However, this is only achieved via an extreme extrapolation of low-redshift evolution; the crucial range $z \simeq 1$ has not yet been probed directly.

3.   COMPARISON WITH OPTICAL STUDIES

The behaviour of powerful radio quasars provides an interesting contrast with optically selected quasars. There are results in the literature on the high-redshift behaviour of the optical LF at high and low optical luminosities. At the high end, Osmer & Smith (1980) presented a V/V_{max}

test over the redshift range 1.9 - 3.25 for 14 luminous objective-prism
quasars with M_B > -28.5 (Ω_0=0). Their (corrected) result was 0.651
(Ω_0=1) or 0.595 (Ω_0=0) - indicating an increasing LF over this redshift
range, Conversely, using faint quasars selected by non-stellar colours,
Koo (1985) argues that, for M_B > -26 (q_0=0), the LF at z > 2 is either
constant (q_0=0.5) or decreasing (q_0=0). There are several interesting
points to make about this differential behaviour:

First, the behaviour of optically weak quasars is similar to that
suggested above for radio-loud quasars (the match is better for low
Ω_0). Given that radio-loud quasars generally have M_B > -27, radio and
optical studies 'agree' on the redshift cutoff, with high-luminosity
quasars being a special case. However, note that there is no reason why
the behaviour of LFs in different wavebands should agree.

Second, the behaviour at low optical powers may be affected by
incompleteness. Dunlop et al. (1985) show that objects with weak lines
like 1351-018 will tend to have stellar UJF colours at z $\gtrsim$ 2.5, and so
be missed. We must await confirmation of Koo's result from CCD grism
surveys sensitive to Lyα equivalent widths W < 100Å.

Finally, the most luminous optical quasars may be affected by
gravitational lensing. Lensing by galaxies is insufficient to cause
important perturbations to the LF (Peacock 1982), but for optically-
selected quasars minilenses with M << $M_\odot$ can be important (e.g. Vietri
1985). We report here some calculations by Peacock (in preparation)
which modify those of Vietri by consistent inclusion of flux
conservation. As an illustration, we show in Figure 4 a Schechter-type
LF and the results of mini-lensing at z = 1.9 and 3.25 for Ω_0 = 0.1
(all in minilenses). The Schechter parameters were chosen to agree with
Koo's results at the lower redshift.

z=0, 1.9, 3.25

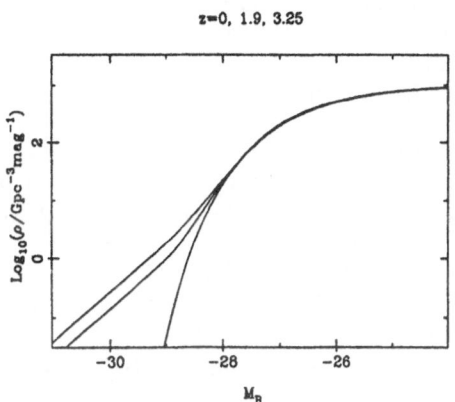

Fig. 4

At Osmer-Smith luminosities, the change in the LF is a factor 2 over
this range. We can convert this to < V/V_{max} > by the approximate
relation

$$< V/V_{max} > \simeq 0.5 + \frac{\Delta \ln \rho}{12}$$

the effect is large enough to make quasars of all luminosities
consistent.

4. CONCLUSIONS

We end with a few points about interpretation of these results. The
evolution of LFs is governed by a conservation equation:

$$\dot{\rho} + \frac{\partial}{\partial P} \; (\dot{P}\rho) \;\; = Q$$

Physically, what we would like to know are the terms $\dot{P}$ (invididual light
curves) and Q (rate of creation). It has been conventional to relate a
lack of quasars at high redshift to quasar (or galaxy) formation - i.e.
low Q at high z - but it is equally possible that $\dot{P}$ is high - i.e.
objects go through their life cycles faster at high z. Similarly, there
is a tendency to invest some significance in 'Luminosity Evolution'
models with $\rho(P,z) = \rho(P/F(z))$. However, such a form need not satisfy
the conservation equation with Q = 0. Indeed, unless the individual
source lifetimes are ~ the Hubble time (which is certainly not the case
for radio sources) Q must be non-zero. Similar comments apply to the
model of Condon (1984): he fits $\rho(P,z) = g(z)\rho(P/F(z))$ so that sources
of all luminosities 'evolve' the same way. Nevertheless, this says
nothing about the luminosity dependences of $\dot{P}$ and Q.

Although a start to physical LF modelling has been made (Cavaliere
et al. 1983), perhaps the best way of interpreting the redshift cutoff
is the direct one: if the radio galaxies at z ≃ 2 are truly young, we
should be able to tell by observations of their colours and dynamics.
This will be an exciting challenge for some years to come.

REFERENCES

Avni, Y. & Bahcall, J.N., 1980. Astrophys. J., 235, 694.
Benn, C.R., Wall, J.V., Grueff, G. & Vigotti, M., 1984. Mon. Not. R.
      astr. Soc., 209, 683.
Cavaliere, A., Giallongo, E., Messina, A. & Vagnetti, F., 1983.
      Astrophys. J., 269, 57.
Condon, J.J., 1984. Astrophys, J., 287, 461.
Djorgovski, S. & Spinrad, H., 1985. Astrophys. J., in press.
Downes, A.J.B., Peacock, J.A., Savage, A. & Carrie, D.R., 1985. Mon.
      Not. R. astr. Soc., in press.
Dunlop, J.S., Downes, A.J.B., Peacock, J.A., Savage, A., Lilly, S.J.,
      Watson, F.G. & Longair, M.S., 1985. Nature, in press.
Koo, D.C., 1985. Proc. Trieste AGN meeting.
Lilly, S.J., Allington-Smith, J.R. & Longair, M.S., 1985. Mon. Not. R.
      astr. Soc., 215, 37.
Osmer, P.S. & Smith, M.G., 1980. Astrophys. J. Suppl., 42, 333.
Peacock, J.A., 1982. Mon. Not. R. astr. Soc., 199, 987.
Peacock, J.A., 1985. Mon. Not. R. astr. Soc., in press.
Peacock, J.A. & Gull, S.F., 1981. Mon. Not. R. astr. Soc., 196, 611.
Vietri, M., 1985. Astrophys. J., 293, 343.
Wall, J.V., 1983. "The origin and Evolution of Galaxies", ed. Jones &
      Jones (Reidel), p295.
Windhorst, R.A., 1984. PhD Thesis, Univ. of Leiden.

DISCUSSION

Kapahi : When the luminosity function is expressed as a polynomial it may be dangerous to extrapolate it outside the range of parameters of the data that went into fitting the polynomial. I wonder if the cutoff in z implied by your models results merely from the fact that the input data contains few large redshifts.

Peacock : I completely agree. Using one polynomial would be meaningless. When you try 5 different ones and get the same behaviour, however, this may be the data telling you something. In any case, I have only used the models as a hint of what aspects to examine in a model - independent way.

Barthel : Basically we differ in approach. You study the high z behaviour statistically, I am trying to 'see' what is (or better : was) going on. You are right that selection effects have an influence on my conclusions, but it remains to be seen to what extent. I am convinced that the truth lies in between our two results. Meanwhile, until further notice I still offer a beer for every new QSR having $\alpha^3_{1.4}$ or $\alpha^5_{.408}$ steeper than 0.6 at z exceeding 2.5.

Peacock : My point is that steep-spectrum quasars at high redshift are going to be rare anyway through selection effects. Therefore, I believe it is hard to be sure that your potentially very important physical effect is operating. We need larger samples to see if you keep your beer.

Shanks : In the radio samples where you have applied the V/V_{max} tests, what is the redshift completeness ?

Peacock : About 80 percent. There is then a worry if the remaining ones are at redshifts high enough to remove the result. We have looked into this and find that most of them would need to have $z \gtrsim 3$. Since the objects are generally neither faint nor red, I think this is very unlikely.

Swarup : Could you comment on your results for redshift cutoff of steep spectrum quasars vis-a-vis those deduced by Windhorst for steep spectrum radio galaxies based on deep surveys at 1.4 GHz.

Peacock : Windhorst claims in his thesis to show that low power radio galaxies cutoff at $z \sim 1$. This is clearly important if true, but I do not think it is firm yet. His argument is that extrapolation of evolution from $z < 0.6$ to > 1 exceeds the number counts. However, he only does one rather extreme extrapolation. I think we need to wait until some redshifts around 1 are measured.

Baldwin : The discrepancy does not, I believe, exist. Windhorst's cutoff referred to radio galaxies of <u>much</u> fainter radio luminosity (by a factor of 10^2-10^3). His cutoff for high power sources is also at $z \gtrsim 2$.

Peacock : I agree there is no contradiction. However, _if_ you think the cutoff is related to galaxy formation, you might expect it to be independent of radio power.

Owen : The redshift estimates by Windhorst are also based on infrared colors as well as direct redshifts.

Peacock : These must be new data. I was talking about the arguments that have been written up.

Roberts : Burke, Turner, Gott, and I observed a sample of the most-optically-luminous radio quasars with the VLA at 0.5 resolution in a search for multiple images which could show that some apparently-luminous QSOs were the result of gravitational lens amplification. No multiple images were found.

Peacock : This interesting observation does not affect my point for two reasons. First, you have looked at radio-selected quasars : the radio luminosity function is less steep than the optical and the selection effects are reduced. Second, the optical depth for lensing by galaxies is known to be small : I was considering a low-mass component of the dark matter which will lens up the optical continuum source only, producing unobservably small splittings.

NONCOSMOLOGICAL REDSHIFTS

Jayant V. Narlikar
Tata Institute of Fundamental Research
Bombay 400 005
India

ABSTRACT. Although the majority of astronomers interpret the redshifts
of quasars according to Hubble's law, there are isolated cases as
well as detailed studies indicating that the redshifts of quasars
and even some galaxies may have noncosmological components. The
evidence discussed here will be on two counts: (i) the periodicities
in redshift distributions, (ii) the associations between discrepant
redshifts objects. In addition, theoretical constraints on physical
models based on the cosmological hypothesis will be briefly mentioned.

1. INTRODUCTION

The fact that a conference on quasars cannot find more than two percent
of total available time for a discussion of the question as to how
far these objects are located, indicates that most astronomers have
already made up their minds about the answer. To many, questioning
the cosmological hypothesis now is like questioning the validity
of special relativity [1]. To a small minority, however, the issue
still remains open; especially because fresh evidence keeps coming
in, throwing doubt on the universal validity of the cosmological
hypothesis. For some of the earlier reviews of the quasar distance
controversy see Refs [2 - 5].

In the interest of brevity, the present review will concentrate
mainly on the more recent evidence. This is not to say that all of
the earlier evidence has gone away. Rather, there has always been
a reluctance on the part of the astronomical community to face it
on the curious grounds that 'the idea must be wrong because we have
no respectable theory to explain it.' To avoid getting into such
arguments the ground rules to be followed here are stated below.

1. The cosmological hypothesis (C.H. in brief) assumes that the
redshift z of an extragalactic object is due entirely to the expansion
of the universe. In other words, given a cosmological model there
exists a unique Hubble relation

$$z = f(DH_o/c) \qquad (1)$$

463

G. Swarup and V. K. Kapahi (eds.), Quasars, 463–473.
© *1986 by the IAU.*

between the redshift and the (luminosity) distance D of the object.
In this relation c and H$_o$ are respectively the speed of light
and the present value of Hubble's constant. For example, the form
of f in the standard dust-dominated Friedmann models is well known
[4].

2. The C.H. may be granted a certain leeway through the formula

$$(1 + z) = (1 + z_C)(1 + z_{NC}),   \qquad (2)$$

where z is the total observed redshift, z_C the cosmological component
obeying relation (1) and z_{NC} the 'noncosmological' part $\lesssim 3 \times 10^{-3}$.
This limit on z_{NC} allows for random motions relative to the cosmological substratum, of upto $\sim 10^3$ km s^{-1}.

3. Evidence in support of the C.H. should ideally demonstrate
the validity of relation (1) in which independent measures of D
are used: that is measures not based on (1). Indirect evidence claiming
that quasars are distant but not providing a measure for D can
at best be considered consistent with (1) but not a proof of it.

4. If the C.H. is claimed to have universal validity, such as
its applicability to galaxies and quasars then it is sufficient to
produce one counter-example where (1) does not hold, in order to
disprove it.

5. A counter-example should demonstrate that the z_{NC} in (2) is
substantially greater than what could be attributed to random velocity
Doppler shifts in clusters of galaxies. It is not necessary to show
that the entire redshift is of noncosmological origin.

6. Finally, to accept the existence of a noncosmological redshift
it is not essential that there should already be a theory available
to explain it. Rather the reverse: if no existing theory can explain
an established phenomenon the onus is on the theoretician to find
a new theory.

Within the above framework I will present recent evidence indicating
that the cosmological hypothesis may not apply to all extragalactic
objects.

2. PERIODICITIES AND PEAKS IN REDSHIFT DISTRIBUTIONS

2.1 Groups of Galaxies

In 1976 Tifft [6] reported a peculiar result that there is a correlation
between the nuclear magnitudes of galaxies in clusters and their
differential redshifts Δz, and that the values of cΔz appear to be
bunched near multiples of 72 km s^{-1}. If CH holds for galaxies, then
the bunching of Δz translates to bunching of ΔD and hence
suggests some discrete quantized structure in the space distribution
of galaxies.

It is hard to understand such a result within the conventional
cosmology and it was hoped that the result would 'go away' as more
accurate data accumulated. The opposite happened. By 1982 high
quality 21cm redshift measurements became available for galaxies in
small groups with cz values known to an accuracy of at least 9 km s^{-1}.

In 1983 Cocke and Tifft [7] analyzed the new data and found that
the effect still holds quite unambiguously.

More recently Arp and Sulentic [8] have reported results of redshift
measurements from over 260 galaxies. Of these over 100 galaxies
in 40 different groups have redshifts measured at 21 cm with accuracy
of $\lesssim 4$ km s^{-1} while over 160 others are taken from the Rood Catalogue
[9] where the accuracy in cz is $\lesssim 8$ km s^{-1}. The entire sample
typically shows systems wherein a large central galaxy is surrounded
by smaller fainter companions. Also, the companion galaxies tend
to have systematically higher redshifts than the brightest galaxy
in the group. The redshift differentials $\Delta z = \{z$ (companion)
$- z$ (dominant)$\}$ for these galaxies show peaks for $c\Delta z$ at 70,140 and
210 km s^{-1}. The Tifft effect therefore persists.

Although Byrd and Valtonen [10] have attempted to explain the
observed values of Δz in the above work in terms of expansion of
companions away from the main galaxy, the morphology of the groups,
the preponderance of positive z-values and the observed peaking
effect remain unexplained.

2.2 Quasars

Geoffrey Burbidge has highlighted the peaks in the redshift distribution
of quasars [3]. In fact the peaks have also shown persistance in
spite of the conventional expectation that with bigger and bigger
samples of quasars, newer methods of their detection the supposed
effect will disappear. In this conference Hewitt and Burbidge [11]
have presented histograms showing the numbers of quasars in different
redshift ranges, for the largest known compilation of data on quasars
- the revised Hewitt-Burbidge catalogue containing about 3000 sources.

For these quasars sharp peaks are present at $z \simeq 0.3, 1.4, 1.95$
as in earlier smaller samples. To counter the criticism that a selection
effect is responsible for it, Hewitt and Burbidge have also given
separate histograms for quasars discovered by the objective prism
method as well as for quasars discovered by other methods. The peaks
persist in spite of such a separation.

Power spectrum analysis of the data by Depaquit et al [12] shows
clear indication of periodicity in the values $\ln(1+z)$.

I leave you with these findings to figure out how they could
be explained in conventional terms.

3. ASSOCIATIONS BETWEEN DISCREPANT REDSHIFT OBJECTS

The main approach here is to demonstrate that two extragalactic objects
with very different redshifts are physical neighbours. Note that
if two objects are shown to be real neighbours it is not necessary
to know their actual distance D from us. That two different values
of z are found at the same D is sufficient to invalidate the Hubble
relation (1).

3.1 Galaxy–Galaxy Associations

Although it is commonly assumed that only quasars (if at all) may
be accused of noncosmological redshifts, even galaxies are not immune!
Indeed, I have already discussed the findings of Arp and Sulentic
[8] that companion galaxies have larger redshifts than the dominant
galaxies. Even if one discounts periodicities in $c\Delta z$, the observation
that $c\Delta z > 0$ remains unexplained. Although I have described the
1985 paper, Arp has been reporting such results from earlier surveys
also [13, 14, 15]. In a recent study Sulentic [16] finds statistically
significant excess redshifts in a sample of 196 companion galaxies
in 60 groups where the dominant galaxies are spirals. (A sample
of 62 companions in 21 E/S0 dominated groups, however, shows no such
excess.) As discussed by Sulentic, the effect, if real, cannot be
explained in conventional terms.

One conventional defence is to discount the companionship of
the companion galaxies and to argue that they are considerably father
away as required by (1) and just happen to be projected near the
dominant galaxies. There are cases, however, where luminous filamentary
connections have been demonstrated between a pair of galaxies of
discrepant redshifts. In 1982 Arp [17] has discussed four such cases,
with $c\Delta z$ ranging from 4,800 km s^{-1} to 26,200 km s^{-1}. However, Sharp
[18] has pointed out that in the case of 0213-2836 discussed earlier
by Arp [15] there is no dynamical evidence of interaction between
the main galaxy and its excess redshift companion.

More recently Sulentic and Arp [19] have found two pairs of galaxies
4030-11 and 4151-46 with luminous connections and perturbations in
the form of asymmetric internal structures of the higher redshift
companions. Both the direct and spectroscopic studies indicate that
the pair members are dynamically interacting. Yet the $c\Delta z$ values
in the two cases are 4859 km s^{-1} and 28661 km s^{-1} - too large for the
pair members to be neighbours according to the conventional view.

3.2 Quasar–Galaxy Associations

Taking the conventional view that bright galaxies are nearby and
quasars with their larger redshifts farther away, we do not expect
any correlation between their respective distributions. Correlations
have, however, been found in several studies, ranging from isolated
groups to large populations.

3.2.1. Large associations.
The most recent study was done by Chu
et al [20] who compared the quasars in the Hewitt-Burbidge catalogue
of 1980 [21], with the bright galaxies in the Second Reference Catalogue
of de Vaucouleurs et al [22]. Using the statistical techniques of
cross-correlation function and the nearest neighbour separation these
authors find that quasars and bright galaxies are associated. For
example, there are $0.36 \pm .09$ more bright galaxies within 40 arc min
of quasars than would be expected under the cosmological hypothesis.

Other investigations of large scale inhomogeneities of quasar
populations have been carried out by Arp [23, 24]. To avoid the

usual criticism of 'selection effect' Arp looked for inhomogeneity
of distribution in the Parkes and 3CR quasars. These are complete
samples of radio loud quasars. Shastri and Gopal Krishna [25] had
also earlier reported inhomogeneity in the distribution of these
quasars with redshifts $z > 2$. Arp found that quasars in the redshift
range $1.4 < z < 2.7$ and magnitude range $17.5 < V < 19$ are distributed
in a non-random fashion. The distribution of these quasars looks
like an elongated jet extending from R.A. $\sim 1^h 30^m$, $\delta \sim 30°$ to
R.A. $\sim 23^h$, $\delta = 0°$. Arp noticed that the companion galaxy M33 in the
Local Group lies at one end of this line. Are these quasars ejected
from the galaxy? If they are distant background quasars then the
above anisotropy implies inhomogeneities in the universe on the scale
of $\sim 10^3$ Mpc. Subsequently Arp [26] has reported jets and filaments in
the distributions of low redshift quasars as well as high velocity
hydrogen clouds associated with galaxies in the Sculptor Group, NGC
55 and NGC 300.

Before one worries about a theory of ejection of objects from
galaxies in jets and filaments one has to be convinced that the observed
anisotropy is statistically significant. The usual statistical tests
of nonrandomness are not applicable to such distributions where the
claim is for an elongated large scale structure. Narlikar and Subramaniam
[27] have devised new tests and applied them to these radio samples.
It is found that Arp's claim for M33 is significant although for
quasars in other redshift or magnitude ranges there is no significant
nonrandomness on a large scale.

3.2.2. Quasars lying close to galaxies. I wish to highlight three
recent investigations.

a) NGC 4319 and Markarian 205. This pair with redshifts 0.006 and
0.071 has a long history dating back to 1971 when Arp [28] announced
a luminous connection between the spiral galaxy with the higher redshift
object. The existence of the connecting filament was finally accepted
(I hope!) after Sulentic's detailed study of the plates [29], although
conventional interpretation of it was not long to follow! Cecil
and Stockton,[30] while admitting the reality of the connection,
have argued that the filament connects M 205 with a companion galaxy
at the same redshift. Recently Sulentic has claimed (private communication)
that the optical spectroscopy of the connection by him with Arp invalidates
this conventional interpretation.

Sulentic [31] has also studied NGC 4319 and its surrounding region
by VLA and finds that there is no corresponding radio connection.
However, NGC 4319 is a double lobed radio source. Thus the fact
that it is an active galaxy capable of ejecting matter lends circumstantial
evidence that M 205 may also have been ejected from NGC 4319.

b) 1E 0104.2 + 3153 and the neighbouring elliptical galaxy. This
example was reported in 1984 by Stocke et al [32] in connection with
the optical identification of the above X-ray source arising from
the Einstein Observatory Medium Sensitivity Survey. It has a quasar
with $z = 2.027$ about 10 arc sec away from a giant elliptical galaxy

at z = 0.111. The authors chose to describe this as a gravitational lens.

In earlier times such a pair would have been dismissed as chance coincidence even though its probability were small. Multiple imaging with same redshifts which is the hallmark of gravitational lensing is absent in this case, however. However, the next case is even more striking.

c) The galaxy 2237 + 0305 and the neighbouring quasar. If I were to pick up only one counter example to CH, I would pick up this one. Not only does it provide strong evidence for noncosmological redshifts - it also illustrates the present sociology of science. The quasar with redshift z = 1.695 was discovered by Huchra et al [33] within 0.3 arc sec of the centre of the galaxy 2237 + 0305, with z = 0.0394. The probability of a 17^m background quasar lying by chance within 0.3 arc sec of the centre of the galaxy is $\sim 10^{-5}$, while if one allows the quasar to be anywhere within the larger slit area used in observing galaxies then the probability is $\sim 10^{-3}$.

By standard statistical practice a hypothesis leading to an outcome of low probability (- even 10^{-3} is normally considered low!) would be rejected and the conclusion should be that the quasar and galaxy are associated. Indeed a number of low redshift quasars have been found with fuzz around them which are conventionally interpreted as galaxies hosting the quasars in their nuclei. Following this line of argument one could have concluded that 2237 + 0305 was hosting the quasar. However, such a conclusion would clearly have been embarrassing for CH. So some argument had to be found to get round the difficulty.

The argument involves gravitational lensing. Make the quasar a background quasar and fainter than the observed 17^m, and then amplify its brightness by invoking the galaxy as a lens. The lensing argument apparently increases the probability of the observed event to 2×10^{-4} from 10^{-5} and to 3×10^{-2} from 10^{-3} (in the case of the larger slit area). However, as pointed out by M. G. Edmunds [34] these probabilities are further reduced if one takes into account the dimming of the quasar light by 3^m through interstellar extinction as implied by its $\nu^{-3.5}$ spectrum.

Thus, even prima facie, gravitational lensing does not improve the situation for CH, it only adds more epicycles. In a comment on the paper by Huchra et al Burbidge [35] has rightly called the title of the paper '2237 + 0305: A New and Unusual Gravitational Lens', a little unfair.

Tyson and Gorenstein have since claimed that they have indeed found other images of the quasar produced by the lens [36]. It is my prediction that these claims would be accepted readily and uncritically while Arp's 1971 claim of a connection had to wait fourteen years for a grudging acceptance. I will therefore base my remaining comments on the assumption that the claim is correct.

The observed image configuration imposes certain limitations on the lens. For example, the lensing galaxy must have a bar-component or it must be a highly eccentric spheroid. Neither possibility belongs to a common category of galaxies. Further, the anisotropy of the lens means it has to be suitably oriented. K. Subramanian (private communication) who has studied the diferent aspects of this lensing system finds that the probability of the lensing configuration is as low as $\sim 2 \times 10^{-5}$.

Schneider [37] has argued that it is improbable that gravitational amplification can sufficiently account for the puzzling observations of quasar galaxy associations. If further observations rule out the lens alternative, or if the lensing probability remains embarrassingly small then the conventional defence would fall back on the oft-expressed view that it is dangerous to calculate a-posteriori probabilities.

3.3 Close Pairs of Quasars

Another test of the cosmological hypothesis comes from the observed occurrence of close pairs of quasars. If the test is to be applied fairly and ideally, we should first decide whether the members of the pair are physical neighbours. If the answer is in the affirmative we should then measure their redshifts. If the redshifts are equal then the CH survives. If the pair has discrepant redshifts the CH is falsified.

How do we decide if the pair members are really physical neighbours? Since we are not permitted to use (1) to estimate their distances we have to rely on other indirect means. These methods are statistical. Given the rules under which the sample of close pairs is to be selected can we estimate the probability of their occurrence by chance?

Burbidge et al [38] in 1974 outlined this procedure soon after two close pairs Ton 155, 156 and 1548 + 114 A, B were found. The aim of this paper was to lay down a formula for computing probabilities that did not have the label 'a-posteriori' attached to it. Simply stated, the expected number of quasars lying within θ arc sec of an arbitrary search centre is given by

$$\langle n \rangle = 2.4 \times 10^{-7} \Gamma(<m)\theta^2 \qquad (3)$$

where $\Gamma(<m)$ is the sky density of quasars brighter than magnitude m expressed in (arc deg)$^{-2}$.

In practice the situation is usually inverted. Observers tend to classify close pairs after measuring their redshifts. Thus those with identical redshifts are supposed to be gravitationally lensed images of a single object. Those with nearly equal redshifts are supposed to belong to the same supercluster, while those with discrepant redshifts are considered to be chance projections without any significance. Thus according to this point of view, the CH is taken to be valid no matter how many and how close are the pairs of the third kind.

Burbidge et al [39], however, have attempted to compute the probability of occurrence of close pairs. They have carefully laid down the values of the parameters that go into the formula (3) and found that the probability of finding the observed number of 6 close pairs with one member a radio source, both brighter than V = 18.5 and lying within 2 arc min of one another to be $\sim 10^{-4}$. As these authors admit, the values of the parameters chosen by them are open to

modification as the data on quasars improve. But if their values
are accepted the close pairs are occurring much too frequently to
be ascribed to chance.

4. THEORETICAL CONSIDERATIONS

I end with three brief theoretical comments. The first is that there
are severe theoretical constraints on quasar models if the cosmological
hypothesis is accepted. Burbidge has outlined some in a recent
review [5]. The general difficulty is as follows. At cosmological
distances the quasars must be very luminous objects. At the same
time the apparent superluminal motions and rapid flux variations
require these objects to be very compact. Whatever energy machine
(including the omnipotent massive black hole) is thought of as the
source of quasar's energy, it has to function with extremely high
efficiency ($\gtrsim$ 10%) in delivering the observed luminosity of a quasar.
Further, the models require large scale relaativistic bulk motions
in a highly ordered form, a concept that has not yet been put on
sound theoretical footing [40]. Both these aspects are novel so
far as the rest of astrophysics is concerned. Those who insist
on conventional physics being applicable throughout the universe
are already stretching the limit of conventionality beyond what
is familiar elsewhere.

My second comment is that the generally held belief that gravitational
lensing has demonstrated that quasars are distant is not quite correct.
The lensing configuration can be scaled down without difficulty.
In fact in a recent paper Chitre and Padmanabhan [41] have shown
that Population III stars or other units of dark matter with masses
$\sim 10^6_\odot$ in the halo of our Galaxy can act as lenses for distant as
well as local quasars, reproducing all the observed features.

Finally, if noncosmological redshifts are present, what is
the physics behind them? Conventional physics offers two alternatives,
gravitational and Doppler redshifts. Do viable and detailed models
exist with these redshifts in which the quasars are considerably
closer than implied by CH?

To some extent this is a chicken and egg problem. Because
it is generally believed that there is no need for noncosmological
redshifts detailed models are not many. At the same time because
there are no detailed investigations of noncosmological models it
is assumed that there are no serious alternatives to CH. Just to
draw attention of this community to recent models of noncosmological
redshifts I mention the gravitational model of Hoyle [42] and the
Doppler model of Narlikar and Subramanian [43].

Should the conventional alternatives to CH also fail then 'new
physics' will have to be invoked. In the last analysis, it is fear
of this prospect that, I suspect, makes most astronomers hold on
tenaciously to the cosmological hypothesis and to ignore the contrary
evidence. This attitude is surprising from workers in a subject
that contributed such new physics as the law of gravitation, spectroscopy
and thermonuclear fusion. Personally, I find the opposite more
dreadful - that astronomy will make no more fundamental contributions

ACKNOWLEDGEMENTS

I wish to thank Geoffrey Burbidge, Chip Arp, Jack Sulentic and Kandaswamy Subramanian for providing me with suitable ammunition for this talk.

References

(1) Smith, D.H., 1983, Sky and Telescope, **66**, 107
(2) Arp, H., 1973, in The Redshift Controversy, ed. G.B. Field (Benjamin, Reading)
(3) Burbidge, G.R., 1981, Ann. N.Y. Acad Sci, **375**, 123
(4) Narlikar, J.V., 1963, Introduction to Cosmology, (Jones and Bartlett, Boston)
(5) Burbidge, G.R., 1985, in Proceedings of the Trieste Conference on Structure and Evolution of Active Galactic Nuclei
(6) Tift, W.J., 1976, Ap. J., **206**, 38
(7) Cocke, W.G. and Tifft, W.J., 1983, Ap. J., **268**, 56
(8) Arp, H. and Sulentic, J.W., 1985, Ap. J., **291**, 88
(9) Rood, H., 1982, Ap. J. Supp., **49**, 111
(10) Byrd, G. and Valtonen, M., 1985, Ap. J., **289**, 535
(11) Hewitt, A. and Burbidge, G.R., 1985, Preprint
(12) Depaquit, S., Pecker, J.-C. and Vigier, J.-P., 1985, Astron. Nach., **306**, I, 7
(13) Arp, H., 1970, Nature, **225**, 1033
(14) Arp, H., 1977, in L'Evolution des Galaxies et ses Implications Cosmologiques p. 377, eds. C. Balkowski and B.E. Westerlund (CNRS, Paris)
(15) Arp, H., 1982, Ap. J., **256**, 54
(16) Sulentic,J.W., 1984, Ap. J., **286**, 442
(17) Arp, H., 1982, Ap. J., **263**, 54
(18) Sharp, N., 1985, Ap. J., 1985, **297**, 90
(19) Sulentic, J.W. and Arp, H., 1985, Ap. J., **297**, 572
(20) Chu, Y., Zhu, X., Burbidge, G.R. and Hewitt, A., 1984, Astron. Ap., **138**, 408
(21) Hewitt, A. and Burbidge, G.R., 1980, Ap. J. Supp., **43**, 57.
(22) De Vaucouleurs, G.H., De Vaucouleurs, A.P. and Corwin, H.G., 1976, Second Reference Catalogue of Bright Galaxies (University of Texas at Austin)
(23) Arp, H., 1984, Ap. J., **277**, L27
(24) Arp, H., 1984, J. Ap. and Astron., 5, 31
(25) Shastri, P. and Gopal Krishna, 1983, J. Ap. and Astron., 4, 109
(26) Arp. H., 1985, A.J., in press
(27) Narlikar, J.V. and Subramaniam, 1985, Astron. Ap. 151, 264
(28) Arp, H., 1971, Astrophys. Lett., **9**, 1
(29) Sulentic, J.W., 1983, Ap.J., **265**, L49
(30) Cecil,G. and Stockton, A., 1985, Ap. J., **288**, 201
(31) Sulentic, J.W., 1985, Preprint
(32) Stocke, J.T., Liebert, J., Schield, R., Gioia, I.M. and Maccacaro, T., 1984, Ap. J., **277**, 43
(33) Huchra, J., Gorenstein, M., Kent, S., Shapiro, I., Smith, G., Horine, E. and Perley, R., 1985, A.J., **90**, 691

(34) Edmunds, M.G., 1985, Nature, **316**, 102

(35) Burbidge, G.R., 1985, A.J., **90**, 1399

(36) Tyson, T. and Gorenstein, M., 1985, Sky and Telescope, **70**, 319

(37) Schneider, P., 1985, Preprint (Max Planck Institute)

(38) Burbidge, E.M., Burbidge, G.R. and O'Dell, S.L., 1974, **248**, 568

(39) Burbidge, G.R., Narlikar, J.V. and Hewitt, A., 1985, Nature, **317**, 413

(40) De Young, D., 1984, Science, **225**, 677

(41) Chitre, S.M. and Padmanabhan, T., 1985, Preprint (T.I.F.R.)

(42) Hoyle, F., 1983, Astrophysics and Relativity, Preprint No. 88 (University College, Cardiff)

(43) Narlikar, J.V. and Subramanian, K., 1983, Ap. J., **273**, 44

DISCUSSION

Lawrence : I hope we all recognise the importance of counter examples in science. But the Universe is a messy place, and there will always be puzzles. In the spirit of a question earlier in the week, I would like to ask on the basis of what evidence would you accept the cosmological hypothesis as a working hypothesis ?

Narlikar : Like any scientific hypothesis the CH is also subject to scientific tests. A more valid question therefore should be posed to the supporters of CH : "What observational test, in your opinion would constitute a disproof of CH ?"

However, in the spirit of the question asked, I may mention that CH for galaxies emerged as a tight Hubble relationship in a natural way. I would have taken CH as a natural working hypothesis if a Hubble law had emerged for quasars. After two decades we still have a scatter diagram.

Burbidge : In reply to the question just asked I would be more inclined to believe in cosmological redshifts if much of the evidence for non-cosmological redshifts had been shown to be wrong. Infact, this has not happened in any critical case.

Birkinshaw : To take up the challenge, I would accept as proof of the non-cosmological nature of quasar redshifts an example of a low-redshift quasar being lensed by a substantially higher redshift object, or showing absorption features from the higher-redshift object in its spectrum.

Narlikar : So far as lensing is concerned a supporter of CH would dismiss the high redshift galaxy as a 'chance coincidence' and he would argue that lensing (if at all) is being done by a faint massive object between the quasar and the observer. The absorption features would be harder to explain away but given sufficient ingenuity, I am sure somebody will find an explanation for it too - because so much is at stake !

Peacock : Do you believe that <u>any</u> quasar has a cosmological redshift ? One quasar in (say) 1000 with a discrepant redshift would not affect conventional statistical studies.

Narlikar : In my talk I have taken the view that there are some genuine counter examples to CH, i.e., <u>a few</u> cases wherein a part of the redshift (say $\gg 10^3$ kms^{-1}) is noncosmological. Rather than statistical studies, a genuine point of fundamental importance is involved here. For, if, noncosmological components are present in the redshifts of a few quasars, then one automatically becomes suspicious about all quasars. The suspicion is reinforced because <u>prima-facie</u> no tight Hubble relationship is seen for quasars.

Leahy : I disagree with your ground rule that "no new theory is required". In the history of science no widely successful theory has been rejected on the basis of a counter-example until a new theory was available which accounted both for the counter-example and for the success of the old theory. (For instance, Maxwellian electrodynamics was not abandoned after it was apparently undermined by the Michelson-Morley experiment.)

Narlikar : I was objecting to the fact that many physicists doubt the evidence against CH on the grounds of lack of theory for the evidence. It was not Maxwellian electrodynamics but the ether theory that was threatened by the Michelson-Morley experiment. Attempts were made, e.g. by Lorentz and Fitzgerald to patch up - but these did not succeed. A new theory came eventually, but nobody doubted the Michelson-Morley result till then.

There can be such a situation that <u>no</u> theory is available to explain observed phenomena. Spectral lines discovered in the last century had no theory to explain them : but this did not make physicists doubt their existence. The situation here is the reverse : physicists doubt the existence of counter examples to CH because there is no alternative theory for them. I consider this attitude unscientific.

"Do you believe that <u>any</u> quasar has a cosmological redshift ?"

John Peacock (p.473)

QUASAR PAIRS

P. A. Shaver
European Southern Observatory
Karl-Schwarzschild-Str. 2
D-8046 Garching bei München
Federal Republic of Germany

ABSTRACT. This paper summarizes various uses which have been made of
QSO pairs. Statistical studies using pairs show that the sky
distribution of QSOs of arbitrary redshifts is random, and that QSOs
of similar redshifts are clustered on a linear scale similar to that
of galaxies today. Close pairs are used to set limits on the masses of
QSOs. And absorption line studies of QSO pairs support the extrinsic
hypothesis for the origin of the narrow absorption lines and the
cosmological interpretation of the redshift, they demonstrate the
presence of absorbing matter around QSOs, and they set limits on the
sizes of the absorbers.

1. INTRODUCTION

An angular separation of one arcmin corresponds to 300-400 h^{-1} kpc at
redshifts of ~1-2, for H_0 = 100 h km s^{-1} Mpc^{-1} and q_0 = 0 (used
throughout this paper). Comparison with typical scales for galaxies
and clusters of galaxies then shows that QSO pairs with angular
separations up to several arcmin can be useful in a variety of
astrophysical and cosmological studies. There is no scarcity of such
pairs: at magnitudes brighter than 20th there is roughly one QSO pair
of separation <1 arcmin for every 100 QSOs, and 10,000 such pairs over
the whole sky. In the following we summarize briefly some of the uses
that have already been made of such pairs.

2. SKY DISTRIBUTION OF QSOs

The use of QSO pairs can by-pass many of the problems encountered in
studying the sky distribution of QSOs from variations in their surface
density, a difficult quantity to measure accurately. The cumulative
number of QSO pairs should increase as the square of the angular
separation if QSOs are distributed randomly on the sky, regardless of
the limiting magnitude of the sample used. Fig. 1 shows that this is
indeed the case, up to angular separations comparable to the field

475

G. Swarup and V. K. Kapahi (eds.), Quasars, 475–480.

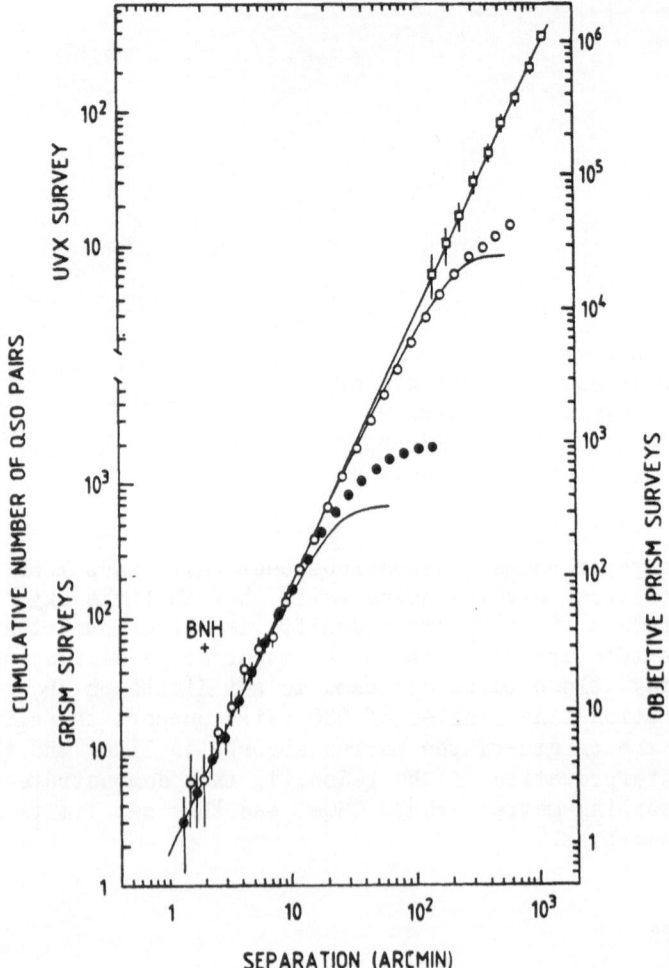

Figure 1. Cumulative frequency distribution of QSO pairs as a
function of angular separation. The open circles, filled circles, and
open squares are from the objective prism, grism, and UVX surveys
listed in Shaver (1985, 1986) respectively. The error bars correspond
to $\sqrt{N}$. The straight line has a slope of +2, the theoretical slope for
a random distribution; the curves are Monte Carlo simulations for
single fields. The cross marks the position expected for the data
point at $\theta = 2'$ if QSOs were clustered as suggested by BNH.

size. Burbidge, Narlikar and Hewitt (BNH, 1985) find an excess of
pairs of small angular separation in a study of published radio QSOs,
but their result is clearly in conflict with the results in fig. 1,
and it is not reproduced in an unbiased sample of radio QSOs studied
by Wills (1978; 1986); Shaver (1986) has shown that the BNH result is
heavily biased by the "publication selection effect" (cf. Wills,
1978). The sky distribution of QSOs of arbitrary redshifts appears to
be random on all scales from arcminutes to tens of degrees.

3.   PHYSICAL CLUSTERING OF QSOs

Homogeneous samples can in principle be used to study the physical
clustering of QSOs, but they must be either very large or very deep
because of the low space density of QSOs. An alternative approach is
to use very large (but heterogeneous) catalogues of QSOs (in this case
that of Véron-Cetty and Véron, 1985). Physical clustering should show
up amongst QSO pairs of small redshift difference and projected linear
separation, but not amongst those of large redshift difference and/or
projected separation, and so can be studied by simply comparing these
two groups. The observational selection effects are the same for both,
and cancel out in the comparison.

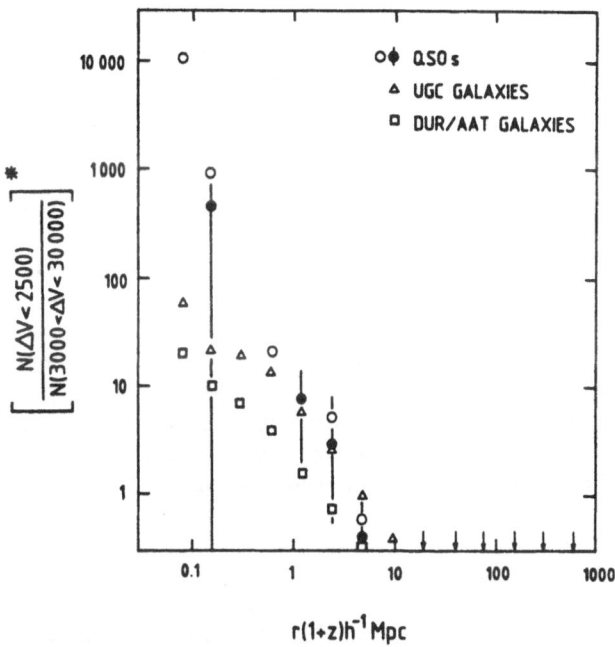

Figure 2. Number of pairs with small redshift difference (<2500
km s[-1]) divided by the number with large redshift difference, as a
function of projected linear separation and normalized to the largest
separations, for QSOs and two samples of galaxies. A random
(unclustered) distribution produces a normalized ratio of zero at all
separations. The filled circles exclude those QSO pairs thought to be
gravitational lenses or subject to the "publication selection effect".

Fig. 2 shows such a comparison using exactly the same technique for
QSOs and two samples of galaxies. After allowance is made for those QSO
pairs thought to be gravitational lenses and a few others possibly
biased by the "publication selection effect", the agreement between the
QSOs and the galaxies is striking: clustering is evident in all three
samples below 10 h[-1] Mpc, and at about the same amplitude, as first

noted by Shaver (1984). If QSOs can legitimately be compared with galaxies, this implies that clustering on these scales has evolved very little since z ~ 2. Furthermore, the fact that QSO clustering shows up at all in this analysis provides further support for the standard cosmology, and further analysis may conceivably help to discriminate between different cosmological and evolutionary models. A more detailed analysis of this work will be published shortly.

4. GRAVITATIONAL LENSING

If the gravitational lens pairs are left in the above analysis, the clustering derived for QSOs exceeds that of galaxies today. As clustering is supposed to increase with time, this could be construed as statistical evidence for the reality of gravitational lensing. It is difficult to distinguish between close physical pairs and gravitational lenses, even for separations of arcminutes which could conceivably be produced by cosmological strings (e.g. Vilenkin, 1984; Hogan and Narayan, 1984; Gott, 1985). One possibility is to compare the redshifts of low-excitation forbidden lines, which should be relatively constant; if they differ in the two spectra, the images are almost certainly distinct QSOs.

　　　Very close pairs of distinct QSOs with different redshifts are also of interest in connection with gravitational lensing. The foreground QSO acts as a lens, and this provides information about its mass. The pair Q1548+114AB, with separation 4″.8 and redshifts 0.4 and 1.9, has been studied by Gott and Gunn (1974) and Iovino and Shaver (1986); the absence of a secondary image of the background QSO B implies an upper limit of $2 \cdot 10^{12}$ $M_{\odot}$ for A, and an ST search for the secondary image will give information on the central mass concentration of A.

5. ABSORPTION LINES IN QSO PAIRS

The twin lines of sight of a QSO pair probe the intervening medium on useful scales, and make possible a variety of studies. The line of sight to the background QSO passes close by the foreground QSO, and the spectrum of the former may contain absorption at the redshift of the latter ("associated absorption"). This has been found in a number of cases (e.g. Shaver, Boksenberg, and Robertson, 1982; Robertson et al., 1986, and references therein), demonstrating that at least some of the high-redshift, high-excitation, narrow-line absorption systems are due to intervening material. Such observations also show that the foreground QSO has the lower redshift, in accordance with the cosmological interpretation (Shaver and Robertson, 1983). The implied excess of absorbing material in the vicinity of QSOs could be due to clusters of galaxies, or possibly large (10^5-10^6 pc) gaseous halos around the QSOs such as that found around MR2251-178 by Bergeron et al. (1983).

　　　The nature of intervening absorbing material, located in front of both QSOs, can be studied by examining the incidence of absorption

systems which show up in both spectra at the same redshift as a
function of projected linear separation. In this way it has been found
that the absorbers, both metal-rich and Lyα systems, have sizes in the
range ~10-400 h^{-1} kpc. This is consistent with the metal-rich systems
originating in galaxy halos, but does not yet distinguish between
various possibilities for the origin of the Lyα systems - more QSO
pairs with a range of separations must be studied for this purpose.

6. CONCLUSIONS

This paper summarizes several uses of QSO pairs, and results obtained
so far. Clearly much more can be done along these lines as the number
of useful pairs increases. Nor are these applications exclusive; a
close pair of radio QSOs has already been used to provide mutual phase
reference in a VLBI study of their structure (Marcaide and Shapiro,
1984), and with relative positions accurate to microarcseconds, proper
motion studies on interesting scales are now feasible. More
possibilities using QSO pairs will undoubtedly be conceived in the
future, further enhancing the value of QSOs in astrophysical and
cosmological studies.

ACKNOWLEDGEMENTS

I am grateful to Ronaldo de Souza and Tom Shanks for providing the
catalogues of galaxies used in section 3.

REFERENCES

Bergeron, J., Boksenberg, A., Dennefeld, M., Tarenghi, M. 1983,
    M.N.R.A.S. 202, 125.
Burbidge, G.R., Narlikar, J.V., Hewitt, A. 1985, Nature 317, 413.
Gott, J.R. III, 1985, Astrophys. J. 288, 422.
Gott, J.R. III, Gunn, J.E. 1974, Astrophys. J. 190, L105.
Hogan, C., Narayan, R. 1984, M.N.R.A.S. 211, 575.
Iovino, A., Shaver, P.A. 1986, Astr. Astrophys. (in press).
Marcaide, J.M., Shapiro, I.I. 1984, Astrophys. J., 276, 56.
Robertson, J.G., Shaver, P.A., Surdej, J., Swings, J.P. 1986,
    M.N.R.A.S. (in press).
Shaver, P.A. 1984, Astr. Astrophys. 136, L9.
Shaver, P.A. 1985, Astr. Astrophys. 143, 451.
Shaver, P.A. 1986, Nature (in press).
Shaver, P.A., Boksenberg, A., Robertson, J.G. 1982, Astrophys. J.
    261, L7.
Shaver, P.A., Robertson, J.G. 1983, Nature 303, 155.
Véron-Cetty, M.-P., Véron, P. 1985, ESO Scientific Report No. 4.
Vilenkin, A. 1984, Astrophys. J. 282, L51.
Wills, D. 1978, Phys. Scr. 17, 333.
Wills, D. 1986, these proceedings.

DISCUSSION

Kunth : What happens to your evidence for QSO clustering at small scales
if one redistributes the redshifts among all QSO from the Veron catalogue by Monte Carlo Simulation ?

Shaver : Scrambling the redshifts completely destroys the clustering.
Furthermore, the clustering only shows up only amongest QSO pairs with
redshift differences less than 2500 km s^{-1}, and not, for example, amongst QSO pairs with redshift differences between 5000 and 7500 km s^{-1}
(using the same technique) which are not expected to be physically
clustered.

Roberts : Do you find any differences between the "associated absorption
systems" and the typical absorption systems seen in other quasars ? Have
you detected Lα in absorption in the former ?

Shaver : No pronounced difference has been found. The associated absorption systems may be of somewhat higher excitation, but Lα is still
seen.

THE ASSOCIATION OF GALAXY CLUSTERS WITH MODERATELY-HIGH-REDSHIFT QUASARS

H. K. C. Yee
Département de Physique,
Université de Montréal

Richard F. Green
Kitt Peak National Observatory,
National Optical Astronomy Observatories

ABSTRACT. CCD images of quasars having redshifts between 0.3 and 0.65 are analyzed to study the association of galaxies with quasars. Average luminosity functions (LF) of the excess galaxies associated with the radio-loud quasars are determined. It is found that for the sub-sample with z>0.55, there is a significant brightening of the characteristic magnitude M^*, if q_0 is assumed to be 0. Comparing computed quasar-galaxy spatial-covariance amplitudes, we can conclude, at the 0.025 significance level, that the spatial-covariance amplitudes of the sub-sample with z>0.55 are greater than those of the lower redshift quasars. This indicates that there has been a strong evolution of preferred sites for bright radio-loud quasars, implying some number-density evolution of quasars has taken place, and that some rich clusters at z~0.6, in comparison with the local rich clusters, have significantly different physical conditions.

In this paper, we present the analysis of the data collected from the Steward 2.3 m survey which provides preliminary evidence of the evolution of the global environments of quasars. Images taken through Gunn r (6500±450Å) and i (8200±600Å) filters of small fields around quasars were obtained using an RCA CCD camera on the Steward 2.3 m telescope. A semi-automatic procedure was used to identify, photometer and classify the detected objects. The sample consists of bright radio-loud quasars having 0.30<z<0.65 chosen from the 3C, 4C and Parkes radio surveys, and optically-selected quasars from the Palomar-Bright-Quasar Survey (PG) in the same redshift range. The completeness magnitudes for detected galaxies range from r=22.0 to 22.9. The field of view of the images is 91"x144". Observations through the r filter of control fields 1° N of the quasars were also made to provide self-consistent background/foreground galaxy corrections.

For analysis, we shall assume cosmological redshifts for quasars, and derive properties of the excess galaxies in the quasar fields by assigning the redshifts of the quasars to them. Unless specified otherwise, H_0=50 km sec^{-1} Mpc^{-1} and q_0=0 are used throughout.

G. Swarup and V. K. Kapahi (eds.), Quasars, 481–487.

To derive the differential luminosity function (LF) of the apparently associated galaxies, we use the following method: galaxies in each quasar field brighter than the completeness magnitude of the field are binned according to their calculated absolute luminosity with the assumption that they are located at the same distance as indicated by the redshift of the quasar. The K-correction term used is a combination of the K-corrections for elliptical-S0, SBc and Scd galaxies as given by Sebok (1982) for Gunn r. For each absolute magnitude bin obtained from each field, the corresponding apparent magnitude background counts are subtracted using the control field counts.

We can determine luminosity functions only for the radio-loud sample, as the PG sample quasars do not have enough excess galaxies to give useful results. The radio-loud sample is divided into two subsets according to redshift: 10 quasars with $0.3 < z < 0.5$ and 9 with $0.55 < z < 0.65$, with median redshifts of 0.41 and 0.61, respectively. For each redshift sample, an average luminosity function is calculated.

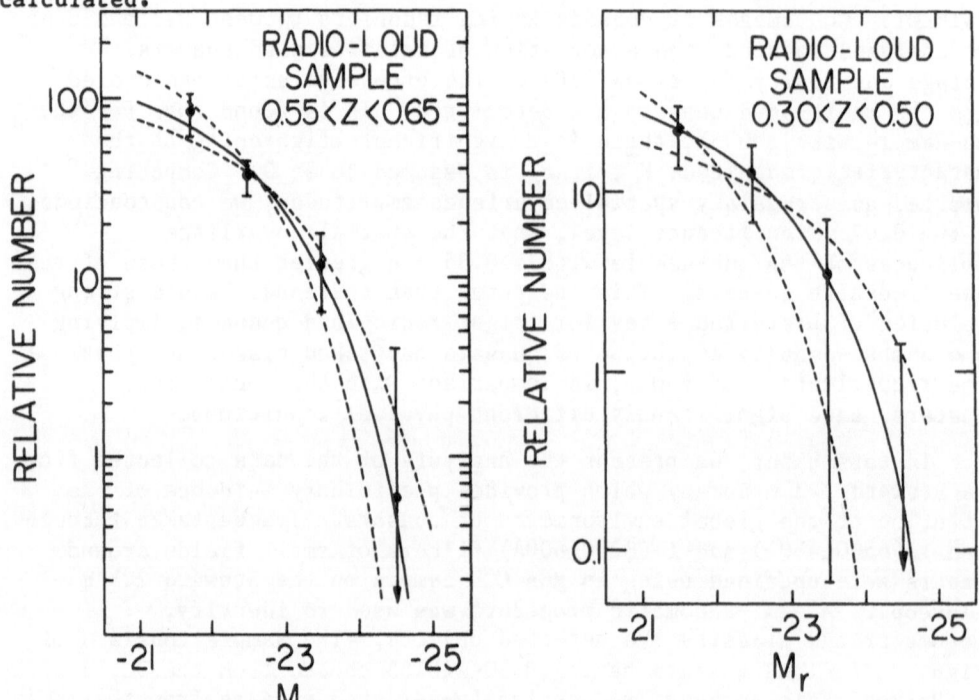

FIGURE 1: Luminosity functions derived for excess galaxies in fields of radio-loud quasars with (a) $0.55 < z < 0.65$ and (b) $0.3 < z < 0.5$ for $q_0 = 0$. The solid and dashed lines represent best fitting and limiting Schechter functions, respectively (see Table 1).

The results are shown in Figures 1a and 1b. To test for a possible evolution in M*, Schechter's (1976) analytical form for the galaxy luminosity function is fitted to the data by minimizing chi-square. By assuming the faint-end slope, α, of -1.2 (Sebok, 1982), we obtain M^* of -22.8 ± 0.9 and -22.8 ± 0.4 for the $\langle z\rangle=0.41$ and 0.61 samples, respectively. These are compared with an M^* of -21.96 for local galaxies (Sebok, 1982) and -22.3 ± 0.3 obtained for the excess galaxies found in fields of radio-loud quasars with $0.15<z<0.3$ (Yee and Green 1984).

We conclude that for $q_0\sim0$, M^* for the $0.55<z<0.65$ sample is significantly brighter than that of local galaxies. An evolution in M^* of up to ~0.8 mag may have taken place. We note that this amount of luminosity + color evolution is comparable to that calculated by Tinsley (1980). If we assume $q_0=1$, the best fitting M^* is still brighter than that of the local galaxies, but the difference is not significant.

We derive the quasar-galaxy spatial covariance amplitude (B_{gq}) by assuming a standard power-law covariance function with index $\alpha=-1.77$. The method as prescribed by Longair and Seldner (1979) is outlined in Yee and Green (1984). To calculate B_{gq}, a proper LF must be chosen, since:

$$B_{gq} \propto A_{gq}/\psi(\langle M(m,z))$$

where A_{gq} is the angular covariance amplitude (which is proportional to the fractional excess of galaxies in the field), and ψ the LF of galaxies. We use two models for the LF: with and without evolution in M^*. The evolving model is based on the M^*'s derived for the LF of the associated galaxies themselves. We adopt an evolution of M^* of -0.5 mag at $z=0.41$ and -0.8 at $z=0.6$. Values for other redshifts are obtained by linear interpolation. The normalization constant for each LF model is determined by matching the counts derived by integrating the LF models in z-space, i.e.:

$$Ng(m)=1/4\pi \int dV \; \psi(\langle M(m,z))$$

to those of the observed control field counts. We note that the evolving M^* model with $q_0=0$ gives the best fit to the observed counts.

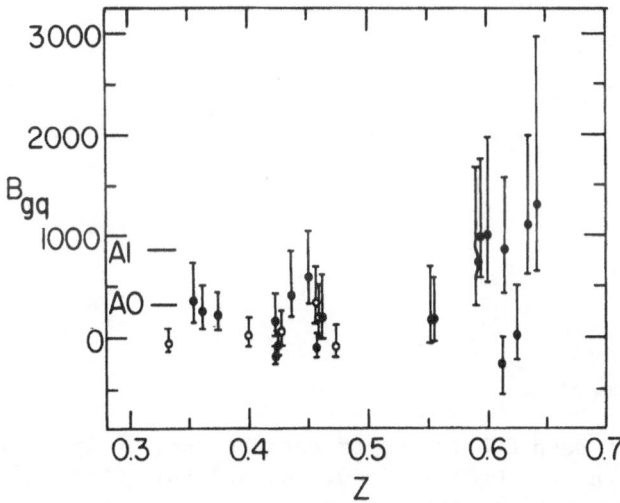

FIGURE 2: Spatial covariance function amplitude versus redshift, assuming an evolution in M^* as described in the text.

The resultant B_{gq}'s for the evolving M^* model are plotted against redshift in Figure 2. For both cases, using the one-sided Kolmogorov-Smirnov test, we can conclude, at a 97.5% confidence level, that the average spatial covariance amplitude for the $0.55 < z < 0.65$ sample is higher than that of the lower redshift sample. The results for the various redshift bins are summarized in Table 1 in the form of $\langle B_{gq}/B_{gg} \rangle$, where $B_{gg} = 67.5 \text{ Mpc}^{-1.77}$ is the average galaxy-galaxy covariance amplitude for normal galaxies (Davis and Peebles 1983). We also included B_{gq}/B_{gg} for quasars with $z < 0.3$ from Yee and Green (1984). These values are scaled by a factor of 0.8 because of a different method used in deriving the angular covariance function.

TABLE I

Average Spatial Covariance Amplitudes

$\langle B_{gq}/B_{gg} \rangle$

Redshift	Radio-loud Sample		Optically-selected-sample
	$M^*=-21.96$	Evolving M^*	$M^*=-21.96$
0.00-0.15	—	—	1.51±0.58
0.15-0.30	2.76±1.01	—	1.57±0.74
0.30-0.50	2.78±1.27	2.54±1.13	1.32±1.14
0.55-0.65	20.43±6.34	9.64±2.69	—

For comparison, normalized covariance amplitudes of galaxies near the center of normal Abell classes 0 and 1 clusters are ~4.6 and 12.6, respectively (Longair and Seldner 1979). Thus quasars at $z > 0.55$ have an average environment approaching that of the centers of Abell 1 clusters.

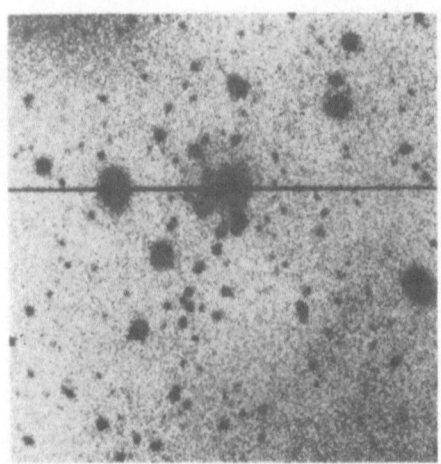 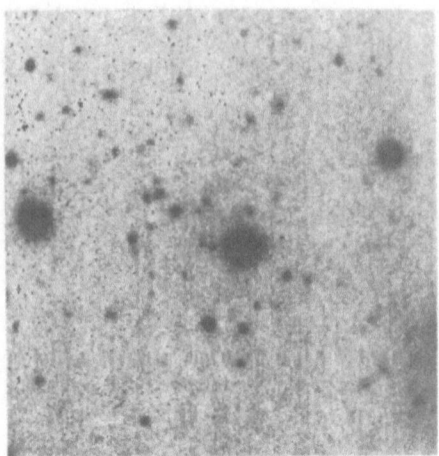

FIGURE 3: Example of deep CCD images of quasars apparently associated with rich clusters: (a) Pks 0405-12 (z=0.574) and (b) 3C263 (z=0.652). Images taken with prime-focus CCD camera at the CTIO 4 m and the CFH 3.6 m telescopes, respectively.

From deeper CCD pictures that we obtained at the CFH 3.6 m and the CTIO 4 m telescopes (Figures 3a and 3b), this conclusion is visually obvious. So far, we have found over 10 quasars apparently associated with such rich clusters.

The apparent evolution of preferred quasar sites has two implications:

1) The physical conditions within some rich clusters at a redshift as small as 0.6 are already significantly different from those of nearby clusters, enabling them to support a bright quasar. Various models of triggering and fueling of quasars have predicted the existence of quasars in clusters at high redshifts (e.g., Stocke and Perrenod 1981, De Robertis 1985 and Roos 1985).

2) The fact that at higher redshifts there are more quasars existing within rich clusters indicates that some number density evolution of quasars must have taken place. This result suggests that the study of the evolution of the quasar luminosity function may have to take into consideration the physical environments of the quasars.

REFERENCES

Davis, M. and Peebles, P. J. E. 1983, Ap. J. 254, 437.
De Robertis, M. 1985, preprint.
Longair, M. S. and Seldner, M. 1979, M.N.R.A.S. 189, 433.
Roos N. 1985, Ap. J. 294, 486.
Sebok, W. L. 1982, Ph.D. Thesis, California Institute of Technology.
Schechter, P. L. 1976, Ap. J. 303, 297.
Stocke, J. T. and Perrenod, S. C. 1981, Ap. J. 245, 375.
Tinsley, B. M. 1980, Ap. J. 241, 41.
Yee, H. K. C. and Green, R. F. 1984, Ap. J. 280, 79.

DISCUSSION

Burbidge : How many redshifts have you got for the supposed galaxies in your sample ?

Green : We have 2-10 confirmations per field for about a dozen objects with z < 0.45. Even the brightest associated objects at z = 0.6 have R > 21 mag; we have not yet been successful in getting an absorption-line redshift.

Peacock : To prove you are seeing evolution of the environment, you must be sure the mean radio powers of the high and low-redshift bins are identical. Otherwise it could be that more powerful quasars lie in richer environments.

Blandford : How did you adjust the normalisation of your evolving galaxy luminosity function ?

Green : We normalised the evolving galaxy luminosity function by fitting to the number counts in the control fields, on the assumption that the field and the quasar host clusters evolve in the same way.

Tyson : This remarkable trend appears to continue to much higher quasar redshifts. A few years ago I made a CCD (unfiltered, back-illuminated RCA CCD) survey of 47 quasar fields. For the quasar field sample with quasar redshifts between 1 and 1.5, there is a statistically significant excess of galaxies (to $R = 22$ mag) nearby on the sky. The two-point angular correlation function implies extreme luminosity evolution of galaxies in these environments at $1 < z < 1.5$.

Van Heerde : By assuming luminosity function evolution you could reduce the excess of galaxies at the high redshift bin significantly. How much evolution do you have to assume to remove the excess completely ?

Green : To bring the excess number of galaxies down to the average value for the low redshift sample requires a change in m^* of > 3 magnitudes over a redshift interval as small as 0.3, with little change in intrinsic colour.

Alighieri : How did you consider the implication on your results of the fact that some of the galaxies around the quasar (after the background galaxies subtraction) might not be at the same redshift of the quasar ? The probability of getting galaxies at higher and lower redshift might not be the same and this might influence your conclusions on the luminosity function.

Green : To first order, any systematic biases from selection or measuring errors are cancelled because the control fields and the object fields are processed identically. If the quasar host clusters produce any gravitational lens amplification, then the background counts in the quasar fields would be slightly enhanced.

Hutchings : Do you see a dependence of your galaxy count excess on the luminosity of the quasars, either optical or radio ? Also, could you remind us of whether there is a difference in environment between radio and optical quasars in your $z < 0.4$ sample ?

Green : The abrupt change of clustering properties with redshift suggests that the dependence on optical luminosity is weak, given the small change in luminosity in this magnitude-limited sample for z going from 0.4 to 0.6.
　　There was a difference in environment for the lower redshift sample, in that the radio quasars showed a higher average clustering amplitude, because of brighter associated objects.

Turnshek : This is a comment. Foltz et.al. have just completed a spectroscopic survey at the MMT of both radio loud and radio quiet QSOs. The redshift range is about 1.4-1.9. They find that about 25% of the radio loud QSOs have $z_{abs} \simeq z_{em}$ systems, but that very few of the radio quiet QSOs do. This is evidence that radio loud QSOs have an excessive amount of material around them compared to radio quiet QSOs. This could be interpreted as radio loud QSOs being embedded in clusters, whereas radio quiet ones are not.

"Untill we reach a deeper understanding of the physical evolution of quasars, we should be very cautious in drawing physical conclusions from successful mathematical representations of the quasar population ensemble properties."

- Richard Green (p.435)

CONSTRAINTS ON THE COSMOLOGICAL EVOLUTION OF AGNs FROM X-RAY DATA[*]

L. Danese , G. De Zotti , G. Fasano , A. Franceschini
Istituto di Astronomia
Vicolo dell'Osservatorio, 5
I-35122 Padova
Italy

1. THE DATA

Both the soft (Maccacaro et al., 1982; Gioia et al., 1984; Giacconi et al., 1979; Griffiths et al., 1983) and the hard (Piccinotti et al., 1982) X-ray surveys provide crucial information on the evolution of AGNs in the X-ray band. Before combining the two data sets, however, the consistency of the two flux scales must be checked. To this end we have compared the soft (SX) and the hard (HX) X-ray luminosities of all powerful (log(SX)>43.5) AGNs for which both IPC and hard X-ray fluxes are available in the literature (lower luminosity objects frequently exhibit large intrinsic, luminosity dependent, absorbing columns which cannot be accurately accounted for). We find

$$<log(HX)/(SX)> = 0.17 \pm 0.06 ,$$

to be compared with the expected value $<log(HX)/(SX)> \simeq 0$ for a typical X-ray energy index $\alpha_x \simeq 0.7$. Thus, an average correction factor of $\simeq 1.5$ needs to be applied to the IPC fluxes to bring them into the hard X-ray flux scale.

Strong additional constraints on evolution models come from the observed properties of the X-ray background (XRB). Its intensity implies, anyway, a fast convergence of the counts of faint AGNs (Setti and Woltjer, 1979). Moreover, as shown by De Zotti et al. (1982), its spectral shape requires that the dominant contributors have a rather narrow range of spectral indices centered around $\alpha_x = 0.4 \div 0.5$; also, the remarkable smoothness of its spectrum (Marshall et al., 1980) strongly argues against a $\geqslant 30\%$ contribution from sources with "wrong" spectra, such as active galaxies which apparently have a "universal" slope $\alpha_x \simeq 0.7$. If indeed QSOs have similarly steep or even steeper slopes and/or a broad range of spectral indices, as suggested by several recent studies (see, e.g., Wilkes and Elvis, 1985) , tight constraints on the evolution properties of AGNs ensue.

2. THE EVOLUTION MODELS

The following evolution models were tested against these data.

[*] Discussion on p.513

G. Swarup and V. K. Kapahi (eds.), Quasars, 489–490 .

1) Luminosity Dependent Density Evolution (**LDDE**) models, as proposed
by Wall et al. (1980) and by Schmidt & Green (1983). The evolution time
scales are in this case proportional to log L_x .

2) Pure Luminosity Evolution (**PLE**) models.

3) Luminosity Dependent Luminosity Evolution (**LDLE**) models, of the
kind proposed by Cavaliere et al. (1985), which allow only high
luminosity sources to undergo strong luminosity evolution.

3. RESULTS

Various statistical techniques were applied to contrast the model
predictions with the data. Minimum χ^2 and Maximum Likelihood methods
were used as parameter point and interval estimators; a 2D Kolmogorov-
Smirnov test (Peacock, 1983; Fasano and Franceschini, 1985) was also
applied to further check the goodness of fit.

The main conclusions can be summarized as follows.

If the X-ray spectra of AGNs are significantly steeper than that of
the 2-10 keV XRB (as current data suggest), a faster convergence of
their counts is implied than can be achieved with the simplest LDDE
models. Luminosity evolution models can meet the XRB constraints on the
condition that only high luminosity sources (log $L_x \gtrsim 43.5$) are allowed to
evolve. Consideration of the Einstein Medium Sensitivity Survey
luminosity distribution, on the other hand, suggests that the transition
between non-evolving and evolving sources occurs just close to log
$L_x \simeq 43.5$.

Further details can be found in Danese et al. (1986).

We acknowledge helpful exchanges with A. Cavaliere, E. Giallongo and
T. Maccacaro. Work supported by CNR (through GNA and PSN) and MPI.

4. REFERENCES

Cavaliere A., Giallongo E., Vagnetti F., 1985, Ap. J. **296**, 402.
Danese L., De Zotti G., Fasano G., Franceschini A., 1986, Astron.
    Astroph. in press.
De Zotti G., et al., 1982, Ap. J. **253**, 47.
Fasano G., Franceschini A., 1985, submitted to Monthly Notices Roy.
    Astron. Soc.
Giacconi R., et al., 1979, Ap. J. **230**, 540.
Gioia I.M., et al., 1984, Ap. J. **283**, 495.
Griffiths R.E., et al., 1983, Ap. J. **269**, 375.
Maccacaro T., Gioia I.M., Stocke J.T., 1984, Ap. J. **283**, 486.
Peacock J.A., 1983, MNRAS **202**, 615.
Piccinotti G., et al., 1982, Ap. J. **253**, 485.
Schmidt M., Green R.F., 1983, Ap. J. **269**, 352.
Setti G., Woltjer L., 1979, Astron. Astroph. **76**, L1.
Wall J.V., Pearson T.J., Longair M.S., 1980, Monthly Notices Roy.
    Astron. Soc. **193**, 683.
Wilkes B., Elvis M., 1986, these Proceedings, p.261.

BL LACERTAE OBJECTS AS AN EVOLUTIONARY POPULATION *

A. Cavaliere, Astrofisica, Dip. Fisica, II Univ. Roma, Italy
E. Giallongo, Ist. Astronomia, Univ. Padova, Italy
F. Vagnetti, Ist. Astronomico, I Univ. Roma, Italy

If the BL Lac Objects are active nuclei with a beamed component that is dominant when directed at us, their observed luminosity function must comprise a flat faint branch: $N(L)dL \propto L^{1+1/p}dL$ with p=4.5 (Urry and Shafer 1984). If this is flatter than the LF $N_p(L)$ of the parent objects at equal observed L, then we expect the counts of BL Lacs to <u>flatten</u> out in turn at fluxes quite <u>higher</u> than the counts of the parents, even when both populations evolve strongly and uniformly with comparable timescales (Cavaliere, Giallongo and Vagnetti 1985).

Figs 1 and 2 illustrate the case when the parent population is a subset of the optically selected AGNs and QSOs, undergoing a luminosity evolution with a timescale $\tau \cong 0.2 \; H_0^{-1}$. A fraction f≪1 of their isotropic power is assumed to be channelled into a randomly oriented beam of emitters with a bulk Lorentz $\Gamma \lesssim 4$; the adopted distribution $F(\Gamma) \propto \Gamma^{-1}$ minimizes the flattening to be demonstrated, yet retains an appreciable proportion of strongly beamed objects. The result is compared with a non-evolving population derived from the bright E galaxies, with a stronger boosting to compensate for the assumed absence of evolution and for the apparent weakness of any isotropic optical emission from these nuclei.

The final convergence at very low fluxes may be enhanced by differential evolution (slower at low L) in the parent population, or by a shorter lifetime of the beamed compared with the isotropic activity.

Two general conclusions ensue. A flat <u>overall</u> behaviour of the counts is not necessarily indicative of weak evolution: a flat LF anyhow allows a fast, early convergence. Conversely, a high degree of <u>order</u> (i.e., LF steeper than $L^{-3.2}$ over most of the observed range) must underly the counts of the optically selected QSOs if they are in fact so steep {$N(<S) \propto S^{-2.2}$ up to $m_B \cong 19$} as indicated by present samples and photometry.

As for the BL Lacs specifically, the statistics of beamed and evolutionary sources predict optical counts <u>flat on average</u>, complying with the present observational bounds; but Fig 3 illustrates the prospects of eliciting their <u>intrinsic</u> evolutionary behaviour, from <u>lower</u> bounds to the counts at $m_B \cong 16.5$ over $\lesssim 1/2$ sky.

A similar prospect holds for the X-ray selected BL Lacs, with the expected improvement of their statistics over the bounds set from the published MSS (Maccacaro et al 1984).

* Discussion on p.513

491

G. Swarup and V. K. Kapahi (eds.), Quasars, 491–492 .
© 1986 by the IAU.

REFERENCES
Cavaliere,A., Giallongo,E., Vagnetti, F. 1985, Astr.Ap., in press (CGV).
Cheng,F.Z. et al. 1985, M.N.R.A.S. 212, 857.
Maccacaro,T. et al. 1984, Ap.J. (Letters) 284, L23.
Setti,G., and Woltjer,L. 1982,in 'Astrophysical Cosmology', Pont.Acad.
    Scientiarum Scripta Varia 48, 315.
Urry,C.M. 1984, Ph.D. thesis, Johns Hopkins University.
Urry,C.M. and Shafer,R.A. 1984, Ap.J. 280, 569.

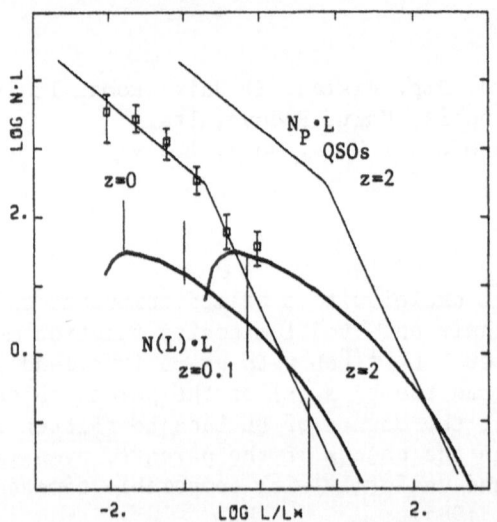

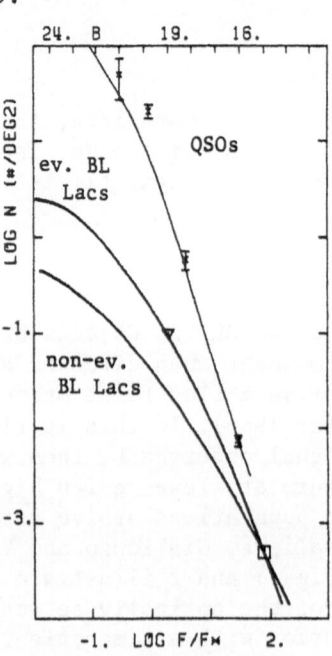

FIG.1. z-dependent LFs: Optical AGNs (local data from Cheng et al. 1985, $\tau=0.17H_0^{-1}$). Evolutionary BL Lacertae Objects derived from the former, boosted with parameters: Γ ranging from 1 to 4, distributed $\propto \Gamma^{-1}$; f= 10^{-2}; normalization to 7% of the AGNs. Bars reproduce estimates of the BL Lac LF (Urry 1984). $H_0=50$ km/s Mpc, $\Omega_0=0.2$, $L*=10^{30}$erg/s Hz.

FIG.2. Predicted optical counts for evolutionary and non-evolutionary BL Lacs. Ev. BL Lacs are given in Fig 1. Non-ev. BL Lacs derive from a population constituted by 5% of the bright E galaxies, boosted with $\Gamma=7$ and $f=5$ 10^{-3}. Observed count (square) and upper limit (triangle) from Setti and Woltjer 1982. Also given are the counts of the evolutionary parent AGNs, from the LF in Fig.1 (cf CGV 1985).

FIG.3. The BL Lac counts of Fig.2 are given in differential, normalized ($\times S^{2.5}$) form. Bars give 1σ statistical uncertainties expected over 1/2 sky.

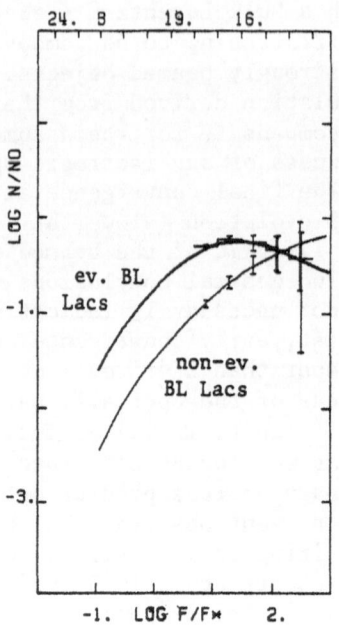

COMPLETE QUASAR SAMPLES AND COMPARATIVE COSMOLOGY

I. E. Segal
Department of Mathematics
Massachusetts Institute of Technology
Cambridge, Massachusetts 02139
USA

ABSTRACT. Comprehensive testing of alternative cosmologies on the basis
of complete samples of discrete sources is described. Application is
made to the comparison of the Friedman and chronometric cosmologies
(denoted C1 and C2) using quasar samples. C2 predicts the results of
analyses predicated on C1 that represent evolution in the C1 framework,
but is itself nonevolutionary, nonparametric, and consistent with
observation.

Statistically efficient methods of estimation of luminosity functions
and related quantities, making due allowance for the observational cutoff bias, in conjunction with Monte Carlo methods, make possible rigorous objective estimates of the probabilities of the observed discrepancies between theoretical cosmological prediction and empirical measurement, assuming correctness of the theory and fairness of the sample
(apart from specified cutoffs). At the same time, the statistical framework makes possible the separation of the question of the conditional
distribution of magnitudes at given redshifts, which is unaffected by
spectroscopic selection, from the statistically independent question of
the spatial distribution, which is extremely sensitive to spectroscopic
selection.

Using the BQS and other complete samples treated by Schmidt and Green
(1983), parallel systematic analyses were made of C1 and C2 as the basis
for equitable comparison. This involved not only direct tests of each
cosmology per se, but also tests of the capacity of each cosmology for
explanation of the results of analyses predicated on the alternative
cosmology. Application to the BQS indicates not only the statistical
acceptability of C2 but shows that the deviations from observation of
the predictions of C1, representing apparent evolution in the C1 framework, are in statistical agreement with the predictions of C2 for the
results of analyses predicated on C1 (Segal and Nicoll, 1986).

Simple theory shows that with suitable frequency-independent pure luminosity and number evolution (LE and NE), C1 will make statistical pre-

G. Swarup and V. K. Kapahi (eds.), Quasars, 493–494 .

dictions identical to those of C2. Nevertheless, LE does not provide as
close a fit to observations as C2 when the evolution is fitted to the
data in question, assuming it naturally related to the Friedman lookback
time, in the cases of optical and X-ray samples studied by Marshall et
al. and Maccacaro et al. Their fitted evolutions are however similar to
the difference between the magnitude-redshift laws for C1 and C2.
Moreover, at small redshifts (< 0.1), LE models are highly deviant from
observed flux-redshift relations, which C2 fit closely; while at large
redshifts (> 2.7) they predict many more quasars than are observed in
faint samples , while C2 predicts that $N(2.7 < z < 4.7) = (0.1)N(< 2.7)$
consistently with observation. Significantly, quasars appear as
'standard candles' in C2, having a bias-corrected dispersion in absolute
magnitude of less than $\frac{1}{2}$ mag. in the range $z > 0.1$ for the BQS, with or
without other complete samples reported in loc. cit. (Segal and Nicoll,
1985), and/or samples of Marshall et al., Mitchell et al.; X-ray selected samples of Maccacaro et al. and Margon et al. inclusive of optical
data but presumptively incomplete fit correspondingly.

The present comparative results are similar to those from earlier studies
of complete homogeneous galaxy samples, which however are quite limited
in redshift, in contrast to the cosmic range of this study. Besides its
statistical effectiveness in fitting systematic observations on discrete
sources, C2 provides wholly natural explanations (devoid of ancillary
parameters or scenarios, such as are used in connection with C1) for a
variety of anomalies within the framework of C1. Among these are the
extreme absolute brightness of quasars, apparent superluminal velocities,
the close connection in all but absolute magnitude of Seyfert I AGNs and
quasars, the missing mass in galaxy clusters, the failure of the Hubble
law in homogeneous complete samples even when allowance is made for
possible local perturbations (peculiar velocities, motion of the Galaxy,
the Supercluster, small-scale clustering), the isotropy of the microwave
and X-ray backgrounds, the flatness of the spectrum of the XRB relative
to those of major discrete sources, the absence of evidence for initial
fluctuations from which galaxies originated, and looming problems in
primeval nucleosynthesis. See Proc. Natl. Acad. Sci. USA **75**, 535; Ap. J.
227, 15; A. & A. **68**, 353; A. & A. **123**, 151; A. & A. **115**, 398 and **118**,
180; Phys. Rev. **D28**, 2393; submitted and forthcoming studies; and references in the foregoing.

Summarizing regarding evolution, since (a) this is incapable of direct
observational substantiation, (b) the deviations from C1 predictions
represented as evolution are precisely as predicted by C2, (c) C2 is itself consistent with statistically appropriate observations, and otherwise acceptable (in terms of theoretical economy and explanatory power),
I submit that the evolution hypothesis can not properly be considered a
truly scientific one.

M. Schmidt and R. F. Green (1983), Ap. J. **269**, 352.
I. E. Segal and J. F. Nicoll (1985), A. & A. **144**, L23.
I. E. Segal and J. F. Nicoll (1986), Ap. J., in press.

DETERMINATION OF THE LARGE-SCALE AND SUPER-LARGE-SCALE STRUCTURE BY MEANS OF THE SPACE DISTRIBUTION OF QUASARS

Y.Y.Zhou[1], D.P.Fang[1], Z.G.Deng[2]and X.T.He[3]

[1] Centre for Astrophysics, Univ. of Science & Technology of China, Hefei,Anhui,China
[2] Dept. of Physics, Chinese Academy of Sciences, Beijing, China
[3] Dept. of Astron., Beijing Normal Univ., Beijing, China

It is very important to study the large-scale structure by means of the space distribution of quasars. Using this method, one may search for the more distant superclusters and explore the super-large-scale structure, i.e., the existence of super-superclusters. The answer to the problem would be of great interest. It is related to the question about the transition of the clustering of galaxies on about 100 Mpc to the uniformity of the universe. Recently Oort et al and de Ruiter et al suggested that the quasars are located in superclusters. So we soppose that analysing the space distribution of quasars might give us some information about the super-large-scale structure of the universe. But up to the present the study of the clustering of quasars has not obtained universally accepted conclusions; in fact, some of them, including grouping and clustering (Arp; Chu and Zhu), no clustering (Chu and Zhu; Osmer; Webster), clustering for $z < 2$ and no clustering for $z > 2$ (Fang et al) and stringing (Deng et al), are contradictory.

In order to obtain more reliable conclusions we should improve the statistics on the clustering of quasars.

Firstly, we choose five independent samples for testing. They are two samples in 25 deg^2 field (Savage and Bolton 1979), one sample of quasar candidates in 25 deg^2 field with redshift value determined by using the slitless spectrum (Savage et al 1984), and two samples of candidates without redshift value in 64 deg^2 field with sampling identifications rates 80% and 83% respectively.

Secondly we tested the five samples with various statistical methods, including the two-point correlation function $\xi(r)$, the cluster analysis, percolation parameter B_c calculation, multiplicity function $F(x)$ analysis and others. Among them the cluster analysis is very effective for determining the essential geometrical structure of the agglomeration region of quasars. Besides the size $L_m(r)$ of the maximum connectea region we use the number $N_m(r)$ of quasars in the maximum connected region to derive the geometrical structure.

Thirdly, we consider the influence of the redshift distribution of the quasars on the clustering of quasars. As is well known, some peaks and dips appear in the redshift distribution. Thus when comparing the considered sample with the uniform Monte Carlo samples, we may get some

G. Swarup and V. K. Kapahi (eds.), Quasars, 495–496.

false clustering or cancel some true clustering owing to the peaks and dips.So we adopt the quasars around the peak z~2 as our sample in order to reduce the statistical fluctuations and increase the number density in the space.

We have computed the two-point correlation $\xi(r)$, the size function $L_m(r)$ and quasar number function $N_m(r)$ of the maximum connected region, percolation parameters $B_{c\alpha}$, $B_{c\delta}$, B_{cz} in the right ascension direction, declination direction, the line of sight, multiplicity function $F(x)$ etc, and compared them with the corresponding function or quantities of the Monte Carlo sampling. We obtain the following results and conclusions.

1. For all five samples either two-point correlation function $\xi(r)$ or $L_m(r)$ and $N_m(r)$ curves are near to the corresponding average curves of the Monte Carlo sampling. The differences between them are within the 2σ range. That means that the clustering and the stringing of the samples are not obvious But some excess of $\xi(r)$ and $N_m(r)$ has been found between σ and 2σ curves (see Figs), i.e., there is possibly some very weak clustering, when the distance between quasars is about 100 Mpc.It is compatible with the view that the quasars exist in supercluster.

2. From $\xi(r)$, $L_m(r)$ and $N_m(r)$, the distribution of quasars is near to the random one up to 400 - 1000 Mpc. It indicates that the universe above the superclusters is uniform and structureless.

3. The computational percolation parameters $B_{c\alpha}$, $B_{c\delta}$, B_{cz} have $|B_c - B_c^{(R)}| < \sigma^{(R)}$. It supports the above mentioned conclusion.

4. The multiplicity function $F(x)$ of the five samples are similar to those of the corresponding Monte Carlo samples, but there is some excess of groups at 160 Mpc.

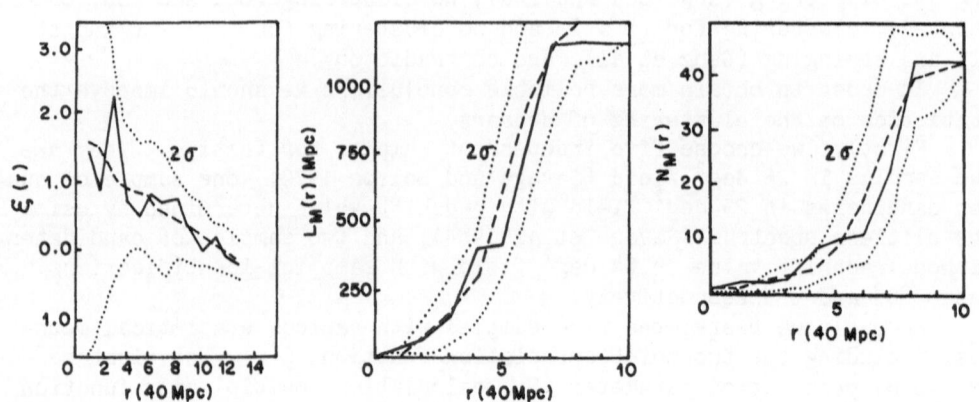

Figs. $\xi(r)$, $L_m(r)$, $N_m(r)$ curves of sample A;
———— sample;   - - - - Monte Carlo

COSMOLOGICAL IMPLICATION OF THE EMISSION LINE REDSHIFT DISTRIBUTION OF QUASARS*

Y.Y.Zhou[1], Y.Gao[1], Z.G.Deng[2] and H.J.Dai[3]

[1] Centre for Astrophysics, Univ. of Science & Technology of China,Hefei, Anhui,China
[2] Dept. of Physics, Chinese Academy of Sciences, Beijing, China
[3] Dept. of Physics, Harbin Inst. of Technology, Harbin, China

The peaks and dips in the quasar redshift distribution seem to be incompatible with the cosmological principle. This lays the cosmological redshift hypothesis under suspicion and censure, and has been considered by some investigators as a manifestation of the intrinsic nature of quasar's redshift. So it is worthwhile studying whether the redshift distribution of quasars could be explained in the framework of cosmological redshift. As we know, this distribution is affected not only by the possible physical origin of redshift but also by the selection effects in the observations (Zhou, Deng, Zhou 1983). From this we have the redshift distribution function

$$f(z) = P(z)R(z),$$

where $P(z)$ is the real distribution function which depends on the evolutionary properties of quasars and the space-time structure of the Universe, and $R(z)$ is the factor caused by the selection effect in the line identification.

For the standard model of the universe with $\Lambda = 0$, we have

$$P(z) = \int_{\tilde{L}(z)}^{\infty} \frac{4\pi c^3}{H_0^3} \frac{[zq_0+(q_0-1)(-1+\sqrt{2zq_0+1})]^2}{q_0^4 (1+z)^6 (1+2zq_0)^{1/2}} n(z,L)dL ,$$

where $n(z,L)$ is Schmidt-Green function of quasar evolution and $\tilde{L}(z)$ satisfies

$$z = q_0 \sqrt{\tilde{L}(z)/L_0} +(1-q_0)[-1+(1+2\sqrt{\tilde{L}(z)/L_0})^2],$$

where $L_0 = 4\pi \mathit{l}c^2/H_0^2$, and l is the flux corresponding the limit apparent magnitude in the observation. According to our previous paper (Zhou, Deng, Dai 1985) $R(z)$ is

$$R(z) = \sum_i R_i(z) + \sum_{i<j}\sum R_{ij}(z) + \sum_{i<j<k}\sum\sum R_{ijk}(z) + ... ,$$

and

$$R_i(z) = \frac{N_i}{N} \frac{S[(1+z)\lambda_{oi}]}{\int P(z)S[(1+z)\lambda_{oi}]dz}$$

* Discussion on p.514

497

G. Swarup and V. K. Kapahi (eds.), Quasars, 497–498.

$$R_{ij}(z) = \frac{N_{ij}}{N} \frac{S[(1+z)\lambda_{oi}]S[(1+z)\lambda_{oj}]}{\int P(z)S[(1+z)\lambda_{oi}]S[(1+z)\lambda_{oj}]dz} \quad , \quad etc.$$

where N_i , N_{ij} , ... are the numbers of quasars identified by the ith line, the ith and jth lines, ... in the sample, respectively, and N is the total number of quasars. $S(\lambda)$ is the total normalized response function. For the αth subsample, which has the common optical measurement window $\lambda_{min}^{(\alpha)} - \lambda_{max}^{(\alpha)}$, from some observational results we assume that

$$S^{(\alpha)}(\lambda) = \begin{cases} A^{(\alpha)}(\lambda-\lambda_{min}^{(\alpha)})/250, & \lambda-\lambda_{min}^{(\alpha)} \leq 250 \text{ Å} , \\ A^{(\alpha)}, & \lambda-\lambda_{min}^{(\alpha)} > 250 \text{ A}, \lambda_{max}^{(\alpha)} - \lambda > 250 \text{ Å} \\ A^{(\alpha)}(\lambda_{max}^{(\alpha)} - \lambda)/250, & \lambda_{max}^{(\alpha)} - \lambda \leq 250 \text{ Å} \\ 0 , & \lambda < \lambda_{min}^{(\alpha)}, \lambda > \lambda_{max}^{(\alpha)} \end{cases}$$

Comparing the calculational redshift distribution histograms with the observational historgram, we come to the following conclusions.

1. The selection effect in redshift identification plays a decisive role in the appearance of peaks and dips in the redshift distribution.

2. For 349 quasars which were discovered by object prism or grating prism technique alone, the calculational distribution histograms are almost as same as the observational one. The four peaks near to z = 0.4, 1.0, 1.4, 2.0 in the distribution appear in the calculational histograms and the correlation coefficients are larger than 0.95 (Fig.1).

3. For 653 quasars which were discovered by the positional method or colour technique, most of the peaks and dips are near to those in the calculational histogram (Fig.2). If we take the limit apparent magnitude $m=20^m$, the correlation coefficients between the observational and calculational histograms are 0.92 for q_o=0.5 and 0.89 for q_o=0.1. Through χ^2-test we confirm that under the significance level α =0.001 the calculational redshift distributions belong to the parent population determined by the observed distribution.

4. To determine q_o by means of the redshift distribution of quasars is a new tentative method, but not a sensitive one. The result for q_o=0.5 may better explain the observational redshift distribution than that for q_o=0.1. This may put some restrictions on dark matter and a more exact value of q_o will be determined in our further calculations.

REFERENCES

Schmidt, M. and Green, R.F. 1983 Ap. J. 269 352
Zhou, Y.Y., Deng, Z.G. and Zhou, Z.L. 1983 Ap. S. S. 97 63
Zhou, Y.Y., Deng, Z.G. and Dai, H.J. 1985 Ap. S. S. 112 93

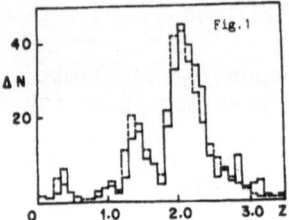

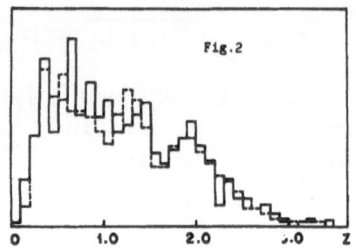

PAIRING, CLUSTERING, OPTICAL VARIABILITY AND GRAVITATIONAL LENSING
IN A SAMPLE OF ABOUT 150 FLAT-SPECTRUM RADIO-SELECTED QSOs

D. Wills and Beverley J. Wills
McDonald Observatory and Department of Astronomy,
University of Texas at Austin, Austin, Texas U.S.A.

Some years ago, Bolton, Peterson, Wills and Wills (1976, BPWW) reported
a statistically-significant excess of QSOs within 2' arc of flat spectrum radio-selected QSOs (specifically, they found 5 such objects while
only 1 was expected by chance). None of the QSO pairs had the same redshifts so the result could be regarded as evidence for non-cosmological
redshifts, non-uniform QSO distributions, or a statistical fluctuation.
Because of the potential importance of BPWW's result, we later used the
same technique to examine an independent sample of about 150 QSOs with
flat radio spectra and found results that are consistent with the known
surface density of QSOs. If we assume a surface density of 3.3 QSOs per
square degree brighter than B = 19 (e.g. Marshall et al. 1983), the 150
fields each of 2' arc radius should contain 1.7 random QSOs above this
limit and we found 2 (e.g. Wills 1978).   A third object is very close
to B = 19, and another field contains an object that is extremely blue,
with B < 19, on the Palomar Sky Survey, but below our plate limit (i.e.
B > 21); we have so far been unable to obtain a spectrum for it. One of
the 3 confirmed QSOs is 119" arc from the radio-emitting one, so rather
small changes in the magnitude limit, search radius and choice of epoch
at which the magnitudes are measured can result in there being anywhere
between 1 and 4 secondary QSOs (or 1-3 if the variable object is not a
QSO), compared with 2 expected by chance, and perhaps 10 that would be
predicted by BPWW's results.

This initial search was limited to B = 19 for comparison with BPWW's,
but most of our two-colour image-tube plates reach B > 21, and we later
obtained spectra of some of the fainter UV-excess objects at McDonald
Observatory. All but one of the objects are within 2' arc of the radio
source positions on which the plates were centred. The detailed results
will be reported elsewhere, but in summary we now have redshifts for 15
of the secondary objects; 3 are low-luminosity galaxies (z < 0.1) and
the remainder are typical QSOs (0.5 < z < 2.2).   There are a few more
objects that have not yet been observed and some whose spectra have so
far given inconclusive results.   Restricting the sample to B = 20, we
have confirmed 8 UV-excess QSOs within 2 arcmin of the radio positions
and there are about 6 more objects to be observed.   Adopting a surface

G. Swarup and V. K. Kapahi (eds.), Quasars, 499–500.

density of 27 per square degree at this magnitude (from Marshall et al. 1983), the expected number of random QSOs on our plates to B = 20 is 14 and our results are therefore quite consistent with these fainter QSOs all being random ones, unrelated to the radio-detected QSOs.

In no case are the redshifts of the members of a pair (or triplet) the same, so no spatial clustering has been found; this does not contradict Shaver's (1984) results, since fewer than 1% of the QSOs in his homogeneous sample of QSOs are clustered in redshift.

Because of the recent interest in discovering new cases of gravitationally-lensed QSOs, we have carefully examined all the QSO images on our plates (7.9 arcsec per mm) for signs of noncircularity or multiplicity. The seeing was typically 1-2" arc and on most of the plates an object 2 arcsec away with B < 21 would easily have been noticed. Only one such case was found, and we are investigating it further.

Finally, we note that 13% of the flat-spectrum radio QSOs show changes of at least 1 mag between the epochs of the Palomar Sky Survey and our image-tube plates (20-25 yr timescale); this is higher than the incidence of variables among steep-spectrum radio QSOs (e.g. Uomoto, Wills and Wills 1976). There is a statistically-significant negative correlation between variability and redshift in our sample, in the same sense as that noted by Uomoto et al. This could be a luminosity effect, but it could also be due, at least in part, to time dilation if the timescale of the variability is comparable to the 20-25 yr baseline of the observations.

In summary, our recent spectra of the fainter UV-excess objects in the fields of flat-spectrum radio QSOs show that their numbers are consistent with expectations based on the known surface densities of QSOs at B = 20, and the new search does not support earlier evidence for excess pairs of QSOs with different redshifts. The recent claim by Burbidge, Narlikar and Hewitt (1985) that the significance of QSO pairing has increased with the addition of more data has been effectively refuted by Shaver elsewhere in these proceedings.

We thank Alan K. Uomoto for his help in obtaining the image-tube plates and acknowledge support of this work by the National Science Foundation (grant AST-8215477).

REFERENCES

Bolton, J.G., Peterson, B.A., Wills, B.J. and Wills, D. (1976).
    Ap.J. Letters 210, L 1.
Burbidge, G.R., Narlikar, J.V. and Hewitt, A. (1985). Nature 317, 413.
Marshall, H.L., Tananbaum, H., Zamorani, G., Huchra, J.P.,
    Braccesi, A. and Zitelli, V. (1983). Ap.J. 269, 42.
Shaver, P.A. (1984). Astr. Ap. 136, L 9.
Uomoto, A.K., Wills, B.J. and Wills, D. (1976). Astr. J. 81, 905.
Wills, D. (1978). Physica Scripta 17, 333.

QUASARS IN THE VIRGO CLUSTER REGION*

X.-T.He
Astronomy Department,Peking Normal University
Peking,China

M.G.Smith
R.D.Cannon
J.A.Peacock
Royal Observatory,Blackford Hill,Edinburgh EH9 3HJ
Scotland

J.B.Oke
C.D.Impey
Astronomy Department,California Institute of Technology
Pasadena,California 91125,USA

1.INTRODUCTION

The region of the Virgo cluster is the subject of an intensive quasar search,since quasars mill provide valuable probes of absorption due to galaxy haloes or the intracluster medium. It is also interesting to test the quasar-galaxy association in such a region of the nearest major cluster of galaxies. More than 200 quasar candidates were found in the Virgo cluster region using both slitless technique and machine search,32 of them have been confirmed as quasars.

2.QUASAR SURVEY AND SPECTROSCOPIC OBSERVATION

The quasar candidates for this study were found on the plates taken with the UK 1.2m Schmidt telescope at Siding Spring in Australia.Three objective-prism plates and one direct plate centred on 12^h27^m, $+13°30'$ (1950.0) in the Virgo cluster were used. We have found 53 emission line quasar candidates and 18 ultraviolet excess object (possible low redshift quasars) in the 5*5 square degrees. Initial results from automatic search with the APM suggest that there may be about 200 quasar candidates in this area. 18 of the candidates were observed in 1982 using the double spectrograph on the Palomar 5m telescope, 17 of the observed objects proved to be quasars (He et al 1984). 17 quasar candidates were observed in 1983 using a grism spectrograph on the University of hawaii 2.2m telescope on Mauna Kea, 15 of them were confirmed as quasars (Impey & He 1985). 4 of 5 candidates proved to be quasars in 1985 using the Palomar 5m telescope (He et al).

* Discussion on p.514

G. Swarup and V. K. Kapahi (eds.), Quasars, 501–502.

3. DISTRIBUTION OF QUASARS

The Kolmogorov-Smirnov method has been used to test the quasar association with galaxies. General speaking, there is no conclusive evidence statistically for quasar-galaxy association in this field, but
a weak evidence for quasar-galaxy association on about 1 degree may
exist. A simple analysis by Arp (Arp 1985) shows significant associations of the quasars with core galaxies in the Virgo cluster, which
may merit further investigation. Although there exist non-uniformities
in the distribution of the quasar candidates, the application of Power
Spectrum Analysis shows no evidence of clustering among them.

4. STUDY OF INTERGALACTIC MEDIUN

One of the aims of this project is to provide a complete sample of
bright quasars for absorption line studies with the Space Telescope.
Two quasars pass very close to bright galaxies, the respective projected quasar-galaxy separation are only 4 and 11 kpc at assumed
distance of the Virgo cluster. Ten quasars have lines of sight that
pass through halo material, if haloes around cluster galaxies have the
size indicated by the statistics of quasar absorption line systems.

References

Arp,H.,Quasars in the Central of the Virgo Cluster, Preprint
He,X.-T., Cannon,R.D., Peacock,J.A., Smith,M.G.,and Oke,J.B.,
      1984, M.N.R.A.S. 211, 443
He,X.-T., Cannon,R.D., Smith,M.G., Oke,J.B., and Impey,C.D.,
      Quasars in the Virgo Cluster Region (2), in preparation
Impey,C.D., and He,X.-T., Quasars and the Virgo Cluster, preprint

A Survey of Quasar Clustering; Initial Results at the SGP *

M.J. Drinkwater
Institute of Astronomy
University of Cambridge
Cambridge CB3 OHA
England

ABSTRACT.   Several U.K.Schmidt objective prism plates are being
searched for quasar clustering. The spectra are measured by the
automated plate measuring facility (APM) and emission line objects
selected for which redshifts can be estimated. Most redshifts tested
(based on Lyman-α) are accurate to 2%. Initial results from the South
Galactic Pole (SGP) field show little evidence of clustering on scales
between 60 and 500 Mpc (H_0=50, q_0=0, at z=2.3).

The use of automated techniques to find quasars has increased the
surface density of candidates by more than an order of magnitude over
that in visual surveys. The first plate in this survey (the SGP) has a
density of 6 objects deg^{-2}, compared with 0.4 deg^{-2} for the CTIO
survey (Osmer 1981). The quasar candidates were selected from spectra on
the U.K. Schmidt objective prism plate UJ3682 of the SGP, measured using
the APM (see Hewett et al. 1985). To obtain the redshift estimates
necessary for a clustering analysis, the candidate selection was based
on the presence of emission lines with a signal-to-noise greater than 3.
    The redshift of each object was assigned by identifying the
strongest emission feature with Lyman-α. The APM spectrum was then
inspected visually to confirm or alter this identification. This
procedure was calibrated using 33 redshifts from Savage et al. (1985);
28 were correct to within Δz=$\pm$0.036. (All the redshifts in the range
1.9<z<2.6 were correctly identified so this range was used to define a
clustering sample of 75 objects.) This accuracy is better than that
claimed for objective prism data by Savage et al. (1985) and is near
the resolution (Δz=$\pm$0.03) they require for clustering studies. In the
above test the 1σ redshift uncertainty was Δz=$\pm$0.02, which
corresponds to 38Mpc at z=2.25. This limits the tests of clustering to
scales larger than about 60Mpc. It was noted that the distribution of
redshifts in the candidate list had an excess at z=2.3, although there
was a significant number of objects at z>2.5. This reflects the greater
ease of detecting Lyman-α emission compared to other lines, but it may
also be caused by the inclusion of stars in the sample. In some stellar
spectra the 4000A break, if combined with strong G-band (4300A)
<u>absorption may resembl</u>e Lyman-α emission at z=2.35.

* Discussion on p.514

G. Swarup and V. K. Kapahi (eds.), Quasars, 503–504 .

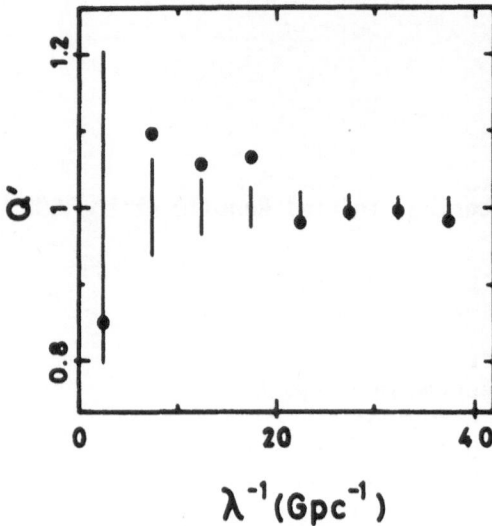

$$\lambda^{-1}(Gpc^{-1})$$

Figure 1. Histogram of Q' for the 3 dimensional PSA. The bars show the ±1σ range of expected values for a random distribution.

The sample of quasar candidates was tested for clustering using the power spectrum analysis (PSA) method of Webster (1976). The two dimensional results do not show evidence of clustering. The PSA results have a higher signal-to-noise when the test is applied in three dimensions. To do this the coordinates of each object were projected into a Euclidian box as in Webster (1982) using a Robinson-Walker metric with H_0=50 kms^{-1}/Mpc and q_0=0 (see Osmer 1981). To allow for the non-uniform redshift distribution all terms in the z direction were discarded in the calculation of Q'. The resulting histogram in Figure 1 shows some evidence of clustering in the first four bins. The statistic Q, calculated under the hypothesis of clustering on a scale of 50Mpc, was 1.065, just outside the corresponding 3σ range of 1.00±0.06, although the degree of clustering is very weak. A close pair noted in the sample (1h3m11s, -29°25'18'' and 1h3m15s, -29°27'10'' both at z=2.25; a separation of 12Mpc) is not responsible for this result.

This initial study has established that it is possible to measure the clustering of quasars on scales between 60 and 500 Mpc although there is little evidence for clustering in the SGP on such scales. The completion of the survey with about 10 plates (some 1000 objects in 224deg^2) will increase the signal-to-noise by a factor of three making the detection of even very weak clustering possible.

REFERENCES.

Hewett,P.C., Irwin,M.J., Bunclark,P., Bridgeland,M.T., Kibblewhite,E.J., He,X.T., & Smith,M.G., 1985. Mon. Not. R. astr. Soc., **213**, 971.
Osmer,P.S., 1981. Astrophys. J., **247**, 762.
Savage,A., Clowes,R.G., Cannon,R.D., Cheung,K., Smith,M.G., Boksenberg, A., & Wall,J.V., 1985. Mon. Not. R. astr. Soc., **213**, 485.
Webster,A.S., 1976. Mon. Not. R. astr. Soc., **175**, 61.
Webster,A.S., 1982. Mon. Not. R. astr. Soc., **199**, 683.

QUASAR CLUSTERING AROUND THE SCULPTOR REGION

D. Kunth
Institut d'Astrophysique
Paris, France

W.L.W. Sargent
California Institute of Technology
Pasadena, USA

ABSTRACT. The Sculptor Region has been searched for quasar candidates: three IIIa-J objective prism plates from the UK Schmidt-telescope centred around the NGC55 and NGC300 galaxies have been used. 76 quasars have been found doubling the number of quasars previously known in this field. A test using Monte-Carlo simulations to reshuffle redshifts of the quasar sample shows that a simple eye examination of the simulated distributions produces as many quasar pairs or associations than in the original field. A further preliminary quantitative study using the autocorrelation function shows that quasar clustering is absent on any scale from 10 to 3500 Mpc ($H_0=100$).

THE QUASAR SURVEY

Our initial goal has been a visual search of objective prism plates to discover bright quasars around NGC55 and NGC300. Further sensitive Ca absorption studies are planned to map metallic absorption in quasar spectra out to large radii around these nearby galaxies. Three plates have been taken at low dispersion and among hundreds of candidates we have spectroscopically observed about 170 objects and found 76 new quasars. Spectra were taken with the Las Campanas Intensified Reticon at the 100-inch Cassegrain spectrograph at low dispersion giving redshifts to an accuracy of ±0.01 except for few BAL QSOs for which line profile fitting is needed. A full account of the survey is given in Kunth and Sargent (1986).

A visual search is very subject to selection effects because it remains very hard to maintain the same search criteria all along the plate examination. The distribution of the objects is thus very clumpy and non-uniform as it is shown in Figure 1 and varies from field to field according to the quality of the plate. The distribution of magnitudes has a broad peak at about 18.8 similar to the distribution found by Osmer and Smith (1980). However while the limiting magnitudes are roughly the same, the number density of quasars found in this work largely exceeds (76 quasars in about 110 deg^2) the one reported by Arp (1984a) and Osmer and Smith (1980) for the same fields. This shows the limitations of the visual technique whenever aimed to discuss statis-

G. Swarup and V. K. Kapahi (eds.), Quasars, 505–508.

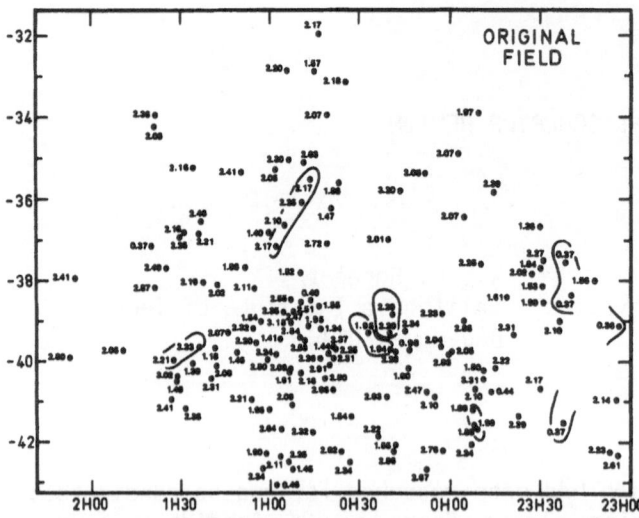

Figure 1 :
Plot of all known quasars -including
those of this work- in the Sculptor
Group found from OP Surveys. Declinations
are in degrees. Some obvious pairs and
groupings are encircled.

tical distribution pro-
perties. Most of the
selection effects have
been quantified by
Clowes (1981) and that
reproduces quite well
the observed continuum
magnitudes at 5000 Å
against the equivalent
widths of Lyα + NV
found in our sample.
From a measure of the
CIV equivalent widths
we do not support Arp's
suggestion (1984b) that
the quasars in the
Sculptor Group follow a
well defined Baldwin
relation. For this
reason, and because of
the non-uniformity of
our survey, as well as
the previous ones, we
find no compelling
evidence that quasars
in the Sculptor Region
have different distri-
bution and intrinsic
properties than field
quasars.

THE CLUSTERING

Clustering at z=2 has been investigated by Osmer (1981) using a large
set of objective prism plates centred in the Sculptor Group. We have
performed such a preliminary analysis with our new set of quasars
which about doubles the number of quasars over roughly 450 deg^2. Moreover the subareas around NGC55 and NGC300 were used to test clustering
on small scales with higher sensitivity. This analysis has been performed by computing the observed autocorrelation function, corrected
for cosmological effects and comparing it with the calculated autocorrelation function for a random distribution generated by Monte Carlo
techniques. The details of the methods will be treated in a forthcoming paper. We found no clustering in all scales up to 3000 Mpc (with
H_0=100 km sec^{-1} Mpc^{-1}) but we still undergo a critical analysis of
the method -using fake fields- before coming to firm conclusions about
quasar clusterings.
    The random distributions were obtained by preserving the same
positions for the quasars on the sky but randomly shuffling the
redshifts. The reason to keep the coordinates unchanged is that they
are subject to the largest uncontrolled selection effects whereas

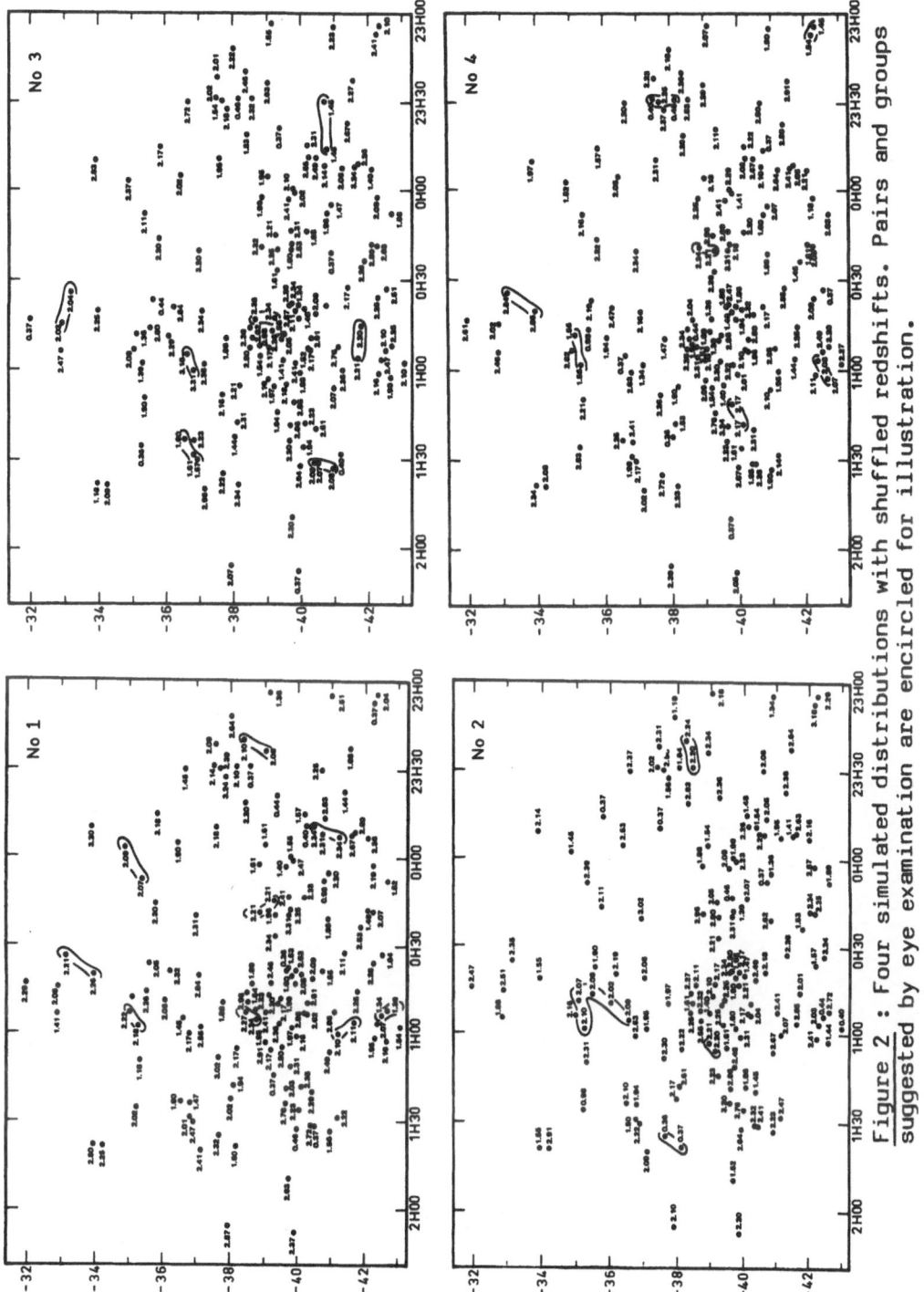

Figure 2 : Four simulated distributions with shuffled redshifts. Pairs and groups suggested by eye examination are encircled for illustration.

there is no reason to find any preferential redshift value within the range set by the objective prism technique. We have made the exercise of plotting some of the simulated distributions for illustration purpose and reproduced some of them (chosen at random) in Figure 2. It is interesting to see that as many pairs and groupings are found in the test fields than in the original one. This gives a visual illustration for the need to step on more quantifying estimates of the tendancy of quasars to cluster.

References

Arp, H., 1984a, Ap.J., **285** , 547
Arp, H., 1984b, Ap.J., **285** , 555
Clowes, R.G., 1981, M.N.R.A.S., **197** , 731
Kunth, D., Sargent, W.L.W., 1986, A.J., in press
Osmer, P., 1981, Ap.J., **247** , 762
Osmer, P., Smith, M.G., 1980, Ap.J. Suppl., **42** , 333.

INFLATIONARY UNIVERSE MODELS AND CLUSTERING OF QUASAR RED SHIFTS

C.Sivaram
Indian Institute of Astrophysics
Bangalore 560034
India

ABSTRACT. Recently it has been shown that many of the puzzling features of conventional cosmological models (such as the horizon and flatness problems) could be explained by invoking inflationary models of the early universe with an exponential expansion phase at very early epochs. These models have the added advantage that they are able to make a definite prediction about the present matter density in the universe, i.e. they require that the density be exactly equal to the closure density which in turn can be easily estimated from the Hubble constant now known to within a factor of two. Now if one goes back to an earlier idea that explored the possibility of unusual clustering of quasar redshifts around z = 2 or 3, we get an example of another cosmological model with a definite prediction for the present overall matter density. This is a modified version of the Eddington-Lemaitre type of model which naturally accommodates such features as a clustering of quasars at certain epochs. From these models one can get a prediction for the present matter density which would be an involved function of the Hubble constant and the redshifts at which such clustering occurs. It can be shown that if such clustering had occurred at any z, the present matter density predicted would be substantially smaller than the corresponding closure density. The conclusion is that any clustering of quasar redshifts is incompatible with inflationary universe models, indirectly providing observational support for these new theories.

The overall matter density of the universe is usually specified by the dimensionless parameter expressed as a ratio $\Omega = \rho/\rho_c$. Here ρ is the actual density and ρ_c is the so called critical density required to just close the universe. ρ_c can be related to the measurable quantity called the Hubble constant H_0 as : $\rho_c = 3H_0^2/8\pi G$ where $H_0 = (\dot{R}/R)_0$ is the present value of the expansion rate of the universe, R being the scale factor and $\dot{R}$ its time derivative, (o denoting present values) G is the gravitational constant.

Now big bang nucleosynthesis of the elements deuterium (D), Helium -4 (He-4), He-3, etc. as well as the dynamics of galaxies and clusters of galaxies are all consistent with $\Omega \simeq 0.1$ to 0.2. However the conventional big bang model despite its success in naturally accounting for both the microwave background and the abundance of the light elements

509

G. Swarup and V. K. Kapahi (eds.), Quasars, 509–510.

faces several serious problems when extrapolated to early epochs such as the horizon, flatness and monopole problems. Inflation models involving an early exponential expansion phase appear to resolve most of these problems. But these models require as a prediction that $\Omega = 1$. One other model which makes a definite prediction for the present density ρ_0 of the universe is a variation of the Eddington-Lemaitre model which can also account for any clustering of quasar red shifts at any z if indeed such clustering does occur. Such models involve a 'cosmological constant or Λ-term' which effectively increases the age of the universe thus :

$$t_H = H_0^{-1} \int_0^\infty \frac{dz}{(1+z)[(1+z)^2(1+\Omega z)-(\Lambda c^2/8\pi G\rho_c)z(2+z)]^{1/2}}$$

If clustering of quasars had occurred at some z(say z = 2.2) when the scale factor of expansion was R_z : then we obtain the following equations for the <u>present</u> scale factor, R_0, H_0 and density ρ_0 (all <u>present</u> values)

$$R_0 = R_z(1+z) \; ; \quad R_z \sim \Lambda^{-1/2}$$

$$H_0 = (\dot{R}/R)_0 = (\frac{1}{3}\Lambda c^2 - \frac{c^2}{R_0^2} + \frac{2c^2}{3\Lambda^{1/2} R_0^3 G})^{1/2}$$

$$\rho_0 = (4\pi G R_0^3 \Lambda^{1/2} / c^2)^{-1}$$

For a given z (where redshifts are thought to cluster), we can solve these equations to obtain both H_0 and ρ_0. Thus for z = 2, we have $H_0 = (20/81\Lambda)^{1/2}$ C, and from the observed $H_0 = 50$ km/s/Mpc we have the (predicted) $\rho_0 \leq 5 \times 10^{-31}$ g/cc $<<\rho_c$. For z = 3, $H_0 = (27/96\Lambda)^{1/2}$ C again $\rho_0 << \rho_0$. For all z between 2 and 3 where observations have now and then reported clustering we have $\Omega < 0.1$ as a prediction. As inflation requires $\Omega = 1$, this is not in agreement with the presence of vast amounts of dark matter in the universe.

ON THE CONCENTRATION OF HIGH REDSHIFT QUASARS IN THE REGION OF M33

Vijay K. Kapahi
Tata Institute of Fundamental Research
Post Box No. 1234, Bangalore 560012, India

ABSTRACT. We have investigated the strong concentration of high redshift radio quasars in a large area of the sky near M33 reported by Arp. Selection effects appear to be important in determining the significance of the inhomogeneity.

A striking example of an inhomogeneous distribution of quasars pointed out by Arp (1984, J.A.A. 5, 31 and Ap.J. 277, L27) is the apparent concentration of high redshift ($z > 1.4$) Parkes and 3C quasars in a $\sim 70^0 \times 20^0$ region extending to the south-west of M33. We have investigated the concentration by using the Veron & Veron Catalogue of Quasars (ESO publication, 1985). Since the exact demarcation of the region near M33 in somewhat arbitrary we have considered a slightly larger area made up of two rectangles (for ease of selection of quasars) as shown in Fig.1. The comparison area (also shown in Fig.1) is obtained by shifting the RA by exactly 12^h. Because redshift measurements are by no means complete for the Parkes catalogues, any possible biases in the regions selected for redshift measurements can be minimized by including quasars from other catalogues, which cover the same areas of the sky in the two regions. We have therefore included quasars from the 4C, B2 and Ohio surveys as well. Although the z distribution for all such quasars in the $21-03^h$ region does appear to peak at somewhat higher redshifts than that for the $09-15^h$ region, (Fig.2) the difference is not statistically significant even at the 10% level.

There are 41 quasars with $z > 1.4$ in the $21-03^h$ region and 40 in the $09-15^h$ region. Counting the Parkes and 3C quasars alone, the numbers are 31 and 21 respectively. The difference is clearly not significant. We do not also find any significant difference in the distributions of V magnitudes of the quasars in the two regions as claimed by Arp. The smaller number of high z quasars in the $09-15^h$ region could arise partly from the fact that many quasars in this region were first identified from Parkes observations of 4C sources selected at 178 MHz (Wills & Bolton 1969, A.J.P. 22, 775). These observations excluded most of the region between 21^h and 03^h where many quasar identifications have been made from surveys at 2.7 GHz. At high and intermediate flux levels high redshift quasars are more likely to be found in high frequency surveys (Kapahi & Kulkarni 1986, this symp.).

511

G. Swarup and V. K. Kapahi (eds.), Quasars, 511–512.

   It should be noted however that we have not exactly verified Arp's
claim of a 14 σ difference (22 quasars versus 2 in the two regions) because his comparison was based on the following additional restrictions,
i) the regions compared had the outline shown by the dotted line in Fig.1,
and the comparison sample was obtained by shifting the outline by 13^h in
RA, (ii) Only Parkes and 3C quasars were considered, (iii) z and V
ranges were restricted to $1.4 \leq z \leq 2.4$ and $17 \leq V \leq 19$ respectively.
If we duplicate his analysis as closely as possible we do indeed find a
large difference (23 quasars versus 5). But then it is not clear if the
regions used and the restrictions imposed have been chosen in order to
emphasize the difference.

   Arp has also noted that the inhomogeneity is supported by the number of radio sources in a Parkes survey of the zone $4^o < \delta < 25^o$, complete to 0.5 Jy at 2.7 GHz; the number of quasar candidates (before spectroscopy) is 3 times larger and the number of all radio sources about 2
times larger in the $21-03^h$ region than in the $09-15^h$ region. The Parkes
survey (Shimmins et al. 1975, A.J.P. Ap.Supp. 34, p63) referred to by
Arp is, however, not a complete sample by itself; it is only a supplementary list of sources that should be added to those already published in
that declination zone. The previous lists are i) the main listing of
the Catalogue for $\delta < 20^o$ (Ekers 1969, A.J.P. Ap.Supp.7); ii) the
$20^o < \delta < 27^o$ region (Shimmins & Day, 1968, A.J.P., 21, 377) and
2.7 GHz observations of 4C sources with $4^o < \delta < 20^o$ (Wills & Bolton,
1969, A.J.P. 22, 775). If the comparison is restricted to all radio sources with $S_{2.7} \geq 0.5$ Jy, the total number in all three lists in the region $4^o < \delta < 25^o$ is found to be 177 in the $21-03^h$ region and 151 in the
9-15 region. The difference is clearly not significant, nor is there any
significant difference in the numbers of all quasars (including those
without measured redshifts).

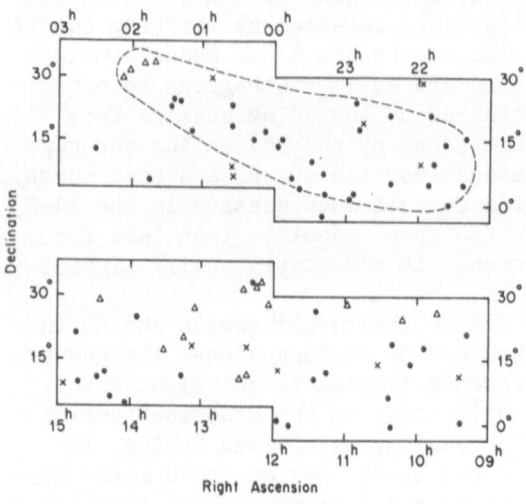

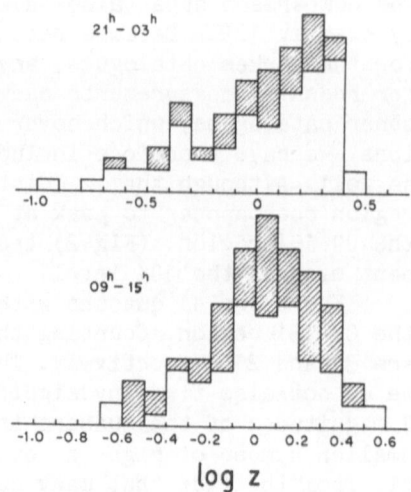

Fig.1. Quasars with z > 1.4.
● 3C & Parkes; x 4C; Δ B2 & Ohio

Fig.2. Redshift distributions.
Hatched region refers to
4C, B2 & Ohio sources

DISCUSSION ON THE PAPER BY **DANESE** ET AL. (p.489)

Cristiani : Why did you choose to normalise the soft flux to the hard and not viceversa ?

Franceschini : Simply because both measurement of the XRB spectral intensity and the HEAO-1 hard X-ray sample are part of the same experiment. On the other hand, we have also verified that the HEAO-1 flux scale is essentially consistent with that of all other hard X-ray experiments (Ariel V, Einstein MPC and UHURU).

Schmidt : Since the HEAOI-A2 contains only one quasar (3C 273) with M_B < -23, the Schmidt-Green LDDE cannot be really applied.

Franceschini : Note, however, that our analysis also takes into account the Medium Sensitivity Survey sample, which contains a large fraction ($\sim$ 50%) of true Quasars.

DISCUSSION ON THE PAPER BY **CAVALIERE** ET AL. (p.491)

Schmidt : In your model of BL Lacs, would a uniform distribution of these objects in Euclidean space yield $N \propto S^{-3/2}$?

Vagnetti : Literally it would, since at high luminosities the LF (however for beamed sources) cuts off or declines parallel to the parents'. In a realistic cosmology, the Euclidean limit may apply at high (possibly statistically irrelevant) fluxes and the counts bend over toward low fluxes. This tends to occur faster and at higher fluxes for the beamed population than for the parent one, as the former's LF is flat up to higher luminosities than the latter's.

Bregman : Do violently variable quasars have the same distribution as the BL Lac sources ?

Vagnetii : If the OVVs are a beamed subset of the QSOs, one expects flatter counts than the latter's, provided the evolutions (including activity lifetimes) are comparable. Quantitatively, the counts may differ from BL Lac', because of possibly different boosting parameters. Both facts might be interestingly tested, if reasonable sample completeness could be defined.

G. Swarup and V. K. Kapahi (eds.), Quasars, 513–514.
© *1986 by the IAU.*

DISCUSSION ON THE PAPER BY **ZHOU** ET. AL (p.497)

Cristiani : Extracting from catalogues the fraction of quasars discovered by means of their radio emission (see poster Barbieri et. al) we
may hope to have a largely unbiased sample from the point of view of
preferential redshifts. From this sample there is no indication of
statistically significant redshift periodicities.

Zhou You-Yuan : The radio quasars still have some peaks, although they
are lower. It may be caused partially by the selection effect of the
line identification in the redshift measurement. As to the periodicities
we should confirm or deny them only after considering the selection
effects.

Burbidge : I showed in 1977 that while there were some selection effects
which can affect the z distribution, comparatively sharp peaks
($\Delta z \lesssim 0.1$) cannot be explained in this way.

Zhou You-Yuan : Our results show quantitatively that both the $p(z)$ and
the selection effect play important roles in the redshift distribution.
I expect to explain possibly the peaks with $\Delta z \sim 0.1$ in the further
calculation. I agree that the problem about peaks with $\Delta z < 0.1$ is
still open.

DISCUSSION ON THE PAPER BY **HE** ET AL.(p.501)

Drinkwater : The surface density of your quasars increases towards the
position of the centre of the Virgo Cluster. I expect that this corresponds to the galaxy-quasar clustering you observe. Could this be a consequence of an increase in sensitivity to quasars towards the centre
of the Schmidt plate ?

X.T.He : No, infact the average surface density of quasars does not
systematically increase towards the position of the plate centre. There
are only a few bright galaxies in the centre of the cluster associated
with quasars.

DISCUSSION ON THE PAPER BY **DRINKWATER** (p.503)

Peterson : What is the limiting magnitude of your survey ?

Drinkwater : The prism spectra go down to 19.5 but few of the selected
candidates would be this faint. I have not yet put an accurate magnitude limit to the candidate list but would estimate it to be 19.

VI

QUASARS AS PROBES OF THE INTERVENING MEDIUM

"The evidence is now at hand to show that the lensing phenomenon is common enough to be a reasonable subject of study and not just a curiosity."

— Bernard Burke (p.525)

Bernie Burke and Vinita Kapahi

GRAVITATIONAL LENSES: OBSERVATIONS

Bernard F. Burke
Department of Physics, Room 26-331
Massachusetts Institute of Technology
Cambridge, Massachusetts 02139

ABSTRACT: The gravitational lens phenomenon is shown to be not uncommon.
A search is now underway to find a larger number of examples, using the
VLA to search for candidates, following up with optical observations to
establish equality of redshifts in the images. Most of the present lensing examples exhibit anomalous behavior that is most easily understood
as evidence for large quantities of intervening non-luminous matter.

1. PROPERTIES OF GRAVITATIONAL LENSES

In 1919, Sir Arthur Eddington led an eclipse expedition to verify
Einstein's prediction of the deflection of starlight by the sun's
gravitational field. In the years that followed, it was widely recognized that the effect was an optical one, and could be observed, in
principle, on an interstellar scale. Eddington (1923) and Einstein
(1936) both recognized that magnification and multiple-imaging of a
distant star by a foreground star was improbable, but Zwicky (1937)
recognized that the effect might be observable in an extragalactic
context. He proposed that the images of compact N-type galaxies might
be significantly modified by foreground galaxies, and he proposed that
the effect, being solely gravitational, could be a new tool for studying the distribution of matter in galaxies.

The discovery of quasars revived the discussion [see Greenfield
et al (1985) for historical references], and the predictions were confirmed when Walsh, Carswell, and Weymann discovered that the radio
source 0957+561 was double, and displayed the characteristics of a
multiply-imaged quasar. The two quasar images were separated by 6",
and were designated A and B (B being southernmost), with both having
similar spectra and equal redshifts. After some puzzles raised
initially by radio and infrared observations [Greenfield et al (1979)
Walsh et al (1979)] the central issues were resolved by the observations
of Stockton (1980) and Young et al (1980) who showed that there was a
foreground cluster of galaxies, and that the brightest member was a
giant cD galaxy only 1" north of the southern image. Any reasonable
model of that galaxy would predict multiple images of some sort, given

G. Swarup and V. K. Kapahi (eds.), Quasars, 517–527.

the proximity of the B image. The major surprise, as demonstrated by
Young et al (1981) and by Greenfield et al (1985), was the inadequacy
of interpreting the imaging in terms of conventional Mass-to-Luminosity
ratios for the visible matter. A substantial quantity of dark matter,
presumably associated with the cluster as a whole, has to be invoked to
explain the observations.

The idealized cases of a point mass and a singular isothermal
sphere have been treated by a number of authors (e.g. see the review
by Peacock, 1983), and these illustrate the scale of the phenomenon.
If there is a 10^{12} M_Q point-mass lens at 1.5 Gpc, a quasar at 3 Gpc,
and one assumes Euclidean geometry, then r_0 = 2" in the point-mass case;
for the isothermal sphere, with σ_v = 300 km/s, r_0 = 0".4. More realistic
cases show a greater complexity of images. Bourassa and Kantowski
(1975) treated ellipsoidal mass distributions and showed that a variety
of caustic surfaces between singly-imaged, triply-imaged, and quintuply-
imaged domains existed, and a wide variety of investigations have been
carried out since then. The orders of magnitude of the lensing effects
remain the same as in the simple cases but the complexity is substantial.

2. THE OBSERVATIONAL STATUS

Both intrinsic and extrinsic astrophysical uses can be expected from
gravitational lens observations. One intrinsic interest derives from
the magnification itself: since the magnification can be substantial -
of the order of 100 in some cases - greater angular resolution is
achieved, an advantage that should be of particular interest for VLBI
observations at radio wavelengths, and for HST observations at optical
wavelengths. Another application has already been noted by Gott and
Gunn (1974) before the lensing phenomenon had been established; they
considered 1548+115, an accidental quasar pair, which allows one to
probe the distribution of matter in the closer quasar by looking for
extra images. Finally, there is the potential problem that might arise
if distant quasars have a high probability of being lensed. In such a
case, the high luminosity, high redshift part of the luminosity function might be subject to correction. In order to check the importance
of such corrections, Roberts, Turner, Gott, and Burke (unpublished) used
the VLA to examine twenty-five high-luminosity quasars, all having absolute B-magnitude -29 or brighter, looking for the multiple images
that might be expected if lensing were a frequent occurrence in the
sample. The results were largely negative: only one quasar in the
sample showed evidence of structure that might be due to lensing, and
that example is marginal. They concluded that lensing does not make an
important perturbation to the observed quasar luminosity function.

The extrinsic uses of gravitationally lensed quasars have received
the most attention, since the phenomenon, being purely gravitational is
an informative probe of the intervening matter in the universe. In the
case of 2237+050, discovered by Huchra et al (1985), one is probing the
matter distribution close to the center of a nearby galaxy, but the
other lens cases, as will be seen, probe both galactic and cluster mass
distributions, with strong evidence for the presence of substantial
quantities of dark matter. The large-scale matter distribution in the

universe may also have an observable effect and one must stay alert to
more exotic possibilities such as strings and black holes. An even
more extreme conjecture, the existence of a "shadow universe" observable
only through the gravitational interaction, has been raised by recent
speculations in particle theory, and such a presence would probably be
manifest primarily through lensing effects.

There are now seven cases of lensing known. Their principal properties are summarized in Table I, which shows the principal features
as they were known in December 1985. The maximum separation between
components is given in Column (2), the flux ratio of the two brightest
components is given in Column (3) with 1115+080 being a special case.
The redshifts of quasar and lens appear in Columns (4) and (5). The
character of the visible lens (if any) is given in (6): galaxy (G),
cluster (C), galaxy plus cluster (B) or no visible candidate (N).
Column (7) indicates radio-loud cases and the number of detected quasar
images is shown in Column (8).

TABLE I

Object	$D_{max}(\pi)$	S_A/S_B	Z_Q	Z_L	Lens	Radio?	No. Images
0957+561	6.1	3,1	1.41	0.39	B	Yes	2
1115+080	2.7	1.1*	1.72	--	G?	No	4
2345+007	7.3	4.0	2.15	--	N	No	2
2016+112	3.4	1.6	3.27	1.0	G,G;No C	Yes	3
1635+267	3.8	4.4	1.96	--	N	No	2
2237+050	2.2	?	1.70	0.04	G	No	2
0023+171	4.8	2.5	0.95	--	N	Yes	2

*Image A is actually double, with nearly equal fluxes for A_1 and A_2.

Three immediate problems are raised by the data of Table I: the
rarity of odd images, the wide spacing between components, and the frequency with which the "N" designation appears in the "Lens" column.
Only one of the seven cases exhibits an odd number of proven images,
contrary to expectation. This problem may be more apparent than real,
since a lens with a compact central component will generally exhibit
one faint image that might easily be below the level of detectability,
since a compact central mass approximates a singularity. The wide
spacing, always greater than an arc-second, is also surprising, especially in view of the model studies of Turner et al (1984), who showed
that conventional assumptions concerning cosmology and the structure of
galaxies would lead one to expect a preponderance of multiple images in
the half-to-one arc-second range. Observational selection may account
for the failure to see this class of object more frequently; 2237+050 is
the only present example, but represents an unlikely coincidence, discovered fortuitously. A systematic search for multiply-imaged quasars,
with image separation of the order of 0.5 to 2 arc-seconds, is urgently
needed.

The final puzzle, the relative frequency with which no visible lens

appears, is an awkward and unexplained anomaly. The most straightforward interpretation is that dark matter is playing an important role,
either in the form of under-luminous galaxies or clusters, or in the
form of intrinsically dark aggregations of matter, perhaps of novel
character.

The position and flux density of the observed images are the principal observables at any given time, but since quasars are frequently
time-variable, there is a further class of interesting phenomenon:
the relative time delay in the time-series of the flux for each image.
A value for the relative time delay between the A and B images in
0957+561 has been published by Florentin-Nielson (1984), but the data
were too incomplete to be convincing. More complete measurements will
be awaited eagerly. The interpretation, as pointed out by Alcock (1985),
is interesting: the Hubble constant and the deceleration parameter are
not determined reliably, despite one's naive expectations, because fluctuations in the matter distribution of the universe can cause major perturbations even if the lens model is correct. Instead, one can take
advantage of this sensitivity; if the Hubble constant can be determined
with precision by luminosity-distance methods, the time delay in the
lens cases will then give a quantitative measurement of matter fluctuations in the universe. If a sufficiently large number of examples can
be measured, the method is also an effective way of measuring the cosmological constant Λ_o.

3. SEARCHING FOR NEW GRAVITATIONAL LENSES

In astrophysics, nearly all observable phenomenon are complex, and only
rarely, if ever, does a single example provide deep understanding. The
Hertzsprung-Russell diagram and the Hubble classification scheme for
galaxies come to mind as instances in which the order underlying a
phenomenon could never have been demonstrated with a handful of objects.
Similarly, it is clear that gravitational lensing is a rich but complex
phenomenon that can be understood and used best if a substantial number
of examples can be found. Both radio and optical searches should be
carried out. Expectations can be estimated from the present data base.
Six examples (the seventh, 2237+050, is exceptional) were recognized by
their multiple character, and all were drawn from reasonably well-
defined samples. There were 486 quasars in the optically-defined
samples; the radio samples include both quasars and radio galaxies, but
for estimating the occurrence of lensing, they are sufficiently well-
defined: about 600 quasars would be found in the radio samples. This
accounting, however, neglects the negative searches, and this number is
more difficult to determine. Private communications from B. and D. Wills
and from E.M. Burbidge give negative results from samples of 250 and 320
quasars, respectively; both groups would surely have seen multiples
having spacings of 2 arc-seconds or greater. Most other negative
searches seem to be incomplete at this stage, and one can conclude,
therefore, that six multiply-imaged quasars have been found in a total
sample size of about 1700. The incidence of multiple imaging, therefore,
can be expected to be about 1 in 300, for high redshift quasars.

A radio search for more examples of lensing has been in progress

for the past few years. The principals in the work have been J. Hewitt,
C. Lawrence, E. Turner, and B. Burke, although much wider help has come
from other colleagues at Princeton and CalTech. The strategy starts
with the raw material of the MIT-Green Bank 300-foot telescope source
study (the MG Survey, Bennett et al 1986); the VLA is used in the snapshot mode to obtain accurate position and source structure, using the
methods described by Lawrence et al 1984). All sources that show multiple, compact structure, except for the obvious cases of double-lobed
radio galaxies, are identified as potential candidates (about one out
of ten sources qualify). Next, optical continuum images are taken, most
commonly by the Kitt Peak 4m telescope, and if there is evidence of
multiple optical structure, the source becomes a candidate for spectroscopic examination (about one in ten of the first-cut candidates, or one
in a hundred of the original sources, meets this criterion). Finally,
spectra are taken, usually with either the Kitt Peak 4m telescope or
the Palomar 5m telescope. If the images have identical spectra and redshifts, this is taken as strong evidence for lensing. Once in a while,
one might accidentally be misled by a galaxy with multiple active nuclei,
but this should not be a common occurrence. The snapshot material now
includes 3000 sources, and we hope to add another 2000 to the list in
the upcoming observing season with the VLA; one might expect, therefore,
between ten and thirty new lens examples when the search is complete.
An examination of the first 1000 snapshots is now nearing completion.
The snapshots should be a valuable resource for other projects, and will
be made available to the astronomical community.

4. THE RADIO-QUIET CASES

Four of the seven examples of lenses in Table I are undetectable at
radio frequencies, or at any rate have fluxes that are below 50 μJy
[Greenfield, Roberts, and Burke, unpublished; Huchra et al (1985)]. As
noted above, 2237+050 is a fortunate alignment of a nucleus of a nearby
galaxy with a distant quasar, and Tyson has reported detection of a
pair of images, plus the nucleus of the galaxy, at this meeting. Two
of the examples, 1635+267 (Djorgowski and Spinrad, 1984) and 2345+007
(Weedman et al 1982), despite the large spacing between images, are
lacking visible foreground matter that could serve as a lens. The
fourth, 1115+080, discovered by Weymann et al (1980), was originally
reported to be triple. Young (1981) predicted that it should be quin-
tuple, with the A image split into two components; the fifth image
would be faint. The lensing material would be an edge-on spiral,
roughly parallel to A_1 and A_2 and situated between these and the B and
C images. Hege (1981) showed by speckle interferometry that the A
image was, indeed, double. Reports of detection of a foreground galaxy
have been indecisive, but a recent image, taken by Wlérick and his
collaborators with the Mauna Kea CFHT, provides clear evidence that
there is no foreground edge-on galaxy at the position predicted by
Young. Whether a galaxy image is merged with the A_1 and A_2 images is
not yet established; the observational proof will not be easy, but is
an intensely interesting result that one hopes will be forthcoming soon
(a preprint by P. Henry presents evidence for such a galaxy).

5.   THE RADIO-LOUD CASES

5.1   0957+561

When a gravitationally lensed quasar is radio-loud, there is usually a
larger set of observable quantities that place more stringent limits on
acceptable lensing models. This, the original example of a multiply-
imaged quasar, provides an illustrative case. A summary of the radio
observation with the VLA has been given by Greenfield et al (1985)and
Roberts et al (1985); the optical data has been summarized by Young et
al (1981). A recent map is given in Figure 1, showing the A and B
images, coincident with the optical quasar images, the G image, nearly

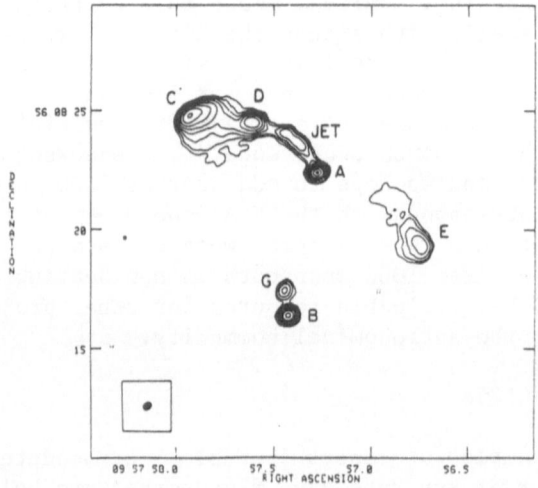

Figure 1.   VLA λ6 cm Map of 0957+561, December 1980

coincident with the galaxy G1 that is the brightest member of the foreground cluster, and the jet-like structures to the NE and SW of A that,
for the most part, are singly imaged. There are a large number of constraints, and Young et al (1981) proposed using a double-ellipsoid model
that, despite the large number of free parameters, does not allow arbitrary freedom in choosing models. The VLBI data of Porcas et al (1984),
Gorenstein et al (1983), Cohen (1984), and Bonometti (1985) give
convincing proof of the general lens properties. Both the A and B
images have jet-like structures extending approximately 50 mas in nearly
the same position angle, and Bonometti's work showed that these jets
each have a spur to one side that exhibits the proper reflection
symmetry predicted by the two-ellipsoid model. In addition, each core
shows a small sub-jet of the order of 0.6 mas length. If these were
moving, the difference in time delay for the two patterns would result
in their presenting different aspects at a given instant. As it turns
out, the motions appear to be slow, and the relative values for all
components of the magnification matrix can thus be derived from observation of the small jets, and independently from the larger (50 mas) jets.

Bonometti gives relative magnifications of 1.236±.007 and -0.46±.01 for the major and minor axes, and position angles of 18°.75±°.06 and 98°±4°. The model values are close to these: Greenfield et al (1985) and Higgs (1984) derived similar double-ellipsoid models that reproduced the data reasonably well.

The observational work to date has established a number of physical facts for 0957+561:

1) No single value of mass-to-luminosity for the foreground cluster of galaxies gives a lens that fits the data, nor does a quadratic law of the form $M = aL + bL^2$.

2) The double-ellipsoid model of the type introduced by Young et al fits the data, but not uniquely. One ellipsoid is centered on the galaxy; the second, more massive ellipsoid is centered to the Southwest approximately 10" from the B image. A large quantity of dark matter, matter not associated with the visible galaxies, is implied by this model.

3) No third image has yet been identified. The models predict that it should be faint, and close to the nucleus of galaxy G1. Gorenstein et al (1983) found a compact VLBI image not far from the center of the radio source G, but their later work (cf. Bonometti) indicates that this is probably not the third image. The G source is extended in the VLA maps, and may have nuclear structure of its own.

4) The forms of the 50 mas jets demonstrate further that gravitational lensing is the definitive explanation for this object.

It is clear that the models are not unique. Falco, Gorenstein, and Shapiro (1985) show that, for any two images, the positions, relative magnification, and difference in propagation time are unchanged under a simple transformation of the mass distribution (generalizations by Gorenstein et al and Falco et al are in preparation). Nevertheless, it is clear from their work that anomalous mass distributions, not related to conventional visible matter, must still be invoked. Radio observations, which impose more restraints, are useful in restricting the allowable models.

5.2   2016+112

The gravitational lens 2016+112, described by Lawrence et al (1984) and Schneider et al (1986) consists of a triple radio structure (Figure 2), with the two upper sources, A and B, being separated by 3".4. These are coincident with faint quasar-like optical objects that exhibit sharp emission lines with identical redshifts of 3.273. The southern radio source, C, is slightly extended. Optically, a faint galaxy D is seen 1".2 from B, in addition to a coupled at C consisting of both a galaxy and a third quasar image [Schneider et al (1986), in preparation]. The galaxy D, in this recent work, exhibits a spectrum that matches closely the spectrum of a giant elliptical galaxy when corrected for redshift, and shows evidence of an H and K absorption break at z = 1.01, probably the largest absorption-line redshift yet. The recent optical work of Schneider et al also shows that there are two faint emission patches NW of A and B. These are shown in Figure 3, which shows a pair of photographs taken with the "PFUEI" camera on the Palomar 5m telescope. The

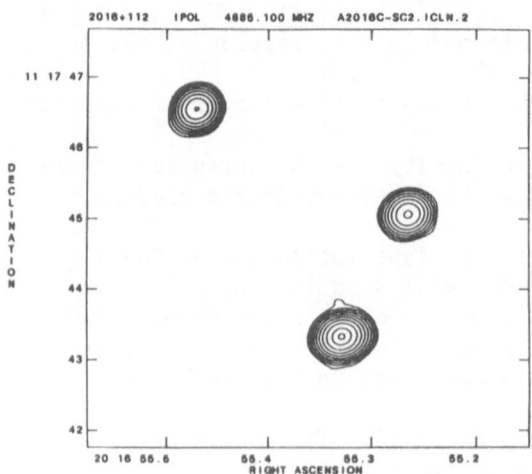

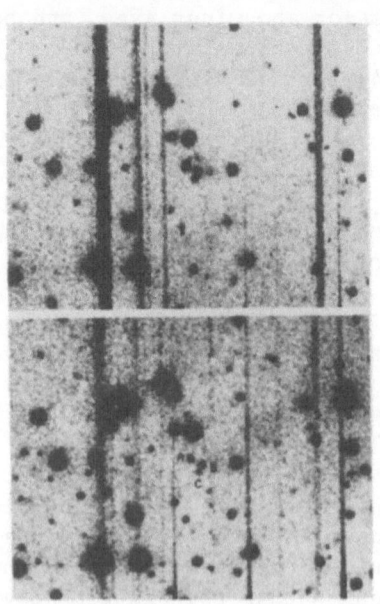

Figure 2.   VLA λ6 cm map of
2016+112.   December 1984 data:
A upper left, B upper right,
C lower; contours .25, .5, 1,   Figure 3.   2016+112 at redshifted Lα
2, 4, 8, 16, 32, 64, 95% of   (upper) and g-band (lower); taken
peak flux 50 Jy.   with PFUEI and Palomar 5m telescope.

upper frame was taken through a narrow filter that passes the redshifted
Lα line at λ5180 Å, and the lower is an image through a g filter. The
relatively bright appearance of C in the Lα filter shows the influence
of the third quasar image, while the two emission patches in the upper
picture are absent in the lower frame.

The lensing properties of the single galaxy D [whose corrected absolute magnitude M_B = -22.5 ± .2 (H_0 = 60, q_0 = 0.5) makes it a typical
bright cluster galaxy] are not well adapted to explaining the observations, as Narasimha , Subramanian, and Chitre (1984) have pointed out.
There is no evidence for an associated cluster in the deep optical photographs that have been taken; if a cluster is present, all of the members
except C and D are fainter than r = 26.

5.3   0023+171

This, the second lens example to be found from the examination of MG
sources described above, is a multiple radio source, as shown in Figure
4.   The optical field shows a faint (r = 23^m) pair of stellar images,
one source being closely coincident with the unresolved radio source to
the NW, and the second falling close to the skewed tongue near the center of the double-lobed radio object.   No galaxies are visible.   The
identification by Hewitt et al (in preparation) should be described as
"probable" in view of the preliminary state of the data, but the indications are convincing that this is another example of a gravitational lens.

The spectra of the visible images are shown in Figure 5. Two lines are evident in both spectra, a strong line at $\lambda7253$ and a weaker line at $\lambda7530$. These correspond to OII $\lambda3727$ and NeIII $\lambda3868$ at a redshift of

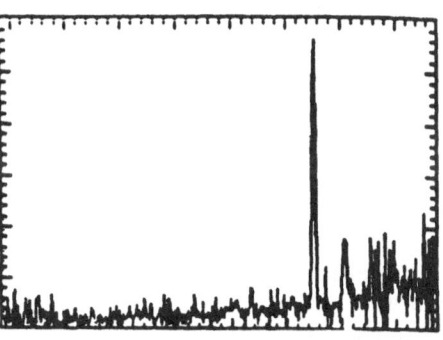

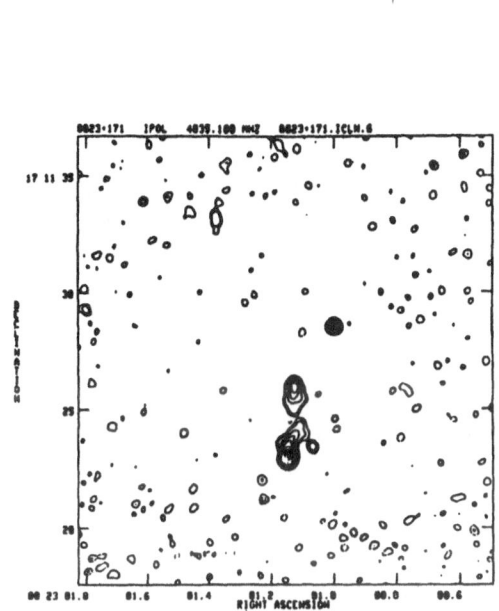

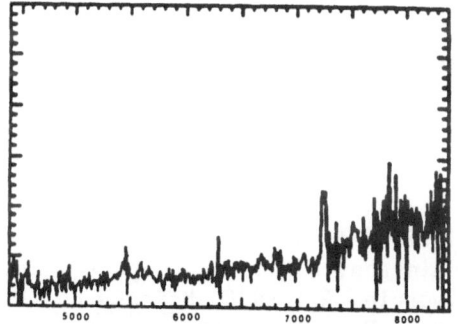

Figure 4. VLA $\lambda6$ cm Map of 0023+171 contours 0.5, 1, 2 4, 8, 16, 32, 64, 95% of peak flux 28 mJy.

Figure 5. 4-Shooter spectra of 0023+171; A (upper) and B (lower).

0.946. The principal difference in the spectra is in the underlying continuum; the object with the fainter lines has the stronger continuum. This may be caused by the presence of a faint galaxy, closely coincident with the quasar, although there is no extended luminosity visible. The absence of visible galaxies in the field implies that once again one has a lens composed of dark matter. The lens redshift is most probably around 0.3, and a giant cD galaxy at that redshift would surely be visible in the deep CCD images taken of the object.

6. SUMMARY

The evidence is now at hand to show that the lensing phenomenon is common enough to be a reasonable subject of study and not just a curiosity. The phenomenon is not so common, however, that it forces major reconsideration of the quasar luminosity function. The apparent lack of close multiples, with spacings of the order of half to one arc-second, may be an observational selection effect, but if there are not at least one or two per thousand quasars, there will be a genuine cause for concern. The number of large-separation pairs is in itself surprising, and provides evidence for the widespread distribution of non-luminous matter in the universe. A striking, and not unrelated phenomenon is the anomalous character of at least six of the seven known images. In the case of 0957+561, rational models can be constructed that fit the data, but it looks like a lucky accident in which the caustic between the one- and three-image loci barely includes the quasar in the multiple-image regime. The sources 0023+171, 1653+267, and 2345+007 all seem to be deficient in foreground material that might form a lens, while 1115+080 and 2016+112 do not have foreground galaxies in the proper place to act as the desired lens. The evidence is building for a large-scale dark component in the universe, not necessarily the same as the dark component inferred to exist in clusters of galaxies.

Thanks are due to J. Hewitt of MIT, E. Turner, J. Gunn, and D. Schneider of Princeton, and to M. Schmidt and C. Lawrence of CalTech. The author was supported in part by a grant from the National Science Foundation, and in part by a Smithsonian Regents Fellowship.

REFERENCES

Alcock, C: 1985, Ap. J. 291 L29.
Bennett, C.L. et al: 1986, Ap. J. Supp., in press;
     see also Lawrence, C.R., ibid.
Bonometti, R.J.: 1985, Ph.D. Thesis, Mass. Institute of Technology.
Bourassa, R.R. and Kantowski, R.: 1975, Ap. J. 195 13.
Cohen, N.L.: 1985, Ph.D. Thesis, Cornell University.
Djorgowski, S. and Spinrad, H.: 1984, Ap. J. 282 L1.
Eddington, A.S.: 1923, Space, Time, and Gravitation 134 (Cambridge).
Einstein, A.: 1936, Science 84 506.
Falco, E.E., Gorenstein, M.V., and Shapiro, I.I.: 1985, Ap. J. 289 L1.
Florentin-Nielsen, R.: 1984, Astron. Astroph. 138 L19.
Gorenstein, M.V. et al: 1983, Science 219 54; 1984, Ap. J. 287 538.
Greenfield, P.E., Roberts, D.H., and Burke, B.F.: 1979, Science 208 495.
Greenfield, P.E., Roberts, D.H., and Burke, B.F.: 1985, Ap. J. 293 370.
Hege, E.K. et al: 1981, Ap. J. 248 L1.
Higgs, W.J.: 1984, M.Sc. Thesis, University of Manchester.
Huchra, J. et al: 1985, Astron. J. 90 691.
Lawrence, C.R. et al: 1984a, Science 223 46.
Lawrence, C.R. et al: 1984b: Ap. J. 278 L95.
Peacock, J.: 1983, 24th Liege Coll. Quasars and Gravitational Lenses 86.
Pooley, G.G. et al: 1979, Nature 280 461.
Porcas, R.W. et al: 1981, Nature 289 758.
Roberts, D.H. et al: 1985, Ap. J. 293 356.
Schneider, D.P. et al: 1985, Ap. J. 294 66.
Stockton, A.: 1980, Ap. J. 242 L141.
Turner, E.L., Ostriker, J.P., and Gott, R.J.: 1984, Ap. J. 284 1.
Walsh, D., Carswell, R.F., and Weymann, R.J.: 1979, Nature 279 381.
Weedman, D.W. et al: 1982, Ap. J. 255 L5.
Weymann, R.J. et al: 1980, Nature 285 641.
Young, P. et al: 1980, Ap. J. 241 507; 1981a,b, Ap. J. 244 723; 736.
Zwicky, F.: 1937, Phys. Rev. 51 290.

DISCUSSION

Peacock : It looks like 0023+171 is being lensed by a double radio galaxy. First, is not this rather unlikely a priori ? Second, if the quasar is at z ≃ 1, you should easily detect the galaxy (unless it isn't at the lower redshift!).

Burke : It is not likely <u>a priori</u> but our selection by using a radio survey would favour radio galaxies being found in the field. In the case of 0023+171 the galaxy is not optically evident, but it might be hiding near the southern quasar.

Rees : I would like to ask about an object which is not claimed to be an instance of lensing, namely the OVV object AO 0235+164. This has a galaxy along the line of sight (within 2 arc seconds). There is no evidence for multiple imaging, and also there is a lot of VLBI data, and evidence of variable 21 cm absorption line profiles. It would be interesting if one could test whether or not this object is displaying magnification. If it definitely isn't, it becomes an object with exceedingly high luminosity, and a prime candidate for optical beaming.

Burke : This OVV was in the list of objects examined by Roberts et.al. with negative results and you are indeed correct in saying that it is important. It should be examined more closely.

GRAVITATIONAL LENSING -- MODELS

Ramesh Narayan
Steward Observatory
University of Arizona
Tucson, AZ 85721, U.S.A.

ABSTRACT. Qualitative features of gravitational lensing are discussed in terms of a scalar framework based on Fermat's principle. The lensing action of galaxy-like models with spherical and elliptical mass distributions are described. The elliptical model has three distinct regimes of lensing, of which two correspond to lensing with three images and one with five images. One of the three-image geometries has been frequently explored in the past. Models proposed for 0957+561 correspond to this. The five-image geometry has been invoked for 1115+080. Some general model-independent properties of gravitational lensing are listed. If image parities were available, it might be possible to make statements about the lensing mass even when it is made up of dark matter.

1. INTRODUCTION

Presently, seven probable cases of gravitational lensing of a high redshift quasar by foreground matter have been identified (see Burke, this volume, for a review of the observations). Attempts to model some of these are briefly reviewed here. A feature to note is that some of the observed cases of lensing (e.g. 0957+561, 2016+112) provide fairly strong constraints on models through a variety of observational details, while in other cases (e.g. 2345+007, 1635+267) the only available information is the angular separation and relative magnifications of the two images. In particular, no trace of the lensing matter is seen in these latter cases. In the light of this wide range of modeling requirements, only a qualitative discussion of lens models will be attempted here. Different regimes of lensing by a galaxy-like elliptical (in projection) lens distribution will be identified and the various known cases of lensing will be discussed within this framework.

Gravitational lensing is usually discussed in terms of a vector formalism (Bourassa and Kantowski 1975, Young et al. 1980) where the position of an image in the sky and that of the undeflected source are related through the vector deflection produced by the lens. This relationship is usually picturised through a bending angle diagram (e.g. Young et al. 1980), but this is possible only when the lensing distribution is circularly symmetric in projection in which case all the vectors in the problem are

G. Swarup and V. K. Kapahi (eds.), Quasars, 529–538.

parallel to one another. For the elliptical lenses that we are interested in here, it is more convenient to consider a scalar formalism in terms of Fermat's principle (Nityananda 1986, Schneider 1985, Blandford, Narayan and Nityananda 1986, BNN).

2. THE TIME SURFACE AND FERMAT'S PRINCIPLE

Consider a source at redshift z_S being gravitationally lensed by matter distributed in a thin slab at redshift z_L. Let θ_S be the angular position in the sky of the source in the absence of the lens. In the presence of the lens, we can define a time-delay surface as follows. Consider a ray that propagates along a null geodesic from the source to the lens plane and then along another null geodesic from there to the observer so as to arrive along the angular direction θ_I. The time delay along this virtual ray path, relative to the direct ray from the source to the observer in the absence of the lens, is given by

$$t(\theta_I) = t_{geom}(\theta_I) + t_{grav}(\theta_I)$$
$$= \frac{(1+z_L)d_{OS}d_{OL}}{2cd_{LS}}(\theta_I - \theta_S)^2 - \frac{2(1+z_L)}{c^3}\int \phi(\theta_I d_{OL}, s)ds \quad (1)$$

The geometrical time-delay t_{geom} due to the extra path-length takes the form of a paraboloid with its minimum at $\theta_I = \theta_S$. The d's are angular diameter distances between the observer, lens and source. The gravitational time-delay t_{grav} arises because light travels slower in the presence of gravitating matter and is proportional to the integral along the ray of the Newtonian potential ϕ. Thus, the surface $t(\theta_I)$ corresponding to the total time-delay is obtained by taking the paraboloidal surface t_{geom} and pushing it up by an amount proportional to the depth of the two-dimensional potential of the projected lens. This time-delay surface has the following properties (BNN):

(a) By Fermat's Principle, true images are located at those θ_I which correspond to extrema in the time-surface. Three types of extrema are possible -- Lows (represented by the symbol L), Highs (H) and Saddle-points (S).

(b) The relative time-delay between two images is directly given by the height difference in the time-surface at the two extrema. Time-delays can be measured if the source is variable.

(c) At a given extremum, the time-surface can be locally described by a curvature tensor $\partial^2 t/\partial \theta_{I1} \partial \theta_{I2}$. If λ_1 and λ_2 are the principal curvatures, then the magnification of the image is proportional to $|\lambda_1\lambda_2|^{-1}$. The relative magnifications of different images are directly observable.

(d) The partial parities of the image along the principal axes are given by the signs of λ_1 and λ_2 respectively (L and H correspond to $++$ and $--$ and S to $+-$ or $-+$, see Figure 1). The total parity is given by the sign of $\lambda_1\lambda_2$ ($+$ for L and H and $-$ for S). Relative parities between different images can be observed if the source has resolved VLBI structure (e.g. a bent jet) or an asymmetric optical fuzz.

In the absence of the lens, the time-surface has a single extremum of type L at $\theta_I = \theta_S$. When the surface is distorted by the presence of lensing

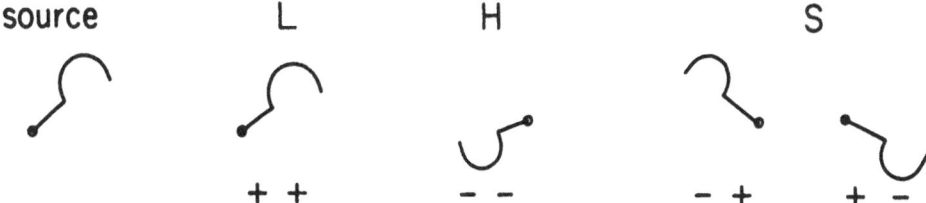

Figure 1. Partial parities associated with different types of extrema

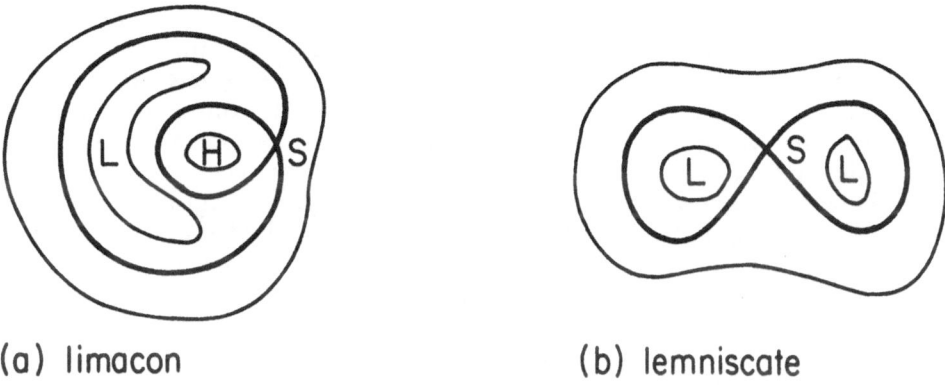

(a) limacon   (b) lemniscate

Figure 2. Time-delay contours for the two possible three-image topologies

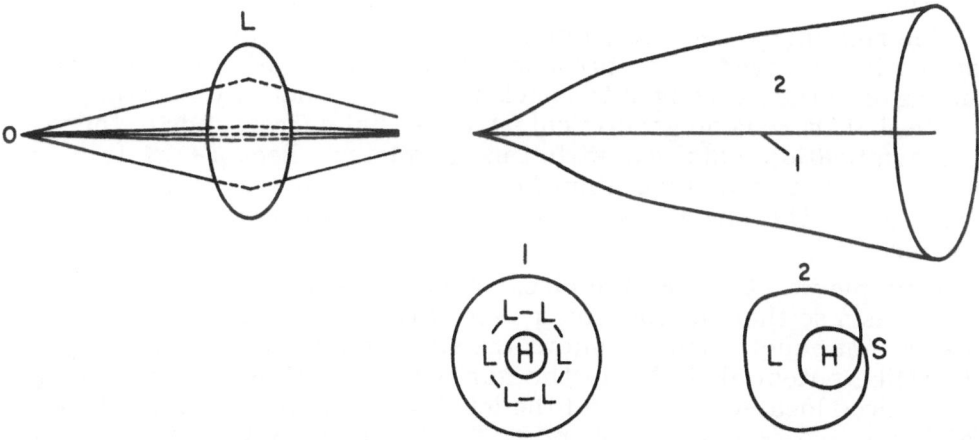

Figure 3. Lensing by a circularly symmetric lens

matter, new images can be produced, but invariably in pairs so that there is always an odd number of images (Burke 1981). There are only two topological arrangements of extrema possible with three images, as shown by the contours of time-delay in Figure 2 (BNN). These topologies are completely defined by drawing the contour that passes through the S image (thick lines in Figure 2). This is the only contour that is not topologically equivalent to a circle and the two possibilities are called the 'limacon' and 'lemniscate' respectively. With five images, there are two S images and hence two non-simple contours. This leads to six possible topologies (BNN).

3. CIRCULARLY SYMMETRIC LENS

We begin with this simple case and consider in Figure 3 the three-dimensional source space behind a typical circularly-symmetric galaxy-like lens with a non-singular core. The conical surface on the right of the figure separates source positions outside it, where only one image is seen by the observer at 0, from positions inside it corresponding to three images. This surface is a caustic sheet since a point source located on this sheet leads to one normal image and a pair of infinitely magnified merging images. In terms of catastrophe theory (Poston and Stewart 1978), the sheet corresponds to a 'fold' catastrophe. For source positions along the axis of cylindrical symmetry, the observer sees one image at the lens center and an infinitely magnified circular ring around it. The time-surface in this case takes the shape of a Mexican hat, with the central image at H in the center and the ring image at the bottom of the surrounding trough. For off-center positions of the source (within the conical fold sheet), the time-surface takes the form of a tilted Mexican hat and so the three images correspond to the limacon topology of Figure 2a.

4. ASTIGMATIC LENS

The ring-image geometry obtained with the circular lens is very special and is destroyed in the presence of any perturbation in the shape of the lens or in the isotropy of the background. Another way of seeing this is to note that there is no generic catastrophe that corresponds to the infinite magnification obtained with this geometry. Because of this non-genericity, studies of high amplification events in gravitational lensing should avoid the circularly symmetric lens, though this is not widely recognised.

A simple way to make the circular lens generic is to introduce some astigmatism so that the focusing power of the lens along its two principal axes are not equal. Figure 4 shows the situation in this case for a typical galaxy-like potential (BNN). In the 'far-field' region of source space (i.e. for a source located far beyond the lens), the ring-image line of Figure 3 'unfolds' into a region corresponding to five images with the characteristic topology marked 3 in Figure 4. This topology is most easily visualized as a squeezed Mexican hat where the bottom of the trough, instead of having a constant height, develops two minima and two maxima along its circumference. The five-image and surrounding three-image regions (marked 2) are separated by a new caustic sheet which consists of four fold sheets connected at four 'cusp' lines. The situation in the 'near-field' region of source space is equally interesting. The single cone of Figure 3 now splits into two sheets that form at the points A and B, corresponding to the different focusing powers along the principal axes of the lens. The region between these two sheets (region 1 in Figure 4) gives the lemniscate three-image topology (Figure 2b), which characteristically forms when the lens is able to split images only along one principal axis and not the other. The two sheets (with two cusp lines each) penetrate each other in an interesting way, forming two 'hyperbolic umbilic' catastrophes at the points marked U.

Thus the average non-singular astigmatic lens produces both possible three-image toplogies and one of six possible five-image topologies. Models of the known cases of gravitational lensing have usually involved elliptic lenses of some sort and therefore correspond to one of the three topologies of Figure 4.

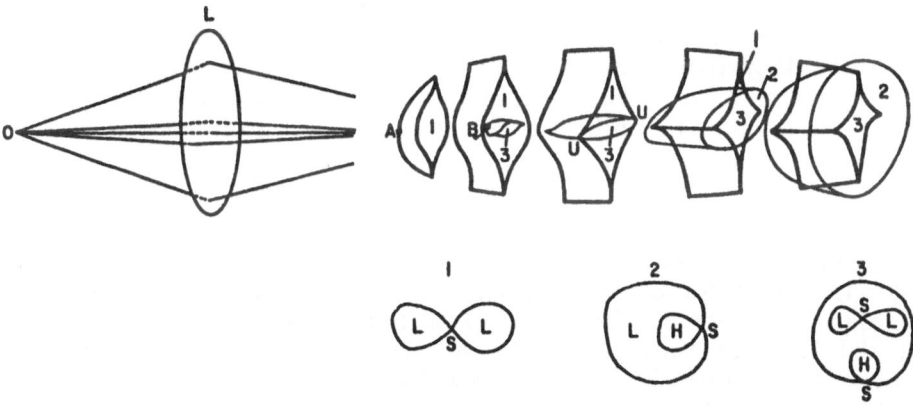

Figure 4. Lensing by an astigmatic lens

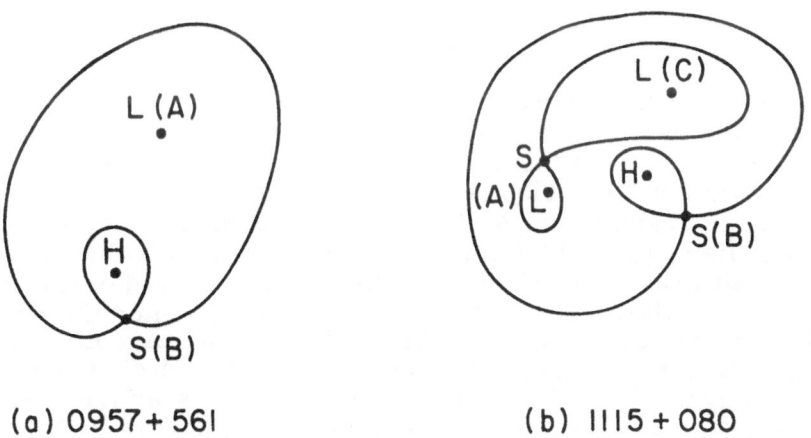

(a) 0957+561   (b) 1115+080

Figure 5. Qualitative features of models of 0957+561 and 1115+080

MODELS OF LENSES

0957+561

Extensive observational data, including VLBI information on parities, are available on 0957+561 and this is consequently the most widely modeled among the known cases of gravitational lensing (Young *et al.* 1981b, Greenfield 1981, Narasimha *et al.* 1984). All the models are similar in their qualitative features, shown in Figure 5a. The topology is that of the limacon, with the A and B images located on L and S respectively. The core of the lensing galaxy produces a third image at H. The galaxy has been detected, but not the image H. The models explain the absence of H by using a small enough core radius for the lensing galaxy, thus making this image very weak. The parity of B is reversed with respect to A. However, the negative parity axis is perpendicular to the line SH, and the VLBI jet points almost exactly along SH (Porcas *et al.* 1981), making it difficult to confirm the reversed parity. More recent observations have apparently succeeded in determining the parities of the images (Burke, private communication, this meeting). It is obvious from Figure 5a that A has a smaller time-delay than B and so it will vary first. Observations seem to confirm this (Florentin-Nielsen 1984). The observed galaxy lens in 0957+561 is not massive enough to produce the 6″ image splitting. Therefore all models invoke the additional effect of a surrounding cluster. This has the effect of introducing extra convergence and shear in the rays from the quasar, but does not affect any of the qualitative features described above.

Most other examples of two-image lensing are probably similar to 0957+561 in their qualitative features. The model of 2345+007 by Subramanian and Chitre (1984) is novel in that it uses two galaxies at different redshifts, but may still have a topology similar to Figure 5a. 2016+112 is a particularly difficult case to model since there are three primary images forming a triangle with angles ~ 90°, 60°, 30°, two additional blobs which may or may not be images, and two possible lensing galaxies (Schneider *et al.* 1986). No model has been proposed so far that fits all the data.

1115+080

Models that have been proposed for 1115+080 (Young *et al.* 1981a, Narasimha *et al.* 1982) use a non-spherical galaxy-like lens in the far-field five-image region (marked 3) of Figure 4. The topology is shown in Figure 5b. The source is assumed to be close to the caustic sheet separating the five- and three-image regions and so two of the five images are close together and magnified. The split A image (distinguished individually as A_1, A_2) is identified with this pair. The B and C images are located on S and L. (The identification of these two images could be reversed since there is a fair degree of symmetry between them.) The missing fifth image is again on H in the core of the lens, somewhere in the middle of the circle formed by the other four images. The fact that no lensing galaxy is seen has been somewhat embarrassing for the models. Nityananda (1986) presents an elegant analysis of this model of 1115+080, showing that the magnifications M of the four images satisfy $M(A_1) \sim M(A_2)$, $M(B) \sim M(C)$ and $M(A_1)/M(B) \sim AB/A_1A_2$ (ratio of angular separations). As regards time-delays, the image C on L (Figure 5b) is the first to vary, the merging pair A varies next, and the image B on S varies last. The time-delay between A_1 and A_2 is very small compared to the other delays.

Lemniscate Models

Models using the lemniscate topology of Figure 2b have rarely been used since they usually require that the lens should under-focus along one of its principal axes, which is unlikely with galaxy lenses that are located 'half-way' to a source at a cosmological distance. Applications include lensing by a supercluster filament (Sanders *et al.* 1984), and lensing by a straight cosmic string (Vilenkin 1984, Hogan and Narayan 1984, Gott 1984), where the missing image is on the string at S. It is tempting to speculate that 2237+031 corresponds to the lemniscate topology since the lensing galaxy is at such a low redshift that it could well under-focus along one axis. In this connection it is important to note that the lemniscate topology can be observationally distinguished from the limacon topology of 0957+561 through image parities.

6. SOME 'THEOREMS' IN GRAVITATIONAL LENSING

There is considerable evidence to suggest that dark matter plays an important role in the observed gravitational lenses. Firstly, in several of the cases there is no visible sign of the lensing matter. Secondly, in both 0957+561 and 2016+112, the two cases with the most information on images and 'lensing' galaxies, there is no successful model where the lensing mass follows the light. Once one concedes the presence of dark matter there is a great deal of freedom in modeling, and it is not clear whether it is worthwhile making very detailed models. An alternative approach is to make general inferences about a particular lens that are model-independent. Some possibly useful 'theorems' (BNN) in this regard are listed below:

(a) In general there are an odd number, $2n+1$, of images, of which n are of type S and $n+1$ are of type L or H, with at least one L among these.

(b) The first image to vary is always of type L and therefore has positive (i.e. majority) parity. If this is violated by observations, then one can conclude that some images have gone undetected.

(c) Because mass is positive, all L images are magnified compared to the unlensed source.

(d) A critical surface density Σ_c can be defined which is the density a sheet placed at the lens plane should have in order to barely focus rays from the source at the observer. It can be shown that rays corresponding to images of type H and L pass through regions of the lens plane with $\Sigma > \Sigma_c$ and $\Sigma < \Sigma_c$ respectively. Nothing can be said about type S images.

(e) Because topologies such as the lemniscate of Figure 2b with no H images are possible, it is possible to obtain multiple images even with $\Sigma < \Sigma_c$ throughout the lens plane.

(f) Images can be infinitely demagnified when they form on singularities in the lens. This can happen only to H and S images, not to L. We conclude the following for three-image topologies: (i) If the missing image is lost on a point singularity such as a black hole, then it is of type H and so the remaining images should have opposite parities (Figure 2a); (ii) If on a line singularity such as a cosmic string, then it is likely to be type S and so the other two images will have the same parity (Figure 2b).

(g) In general, the more magnified the images are (small curvature of the time-surface) the smaller their differential time-delays.

Most of the above theorems require that we be able to distinguish the image type (L, H, or S), for which the pre-requisite is that parities should be measured. Presently, VLBI seems to be the only way to obtain parity information. It has been already used in 0957+561 (Porcas *et al.* 1981, Gorenstein *et al.* 1984), and there are good prospects in 2016+112. For the rest of the lenses, the only hope is that the Hubble Space Telescope may be able to resolve the fuzz around the lensed quasar images. If there is any asymmetry in the fuzz, then this could be used to infer image parities.

ACKNOWLEDGEMENTS. I thank Rajaram Nityananda for sharing with me his pioneering ideas on the application of Fermat's Principle to gravitational lensing much before publication. The work presented here is based largely on collaborative work that I carried out with him and Roger Blandford. Support by the NSF under grants AST 84-15355 and AST 83-13725 is also gratefully acknowledged.

REFERENCES

Blandford, R., Narayan,R., and Nityananda, R. 1986, *Ap. J.*, in press.

Bourassa, R. R., and Kantowski, R. 1975, *Ap. J.*, 195, 13.

Burke, W. L. 1981, *Ap. J. (Letters)*, 244, L1.

Florentin-Nielsen, R. 1984, *A. & A.*, 138, L19.

Gorenstein, M. V. *et al.* 1984, *Ap. J.*, 287, 538.

Gott, J. R. 1984, *Ap. J.*, 288, 422.

Greenfield, P. E. 1981, Ph. D. Thesis, M.I.T.

Hogan, C. J., and Narayan, R. 1984, *M.N.R.A.S.*, 211, 575.

Narasimha, D., Subramanian, K., and Chitre, S. M. 1982, *M.N.R.A.S.*, 200, 941.

Narasimha, D., Subramanian, K., and Chitre, S. M. 1984, *M.N.R.A.S.*, 210, 79.

Nityananda, R. 1986, preprint.

Porcas, R. W. *et al.* 1981, *Nature*, 289, 758.

Poston, T., and Stewart, I. N. 1978, *Catastrophe Theory and its Applications* (London:Pitman).

Sanders, R. H., van Albada, T. S., and Oosterloo, T. A. 1984, *Ap. J. (Letters)*, 278, L91.

Schneider, P. 1985, *A. & A.*, 143, 413.

Schneider, D. P. *et al.* 1986, preprint.

Subramanian, K., and Chitre, S. M. 1984, *Ap. J.*, 276, 440.

Vilenkin, A. 1984, *Ap. J. (Letters)*, 282, L51.

Young, P. *et al.* 1980, *Ap. J.*, 241, 507.

Young, P. *et al.* 1981a, *Ap. J.*, 244, 723.

Young, P. *et al.* 1981b, *Ap. J.*, 244, 736.

DISCUSSION

Subramanian : (1). Fermat's principle is a very nice way of looking at lensing by a mass sheet at a particular redshift but when you have lenses at different redshifts, then Fermat's principle formalism becomes less intuitive and useful.

(2).To add to your list of general theorems on lensing : Stephen Cowling and I have proved a number of theorems on local conditions on the surface density of the lens for multiple imaging by smooth and bounded gravitational lenses. (MNRAS in Press)

Narayan : (1).If the image splitting is produced by only one lens sheet and the others at different redshifts merely modify the images without introducing extra splitting, then some of the single lens-plane ideas can be carried through (BNN). However, if there is strong lensing in more than one plane,the geometrical intuition provided by Fermat's principle seems to be lost (though the principle of extremum time is still valid).

Rees : (1). Even those who advocate cosmic strings as triggers for galaxy formation would only expect them to contribute 10^{-5}-10^{-4} of the critical cosmological density. The probability of observing lensing due to a string, along a given line of sight, would be only of this order, so doesn't this make it unlikely that strings are relevant to observed lensing ?

(2). Could you comment on how "minilensing" might affect your discussion of the number and relative brightness of the multiple images?

Narayan : (1). This is substantially correct. However, if the luminosity functions of quasars were sufficiently steep [$\phi(>L) \propto L^{-\alpha}$; $\alpha > 2$] then the probability of lensing by a particular class of lens could be significantly greater than the Ω in those lenses because of 'amplification bias'. The effect is absent for a straight string since it produces no magnification, but could be important for string loops (see Hogan and Narayan 1984).

(2). Most of my discussion assumed a smooth lens potential. If there ae 'mini-lenses' present, e.g. stars, then there are two possibilities :

(a) If the angular size of the source is greater than the mean angular separation of mini-lenses, the smooth potential approximation is valid.

(b) If not, each image is split into a large number of mini-images which cannot in general be resolved. The net magnification of an image can be substantially modified, and this has been invoked to explain the absence of the odd image in several observed examples of lensing. Time-delays are not affected.

GRAVITATIONALLY LENSED QUASARS AS COSMOLOGICAL PROBES: THE UNIQUENESS
PROBLEM *

M. Birkinshaw
Department of Astronomy, Harvard University,
60 Garden Street
Cambridge, MA   02138 U.S.A.

ABSTRACT. The generalisation from circular to elliptical mass distributions acting as gravitational lenses produces a number of interesting
effects. These effects include an extra indeterminacy in predictions of
the time delay between the two brightest images, if only the relative
brightness and separation of these images is known. This indeterminacy
is removed if further information about the system (such as a measurement of the astigmatism of the lens) is available.

1. THE PROBLEM

Gravitationally-lensed quasars producing multiple images have been
suggested as cosmological probes (Refsdal 1964; Young et al. 1981):
the time delays between variations of the image brightnesses measure
the distance of the quasar, so that the system can be used to determine
the Hubble constant. This method of measuring H_0 is feasible only if
the time delay as a function of distance can be predicted from the
observables of the image system, the relative magnifications and separations of the images. The variables in the problem are the structure
and location of the lensing mass -- since dark matter may contribute to
the lens, the use of the apparent (luminous) lens structure is pre-
cluded.

For a circularly-symmetrical lens, the image positions and brightnesses are determined entirely by the masses enclosed at the image radii
and the surface densities at the images. An uncertainty in the predicted time delay arises to the extent that the distribution of mass
between the images is unknown (Gorenstein et al. 1985), and that the
symmetry transformation of Falco et al. 1985 can be applied.

2. THE MODEL

The properties of elliptical lenses have been explored to investigae
the time delays that arise from a lens model without circular symmetry.
Only one simple case will be discussed here -- that of a uniform elliptical disc mass, with axial ratio b:a = 2:1. It will be assumed that
the observed image structure consists only of the two brightest images
with relative brightness μ_{12} and separation β_{12}. The consistency of
this assumption is assured by discussing solutions only where the third

* Discussion on p.553

539

G. Swarup and V. K. Kapahi (eds.), Quasars, 539–540.

image is fainter than either of the other images by at least a factor 10. The problem of the uniqueness of the time delay τ_{12} is then reduced to that of determining whether observations of (β_{12}, μ_{12}) uniquely define τ_{12}, once the angular size of the lens and its surface mass density have been estimated (by other observations or by physical arguments).

The deflection angles produced by an elliptical disc correspond to an astigmatic focusing lens for images projected within the disc (a circular disc is not astigmatic). Outside the disc, the elliptical lens shows regions where the deflection angles do not tend to zero monotonically as the images move further from the disc centre, so that there are regions where magnification features unlike those present in the circular lens may appear.

3. THE RESULTS

The observed (β_{12}, μ_{12}) for a given circular lens specify a unique τ_{12} at all but one (β_{12}, μ_{12}). The time delays at this equivalent image configuration are of the same sense (the brighter image to fainter image delay is negative). If the third image can be detected, its location resolves the uncertainty in the lens/source configuration produced.

The elliptical lens, on the other hand, produces a single (β_{12}, μ_{12}) at two different source locations over much of the (β_{12}, μ_{12}) plane. For the cases of most interest, the brighter two images are projected outside the disc, and a detection of the third image may not resolve the indeterminacy in the configuration. As an example of the uncertainty in τ_{12} that can be produced, for a particular mass of the elliptical lens model, an equivalent image configuration arises at $\beta_{12} = 2.51a$, and $\mu_{12} = 1.57$. In this case, the time delay between variations in the brightest images can be either -0.56 or $+0.52\tau_0$, where τ_0 is a scale time depending on the distances of lens and quasar. In this example, the geometry of the two configurations is so similar that observations of all three image positions cannot resolve the ambiguity, and the difference in the relative brightness of the third image is only 0.4 mag.

The conclusion from this work is that an elliptical gravitational lens produces an extra indeterminacy in the time delay over that produced by a circular lens. Further information than simply the brightness ratios and separations of the images and the characteristics of the lens are required to remove this indeterminacy, and to predict a unique time delay that can be used to measure H_0. It is of interest to note that the best models of the double quasar of Gorenstein et al. (1985) require the use of an astigmatic lens mass resembling the elliptical model considered here, and that the removal of the indeterminacy in this case is effected by the measurement of all components of the relative magnification matrix of the brightest two images.

4. REFERENCES

Falco, E.E., Gorenstein, M.V. & Shapiro, I.I., 1985. Ap. J., 289, L1.
Gorenstein, M.V., et al., 1985. In preparation.
Refsdal, S., 1964. Mon.Not. R. astr. Soc., 128, 307.
Young, P., et al., 1981. Ap. J., 244, 736.

"MISSING IMAGE" IN GRAVITATIONAL LENS SYSTEMS

D. Narasimha
Department of Physics, University of Calgary
Calgary, Canada
K. Subramanian
Astronomy Centre, University of Sussex
Brighton, U.K.
S.M. Chitre
Tata Institute of Fundamental Research
Bombay, India

ABSTRACT. The gravitational lens models involving extended mass distribution generally predict an odd number of images with one of the images close to the centre of the principal lensing galaxy. In all the observed lens systems only an even number of images have been unambigously detected so far. It is demonstrated that the presence of a compact nucleus at the centre of lensing galaxies would dim the "odd" image significantly without affecting the rest of the image configuration.

It is an intriguing fact that in almost all the observed lens systems only an even number of images have been unambigously detected, although an important property of the lensing action of extended smooth distribution of matter is the generation of an odd number of images (Burke, 1981). In fact, the model calculations which invoke a dominant lensing component contributed by a galaxy predict the existence of an image close to the galactic centre. A number of suggestions have been advanced to explain why this "odd" image has not been detected so far, none of which seem compelling enough (Chang and Refsdal, 1984; Subramanian, Chitre and Narasimha, 1985, Narayan, Blandford and Nityananda 1984).

Here we propose a scenario to account for the "missing odd image" phenomenon by making use of a massive compact nucleus that could possibly exist at the cores of giant galaxies. We argue that such a nucleus with even a small fraction ($\lesssim 2\%$) of the total galaxy mass contained within about 50 pc, would have negligible influence on the properties of the images located outside the core of the galaxy. But, the nucleus becomes gravitationally effective in pulling the odd image closer to the galactic centre, at the same time making it significantly dimmer.

Consider a spherically symmetric distribution of matter for the core of the lensing galaxy. Following the notation of Bourassa and Kantowski (1975), the scattering function that represents the bending

G. Swarup and V. K. Kapahi (eds.), Quasars, 541–543.
© *1986 by the IAU.*

of light rays for an impact parameter distance small compared to the galactic core-radius, may be written as

$$I_G(z_0) = \pi \sigma_G z_0^* \, ,$$

where σ_G is the central surface density of the galaxy and z_0 is the complex number giving the image position in the deflector plane. Suppose, in addition we consider the effect of a nucleus of mass, M_N and radius, R_N with a constant surface density. Then, the scattering function of the nucleus is given by

$$I_N(z_0) = \frac{M_N}{R_N^2} z_0^* \, , \qquad |z_0| < R_N$$

$$= \frac{M_N}{z_0} \, , \qquad |z_0| > R_N$$

The image position z_0 is then related to the source position z by the equation

$$z = z_0 - \frac{4\mathcal{G}D}{c^2} I_G^* - \frac{4\mathcal{G}D}{c^2} I_N^*$$

$$= z_0 \left(1 - \kappa_G - \kappa_N \right) \, , \qquad |z| < R_N$$

Here, $\kappa_G = \frac{4\mathcal{G}D}{c^2} \pi \sigma_G$, $\kappa_N = \frac{4\mathcal{G}D}{c^2} \pi \left(\frac{M_N}{\pi R_N^2} \right) \equiv \frac{R_0^2}{R_N^2}$,

with $R_0 = \left(\frac{4\mathcal{G}D}{c^2} M_N \right)^{1/2}$ being radius of influence of the nucleus and D = (observer-deflector distance) x (deflector-source distance).
                                            (observer-source distance)

For the present problem both κ_G and κ_N exceed unity. In the absence of the nucleus the image would have been located at a position z_0^I given by

$$z = z_0^I \left(1 - \kappa_G \right) \, .$$

Writing $r = |z|$, $r_0 = |z_0|$ and $r_0^I = |z_0^I|$, the "odd" image without the nucleus is located at

$$r_0^I = \frac{r}{(\kappa_G - 1)} \, .$$

After inserting the nucleus the image gets shifted to the position

$$r_0 = \frac{r}{(\kappa_G + \kappa_N - 1)} = \frac{(\kappa_G - 1) r_0^I}{(\kappa_G + \kappa_N - 1)}$$

The amplification A^I of the image in the absence of the nucleus is obtained as

$$A^I = \frac{r_0^I}{r} \frac{dr_0^I}{dr} = \frac{1}{(\kappa_G - 1)^2} \, ,$$

while, with the nucleus, the amplification A is given by

$$A = \frac{r_0}{r} \frac{dr_0}{dr} = \frac{1}{(\kappa_G + \kappa_N - 1)^2} \, .$$

Assume the parent galaxy to have a mass $M \sim 5 \times 10^{11} M_\odot$ within a core-radius 3 kpc and let us suppose $D \sim 600$ Mpc. We then have $\kappa_G \sim 6.5$. Consider now a nucleus with mass $M_N \sim 10^{10} M_\odot$ and radius $R_N \sim 50 pc$ to get $\kappa_N \sim 470$. Clearly $\kappa_N \gg \kappa_G$ and we have

$$r_0 \simeq \frac{\kappa_G}{\kappa_N} r_0^I = \kappa_G \left(\frac{R_N}{R_0} \right) \left(\frac{r_0^I}{R_0} \right) R_N$$

Thus, for the location, r_0^I of the odd image, without the influence of the nucleus, within a critical radius $\simeq \kappa_G^{1/2} R_0$, the image in the presence of the nucleus is pulled in to the position

$$r_0 \lesssim \kappa_G^{3/2} \left(\frac{R_N}{R_0} \right) R_N \lesssim \frac{3}{4} R_N .$$

The amplifications, before and after the introduction of the nucleus are respectively given by

$$A^I \simeq 3 \times 10^{-2} , \quad A \simeq 4 \times 10^{-6}$$

i.e. the influence of the nucleus becomes effective in further depressing the intensity of the odd image by an order of 10^4 (or, equivalently dimmed by some 10 magnitudes), besides being brought closer to the centre of the galaxy.

We added the gravitational effect of the compact nucleus on the image properties by adopting the realistic lens models which we had previously constructed, without altering any of the parameters of those models (Narasimha, Subramanian and Chitre, 1984; Subramanian and Chitre, 1984). The masses of the parent lens galaxy required to reproduce the observed configuration turn out to be in the range $5 \times 10^{11} - 3 \times 10^{12} M_\odot$, with the mass of the nucleus $\sim 1 - 2 \times 10^{10} M_\odot$ contained within ~ 50 pc. The computations show that in the absence of the nucleus the intensity of the "odd" (third) image is about a few per cent of that of the external bright image. But the introduction of the nucleus depresses the intensity of the image by a factor anywhere from 10^2 to 10^4, i.e., the third image is dimmed by some 6 - 9 magnitudes. The properties of the main images like their positions and intensities are left practically unaltered by the presence of the nucleus.

We should perhaps stress that we do not necessarily invoke any singularities like black holes in the density distribution of matter, although such singularities can also play a similar role in explaining the absence of the "odd" image. Rather we attempt to account for the observed features of the gravitational lens system in the framework of a conventional smooth density distribution before appealing to the presence of any singularities. Young et al (1978) have inferred from photometric and spectroscopic observations that the nucleus of the giant elliptical galaxy M87 contains a compact mass $\sim 5 \times 10^9 M_\odot$ with 100 pc. We have simply demonstrated that the presence of such a compact nucleus in the core of the lens galaxy can lead to a signficant dimming of the third image.

References

Burke, W.L., 1981. Ap. J. (Letters) 244, L1
Bourassa, R.R. and Kantowski, R., 1975. Ap. J. 195, 13
Chang, K. and Refsdal, S., 1984. Astron. ¢ Ap. 132, 168
Narasimha, D., Subramanian, K. and Chitre, S.M., 1984. M.N.R.A.S. 210, 79
Narayan, R., Blandford, R. and Nityananda, R. 1984, Nature, 310, 112.
Subramanian, K. and Chitre, S.M. 1984. Ap.J. 276, 440
Subramanian, K., Chitre, S.M. and Narasimha, D. 1985 (preprint)
Young, P.J., Westphal, J.A., Kristian, J., Wilson, C.P. and Landauer,F.P.
    1978, Ap.J. 221, 721.

Ramnath Cowsik, Judith Perry, Peter Sohnius and
Patricia Boeshaar

GRAVITATIONAL LENSES AS PROBES OF THE DISTRIBUTION OF DARK MATTER IN THE UNIVERSE

R. Cowsik and P. Ghosh
Tata Institute of Fundamental Research
Bombay 400 005

ABSTRACT. Studies of the characteristic properties of gravitational lensing by clusters of galaxies suggest that the dark matter in them is probably smoothly distributed on the scale of the cluster itself, rather than being clumped into halos around individual galaxies.

1. INTRODUCTION

Gravitational lensing by clusters of galaxies give information on the distribution of dark matter in the universe. Different properties are obtained from the lens depending on whether (1) dark matter is smoothly distributed on the scale of clusters and individual galaxies are embedded in it, or (2) dark matter clumps into halos around individual galaxies. We show here the difference between these properties, and suggest that observations favour scenario 1. Fig. 1 shows the calculated profiles of surface density (σ) of the galaxies (solid line) and the dark matter (dashed line) according to scenario 1 (Cowsik and Ghosh, submitted to Ap.J.); the data (dots) is also shown.

2. SMOOTHENED POTENTIALS

Lenses are studied through the scalar potential formalism (Schneider 1985, Astr. Ap. 143, 413, Nityananda 1985 private communication). The deflection $\bar{\alpha}$ is the gradient of the scalar potential

$$\psi(\underline{r}) = \int \rho(\underline{r}') \ln|\underline{r} - \underline{r}'| d^2 r'$$

Potentials ψ_1, ψ_2 generated by smoothened distributions of galaxies and dark matter according to scenarios 1 and 2 are shown in Fig.2. A background potential, generated by a (critical) background density is subtracted in each case to avoid double counting. The deflection α_1, α_2 generated by ψ_1, ψ_2 are also shown. The radial scale $1_\nu = [<v_\nu^2>/12\pi G \rho_\nu(o)]^{\frac{1}{2}}$ is the structural length of the cluster.

G. Swarup and V. K. Kapahi (eds.), Quasars, 545–546.

Here $\langle v_\nu^2 \rangle$ is the dispersion in the velocities of the neutrinos, m_ν is the neutrino mass and $\rho_\nu(o)$ is the density of dark matter at the centre of the cluster.

3. FLUCTUATIONS

Individual galaxies cause fluctuations about the smoothened potentials. Fig.3 shows the fluctuations in ψ along the x-axis of a typical cluster. The length scale for the fluctuations if $l_* = [\langle v_*^2 \rangle / 12\pi G \rho_*(o)\ ]^{\frac{1}{2}}$, the structural length of a galaxy. Here, $\langle v_* \rangle^{\frac{1}{2}}$ is the r.m.s. velocity of the stars in the galaxy and $\rho_*(o)$ is the visible density at the centre of the galaxy. Deflection angles obtained from the potentials of Fig.3 are displayed in Fig.4.

4. COMPARISON OF SCENARIOS

First, mass models for the gravitationally lensed quasar 0957 + 561 (Greenfield, Roberts and Burke 1985, Ap. J., 293, 370) have shown that scenario 1 is preferred over scenario 2. Second, the straight lines in Fig.4 correspond to various values of the left-hand side of the lens equation $\underline{r} - \underline{x} = \underline{\alpha}(\underline{r})$; their intersections with α-curves give the positions of the images. Scenario 1 give well-separated images which are sharp and undistorted. In scenario 2, several of these images are actually multiple images with small splitting; at low resolution these appear as diffuse, distorted images. The observed lack of distorted images (Tyson et al, 1984, Ap.J. (Letters), 281, L59) favours scenario 1.

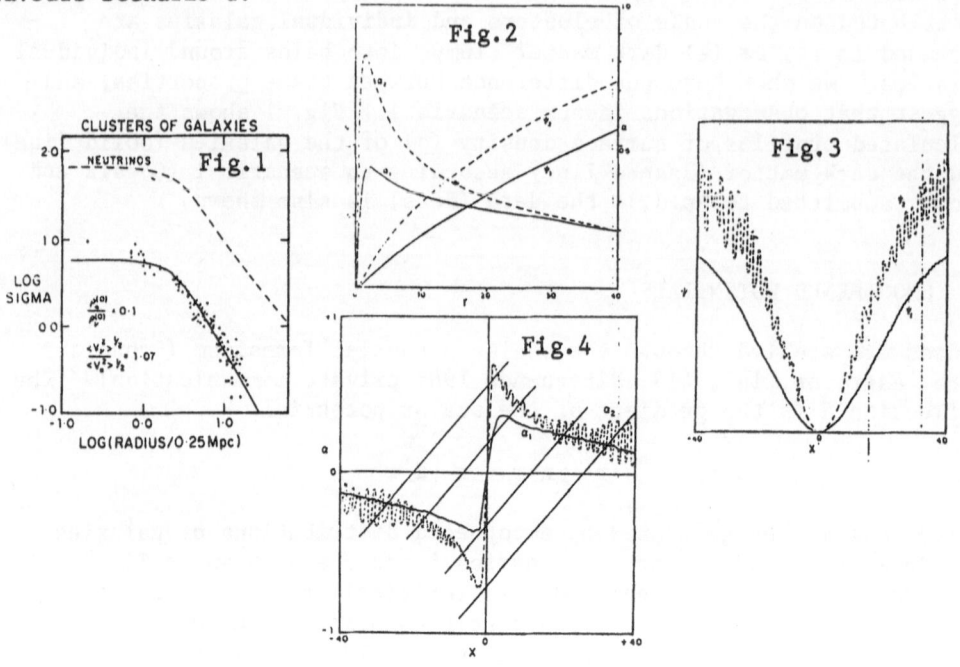

OBSERVATIONS OF CLOSE PAIRS OF FAINT BLUE OBJECTS: TOWARDS MIRAGE OR REALITY*

H. Reboul
Labo. d'Astronomie
U.S.T.L.   34060
MONTPELLIER Cedex
France

A.M. Fringant
Observ. de Paris
61 Av. de l'Observ.
75014 PARIS
France

C. Vanderriest
Observ. de Meudon
92190 MEUDON
France

ABSTRACT, INTRODUCTION and CONCLUSION. We present the results of a preli⁻ minary observational run of the FRV sample (Fringant et al,1983) of close pairs of UVX objects at high galactic latitude . The extragalactic physical pairs are ~30% and most of them do appear as "interactivating" AGNs . The method seems equally adapted to find new gravitational mirages as well as really double QSOs.

RESULTS

15(12 UVX-UVX + 3 mixed) pairs have been investigated at this time . Observations and results are summarized in Table 1 where the following abbreviations are used . Only the initials are mentioned for the observers: Buser, Cayrel, Fringant, Lelievre, Mathez, Reboul, Stockton, Vanderriest . P or O in the last column means : proved Physical or Optical .

TABLE I

Object	θ	B1(color)	B2(col)	TELESCOPE(Observer)	NATURE	REDSHIFT	C
PHL2758-C	3.7	18.2(II)	18.4(III)	CFH3.6(V)/ESO3.6(R)	HS + ?	(HS=HaloStar)	
PHL6657-8	4.	18.1(II)	18.2(II)	ESO2.2(C,B,R)	HII + HII	0.079+0.079	P
PHL925-C	6.	17.3(I)	18.4(II)	ESO3.6(R)	HS + ?		
PB6378-C	8.	17.(II)	20.(I)	CFH3.6(V)/OHP1.9(V)	SY?+LG?	0.087+0.087	P
PB9261-C	2.	17.5(II)	19.(II)	" "	GAL+JET	0.066	P
US2065-6	4.	18.2(I)	18.8(I)	CFH3.6(L,M,V)	HS + ?		
PB3424-5	2.8	18.(I)	18.5(I)	HAWAII2.2(S)	WD+WD		P
PB4053-C	8.	16.5(I)	18.(K)	OHP1.9(V,R,F)	SY1+GAL	0.089+?	
PHL1696-7	2.2	16.7(II)	17.8(II)	CFH3.6(L)	SB GAL		P
PB5062-C	5.5	17.5(II)	19.(G)	ESO2.2(C,B,R)	QSO+HS	1.77	O
PHL241-2	7.	17.(I)	17.9(II)	ESO3.6(R)	WD + ?		
PB7348-C	5.	17.5(II)	18.5(A)	ESO3.6(R)	QSO+ ?	1.33+ ?	
PB7496-7	6.	18.(II)	18.(II)	ESO3.6(R)	WD + ?		
PHL2236-C	6.	18.2(II)	18.4(II)	ESO3.6(R)	HS + ?		
PHL6171-C	2.7	16.2(III)	16.5(III)	ESO3,6(R)	HS + ?		

* Discussion on p.553

547

G. Swarup and V. K. Kapahi (eds.), Quasars, 547–548.

DISCUSSION

The 5 proved physical pairs and the absence of proved optical pairs among
the 12 UVX-UVX associations confirms the predictions of the 2-points autocorrelation function of the Berger-Fringant catalogue (Fringant et al 1983).
It seems possible to give a crude estimation of $\sim$30% of extragalactic physical pairs for such samples (<θ>=4.0",<B1>=18.2,<ΔB>=1.9) . Physical pairs
of Halo Stars and White Dwarfs are respectively ~40% and ~25% .

The projected linear separations (H_0=50,q_0=0.15) are 8 kpc for the
pair of HII regions PHL6657-8, 18 kpc for the pair of active galaxies PB
6378-C and 19 kpc for the pair of probably active galaxies PB4053-C . The
case of PB7328-C is doubtful . The object could be a really double QSO or
a QSO-BLac association (50 kpc) but we cannot dismiss an optical pair .

The fact that quite all the extragalactic pairs appear in strong gravitational interaction and the high proportion of AGNs is an argument for
the hypothesis of "interactivation" i.e. mutual activation by gravitational
interaction (direct or tidal fueling of the accretion disks of the central
massive bodies) . The activation hypothesis has been investigated by some
authors (see e.g. Hutchings et al 1982, de Robertis 1985) .

Due to its low <ΔB> , the FRV sample could reveal symmetrical cases .
If double interactivating QSOs do exist, they must lie in the same observational range than mirages (fig. 1) . In addition of testing the validity
of the activation hypothesis at remote times, really double QSOs could
have on q_0 the same potentiality than mirages have on H_0 .

Figure 1. Projected linear
separations (H_0=50,q_0=.15)
H= sym. pairs of HII regions
     (Shaver and Chen, 1985,
      Halpern et al, 1984)
h= as H but very unsym.
q= QSO-compan. (Stockton, 1982)
G= FRV interactivating gal.
J= FRV Jet-galaxy
H= FRV pair of HII regions
Q= FRV interactivating QSOs??
M= known mirages
A= QSO-Gal. (Gilmore, 1984)

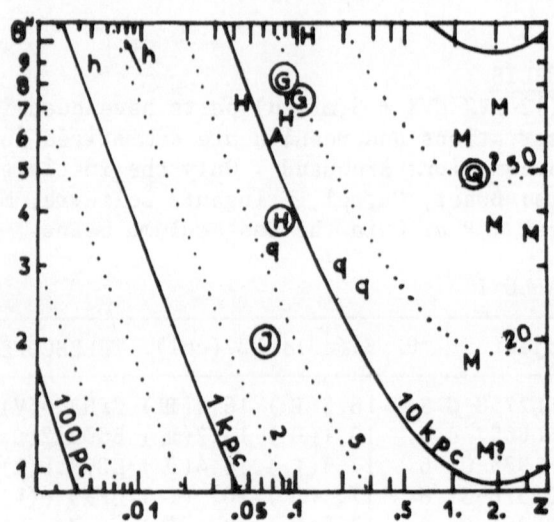

REFERENCES
de ROBERTIS, M.: 1985, Astron. J. 90, 998
FRINGANT A.M., REBOUL, H., VANDERRIEST, C.: 1983, Proc. of the XXIV th
     Liège International Astrophysical Colloquium , p. 155
GILMORE, G.: 1984, M.N.R.A.S. 211, 25 P
HALPERN, J.P., MARSHALL, H.L., OKE, J.B.: 1984, Astron. J. 89, 1802
HUTCHINGS, J.B., CAMPBELL, B, CRAMPTON, D: 1982, Astrophys. J. 261, L23
SHAVER, P. A., CHEN, J.S.: 1985, Astron. Astrophys. 148, 443
STOCKTON, A.:1982, Astrophys. J. 257, 33

OPTICAL BRIGHTNESS MONITORING OF THE TWIN QSO Q0957+561A,B [*]

R. Schild
Harvard-Smithsonian Center for Astrophysics
60 Garden Street
Cambridge, Massachusetts 02138

ABSTRACT. The gravitationally lensed QSO 0957+561A,B has been monitored from 1979 to the present with CCD cameras. We believe that the arrival time difference is 1.2 ± 0.2 years.

Our program of optical brightness monitoring of the twin QSO 0957A,B (TwQSO) has been continued and results as of August 1984 have been published by Schild and Cholfin (1986), who found an arrival time difference Δt of 1.03 years. Observations with photometric accuracy of 1% on 30 nights from the most recent observing season have been added to the data set with the following results:

1.) The cross-correlation of the two light curves now has only a single peak, corresponding to Δt = 1.3 ± 0.3 years (Figure 1).

2.) The light curve for the QSO now covers a span of 6 years (Figure 2). The preceding QSO image (Component A) gradually increased in brightness by 15% from 1979.0 to 1982.0. Between 1982.0 and 1983.0 it increased by an additional 15% and then remained nearly constant until 1984.0, when it dropped by 0.1 mag. This brightening and subsequent fading provided a signature which was seen one year later in the B image.

3.) Observations in the Nov. 1984-May 1985 observing season on 30 nights showed brightness variations of 10%. We plan to watch for these brightness fluctuations in nightly monitoring during the coming observing season with the hope of determining Δt to within 1 day.

4.) From six nights of intensive and protracted observing with a time resolution of 10 minutes, we do not find evidence for 1% or greater brightness fluctuations on 10 minute to 4 hour time scales. We do find evidence for brightness fluctuations on a time scale of a day.

[*] Discussion on p.553

G. Swarup and V. K. Kapahi (eds.), Quasars, 549–550.

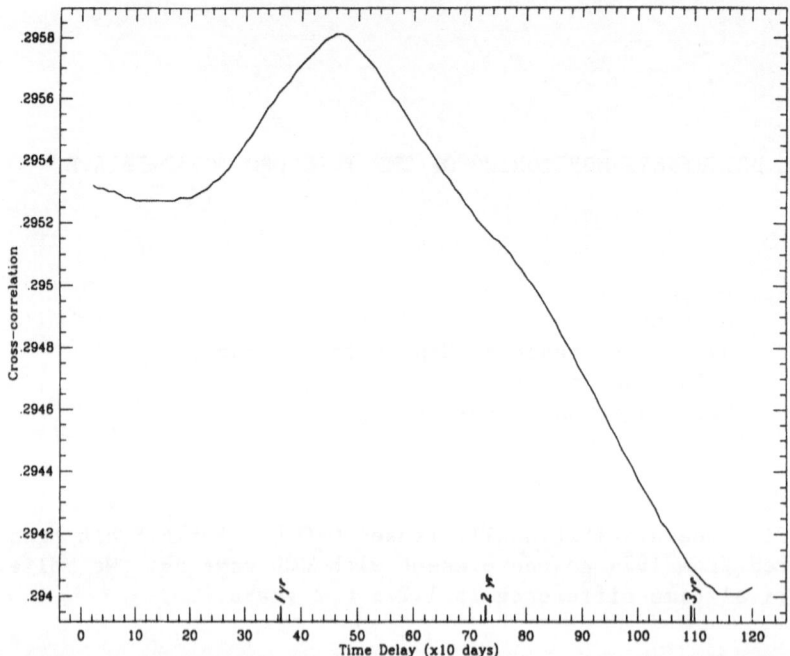

Figure 1. Cross correlation of the light curves of the two QSO images. The correlation peak corresponds to a Δt = 1.3 ± 0.3 years, with the northern image arriving first.

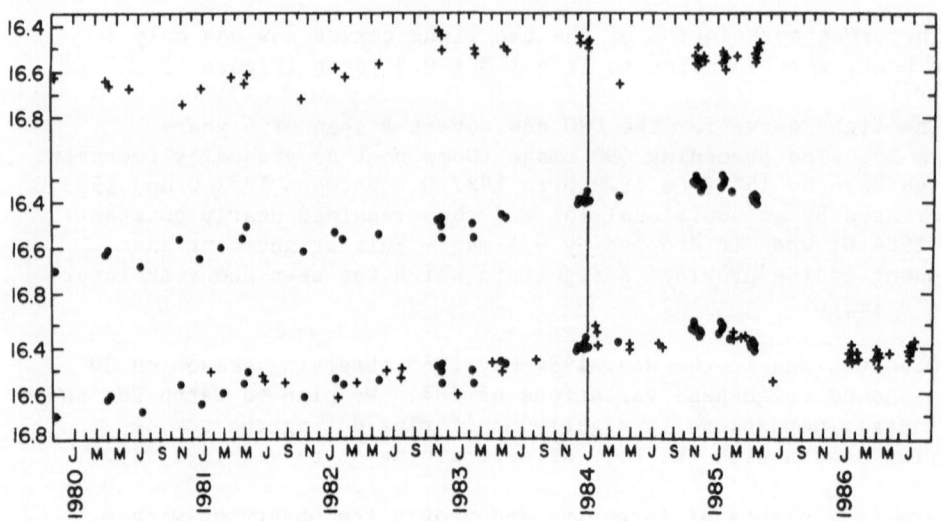

Figure 2. Light curves of the northern (upper) and southern (middle) QSO components. The lower curve shows the two components together with a shift of Δt = 1.2 years.

A CLOSE LOOK AT Q2237+0305 *

J. Anthony Tyson
AT&T Bell Laboratories
600 Mountain Avenue
Murray Hill, NJ 07974

ABSTRACT. CCD images of the lensed QSO candidate 2237+0305 in the blue and near-infrared are examined. At least two blue images of the QSO, separated by about 1 arcsec, are found. The inferred velocity dispersion of the lensing galaxy core of 150 km/sec implies that no dark matter is required for this case. Limits on the mass distribution of the lensing galaxy core and bar can be obtained from the data. Since the galaxy is so nearby, this lensed QSO is a good candidate for assisted lensing by individual stars in the galaxy core.

This sixth known gravitational lens system was discovered by John Huchra and collaborators. It is unlike the other five: the foreground galaxy that acts as a lens is nearby. The galaxy, 2237+0305, is a 15.7 magnitude spiral at a redshift of 0.0394. The discovery was made with spectroscopic observations of 2237+0305 that revealed a galactic spectrum contaminated with light from a quasar having a redshift of 1.695. Huchra, et.al. noted that the galaxy could produce either a single quasar image or a cluster of images which would not have been resolved in the 2 arcsecond seeing.

To confirm 2237+0305 as a strongly amplified quasar image, multiple images must be found, leading to eventual spectroscopy on each separately. Moreover, any multiple image structure would supply needed data on the mass distribution in the lensing object: is the nuclear bulge of the galaxy or a more compact object responsible for the lensed quasar image structure?

CCD frames of 2237+0305 were obtained in excellent seeing at Cerro Tololo Interamerican Observatory using the prime focus CCD camera, on 27 October, 1984. A bright star was included in the CCD field for use later as a comparison star. To avoid saturating on the quasar, four 100 sec exposures were taken through a J (blue) filter, and two 100 sec exposures through a special I filter (.78 - 1.1 micron), with the telescope offset slightly between exposures. The images were re-combined later after removal of any non-reproducible features.

The raw images were processed on the VAX 11/785 at Bell Labs. Examination of the full range of the raw images revealed something strange: the bright peaks of the quasar image on each CCD frame were triangular

* Discussion on p.554

551

G. Swarup and V. K. Kapahi (eds.), Quasars, 551–552.

in shape. But the bright star in the field was circular at its peak.
This was the first indication that lensed quasar multiple image structure had been resolved. Initial image processing removed bad pixels
and cosmic rays, and, for each band, the frames were then registered to
fractional pixel accuracy and averaged by taking the median intensity
of each pixel.

The full-width at half-maximum of the triangle ensemble was larger
than that of the comparison star. To reduce scattered light and sharpen
the images near the quasar, the comparison star was smoothed with a
gaussian filter until it was the same width, and then scaled and subtracted from the centroid of the quasar image combination. Three
"stellar" images are seen.

One of the three "stellar" images has a different color than the
other two. The fainter "third" image is much redder than the other two,
and is presumably the galaxy core. Thus this is a likely example of a
"double" quasar gravitational lens. The separation of the two quasar
image components is about 1.2 arcsec, and is the closest spaced binary
lensed quasar yet discovered. If the lensing mass has an isothermal
distribution then the 1.2 arcsec quasar image separation would imply a
line-of-sight velocity dispersion in the lensing mass of 150 km/sec.
This is slightly smaller than a typical spiral galaxy's core dispersion.
What distribution of mass is consistent with this image two-dimensional
morphology and color data? The center of the galaxy does not lie on
the line joining the two quasar images. In order to produce two mis-
aligned images, the galaxy must have a mass distribution more complex
than spherical. The CCD images show that the galaxy is a barred spiral
with the bar oriented at 45 degrees to the east of north on the sky.
It is possible that this bar represents mass that can account for the
quasar image structure.

A composite color image formed from the blue and infrared CCD exposures shows, in addition to the relatively blue binary quasar images, a
suggestion of a blue contribution near the third (red) component of the
triangle. In collaboration with E. Falco and M. Gorenstein (CFA), we
find that although the observed quasar image separation is easy to get,
the displaced galaxy core (and possible third quasar image) is only consistent with a combination of galaxy core and extra quadrupole, aligned
with the observed bar in the galaxy.

Paczynski has pointed out that 2237+0305 is a likely candidate for
"mini-lensing": additional lensing by a star in the lens galaxy core.
It is highly probable that some fraction of the lensing in this case is
due to a star in the core of the lens galaxy. This predicts that
(1) high resolution ST or speckle images of 2237+0305 may reveal higher
multiplicity QSO images of spacing < 0.1 arcsec, and (2) the relative
flux from the QSO components may vary on month to year timescales.

Based only on the positional coincidence of the relatively high
redshift quasar and the nearby galaxy, 2237+0305 has been cited as a
possible discrepant redshift object. However, its morphological similarity to other confirmed lensed quasars makes any non-cosmological redshift interpretation unlikely. ST offers an excellent opportunity to
examine the spectrum of the two blue images, and search for fainter
QSO images.

DISCUSSION ON THE PAPER BY **BIRKINSHAW** (p.539)

Shanks : How do you rate the prospects for determining H_o from time delays in a gravitational lens ?

Birkinshaw : Where the observations are sufficiently detailed to measure elements of the reltive magnification <u>matrix</u> of the brightest images, there the models are far more constrained than when only the relative brightness of the image are known, so that the indeterminacy that I discussed can be resolved. However, the symmetry transformation of Falco <u>et. al.</u> (1985) means that no <u>measurement</u> but only <u>limits</u> to H_o can be derived from the comparison of model and observed time dealys.

DISCUSSION ON THE PAPER BY **REBOUL** ET AL. (p.547)

Alladin : What is the typical relative velocity for the pairs containing an active component ?

Reboul : The typical redshift is ~ 0.1. There are no very significant differences between the two components due to the low dispersion. Relative velocity is lower than 200 kms^{-1}.

Veron : Do you have cases where both components of a pair of active galaxies are Seyfert galaxies rather than one Seyfert and one more conventional region ionised by young stars ?

Reboul : There is no clear case of pair of good Seyfert galaxies. But the objects appear in strong gravitational interaction. This is in the reason why we put forward the hypothesis of "interactivation".

DISCUSSION ON THE PAPER BY **SCHILD** (p.549)

Hutchings : Do radio flux measurements confirm the 1.2 year time delay in 0957 found in optical data ?

Burke : The radio components A and B of 0957+561 exhibited a steady decrease of 1 mJy/yr for several years. Last year A brightened by several mJy, without a corresponding change in B. We look forward eagerly to the coming year's data.

G. Swarup and V. K. Kapahi (eds.), Quasars, 553–554.
© *1986 by the IAU.*

DISCUSSION ON THE PAPER BY **TYSON** (p.551)

Shanks : Have you any plans to try and obtain individual spectra for the images you resolved in the Huchra quasar, from a ground based observatory ?

Tyson : No. The objects appear to be separated by about one arcsecond. Thus, spectroscopy done on the surface of this planet will be hoplessly contaminated. Space Telescope spectroscopy, using our best positions for A, B and C is hopeful.

QSO ABSORPTION LINES: HEAVY ELEMENTS AND LYMAN-α CLOUDS

BRUCE A. PETERSON
Mount Stromlo and Siding Spring Observatories
The Australian National University
Woden, A. C. T., 2606
Australia

ABSTRACT

The absorption lines in QSO spectra may be produced in material surrounding the QSO, in intergalactic clouds and in the interstellar gas of galaxies along the line of sight to the QSO. The intergalactic clouds produce weak Lyα absorption lines. The intervening galaxies produce absorption line systems with heavy element abundances and ionizations similar to the H I clouds in the halo of our galaxy. The material surrounding the QSO produces broad absorption troughs. Evidence suggests a continuity in spatial correlation and heavy element abundances for the Lyα clouds and intervening galaxies.

1. THE LYMAN-α CLOUDS

High redshift QSOs (those with $z > 1.7$) have many weak absorption lines that can be observed between the short wavelength atmospheric cutoff at 3100Å

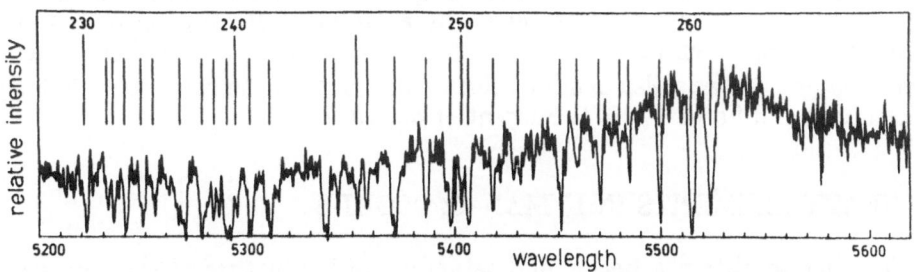

Figure 1. The spectrum of 1442+101 (OQ172) from 5200–5620Å (Peterson *et al.*, 1986). The Lyα emission line is at 5523Å. Almost all of the 261 absorption lines on the short wavelength side of the Lyα emission line are Lyα absorption lines at smaller redshifts, produced in intergalactic clouds along the line of sight to the QSO.

G. Swarup and V. K. Kapahi (eds.), Quasars, 555–561.

and the redshifted Lyα emission line. Figure 1 shows a portion of the spectrum on the short wavelength side of the Lyα emission line of 1442+101 (OQ172) (Peterson *et al.*, 1986).

Counts of the number of Lyα lines per unit redshift interval, as a function of redshift are shown in Figure 2. The data are well represented by the solid line corresponding to

$$\frac{dN}{dz} = K(1+z)^\gamma \tag{1}$$

with $\gamma \approx 2$. The dotted line represents

$$\frac{dN}{dz} = \frac{K}{(1+z)\sqrt{z}} \tag{2}$$

corresponding to the number of uniformly distributed clouds along the line of sight assuming time invariant cloud properties in the chronometric cosmology (Segal 1983). The data are inconsistent with the chronometric theory. The dashed line represents the relation

$$\frac{dN}{dz} = \frac{c}{H_\circ}\sigma\rho_\circ\frac{(1+z)}{(1+2q_\circ z)^{\frac{1}{2}}} \tag{3}$$

corresponding to the number of uniformly distributed clouds along the line of sight assuming time invariant cloud properties in an expanding Universe with $q_\circ = 0.05$ (Peterson 1978). The difference between the observations and Equation 3 can be understood in terms of highly ionized H I clouds (Ostriker and Ikeuchi 1983, Atwood, Baldwin and Carswell 1985) which become progressively more ionized as the Universe expands.

2.   THE HEAVY ELEMENT SYSTEMS

The heavy element absorption systems are associated with the gas in intervening galaxies. The observed heavy element lines are mostly ground state transitions of neutral and low ionization states of the more abundant elements. For example, the QSO 0528—250 (Morton *et al.* 1980) has two heavy element absorption systems with Lyα column densities of $n \approx 10^{21}$ cm^{-2}, and the prominent heavy element ions of C II, N I, O I, Si II, Al II, S II, Fe II, Si IV, and C IV. The column densities measured relative to H I and compared to solar abundances show that O I, N I, Si II, S II, and Fe II are down by a factor of 10, typical of H I clouds in the halo of our own galaxy (Savage and de Boer, 1981). The heavy element absorption line systems also exhibit velocity structure in the range 10–100 km s^{-1} (Boksenberg and Sargent, 1975, Sargent *et al.*, 1982).

3.   HEAVY ELEMENTS IN LYMAN–α CLOUDS

In order to obtain a better understanding of the nature of the Lyα clouds, searches have been made for absorption lines of heavy elements in Lyα cloud spectra. As primeval matter was almost entirely hydrogen and helium, and the nucleosynthesis of heavy elements takes place in stars, the presence of heavy elements in the Lyα clouds would imply that the abundances of the clouds have been enriched by stellar material, and that the Lyα clouds are associated with galaxies.

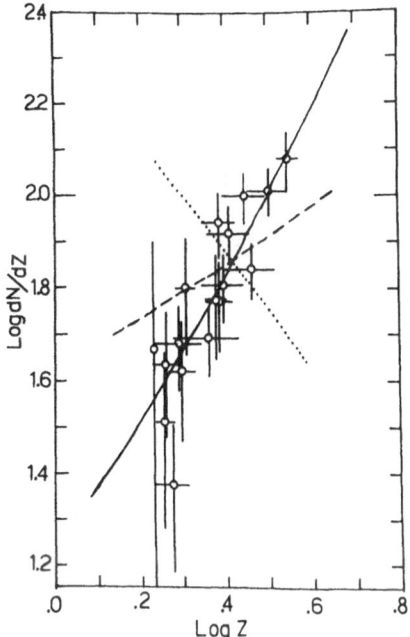

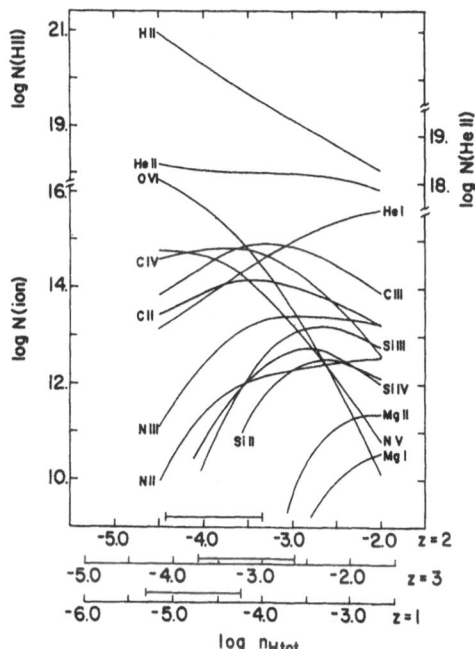

Figure 2. Counts, as a function of redshift, of the number of Lyα lines per redshift interval. The solid line is a power law fit to the data (Eq. 1). The dotted line is the relation expected from the chronometric theory (Eq. 2) for uniformly distributed clouds. The dashed line is the relation expected in an expanding Universe with uniformly distributed, non-evolving clouds (Eq. 3).

Figure 3. Column densities for dominant ions as a function of the total hydrogen density for a plane parallel slab with log N(H I) = 16.5 and [Z/H] = −1.7 (Chaffee et al., 1986). Different epochs are labeled by redshift, and correspond to horizontal shifts in the density axis. The horizontal bars on each density axis show the the limits implied by the observed size constraints on the clouds at $z = 2$.

In order to be detectable in a Lyα cloud, an ion must have a column density greater than about 10^{14} cm^{-2}. In the sun, hydrogen is more abundant than the elements heavier than helium by about 10^6. Thus, in a Lyα cloud with solar abundances, a barely detectable heavy element ion, which was the dominant ionization state for that element, would be accompanied by a Lyα absorption line with a column density of 10^{20} cm^{-2} if the hydrogen was neutral. Low ionization absorption line systems such as these would be similar to those observed in 0528—250, and are associated with intervening galaxies.

For a typical Lyα cloud, the H I column density is about 10^{15} cm^{-2}. If heavy elements were present in these Lyα clouds with approximately solar abundances, they would be impossible to detect unless the Lyα clouds were very highly ionized. However, if the cloud was so ionized that only one part in 10^6 of the hydrogen was neutral, then the Lyα absorption line would be similar in strength to absorption lines produced by highly ionized species of the more abundant heavy elements. This is illustrated in Figure 3, where Chaffee et al. (1985) have plotted column densities for dominant ions as a function of the total hydrogen density as predicted by models where the Lyα clouds are photoionized by the background light from

QSOs, and the heavy element abundances, with respect to hydrogen, is $10^{-1.7}$ times the solar value ($[Z/H] = -1.7$). Note that the galactic stars with the lowest known abundances are CD$-38°245$ with $[Fe/H] = -4.5$ (Bessell and Norris, 1984) and G64–12 with $[Fe/H] = -3.5$ (Carney and Peterson, 1981). Heavy element absorption lines in Lyα clouds with abundances as low as $[Z/H] = -3$ would be practically undetectable.

Two approaches have been used to search for heavy elements in Lyα clouds. In the first, the ability to detect weak lines is enhanced forming a composite Lyα cloud spectrum by summing portions of QSO spectra in the rest frame of each Lyα cloud. In the second, a spectrum of a Lyα cloud with exceptionally high column density is examined for heavy element absorption lines.

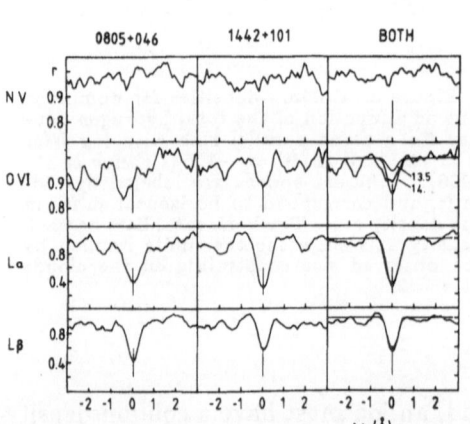

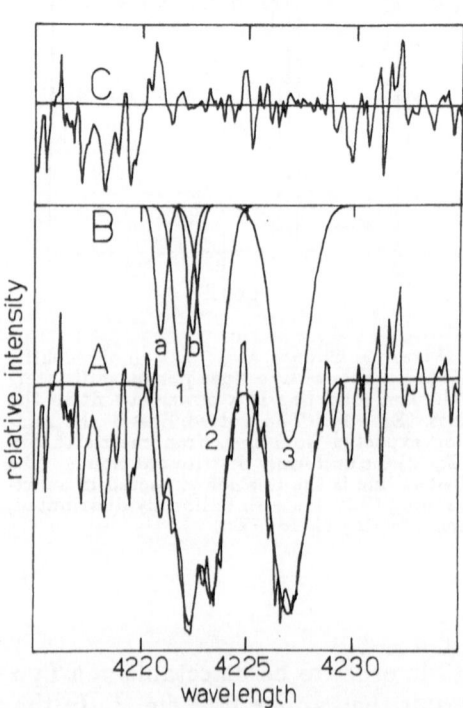

Figure 4. Composite spectra of the QSOs 0805+046 (left), 1442+101 (middle), and a weighted average of both objects (right) in the regions of N v, O vi, Lyα, and Lyβ (Norris et al., 1982). The abscissa represents the distance from the folded line center. The composite spectra were produced by adding, in the rest frame, the spectra of 27 absorption redshift systems identified in the spectrum of 0805+046 by Chen et al. (1982) and 38 systems identified by Peterson et al. (1986) in the spectrum of 1442+101. Each redshift system consisted of a Lyα and Lyβ pair with the Lyα rest frame equivalent width greater than 1.0 Å. No evidence was found for N v. Theoretical profiles are fitted to the O vi, Lyα and Lyβ absorption lines.

Figure 5. A portion of the spectrum of 0014+81 obtained by Chaffee et al. (1986). Panel A – The observed spectrum and computed spectrum. Panel B – The five Gaussian components of the computed spectrum. Components a and b are the C iii lines predicted from a photoionization model assuming log $N(H\ I) = 16.5$ and $[Z/H] = -2.7$. Components 1, 2, and 3 are Lyα absorption lines. Panel C – Residuals for the observed and computed spectra.

Figure 4 shows portions of the resulting composite spectrum for the absorption lines of N v, O vi, Lyα, and Lyβ that Norris, Hartwick, and Peterson (1983) obtained by summing the spectra of 27 Lyα clouds in front of 0805+046 and 38 in front of 1442+101. They detected O vi with 96 percent confidence (2.1σ) and found that the line profiles of the Lyα, Lyβ, and O vi lines were well represented by calculated profiles with a doppler parameter of $b = 30$ km s^{-1}, and column densities of of log $N(\text{H I}) = 14.9$, and log $N(\text{O VI}) = 13.8$. Limits on the column densities of C iv and N v were given as log $N(\text{C IV}) \leq 13.2$, and log $N(\text{N V}) \leq 13.5$. They concluded that their results were consistent with Lyα clouds, with Population II rather than primeval abundances, that were photoionized by the QSO background flux.

However, subsequent applications of this method have been unsuccessful in detecting heavy element absorption lines. Sargent and Boksenberg (1983) report finding no evidence for O vi, in an analysis of the spectrum of 1623+269, and Norris and Peterson (1986) failed to detect O vi in the Lyα clouds in front of 2000-330.

Using the second approach, Chaffee et al. (1985) examined a double Lyα absorption system in the QSO 0014+81 which had H i column densities of $N_a = 5 \times 10^{16}$ and $N_b = 2 \times 10^{16}$ cm^{-2}. They marginally (2.8σ) detected two features which could be Si iii (λ1206), and obtained an upper limit (5σ) for O vi of log $N(\text{O VI}) \leq 13.8$. From their photoionization model they inferred that the heavy element to hydrogen ratio was $10^{-2.7}$ of the solar value.

However, in a following paper, Chaffee et al. (1986) discuss their ionization model, which predicts that if Si iii is detectable, C iii will be even stronger and must also be detectable (see Figure 3), and, on the basis of further observations, they conclude that the 3σ upper limit for C iii is log $N(\text{C III}) \leq 12.9$, five times less than predicted by their model from the strength of previously found Si iii lines.

In Figure 5A, I have re-plotted the spectrum obtained by Chaffee et al. (1986), and fitted the spectrum with the two predicted C iii lines (a and b) and with three arbitrary Lyα lines (1, 2, and 3). The individual fitted components are shown in Figure 5B, and the residual from the fit is shown in Figure 5C. It is the large positive residual at 4220.5Å that argues against the presence of line–a of C iii, and thereby invalidates the previous Si iii identification. Although a positive residual of similar amplitude also occurs at 4232Å, I estimate that there is only a 2 percent chance that line–a of C iii would be filled in by noise greater than or equal to the observed residual.

Sargent and Boksenberg (1983) report that in an analysis of a Lyα absorption system in the QSO 2126 − 158 which had an H i column density of $N = 2 \times 10^{17}$ cm^{-2}, possible O vi lines were found, setting an upper limit of log $N(\text{O VI}) \leq 14.4$, and log $N(\text{O VI/H I}) \leq -2.9$, which is less than log $N(\text{O VI/H I}) = -1.1$ found by Norris et al. (1983) for clouds with H i column densities of $N = 8 \times 10^{14}$ cm^{-2}.

Thus, the original reports of heavy elements in the Lyα clouds have not been satisfactorily confirmed by subsequent observations. On the other hand, the observations are not sufficient to exclude the presence of the lowest heavy element abundances found in the galaxy.

4.   CONTINUITY OF LYMAN-α AND HEAVY ELEMENT SYSTEMS

Recent investigations indicate that there is continuity or overlap between the properties associated with Lyα clouds on the one hand and with heavy element systems on the other.

Tytler (1985) states that the H I column density distribution is a single power law from the weakest Lyα line up to the strongest heavy element system. Webb, Carswell and Irwin (1985) find that the Lyα absorbing clouds show velocity clustering on scales up to 150 km s^{-1}, which is similar to that for the heavy element systems. Bergeron and Boissé (1984) indicate that number of heavy element systems identified by C IV absorption lines increases with redshift as do the Lyα systems, and that the equivalent width distribution of the Lyα lines may overlap with the equivalent widths of Lyα systems inferred from the C IV lines.

REFERENCES

Atwood, B., Baldwin, J. A., and Carswell, R. F. 1985, *Astrophys. J.*, **292**, 58.
Bergeron, J. and Boissé, P. 1984, *Astron. and Astrophys.*, **133**, 374.
Bessell, M. S. and Norris, J. 1984, *Astrophys. J.*, **285**, 622.
Boksenberg, A., and Sargent, W. L. W. 1975, *Astrophys. J.*, **198**, 31.
Chaffee, F. H., Jr., Foltz, C. B., Röser, H.-J., Weymann, R. J., and Latham, D. W. 1985, *Astrophys. J.*, **292**, 362.
Chaffee, F. H., Jr., Foltz, C. B., Bechtold, J., and Weymann R. J. 1986, (pre-print)
Carney, B. W., and Peterson, R. C. 1981, *Astrophys. J.*, **245**, 238.
Chen, J.-S., Morton, D. C., Peterson, B. A., Wright, A. E., and Jauncey, D. L. 1981, *Mon. Not. R. astr. Soc.*, **196**, 715.
Morton, D. C., Chen, J.-S., Wright, A. E., Peterson, B. A., and Jauncey, D. L. 1980, *Mon. Not. R. astr. Soc.*, **193**, 399.
Norris, J., Hartwick, F. D. A., and Peterson, B. A. 1983, *Astrophys. J.*, **273**, 450.
Norris, J. and Peterson, B. A. 1986, (in preparation)
Ostriker, J. P., and Ikeuchi, S. 1983, *Astrophys. J. Letters*, **268**, L63.
Peterson, B. A., Morton, D. C., Chen, J.-S., Wright, A. E., and Jauncey, D. L. 1986, *Mon. Not. R. astr. Soc.*, (in press)
Savage, B. D., and de Boer, K. S. 1981, *Astrophys. J.*, **243**, 460.
Sargent, W. L. W., Young, P. J., Boksenberg, A., and Tytler, D. 1980, *Astrophys. J. Suppl.*, **42**, 41.
Sargent, W. L. W., Young, P. J., and Boksenberg, A. 1982, *Astrophys. J.*, **252**, 54.
Sargent, W. L. W., and Boksenberg, A. 1983, "The Lyα Absorption Lines in QSO Spectra" in the 24th Liège International Astrophysical Colloquium, *Quasars and Gravitational Lenses*, pp. 518–537.
Segal, I. E. 1983, (private communication at the 24th Liège International Astrophysical Colloquium, *Quasars and Gravitational Lenses*).
Tytler, D. 1984, *Bull. American astr. Soc.*, **16**, 1008.
Webb, J. K., Carswell, R. F., and Irwin, M. J. 1984, *Bull. American astr. Soc.*, **16**, 733.

DISCUSSION

Segal : I am not at all clear what you mean by the chronometric prediction, - which I do not see how to make in any model-independent way for
the complex systems you are studying. Aren't you making model-dependent
assumptions without explicitly stating them - the origin of the absorbing clouds and the parametrization of the effect, in particular.

Peterson : The chronometric prediction that I have shown is that given
by you, for the z dependence of non-evolving absorbing clouds that are
uniformly distributed in 3-space.

Mallik : Is there any evidence for molecular Hydrogen absorption in
these clouds ?

Peterson : There is no compelling evidence for molecular Hydrogen. It
is more likely that the HI clouds are highly ionized.

Burbidge : Are there still many unidentified lines in these objects ?

Peterson : In the case of 0528-250, all of the lines are identified that
lie on the long wavelength side of Lyα absorption line of the high
redshift metal line system. On the short wavelength side of this Lyα
line, the number of lines increases dramatically, and these are interpreted as Lyα lines produced by intergalactic Hydrogen clouds. There
are also a few metal lines that can be identified in this wavelength
region. Note that the emission redshift is very close to the redshift
of the high redshift metal line system.

Claude Canizares (?), Bruce Peterson, John Peacock and
Peter Shaver

THE EVOLUTION OF INTERGALACTIC HYDROGEN CLOUDS: CONFUSION RESOLVED

H.S. Murdoch[1], R.W. Hunstead[1], M. Pettini[2] and J.C. Blades[3]

[1]School of Physics, University of Sydney, Australia
[2]Royal Greenwich Observatory, Hailsham, U.K.
[3]Space Telescope Science Institute, Baltimore, U.S.A.

1. INTRODUCTION

It is now widely accepted that a population of intergalactic hydrogen clouds is responsible for the plethora of absorption lines seen at wavelengths shorter than Ly α emission in all high-z QSOs. The question of whether the comoving number density of such clouds evolves with redshift has been somewhat controversial. It is usually assumed that

$$\frac{dN}{dz} \propto (1 + z)^{\gamma} \qquad \text{for } W \equiv \frac{W_{obs}}{1+z} \geq 0.32 \overset{0}{A}.$$

For a non-evolving population of clouds we expect $\gamma \leq 1$ for $q_0 \geq 0$.

The results obtained for γ have varied widely depending on the method of analysis used, particularly where the range of z is limited. This may be seen clearly in Table 1, especially for the results of Carswell et al. (1982). Their last method gives the weighted mean of maximum likelihood (ML) estimates for each QSO separately. We aim to apply rigorous statistical methods to test for evolution and to attempt to explain the variety of results previously obtained.

Table 1:   Various estimates of γ for the SYBT[*] sample

Author(s)	γ	Method
SYBT	+0.48 ± 0.54	ML
Our analysis	+0.45 ± 0.69	ML
Phillipps and Ellis (1983)	−0.1 ± 1.0	Own
Peterson (1983)	+1.6 ± 1.3	Peterson
Carswell et al. (1982)	+1.4 ± 0.7	Peterson
	+0.6 ± 0.6	ML
	−2.1 ± 1.5	$\bar{\gamma}$

[*] SYBT = Sargent et al. (1980); sample of 5 QSOs with 187 lines. Carswell et al. add a further 25 lines from 2 QSOs.

G. Swarup and V. K. Kapahi (eds.), Quasars, 563–567.
© 1986 by the IAU.

2. OUR APPROACH

We form a homogeneous sample of QSOs observed at comparable resolution (∼1 Å) by adding 2000–330 (z=3.78; Hunstead et al. 1986) and 0215+015 (z=1.7; Blades et al. 1985) to the 9 QSOs of the Young, Sargent and Boksenberg (1982; YSB) sample. This gives the widest range of z (1.5 – 3.78) currently achievable. We exclude all lines in known heavy–element systems, including Ly α, since these probably form a separate population.

We then estimate γ using ML since this method gives a minimum variance for the fitted parameter. Next we apply various statistical tests for the assumed parametric form; this is a very important step. Specifically we test for:
- (i)   assumed overall power law
- (ii)   uniformity of dN/dz among QSOs
- (iii) uniformity of dN/dz within QSOs
- (iv)   rank correlation between W and z within individual QSOs for comparison with overall rank correlation (if any).

3. RESULTS AND TESTS

For our sample of 277 lines we find γ = 2.17 ± 0.36, which differs from the case of no evolution by ≳3.25 σ for q_0 ≳ 0. A Kolmogorov test shows an excellent fit to the assumed power law and there is no significant overall rank correlation between W and z.

It is essential, however, to apply the remaining tests for the uniformity of the power law expression for dN/dz both among and within QSOs. In Table 2, the χ^2 test is consistent with a uniform distribution

Table 2. Tests for individual QSOs

QSO	Test for uniformity of dN/dz among QSOs			Uniformity of dN/dz within QSOs	Rank correlation W:z for each QSO
	N_k	$\langle N_k \rangle$	χ^2	$\overline{Q}_k - 0.500$	ρ_k
2000–330	73	67.6	0.4	+0.016 ± 0.030	−0.20 ± 0.12
2126–158	52	56.9	0.4	−0.050 0.036	−0.40 0.14
0002–422	34	32.6	0.1	−0.078 0.048	−0.12 0.17
PHL 957	29	35.7	1.3	−0.085 0.056	−0.02 0.19
0453–423	37	25.7	5.0	−0.041 0.051	+0.09 0.17
1225+317	20	20.5	0.0	−0.090 0.064	−0.09 0.30
0421+019	8	13.5	2.3	−0.15 0.11	+0.14 0.38
0119–046	10	9.7	0.0	−0.04 0.09	−0.22 0.33
0002+051	5	8.6	0.3	−0.04	−0.50
1115+080	2			+0.10	−1.0
0215+015	7	6.2	0.1	−0.19 0.09	−0.68 0.41
	277	277.0	9.9	−0.046 ± 0.017	−0.172 ± 0.061

from one QSO to another but the other two tests each indicate a
significant departure <u>within</u> individual QSOs from the global trend. The
weighted mean values of the test statistics are each significant at a
probability level p $\sim$ 0.003 in a sense consistent with the alternate
test. The Q test (Murdoch <u>et al.</u> 1986) indicates that, compared with the
overall trend, there are relatively fewer absorption lines close to the
emission redshift, while the rank correlation test for individual QSOs
indicates that lines closer to the emission redshift tend to be weaker.

4. THE INVERSE EFFECT

We use this term for the weakening of the global trend for dN/dz within
individual QSOs, as shown by the above tests.   When lines within the
emission line profile are excluded the inverse effect becomes insignificant (and the value of γ is increased). This suggests an effect located
close to each QSO. A probable explanation is the increased ionization
of the hydrogen clouds which occurs within $\sim$5 Mpc from a luminous QSO;
within this range the flux from the QSO is calculated to exceed that
from the general QSO background.   This explanation is also consistent
with the occurrence of high-ionization absorption systems (with N V) at
redshifts relatively close to that of the QSO.

The inverse effect is especially important where the overall range
of z is small and it can explain the divergence of results reported in
Table 1.   From Table 3 the improved concordance in the results for
the various samples can be seen when we progressively i) exclude Ly α
lines in heavy-element systems and then ii) remove the region of
spectrum close to Ly α emission. The inverse effect is small in our
sample because of the wide overall range of z.

Table 3.   Summary of estimates of γ

	SYBT sample	(N)	YSB sample	(N)	Our sample	(N)
Original result :	0.48 ±0.54	(187)	1.81 ±0.48	(214)	...	
<u>Our results:</u> incl Ly α in heavy-element systems	: 0.45 ±0.69	(187)	1.31 ±0.55	(214)	1.79 ±0.35	(298)
excl Ly α in heavy-element systems	: 0.97 ±0.71	(172)	1.72 ±0.57	(197)	2.17 ±0.36	(277)
after removing region under emission line	: 1.43 ±0.88	(131)	1.95 ±0.73	(144)	2.31 ±0.40	(222)

5. WHAT HAPPENS TO THE CLOUDS?

We find that, for $W \gtrsim 0.32$ Å, a good approximation to the combined distribution in z and W is

$$\frac{\partial^2 N}{\partial z\, \partial W} \propto e^{-W/W^*} (1 + z)^\gamma.$$

with W^* approximately constant (Murdoch et al. 1986). Clearly there are fewer clouds with $W(Ly\ \alpha) \gtrsim 0.32$ Å at lower z. Hence more lines have left the sample through becoming too weak but the _form_ of the distribution in W remains essentially unchanged. This implies that the evolution of W with z has the form

$$W = \gamma W^* \ln(1 + z) + \text{const.}$$

Since, in the relevant region of the curve of growth, $W \sim k \ln N(H\ I) +$ const., a consistent expression for the evolution in the column density of neutral hydrogen is given by

$$N(H\ I) \propto (1 + z)^\eta, \quad \text{with } \eta = \gamma W^*/k.$$

For a typical H I cloud with $b \sim 40$ km s^{-1}, this leads to an indicative value for $\eta \sim 6$. This agrees well with the models of Ikeuchi and Ostriker (1986) and Atwood, Baldwin and Carswell (1985) which predict $\eta \sim 4$ to 7 in the relevant range of z. In these models, clouds with mass $\sim 10^6$ to $10^{8.5}$ $M_\odot$ do not collapse to form galaxies but rather expand approximately isothermally (for $z \gtrsim 2$) in pressure equilibrium with a hot intergalactic medium. Our empirically deduced result should encourage the further development of these models and also provide a useful challenge for competing models.

REFERENCES

Atwood,B., Baldwin,J.A. & Carswell,R.F., 1985. Ap. J., **292**, 58.
Blades,J.C., Hunstead,R.W., Murdoch,H.S. & Pettini,M., 1985.
    Ap. J., **288**, 580.
Carswell,R.F., Whelan,J.A.J., Smith,M.G., Boksenberg,A. & Tytler,D.,
    1982. M.N.R.A.S., **198**, 91.
Hunstead,R.W., Murdoch,H.S., Peterson,B.A., Blades,J.C., Jauncey,D.L.,
    Wright,A.E., Pettini,M. & Savage,A., 1986. Ap. J. (in press).
Ikeuchi,S. & Ostriker,J.P., 1986. Ap. J. (in press).
Murdoch,H.S., Hunstead,R.W., Pettini,M. & Blades,J.C., 1986.
    Ap. J. (submitted).
Peterson,B.A., 1983. IAU Symposium 104, 350 (Dordrecht: Reidel)
Phillipps,S. & Ellis,R.S., 1983. M.N.R.A.S., **204**, 493.
Sargent,W.L.W., Young,P.J., Boksenberg,A. & Tytler,D., 1980.
    Ap. J. Suppl., **42**, 41 (SYBT).
Young,P.J., Sargent,W.L.W. & Boksenberg,A., 1982. Ap. J., **252**, 10 (YSB).

DISCUSSION

Malkan : The observation that $\frac{dN}{dz}$ goes the wrong way in an individual QSO spectrum (i.e. the lack of absorption in the blue wings of the quasar emission lines) was previously noted by Dave Tytler. He suggests that this effect may be explained if the absorbing clouds are often too small to cover the emission line region. Then the observed absorption equivalent widths become 2 or 3 times weaker when seen in the emission line profiles, explaining the effect.

Murdoch : We seriously considered this possibility but believed that is was probably ruled out by the work of Foltz et al. on the size of typical clouds.

Turnshek : Concerning Matt Malkan's comment, I have discussed with Foltz and Weymann their observation of a QSO pair in which they detect coincident Ly-α lines. They are concerned that their observations may have been a bit misinterpreted. They make the point that, while there are Lyα lines in common, there are many that are not, and so there is at least a distribution of smaller cloud sizes. The common lines may just be associated clouds or filaments. In either case, the Lyα emitting region may not be completely occulted giving smaller observed equivalent widths near the emission line.

Murdoch : We would be equally happy with the explanation that the clouds close to the QSO may be too small to cover the emission regions. It is possible that both mechanisms operate, since high ionization heavy element systems are also found close to the QSO.

Segal : You have established the consistency of evolution with certain data, but this by no means implies that the same data may not be consistent with quite different hypotheses, - e.g. chronometric theory with a different model for the origin of absorbing clouds than you have implicitly assumed. Moreover, in part your data appears a posteriori, - it includes that from which your hypothesis emerges. Could you comment ?

Murdoch : We have clearly established that N(HI) or alternately W depend strongly on z. We accept the majority view that z is a measure of epoch. Any alternate thoery needs to explain our specific results.

ABSORPTION SPECTRUM OF THE z=3.78 QSO 2000-330 AT HIGH RESOLUTION

R.W. Hunstead[1], H.S. Murdoch[1], J.C. Blades[2] and M. Pettini[3]

[1]School of Physics, University of Sydney, Australia
[2]Space Telescope Science Institute, Baltimore, U.S.A.
[3]Royal Greenwich Observatory, Hailsham, U.K.

The forest of absorption lines observed in all high-redshift QSOs at wavelengths shorter than Ly α emission is generally accepted as being due to a population of intergalactic hydrogen clouds. In order to study the physical properties of these clouds and possible changes in the properties with redshift it is first necessary to establish the true continuum level. The difficulty in defining continuum levels in the Ly α forest region is readily acknowledged in the literature but only rarely do published spectra show the adopted continuum.

Our AAT spectrum of the highest-redshift region of the Ly α forest in 2000-330 is shown in Fig. 1 at a resolution of 0.6 Å FWHM (32 km/s). Despite this high resolution and good S/N ratio there are very few regions which can confidently be recognised as genuine continuum. Conventional methods for continuum estimation are based on the local

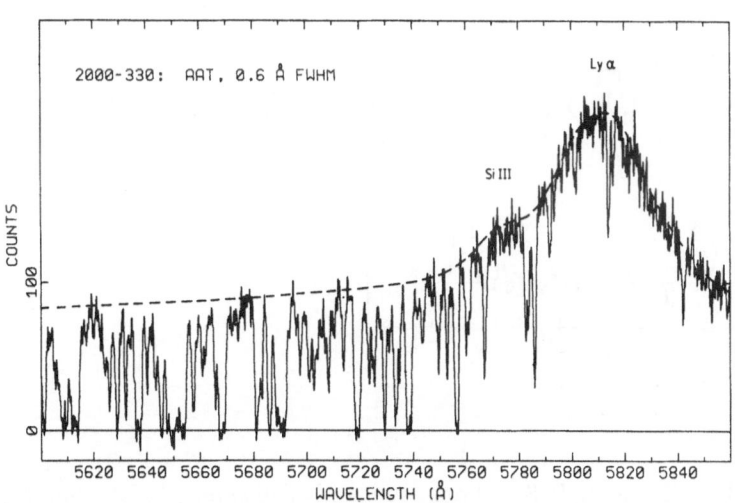

Figure 1. IPCS spectrum of 2000-330 near Ly α emission. The adopted continuum is shown as a dashed line.

G. Swarup and V. K. Kapahi (eds.), Quasars, 569–570.
© *1986 by the IAU.*

variance in the data and fail completely when confronted with such high line densities (Hunstead et al. 1986). The problem is even more severe below Ly β emission where higher order Lyman lines add to the confusion.

Consequently we adopt a pragmatic approach based on the reasonable assumption that the underlying continuum is a power law with constant exponent; in some spectral regions this necessitates the continuum level being set substantially higher than the data. The level is chosen so that positive excursions are consistent with photon statistics. Our adopted continuum is shown in Fig. 1; the evidence for Si III emission is discussed by Hunstead et al. (1986). A full analysis of all our high-resolution data for 2000–330 will be published elsewhere.

Once the continuum level has been defined, the customary approach for identifying lines belonging to QSO absorption systems involves setting up a line list and then looking for wavelength matches with various redshifted ionic species. This method does not utilise fully the information contained in the spectral data, since it demands that every line subjected to scrutiny be independently assessed as real (>5σ) and then abstracts just one parameter, λ, for correlation purposes. Since most spectra contain blends of lines, an alternative approach which uses the entire line profile can be far more revealing. Superimposed spectrum plots are prepared by rebinning the data (normalised by the local continuum) onto a logarithmic wavelength scale, with origins corresponding to the redshifted species being compared. These spectra are then plotted directly on top of one another, preferably using different colours. Alignments and blending can readily be recognised and additional velocity structure can often be inferred. An example is shown in Fig. 2 for the z_{abs} = 3.55 system in 2000–330. The ratio of C IV/Si IV is seen to vary significantly among the components, which span a velocity range of at least 400 km s^{-1}.

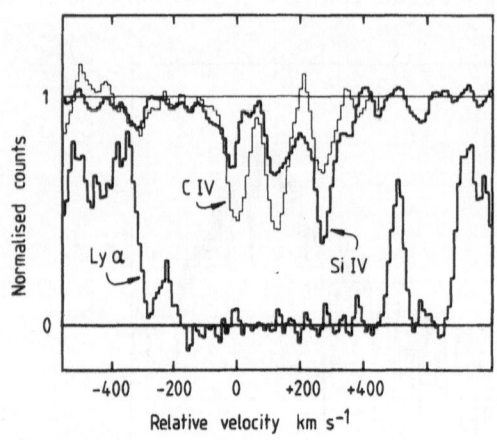

Figure 2. Superimposed spectrum plots of Ly α, C IV 1548 and Si IV 1393 in the z_{abs} = 3.55 system. The velocity origin is at z = 3.5481.

REFERENCE

Hunstead,R.W., Murdoch,H.S., Peterson,B.A., Blades,J.C., Jauncey,D.L., Wright,A.E., Pettini,M. & Savage,A., 1986. Ap. J. (in press)

HIGH RESOLUTION SPECTROSCOPY OF ABSORPTION LINES IN THE Z = 1.7 BL LAC OBJECT 0215+015 *

J. Chris Blades
Space Telescope Science Institute

Richard W. Hunstead and Hugh S. Murdoch
School of Physics, University of Sydney

Max Pettini
Royal Greenwich Observatory

We describe the various absorption systems in 0215+015, and present results from our new, high-resolution studies at 10 km/sec (FWHM).

1. THE ABSORPTION SYSTEMS IN 0215+015

The radio source 0215+015 was shown by Gaskell (1982) to be a highly variable BL Lac object with absorption systems at $z = 1.345$, 1.549, and 1.649. Subsequent work has found 7 separate absorption systems in the optical spectrum of the object as well as a weak Lyman α forest (Blades et al. 1985).

The system at $z = 1.345$ exhibits a strong, mixed-ionization spectrum and is one of the richest found in any QSO—hence it is of considerable interest. We have detected 5 neutral species, H, C, N, O and Mg; 8 singly ionized species, C, Mg, Al, Si, Ca, Mn, Fe and Zn; as well as the higher ionization species, Al III, Si III, C IV and Si IV. This is the first time that Zn II has been found in any QSO absorption system. The ion column densities allows us to deduce depletions of factors between 10 and 4 relative to solar values for Mg, Al, Si and Fe, similar to the gas phase abundances of these elements in local, low-density, interstellar clouds. The Zn strength implies it has solar abundance (Pettini, 1985). In Blades et al. (1985) we attributed this absorption system to an intervening galaxy, with the sight line intercepting both halo and disc material. Indeed, judging by the strength and complexity of the lines, the intercept probably passes through the inner region of the intervening system.

The other six absorption systems in 0215+015 are of higher ionization, with species, C III, C IV, and S IV dominating (Blades et al. 1985; Bergeron & d'Odorico, 1985). When observed at high resolution, the $z = 1.549$ and 1.649 systems reveal highly complex C IV absorption, the latter showing at least nine distinct components over a velocity range of 910 km/sec. The origin of such complex absorption is not at all clear. In Pettini et al. (1983) we postulated two, rich clusters of galaxies at redshifts of 1.549 and 1.649, with the complex absorption occurring in overlapping halos and extended disks of individual galaxies in the sight line to 0215+015.

---

* Discussion on p.577

G. Swarup and V. K. Kapahi (eds.), Quasars, 571–572.

2. THE IMPORTANCE OF HIGH RESOLUTION OBSERVATIONS

In order to disentangle the complex absorption in 0215+015 we have found it necessary to use the highest possible spectroscopic resolution. Such observations yield detailed information on the overall velocity structure and give well-determined values for N and b. For this purpose we have been studying selected absorption lines in 0215+015 at 10 km/sec resolution, and we describe here our first results.

The adjacent figure compares the Fe II lines at 10 km/sec resolution (lower profiles) with an earlier observation of the λ2600 line at ∼ 25 km/sec resolution. Even at the higher resolution we have not satisfactorily resolved the overall complex.

Nevertheless, our detailed model fit to the Fe II pair shows that there are at least 12 different velocity components over the 250 km/sec range, with individual values of the column density $N \sim 1 - 2 \times 10^{13}$ cm^{-2} and b values in the range $4 - 6$ km/sec.

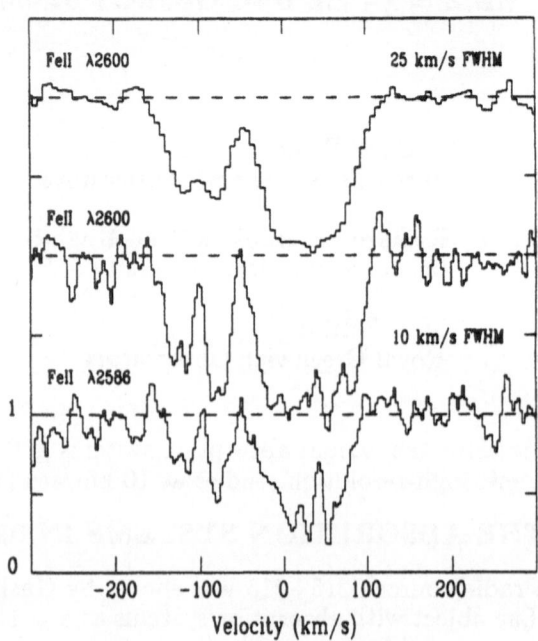

Figure 1. Fe II at $z = 1.345$.

For the case of the $z = 1.549$ C IV system, our new data show that some of the components that make up this redshift complex are exceedingly narrow, with b-values around ∼ 5 km/sec, indicating that the lines must originate in gas with $T \sim 10^4$ K. This implies that photoionization is the principal mechanism for production of C IV, and is consistent with models of our own galactic halo that propose photoionization as the major means of producing highly-ionized species (see York 1982; Hartquist, Pettini & Tallant 1983).

REFERENCES

Bergeron, J. and d'Odorico, S., 1985, preprint.

Blades, J. C., Hunstead, R. W., Murdoch, H. S. and Pettini, M., 1985, *Ap. J.*, **288**, 580.

Gaskell, C. M., 1982, *Ap. J.*, **252**, 447.

Hartquist, T., Pettini, M. and Tallant, A., 1984, *Ap. J.*, **276**, 519.

Pettini, M., 1985, ESO Workshop on Production and Distribution of C, N, O Elements, Munich.

Pettini, M., Hunstead, R. W., Murdoch, H. S. and Blades, J. C., 1983, *Ap. J.*, **273**, 436.

York, D. G., 1982, *Ann. Rev. Ast. Astrophys.*, **20**, 221.

A SEARCH FOR QSOs IN THE FIELDS OF NEARBY GALAXIES

A S Pocock[1], M V Penston[1], M Pettini[1] and J C Blades[2]

[1] Royal Greenwich Observatory
[2] Space Telescope Science Institute

The extent and physical conditions of diffuse gas in the outer regions of galaxies are currently the subject of considerable interest. A very sensitive way to probe the gas is by observing the absorption lines it produces in the spectra of background objects. However, a detailed investigation of the interstellar medium associated with external galaxies requires the availability, in the field of the galaxy under study, of *several* probes (QSOs, Active Galactic Nuclei, supernovae) which are: (a) sufficiently bright for high-resolution spectroscopy (B < 17.5) and, (b) located over a range of projected distances from the galaxy, say from 10 to 200 kpc. As there are very few QSOs in the literature which meet these requirements, we have been carrying out a search of nearby galaxy fields for the specific purpose of finding a number of suitable background probes.

QSO candidates are first identified from visual inspection of objective prism plates obtained on our behalf by staff of the UK 1.2m Schmidt Telescope Unit at Siding Spring, Australia. The plates are short exposures (< 20 minutes), as only QSOs brighter than B ~ 17.5 are useful for our purposes. The criteria used to select promising candidates are ultraviolet excess (objects with long spectra of approximately uniform intensity, unlike the spectra of field stars which show a rapid fall-off in intensity towards shorter wavelengths) and/or the presence of emission lines, which is diagnostic. QSO candidates are then confirmed by follow-up low-resolution slit spectroscopy carried out with the 1.9m telescope of the South Africa Astronomical Observatory, in Sutherland. On average, approximately 30 percent of the candidates turn out to be *bona fide* QSOs.

To date, we have searched and followed-up 13 objective prism plates centred on nearby galaxies (of different morphological types) and clusters which subtend a large angle on the sky. The search area extends over a projected distance of ~ 100 kpc from the foreground galaxy, this being the extent of galactic haloes suggested by the statistics of QSO absorption line systems. So far, we have discovered 37 QSOs and 38 compact or emission-line galaxies (also useful for our

573

G. Swarup and V. K. Kapahi (eds.), Quasars, 573–574.

purposes) in these fields. A high degree of success has been achieved
for galaxies in the Sculptor group, particularly for NGC 253 (see
figure), where we have discovered 7 bright QSOs within ~ 100 kpc of the
galaxy centre. Statistically, however, the concentration of background
QSOs in this field is consistent with random fluctuations from the
average sky density of QSOs.

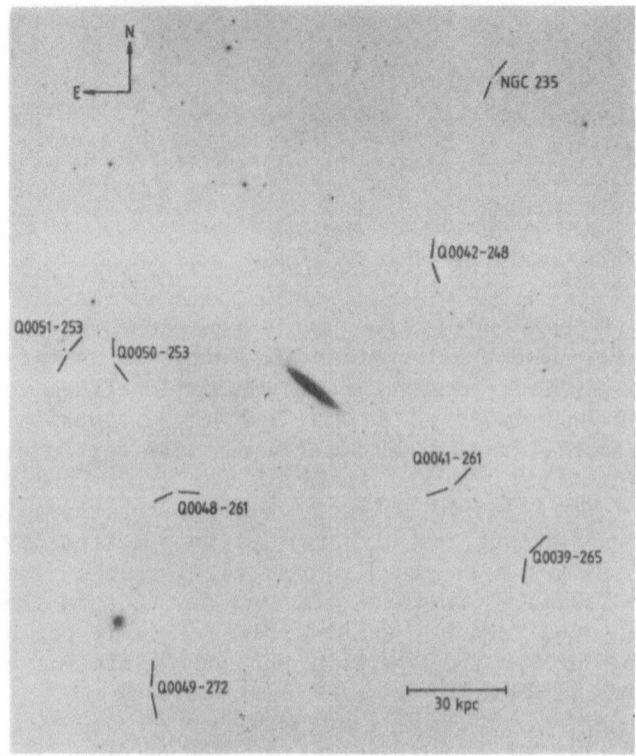

Figure 1.

*The field of NGC 253,
showing the relationship
of the foreground galaxy
to the QSOs we have
discovered (reproduced
from the ESO B survey).*

To augment the list of suitable probes, we have carried out a
literature search, which to our knowledge is complete up to the end of
1984, for all QSOs and AGNs which are noted in their discovery papers
as being situated near galaxies. In addition, we have conducted an
observational search, on Palomar Observatory Sky Survey prints, for
foreground galaxies close (within 10 galaxy diameters) to bright QSOs
discovered in optical and X-ray surveys, and have found a number of new
cases.

Combining all these data, we have compiled a list of all QSO-galaxy and
AGN-galaxy groupings which are suitable for studies of the intervening
interstellar media. The compilation is given in a paper submitted for
publication in Monthly Notices of the Royal Astronomical Society
(Pocock et al. 1986). High resolution optical and ultraviolet
observations of the background QSOs will be carried out in the near
future with the Isaac Newton Telecope on La Palma and the Hubble Space
Telescope.

GAS IN COSMIC VOIDS[*]

P M Gondhalekar[1] and N Brosch[2]

[1]Rutherford Appleton Laboratory, Chilton, Didcot,
Oxon, OX11 0QX, United Kingdom
[2]Wise Observatory, University of Tel Aviv,
Tel Aviv 69978, Israel

ABSTRACT. Absorption lines at the redshifts of cosmic voids in Perseus-Pisces and Böotes have been detected in the ultraviolet spectra of background quasars. The detection of Si IV and C IV lines besides Ly α suggests chemical enrichment of gas in the voids. Additional observations of three sight-lines through the void in the Böotes suggests that the gas in the void is clumped in large clouds.

1. INTRODUCTION

The large-scale structure of the Universe is characterised by filamentary distribution of groups and clusters of galaxies (eg Davis et al 1982). The filaments delimit voids in space in which the density of galaxies is very low. The spatial scale of these voids is $(10 - 40)$ h^{-1} Mpc (h is the Hubble constant in units of 100 kms^{-1} Mpc^{-1}). Krumm and Brosch (1984) searched for H I protogalactic clouds in the voids of Perseus-Pisces and Hercules. They determined that in these voids clouds of mass less than 10^{10} h^{-2} M$_\odot$ were not present.

The advantage of ultraviolet observations to detect small quantities of gas has been demonstrated by Brosch and Gondhalekar (1984). These authors, following the suggestion of Brosch and Greenberg (1983), analysed the ultraviolet spectra of quasars and Seyfert galaxies beyond the voids of Perseus-Pisces and Böotes. Absorption lines of Ly α, Si IV and C IV, at the redshifts of voids, were detected in the spectra of three objects. In order to determine the distribution of gas in the voids additional observations were made along three sight-lines through the void of Böotes. The results of these observations are presented here.

2. DISCUSSION

Three quasars beyond the void in Böotes were observed with the International Ultraviolet Explorer. Two short wavelength (1200Å – 1900Å) spectra of each quasar were obtained.

[*] Discussion on p.577

G. Swarup and V. K. Kapahi (eds.), Quasars, 575–576.

The spectra were analysed with the aid of a suite of programmes available on the SERC VAX Computer at the Rutherford Appleton Laboratory. The two spectra of each quasar were merged, with equal weights, to enhance the signal-to-noise ratio in the spectra and the signal-to-noise ratio was determined in the region of the spectrum free of emission lines. Absorption lines due to Ly α, Si IV and C IV, in the redshift interval of the void in Böotes, were identified in these spectra and only lines stronger than 3 σ were considered. The equivalent widths of these lines and the derived column densities are given in Table I.

TABLE I

Equivalent Widths and Column Densities in the Böotes Void

| QSO | Observed Line | | |
	Ly α	Si IV	C IV
1512+37	[1]0.036 [2]3.09±0.77 [3]2.08±1.07x10^{19}	0.036 1.49±0.37 2.20±0.50x10^{14}	0.039 1.43±0.35 4.82±1.20x10^{14}
1444+41	Only weak lines detected		
1415+45	No absorption lines detected		
1315+64*	0.04 2.55±0.01 1.20±0.06x10^{19}	0.03 2.92±0.02 3.20±0.02x10^{14}	0.04 1.60±0.05 3.90±0.12x10^{14}

[1]redshift of absorption line
[2]equivalent width in A
[3]column density in cm^{-2}

*from Brosch and Gondhalekar (1984)

The presence of absorption lines at the redshift of the void in Böotes demonstrates the presence of gas in this void. The absence of absorption lines along some sight-lines seems to suggest that the gas in the void is concentrated in large clouds. The detection of Si IV and C IV lines confirms the conclusion reached previously that the gas in the voids is enriched.

REFERENCES

Brosch N & Greenberg J M, 1983. Astrophys Sp Sci 90, 457.
Brosch N & Gondhalekar P M, 1984. Astron Astrophys 140, L43.
Davis M, Huchra J, Latham D W & Toury J, 1982. Astrophys J 253, 423.
Krumm N & Brosch N, 1984. Astron J 89, 1461.

DISCUSSION ON THE PAPER BY **BLADES** ET AL (p.571)

Kundt : How did you determine the abundance of zinc in the z = 1.345 system ?

Blades : We determined the HI column density from our IUE spectrum and then deduced the zinc abundance from the respective columns.

Schilizzi : How do you know the sight line passes close to the centre of the galaxy which you claim gives rise to the z = 1.345 system ?

Blades : By an indirect method. The absorption system at z = 1.345 shows CAII. This species is known to have a restricted extent in our galaxy and in some nearby NGC galaxies - its scale height is about 1-2 kpc from the plane in our galaxy, and seems to extend only to ~10 kpc in the plane for the NGC galaxies. If the 1.345 galaxy is similar (an admittedly mind-boggling assumption) to these nearby galaxies, then detection of CAII suggests that the sight-line passes near the centre of the intervening system.

Green : Have you compared the CIV to CII ratio in these absorption systems to that found in our galaxy ?

Blades : This is a major purpose of our work - it takes a lot of telescope time to obtain the required high resolution data for such comparisons and our work is not yet complete.

DISCUSSION ON THE PAPER BY **GONDHALEKAR & BROSCH** (p.575)

Malkan : As you know, I got blue high-resolution spectra with high signal/noise ratios, of 3 of the background QSO's you detected UV absorption in, at the void redshift. To very tight limits (e.g. 50 to 100 milliangstroms) I saw no corresponding CaII H and K absorption lines. However, you felt that CaII may not have been expected, given the apparent ionization of the gas.

Gondhalekar : I suspect the problem is the shear weakness of lines. We are looking at very small columns of gas and $Ly\alpha$, SiIV and CIV are the strongest lines in the absorption spectrum. Not surprisingly these are the lines we detect.

Trimble : If this gas exists in isolation well away from any galaxy, stars, or other sources of ultraviolet radiation, what keeps the gas ionised ?

Gondhalekar : You will have to ask Martin Rees - and theorists on galaxy formation. I believe there is no problem in having ionized gas in voids.

G. Swarup and V. K. Kapahi (eds.), Quasars, 577–578.
© *1986 by the IAU.*

Alloin : What is the signal to noise ratio in the continuum for this spectrum, and consequently, how confident are you about your line measurements ?

Gondhalekar : We only tried to identify lines which were stronger than 3σ. The rms noise was determined from regions of spectra free of known emission lines.

Wandel : In the case you do make a positive detection of absorption systems at the redshift of the voids, are you able to rule out the possibility that the absorbing material is actually at the edge of the void, and has a large velocity relative to the Hubble flow, as for example, in the Ostriker-Cowie explosive shells.

Gondhalekar : We need to look at a number of sightlines. If we find all lines at redshifts corresponding to the edges of void then we have a problem. The problem of Hubble flow is much more difficult; we need to determine the profiles of these lines - may be with ST.

Turnshek : I would simply point out that there are features to the red of your Lα emission line which appear as strong as the absorption features you have proposed an identification for. With the IUE spectrum you would expect the signal-to-noise to be poorer in the blue. Since you would not expect the lines to the red of Lα emission to be real, I doubt that the lines you have identified are real. Also, even if the absorption lines were real, I don't believe that you could even hope to determine column densities with data of that quality.

Gondhalekar : There is no problem in determining column densities. Of course, the accuracy of these column densities is not very high.

SUMMARY : FUTURE PROSPECTS

L. Woltjer
European Southern Observatory
Karl-Schwarzschild-Strasse 2
D-8046 Garching b. München

Before this meeting, it occurred to me that the simplest thing to do would be just to read the summary I gave at Liège two and a half years ago. Fortunately, it turned out that this is not possible, because in some areas substantial progress has been made. With 137 papers presented at this meeting, however, it is impossible to review them all; so I shall just make some general comments.

Much has been said about the basic framework in which the observations are to be discussed. Just two comments about this. First, if one wishes to show that the majority view of cosmological redshifts is incorrect, a few specific cases will carry much more weight than many a posteriori statistics. Second, the very high confidence levels of 10^{-4}–10^{-8} are misleading: a systematic error can easily be equal to a number of r.m.s. statistical errors and the "confidence" levels become so large only in the case of gaussian error distributions. With regard to the first point, close pairs of objects with very unequal redshifts and with absorption features in the spectrum of one at the redshift of the other may be particularly telling. If there were several cases with absorption at the higher redshift in the spectrum of the lower z object, a powerful case could be made for redshifts not related to distance; one case might not suffice, however, since then there would be endless arguments about the correct redshift of the absorption system. The results discussed by Shaver and also the interesting quasar-galaxy superposition reported by Jacqueline Bergeron, however, all seem to give further support to the cosmological interpretation.

Let us now come to some basic statistics, numbers, etc. Relatively little has changed, but the data have become better. To a B magnitude of 21 there appear to be about 40 quasars per square degree. This corresponds to somewhat over a million over the whole sky, and there is therefore an ample supply of objects for more detailed study. Beyond B = 21 the N(>B) curve flattens. The X-ray picture does not seem to be very different. There seems to be agreement that L_x/L_{opt} varies about like $L_{opt}^{-1/4}$ and that it decreases perhaps modestly with z. As a result of all of this, the estimates for

579

G. Swarup and V. K. Kapahi (eds.), Quasars, 579–583.
© *1986 by the IAU.*

the contribution of quasars to the X-ray background are still within the range 40 ± 25 per cent. Probably, it will remain difficult with the kind of data likely to become available in the coming years to do much better than this. However, there is a fundamental problem with a large part of the X-ray background consisting of unresolved quasars, and that is that the incompatibility of the spectra seems to become better established. Various papers at this meeting have reported spectral indices of 0.5 - 1.0 for quasars and around 0.9 for Seyferts, all steeper than the probable value for the background. The cheap way out is to say that the fainter quasars which account for most of the background may have different spectra, but this seems somewhat artificial. I suppose that it is more likely that this means that the contribution of quasars is not large enough to be fully dominant, and this raises the more interesting question of the origin of the remainder.

Quasars have a very steep luminosity function. Evidence was presented that the density of the very bright quasars has a maximum around z = 2, followed by some decrease towards z = 3.8 at which point the distribution abruptly stops. But, of course, there are serious problems remaining in all of this. There are the effects of variability and of photometric errors, and in particular uncertainties of the variation of the equivalent widths of the emission lines with z and L. A few objects at large z appear to have relatively faint lines, and this may reduce the discovery probability for high z objects. Wampler and Ponz made the claim that if all these effects are taken into account, it is even possible that there is no evolution. While most of us might feel that this is a rather extreme point of view, their analysis certainly shows that one should not take too seriously the disagreements about the details of the evolutionary picture. What is needed are samples of high photometric accuracy at the time of discovery - to eliminate variability effects, obtained with sufficient spectroscopic resolution to observe lines of small equivalent width.

There seems to be some evidence that the luminosity function narrows at large redshifts. Schmidt and others have been unable to find significant numbers of large redshifts at magnitudes fainter than about B = 20. If this is confirmed - and the accidental discovery of a very faint quasar with z = 3.2 reported by D'Odorico should perhaps still serve as a warning - it would tell one that there are relatively fewer low luminosity objects at high z, with interesting implications for quasar evolution.

Turning now to the more physical aspects of quasars, let me begin with the question of anisotropies which has been extensively discussed here. Are the observed anisotropies due to Doppler boosting or are they intrinsic ? On the positive side for Doppler boosting are the superluminal velocities that seem to be difficult to get in other ways. Also the relation between the width of Hβ and the relative core luminosity discussed by Beverley Wills seems to point in the same direction. On the other hand, as Porcas and others discussed, the lack of a correlation between the extended radio structures and some of these phenomena seems difficult to understand on this basis. And

then there are the extremely interesting observations by Davis of the
jet in 3C 273 which show an eventual counterjet to be at least
5500 times fainter than the jet – something which would be hard to
achieve with the Doppler models. The not very surprising conclusion
of all of this seems to be that both Doppler boosting and intrinsic
ansisotropies probably play a role, but that their relative importance
still needs to be clarified. Larger samples of VLBI data on
superluminal expansion in different radio sources with different
structures are needed, as are more extensive studies of the
relationship between radio characteristics and the detailed optical
emission line profiles of lines with different ionization stages.

Relatively little has changed in our knowledge of the
relationship between galaxies and quasars. Fuzz has been detected
around most quasars with z < 0.5, and its characteristics are not
incompatible with what would be expected for underlying galaxies. Of
particular interest are the images obtained in [O III] which show much
excited gas in irregular distributions around some quasars. While
there seems to be a general belief that elliptical galaxies are
related to radio galaxies and radio quasars, and spirals to Seyferts
and optical quasars, the situation is still not very clear for the
higher luminosity objects. The issue is important also in connection
with quasar evolution. As Green showed, many evolutionary scenarios
are possible, with perhaps luminosity dependent luminosity evolution
providing the best fit to the data. If there were important
differences in the types of galaxies associated with different kinds
of quasars, further constraints on the possible evolutionary tracks
might become available. In the same context, there are the clustering
effects which indicate connections between quasars, galaxies and the
environment; again, very carefully selected (by luminosity, etc.)
samples of quasars are needed before meaningful conclusions can be
drawn in particular about the redshift dependence.

Numerous papers have been presented about the optical emission
lines and about the flow patterns they indicate. The "Broad Line
Region" in the inner few parsec and the "Narrow Line Region" further
out (kpc scale) have been recognized long ago. The location of the
"Broad Absorption Line Region" seen in some quasars is less certain,
but it may be related to the Broad Emission Line Region. The electron
densities in the Broad Line Region vary over a large range
($10^6 - 10^{11}$ cm^{-3}) indicating that a simple pressure confinement
model is probably inadequate and that a dynamic model is needed.
Judith Perry discussed the various winds and flows that occur and
their connection to stellar mass loss, supernova explosions, etc. The
variability of the emission lines gives detailed information about the
location of some of the emitting regions, and one begins to get an
impression of a certain continuity between them. I think that once
one has realized that all of this takes place in a very dynamic
situation, one also begins to wonder whether pure photoionization
models are reasonable or if energy input from hydrodynamical processes
or from high energy particles also plays a role.

Of particular interest, of course, is the physics of the interior
of the quasar. There seems to be agreement about the ingredients that

are necessary. First, there is in most pictures a black hole. While
the characteristic signature of a black hole has certainly not been
seen, it seems clear that - unless conventional physics is wrong - any
very massive object or set of objects in the center of a galaxy
ultimately will have little choice but to become a black hole. In
addition to this, in most models there is an accretion disk, there are
magnetic fields which facilitate the transfer of angular momentum and
which may cause much flaring activity, and there are in the compact
sources the electron-positron plasmas which, as Novikov showed, may in
a relatively model-independent way explain the spectral properties of
these sources.

The basic power in quasars would come from the rotational energy
of the black hole and/or from gravitational energy released in the
accretion disk. Jets are seen in many cases and while their origin is
still not entirely clear, there appears to be agreement that a
magnetic field is needed to confine the jet and also a confining
medium to keep the magnetic field in place. As was noted by Barthel,
at larger redshifts the medium may become so confining that the jets
do not make it out of the galaxy anymore. In a way, this is
unfortunate, because it may make it more difficult to probe the
intergalactic medium at those redshifts.

The accretion rate, and thereby the luminosity, can be limited
either by the supply of matter to the black hole or by the Eddington
luminosity. In the former case, much irregular variability could be
expected, in the latter more systematic oscillations would also become
possible. An interesting example of such an oscillation in an
accretion disk was given by Abramovicz and applied to the observations
of Alloin and Pelat on the Seyfert galaxy NGC 1566 which seems to show
quasi-periodic behavior. The combination of such results with the
general evolutionary picture of quasars could have far reaching
consequences. Taking the example of this Seyfert galaxy and the fact
that such oscillations require the luminosity to be near the Eddington
luminosity, we would conclude that the central mass would be around
10^6 solar masses, which with an accretion rate of 0.01 $M_\odot$/year
gives a time scale of 10^8 years. This could give difficulties for
some evolutionary scenarios, since as Rees pointed out at the
beginning in the case of pure luminosity evolution central masses of
the order of 10^9 $M_\odot$ are required, because the objects were much
brighter in the past. Hence, if in Seyfert galaxies such oscillations
do, in fact, occur, it might favor a picture in which many galaxies
host a quasar or Seyfert nucleus during a relatively small part of
their life. Perhaps also the very high dynamic range maps in the
radio domain from Westerbork and elsewhere can give further
information on these issues. In models of pure luminosity evolution,
quasars would be more or less eternal, and one could perhaps expect to
find faint residues of past activity around them, especially at lower
frequencies.

Finally, with regard to cosmology, it follows from Burke's
presentation that we have been very lucky: the number of
gravitational lenses does not seem to be so large as to obscure the
overall evolutionary picture, but at the same time, it is large enough

to have many cases in which significant inferences about the distribution of (dark) matter in the universe may be made. Results about clustering of quasars are just beginning to come in, but their interpretation is unclear owing to possible effects of intergalactic absorption on the apparent distribution and of environmental effects on the real distribution. The latter may make the clustering properties of galaxies and quasars rather different. Lastly, the quasars are, of course, useful probes of the intergalactic medium (Ly α clouds) and also of intervening galaxies. As an example, the report by Hunstead of an absorption line due to zinc shows how much may be learned from studies of absorption lines in quasars about nucleogenesis in high redshift galaxies.

To make significant further progress, a new generation of instrumentation will be required. For the absorption line studies just mentioned, observations of faint objects at high spectroscopic resolutions will be needed which will require a new generation of optical telescopes as currently planned in various parts of the world. Optical interferometers will be needed in order to resolve the various emission or absorption line regions; they will almost certainly ultimately have to be located in space to overcome the effects of the earth's atmosphere. To qualitatively improve the imaging of the radio structures in which the superluminal motions are observed, new VLBI techniques are required involving also instrumentation like the proposed Quasat satellite. And, of course, further IR and X-ray satellites are needed, the latter also to study rapid time variations. While EXOSAT, ROSAT and perhaps other satellites will make important contributions, still larger facilities will be required thereafter. It is clear that the study of quasars can provide an important chapter in the scientific justification of most of the observational facilities, both space- and ground-based, currently under consideration. And once these facilities will be available ten or fifteen years from now, an entirely new picture of the nature and evolution of quasars may well emerge.

SUBJECT INDEX